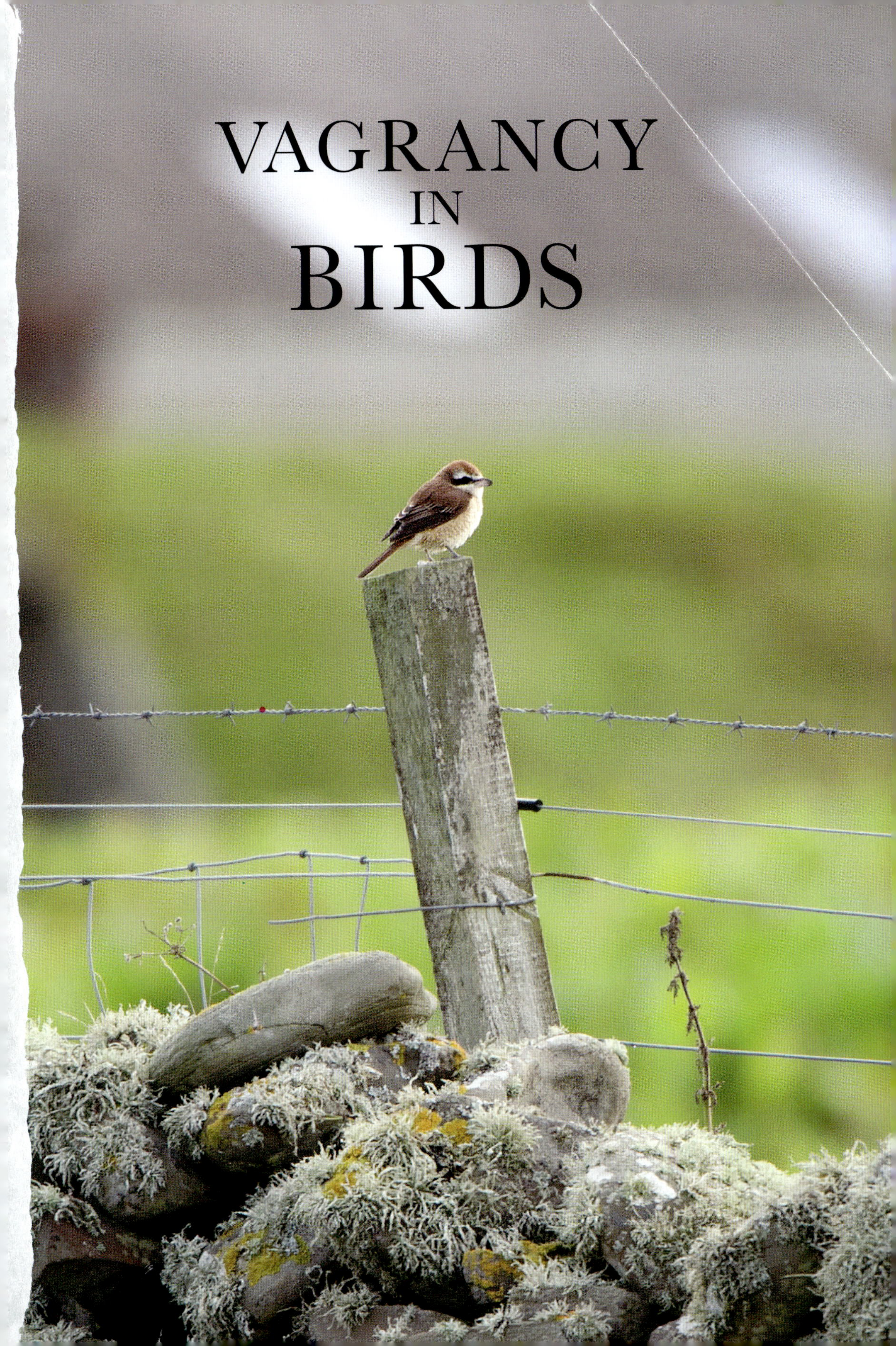
VAGRANCY
IN
BIRDS

VAGRANCY IN BIRDS

Alexander Lees &
James Gilroy

Princeton University Press
Princeton and Oxford

This edition published in the United States, Canada, and the Philippines in 2022 by
Princeton University Press
41 William Street, Princeton, New Jersey 08540

press.princeton.edu

First published in the United Kingdom in 2021 by Bloomsbury Publishing Plc

Recommended citation: Lees, A.C. & Gilroy, J.J. 2021. *Vagrancy in Birds*. Christopher Helm, London.

Library of Congress Control Number 2021942989
ISBN 978-0-691-22488-6

2 4 6 8 10 9 7 5 3 1

Design by Julie Dando, Fluke Art
Printed and bound in India by Replika Press Pvt. Ltd.

Half-title image: Brown Shrike *Lanius cristatus*, Grutness, Shetland, Scotland 14 October 2019 (*Rebecca Nason*).
Title image: Connecticut Warbler *Oporornis agilis*, Southeast Farallon Island, California, United States, 14 September 2010 (*Matt Brady*).
Cover images: Main image Black-browed Albatross with Northern Gannets, Heligoland, Germany (*Vincent Legrand*). Front cover, clockwise from top left: Yellow Warbler, Shetland Islands (*Hugh Harrop*); Squacco Heron, Somerset (*Rich Andrews*); Pallas's Warbler, Norfolk (*James Lowen*); Tengmalm's Owl, Shetland Islands (*Rebecca Nason*); Red-flanked Bluetail, Wiltshire (*Rich Andrews*); Long-billed Murrelet, Devon (*Shutterstock/Agami Photo Agency*); Black-throated Thrush, The Netherlands (*Shutterstock/Agami Photo Agency*); African Crake, Cape Verde Islands (*Shutterstock/Agami Photo Agency*); Siberian Accentor, Shetland Islands (*Stef McElwee*); Sandhill Crane, Japan (*Shutterstock/Agami Photo Agency*); Red-eyed Vireo, Easington, East Yorkshire (*Jake Gearty*); Sharp-tailed Sandpiper, California (*Shutterstock/Agami Photo Agency*).
Back cover, clockwise from top left: Ring-necked Duck, Durham (*Stef McElwee*); Woodchat Shrike, Somerset (*Rich Andrews*); Green Heron, Pembrokeshire (*Jake Gearty*); Siberian Rubythroat, Shetland Islands, (*Hugh Harrop*); Iceland Gull, Shetland (*Rebecca Nason*); Black Redstart, Northumberland (*Stef McElwee*); Black Stork, Tyne & Wear (*Stef McElwee*); Olive-backed Pipit, Shetland (*Stef McElwee*); Black-winged Stilt, Somerset (*Rich Andrews*); Lanceolated Warbler, Shetland (*Shutterstock/Agami Photo Agency*); Nutcracker, The Netherlands (*Shutterstock/Agami Photo Agency*); Blyth's Reed Warbler, Shetland Islands (*Jake Gearty*).

CONTENTS

ACKNOWLEDGEMENTS

This book would not be possible without the work of ornithological giants past and present, and we apologise in advance for missing any key literature.

We thank Ben Sheldon for suggesting we should tackle this subject, and thank Bloomsbury for their patience through various delays, not least those caused by childcare in a pandemic. We thank Chris Batty, Rohan Clarke, Marshall Iliff and Simon Mahood for commenting on different family accounts and catching omissions and errors. Many thanks to the many photographers – Rik Addison, Ciro Albano, Gary Allport, Chris Batty, Adrian Boyle, Matt Brady, Graham Catley, Sayam Chowdhury, Rohan Clarke, Ian Davies, Paul Dufour, James Eaton, Jens Eriksen, Paulo Ricardo Fenalti, Paul French, Alan Henry, Steve Howell, Marshall Iliff, Tom Johnson, Josh Jones, Yann Kolbeinson, Vincent Legrand, Paul Lehman, Wich'yanan Limparungpatthanaki, James Lowen, Rob Martin, Anthony McGeehan, Jay McGowan, Richard Moores, Yann Muzika, Rebecca Nason, Fabio Olmos, Yoav Perlman, Niall Perrins, Stuart Piner, Robson Silva e Silva, Brian Sullivan, Terry Townshend, Eric Vanderwerf, Peter Vaughan, Hugh Venables and Chris Wood for their time both in digging out images and providing further background information on records.

In a similar vein we thank Sayam Chowdhury, Josh Jones, Ralph MacNally and Tom Squires for help with contacting photographers. For sundry information-sharing and discourse about vagrancy and migration we thank Alexandre Aleixo, Tom Auer, Steve Bailey, Alan Ball, Jessie Barry, Stuart Butchart, Mario Cohn-Haft, Martin Collinson, David Edwards, David Fisher, Aldina Franco, Ben Freeman, Steve Gantlett, Jen Gill, Guy Kirwan, Ilya Maclean, Stuart Marsden, Tom McKinney, Oliver Metcalf, Eliot Miller, Vitor Piacentini, Ken Rosenberg, Ray Scally, Tom Schulenberg, Luís Fábio Silveira, Dave Smith, Ian Thompson, Joe Tobias, Bill Underwood, Rob van Bemmelen, Benjamin Van Doren, Andrew Whittaker and Kevin Zimmer.

Alex thanks his wife Nárgila and son Sebastian for their patience throughout a year of working evenings and weekends on the book throughout 2019–21, and his parents and sister for their part in his fostering his early ornithological ontogeny.

James thanks his wife and children for their endless support and patience, and his mother, sister and especially his father for inspiring his lifelong interest in birds and nature.

▲ Hume's Warbler *Phylloscopus humei*, Easington, Yorkshire, England, 10 November 2019, an increasingly recorded vagrant to Europe, usually in late autumn (*Alexander Lees*).

SCOPE OF THE BOOK

'Vagrants', 'accidentals', 'rarities', 'extralimitals' and 'casuals' are all synonyms for unusual records of nominally 'out of range' individuals of a given bird species. Humans have long coveted records of these 'lost' individuals, and there is a rich ornithological literature that describes the various subcultures associated with their pursuit, from 19th-century collectors to 21st-century twitchers (Mearns & Mearns 1998, Wallace 2004). The obsession surrounding vagrant birds has been historically derided by some ornithologists, who argued that records of vagrants are of little biological relevance, but we share the contention of others (e.g. Grinnell 1922, Rose & Polis 2000, Newton 2008) that vagrancy is a powerful biological phenomenon whose study is fundamental to understanding the diversity of life on earth. This book will for the first time systematically explore the taxonomic and geographic patterns of extralimital avian occurrence globally and try to synthesise what we know about the processes that underpin these occurrences, based on the latest scientific discoveries.

Before we can embark on this study of pattern and process in avian vagrancy, we need to define what a vagrant is in the first instance. This is surprisingly more challenging than one might expect. If our starting assumption is that a vagrant of a species is an individual occurring outside of its normal geographic range, then defining vagrants is simply a matter of identifying the range edge – and any individuals that lie beyond it are vagrants. For example, Short-toed Treecreepers *Certhia brachydactyla* do not breed in Britain and are a strictly sedentary species – thus any individuals that appear in Britain are vagrants, even if they are only a few dozen kilometres from regular breeding sites on the near Continent, separated by the English Channel which is a significant barrier to their dispersal. Meanwhile Icterine Warblers *Hippolais icterina*, which breed equally close to Britain, are highly migratory and occur in small numbers regularly – are these individuals vagrants? Most British birders would label them as 'scarce migrants', but they do meet our criteria for being out of range, highlighting the fact that there is a smooth continuum between the rarest of vagrants to true 'scarce migrants' occurring at the edge of their normal migration routes. It is tempting to try and apply some sort of quantitative threshold to define vagrancy, but this is a major challenge, not least because geographic ranges are not fixed in time and space, and can change dramatically very quickly. However, a ballpark definition might be that the geographic range of a species should encompass something like 99.99 per cent of individuals of a species at a given time – anything outside this range, based on the density of the species, might be defined as a vagrant.

▲ Icterine Warbler *Hippolais icterina*, Foula, Shetland, Scotland, 4 June 2008, an expected 'scarce migrant' in late spring in the Northern Isles but still out of habitat and out of range (*James Gilroy*).

In case you are an all-round naturalist picking up this book and thinking, well 'is this just a bird thing?', it would be remiss of us not to point out that vagrancy is not just a bird-specific phenomenon. Vagrancy is most appreciated and celebrated in birds because of their high mobility, but taking Britain again as an example, natural historians have recorded vagrant butterflies, dragonflies and bats from North America, vagrant moths, grasshoppers and sea turtles from the tropics, and vagrant seals and cetaceans from the Arctic.

Most newspaper articles reporting on the latest 'waif' or 'stray', and the 'hordes' of birders who come to see it typically invoke the weather as the root cause of vagrancy – the oft repeated adage of 'blown off course'. Weather is an important driver of vagrancy, but is far from the only one. The aforementioned Short-toed Treecreepers do not migrate and can't realistically be blown 'off course', as they don't have a 'course' to take; their appearance outside their normal range belies one of the many other different mechanisms for vagrancy, the description of which forms a substantial part of this text. Another old adage – sometimes heard from birders – is that 'anything can turn up anywhere'. Although it might sometimes feel like this is the case, particularly at the height of exciting autumn migrations, this is demonstrably not true. Vagrancy is far more frequent in highly migratory species and

▲ Thousands of birders line the seafront at Dawlish, Devon, England, to see the first record for Britain of a Pacific seabird – Long-billed Murrelet *Brachyramphus perdix* – on 12 November 2006 (*Alexander Lees*).

migration span is a good predictor of both vagrancy likelihood (McLaren *et al.* 2006) and the distance that birds may travel in the 'wrong' direction – to reach places vast distances outside of their regular ranges. Sedentary species can occur as vagrants, but the distances involved tend to be far more modest. This in turn varies latitudinally – tropical forest understorey species, for example, may be the least vagrant-prone of all taxa, and many may be physically incapable of flying more than a few dozen metres before collapsing exhausted (Moore *et al.* 2008). As such, it is perhaps possible to encounter a species like Sanderling *Calidris alba* almost anywhere in suitable habitat on Earth, yet finding an Andean *Scytalopus* tapaculo more than a few hundred metres outside of its elevational range limit (and hence geographic range) would be an incredibly rare vagrancy event, as these species are extremely sedentary. The life history characteristics of birds are thus reliable predictors of their vagrancy likelihood, and we explore this variation across the avian 'tree of life' in the second part of the book.

Broadly speaking, vagrancy can arise through both *exogenous* factors (i.e. with an external cause) or *endogenous* factors (an internal cause). The Icterine Warblers that are drifted across the North Sea to eastern England during 'fall' conditions are thus exogenous and might even be termed 'exovagrants'. Vagrancy in Short-toed Treecreeper, although potentially assisted by favourable winds, is more likely to be driven by an endogenous impulse to disperse – making these nominally 'endovagrants'. These categories are simplistic, and in many cases vagrancy may stem from multiple processes occurring simultaneously – Icterine Warblers may appear in Britain primarily during ideal wind drift conditions, but a significant proportion of those individuals are probably following errant compass headings anyway. In this book we will explore in detail the various internal and external drivers of avian vagrancy, first looking at the mechanisms by which migratory species navigate, and how errors in these mechanisms can help explain patterns of vagrant occurrence.

◄ Tschudi's Tapaculo *Scytalopus acutirostris*, Bosque Unchog, Huánuco, Peru, 24 June 2013. Tapaculos are among the most sedentary of all bird species and hence the least vagrancy prone. Their poor dispersal capacity has contributed to the high endemism and species richness in the group, as populations easily become isolated from one another by relatively minor habitat barriers (*Alexander Lees*).

HOW BIRDS NAVIGATE

Birds have remarkable navigation capacities, with many species migrating tens of thousands of kilometres only to return with pinpoint precision to the same nesting site they used in the previous year. Understanding the basics of avian navigation is essential if we are to make sense of vagrancy patterns, both in terms of identifying how endogenous navigatory errors are made, and also understanding how exogenous forces like weather can inhibit avian navigational abilities and lead them to stray off course.

When humans think about navigation, we naturally conjure up maps of the world in our minds. This ability is not innate, however – we are fortunate that our predecessors have meticulously mapped the planet, providing us with visual resources that we can memorise and use to keep track of our location within the world around us. Our capacity for navigation thus depends almost entirely on learned information, much of which has been handed down from previous generations, as well as technological solutions like GPS – all of which are luxuries not afforded to most migratory birds. In the majority of bird species, the ability to navigate with precision to exact localities is not innate but is learned through gradual experience (Wiltshko & Wiltscko 2015). Migratory birds are apparently able to develop 'mental maps' during their early years of life, using cues they experience during their first migratory journeys, to allow them to assess exactly where they are in relation to specific localities such as former breeding locations, stopover sites or winter territories. However, when embarking on their first migration, most juvenile birds cannot rely on such a developed map sense. Unlike more experienced adults, juveniles are generally unable to determine exactly where they are, at any given point in time, relative to their origin and goal.

If we consider a juvenile of a migratory species that does not migrate in flocks – in other words, a species that cannot rely on following other individuals – it is clear that for the first migration at least, navigation must be done using only the innate instincts the bird has inherited from its parents. In many cases, such birds must embark on migration within weeks of fledging if they are to avoid being exposed to unsuitable environmental conditions as the seasons shift. Many migrants must also adjust their migratory headings multiple times over the course of their first migration to avoid barriers such as oceans or deserts. The fact that billions of naive juvenile birds successfully complete these journeys annually is genuinely remarkable. Without a map to guide them, how do they find their way?

Birds have evolved many different solutions to this problem, but research suggests that in general, juvenile long-distance migrants undertake their first migration using just two key tools – a clock (i.e. an ability to track the passage of time) and a compass. These tools are inherited genetically (Helbig 1992, Willemoes *et al.* 2014), and in theory they are all a young bird needs to complete its first migration entirely alone. Juvenile migratory birds typically inherit a migratory 'program' from their parents, which encodes both the direction and timing of

▲ Swainson's Thrush *Catharus ustulatus*, Norwick, Shetland, Scotland, 28 September 2014, the 33rd record for Britain (*Paul French*).

movements necessary to make a return trip to a suitable wintering area. The genetic heritability of this program is quite precise – a series of experiments with Eurasian Blackcaps *Sylvia atricapilla* revealed how interbreeding between individuals with different migratory programs produces offspring with intermediate strategies (Helbig 1992). This was further confirmed in a recent tracking study with Swainson's Thrushes *Catharus ustulatus* from a contact zone between two populations with different migration routes, where some offspring followed routes that were intermediate between their parents, whilst others adopted the orientation of one parent in autumn and the other parent the subsequent spring (Delmore & Irwin 2014).

From the point of view of understanding vagrancy, errors in any part of the inherited migratory program could result in first-year birds moving off in the 'wrong' direction. For example, a bird might inherit a perfectly functioning clock and compass (i.e. the machinery of navigation), but due to some mutation or other natural variation its directional or temporal 'instructions' differ from the rest of the population. Such an individual would theoretically be quite capable of understanding which way is north, but it will move in a very different direction than other individuals of its species. Alternatively, a bird might inherit a perfect set of directional 'instructions' but carry some mutation or variation of its clock or compass apparatus that prevents it from accurately implementing the movements encoded in its genes. Clearly, there are many ways that navigatory errors could lead to vagrancy – and this variability in part explains the vast diversity of observed vagrancy patterns. However, by examining recent advances in the scientific understanding of the mechanisms of bird navigation, it is possible to make some sense of just how and why vagrancy occurs. In the remainder of this section, we will summarise the current state of knowledge of how migratory birds inherit and implement their migratory programs. First and foremost, we will consider arguably the most fundamental component of the migratory apparatus, at least in terms of drivers of vagrancy: the compass sense.

◄ Chinese Leaf Warbler *Phylloscopus yunnanensis*, Zoige, Sichuan, China, 1 June 2011, a tired migrant 'out of habitat' (*James Eaton*).

The four avian compasses

Compasses have just one purpose – to tell the user where north is. Armed with such a tool, a naive juvenile bird should always be able to use north as a reference point, and then orientate itself by rotating its body to a point where it is in line with whatever desired migratory heading it has inherited from its parents. This basic compass orientation is thought to be the cornerstone of migration in naive juvenile birds and has been the subject of exhaustive scientific study. For decades, researchers have performed increasingly complex and elegant experiments to examine how birds 'read' their internal compass, and how they then use this information to orientate themselves. Many of these experiments have involved trapping wild birds (or raising them in the lab) and placing them in a cage apparatus – an Emlen funnel – overnight, where their desired target orientation can be evaluated from the pattern of marks they leave around the edges of the cage. By manipulating the information available to these birds – for example, by preventing them from seeing the stars above – researchers have been able to painstakingly unpick the different cues that birds use to orientate themselves.

Thus far, research has identified four separate avian compass mechanisms, each using information from four different sources: 1) the Sun, 2) the pattern of polarised light, 3) the stars, and 4) Earth's magnetic field (Åkesson & Helm 2020). Interestingly, evidence suggests that most migrants have evolved to use at least two, and perhaps sometimes all of these compasses simultaneously, forming a hierarchy of tools that they can cross-reference to make the best judgements possible. Despite having this arsenal of compass mechanisms at their disposal, it is still clear that the system is fragile and prone to error – and thus potentially vagrancy – with each compass having its own particular weaknesses and complications.

The Sun compass was the first to be identified in birds; Kramer (1952) revealed how Common Starlings *Sturnus vulgaris* would show a very clear pattern of migratory orientation within their cages, but only on days when the Sun was visible. In a follow-up experiment, Kramer used mirrors to change the apparent position of the Sun above the Starlings and found that they duly shifted their orientation in line with this new position. Observing the position of the Sun at sunrise and sunset is the simplest solar compass, but this frame of reference is only available as two snapshots during each 24-hour period, which is clearly insufficient for a migrant that must keep track of its heading throughout the day. Research in 'homing pigeons' *Columba livia domestica* has revealed that they solve this problem using their high-precision internal body clock, allowing them to keep precise track of the Sun's arc of movement (the azimuth, or horizontal position) with respect to the passage of time (Wiltschko *et al.* 2000). In doing so, they can continually adjust their mental compass to ensure that they can always tell north from south.

Whilst this clearly provides an effective compass for diurnal migrants, there are two major drawbacks to reliance on the Sun compass in naive juvenile birds. One is that the time-adjustment method apparently requires a high degree of learning; evidence from pigeons suggests that they are only able to do it from around three months old and some individuals master learning it far better than others. The second issue is that the Sun must be clearly visible – the compass will not work when there is significant cloud cover. Clearly, a migratory bird must be prepared to move in cloudy conditions, and therefore relying solely on a Sun compass for navigation is not a safe strategy.

One way that some species have solved the problem of cloud cover blocking the Sun is to use the pattern of polarised light, rather than the Sun itself, as a compass cue. Polarisation describes the degree to which light waves coming from a source are either scattered or aligned – and helpfully for a migratory bird, light from the Sun always carries a consistent pattern of polarisation across the sky. The alignment of polarised light waves is most intense at an angle 90 degrees from the Sun, and most scattered at points either towards or away from the Sun itself (Figure 1). Importantly, this pattern is present even when the Sun is obscured by clouds or objects, meaning that any organism that can 'see' the polarisation of light can roughly identify the position of the Sun at any time of day.

Many species, including birds, have an acute ability to see patterns of polarised light whenever at least some light is available. In fact, most humans can see polarisation patterns too – although until recently we have been largely unaware of this apparent 'extra sense' (Temple *et al.* 2015). Referred to as 'Haidinger's brush', human

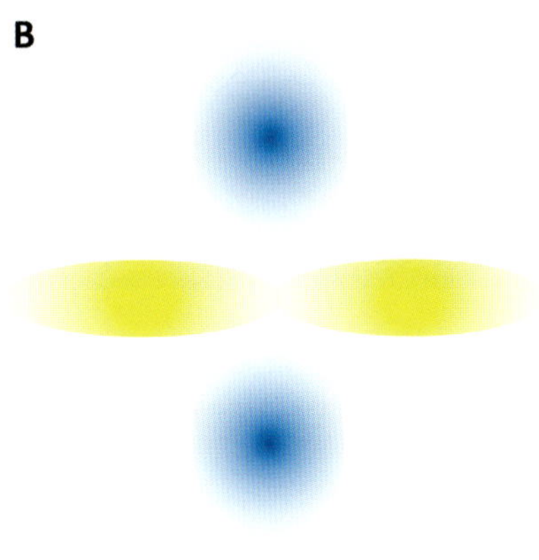

▲ **Figure 1.** A) Birds can find north by viewing the pattern of polarisation across the sky at sunrise (shown here) and sunset. B) 'Haidinger's brush', a pattern of polarised light that is visible to the naked eye in most humans.

eyes have the capacity to see polarisation in the right circumstances – you can try it yourself using an LCD screen (e.g. a laptop or tablet). If you look at a blank white portion of the screen and tilt your head from side to side, you should see faint yellow and blue 'bow-tie' patterns (Figure 1B) appear at certain angles. With practice, you can also see the same pattern in the blue parts of the sky at 90 degrees from the Sun, particularly around sunrise and sunset. The long axis of this yellow 'bow-tie' will point approximately towards the Sun – it is likely that this is something akin to the polarisation compass used by birds to navigate. Although discovery of this capacity in humans was a surprise to modern science, Viking civilisations apparently discovered this more than 1,000 years ago, and even used quartz crystals (the enigmatic 'sun stone' of viking lore) to amplify the polarisation pattern for use as a compass tool (Le Floch *et al.* 2013).

Polarisation can provide a cloud-proof mechanism for identifying north, but its value as a universal compass for migratory navigation is limited. Evidence suggests that polarisation patterns are only likely to be strong enough to use at sunrise and sunset, and therefore may not be used for continual navigation throughout a flight. Moreover, knowing the precise position of the Sun at sunset is not sufficient for an individual to find north and south – the individual must also have a keen sense of time in order to compensate for changes in the Sun's position through the year. Again, this time-compensation ability appears to be a learned skill that is honed by developing birds over time – and thus may not be particularly accurate in a naive juvenile bird, and especially a late-fledged individual in a rush to embark on its first migration. For nocturnal migrants (which includes most passerines and many non-passerines), the stars provide another potential source of compass information. Some of the most elegant experiments conducted on bird navigation were carried out by Emlen (1970) using caged Indigo Buntings *Passerina cyanea* to examine their use of celestial cues. Emlen did this by taking caged buntings to Cornell University's planetarium, where he was able to measure how the migratory orientation of birds varied when he experimentally changed the patterns of stars in the night sky above them. The results were dramatic and conclusive – the birds did indeed use the stars to orientate, and in particular they used the point of axial rotation as their reference point (Figure 2). In the northern hemisphere this corresponds to Polaris, the North Star (Emlen 1970).

◀ **Figure 2.** Many migratory birds apparently memorise the pattern of star rotation around the poles during their first weeks of life, allowing them to identify north and south whenever sufficient clear sky is in view. Image reproduced with permission from the European Southern Observatory (ESO).

Emlen's experiments went on to show in detail how this mechanism worked. By changing the point of rotation in the planetarium from Polaris to Beetlejuice, he was able to prove that it was not the pattern of stars themselves that the birds used, but the point of rotation, as birds exposed to this new rotation point shifted their migratory orientations to reflect this new 'north' around Betelgeuse, moving in a different direction to those exposed to the fixed point of Polaris.

Another key insight from subsequent planetarium experiments was that much like the Sun compass, the star compass requires considerable learning and skill refinement on the part of juvenile birds to be useful. Specifically, individual birds must first identify the point of rotation itself, presumably through careful observation over multiple cloud-free nights. Then, the bird must learn the pattern of stars around the fixed point of rotation, so that they can then check and identify the north point at any moment during their subsequent journey – even if Polaris itself is obscured by clouds (Wiltshchko *et al.* 1987). This clearly represents an acute challenge for a recently fledged bird with only weeks between taking its first flight and embarking on its first full migration. Even after successfully internalising a star map, there are still many further pitfalls to the use of stars as a compass during an ongoing migration. Clearly, during flight itself, individuals must be able to continually assess their heading with respect to their compass – and yet this will not be possible whenever cloudy conditions mask the positions of the stars.

Prolonged cloudy conditions during a peak migration window – or indeed smoke from wildfires – could therefore cause significant problems for any birds relying largely on star-based or solar compasses to orientate. Also, as birds imprint on a star map based on the skies above them in their first weeks of life, the stars themselves will change considerably as the season progresses, and even more so as the bird begins to move across the planet during migration, with some constellations disappearing altogether whilst other new ones emerge on the horizon. The learned star map therefore must be updated continually during the course of a migration in order to remain usable as a compass (Muheim *et al.* 2014) – something that may be further inhibited by cloudy weather during the course of migration. It will also be made more difficult the shorter the timespan within which the migration takes place.

The stars, the Sun and polarised light therefore all appear to require some degree of learning, as well as being highly vulnerable to cloudy weather. The final tool in the avian compass toolkit, however, does not suffer from these weaknesses, and can apparently be used without the need for learning or skill development – the magnetic compass. It appears that most or even all species have evolved to use magnetic cues – and this likely represents the default 'fail-safe' compass among naive juvenile migrants.

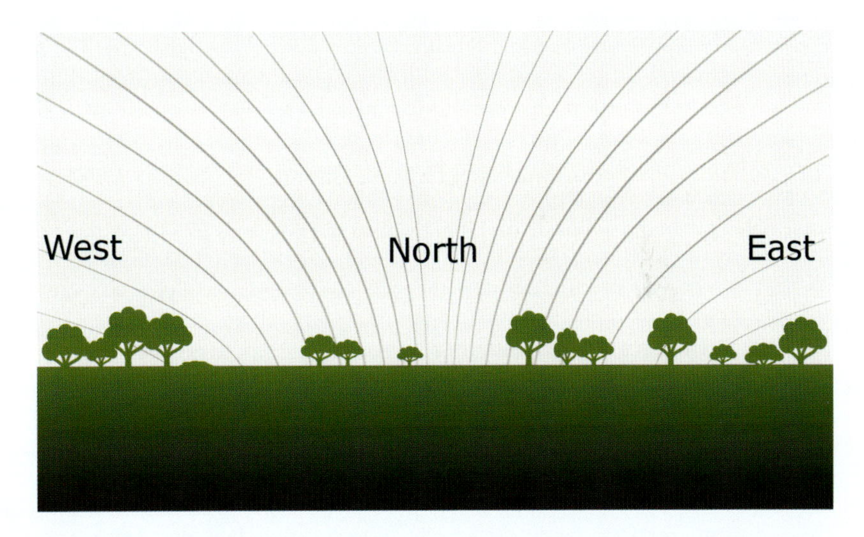

► Figure 3. Hypothetical visualisation of how birds may view Earth's magnetic field, showing the apparent 'dip' towards the pole and away from the Equator, as 'seen' by birds such as European Robin *Erithacus rubecula*.

The avian magnetic compass was first discovered in European Robins *Erithacus rubecula* (Merkel & Wiltschko 1965), and our current best evidence suggests that it primarily works by inclination (the 'dip' of magnetic lines towards the pole) rather than polarity itself (Wiltschko & Wiltschko 1972; Figure 3). Just how birds perceive this 'dip' remains poorly understood, although it appears that at least two different mechanisms have evolved separately – one involving magnetic particles (magnetites) located in the sensory organs of the bird, and the other a light-based system involving specialised receptors in the retina that allow birds to literally 'see' the magnetic field around them. To use the magnetic compass, a bird must evaluate the point at which magnetic 'lines' dip towards the horizon – the steepest angle of this dip always points towards the magnetic pole (Ritz *et al.* 2009). As such, the magnetic compass seems relatively infallible – indeed, one might question why the other compasses are needed, if the magnetic compass can be used in any weather conditions and without any learning or imprinting. The fact that almost all migratory birds appear to have evolved the ability to use other compasses alongside their magnetic sense (Muheim *et al.* 2006) indicates that this cannot be the case – whilst evidently useful, it is apparent that the magnetic compass is not sufficient for orientation by itself.

◄ European Robin *Erithacus rubecula*, Mutsu, Tochigi, Japan, 26 February 2019, thousands of kilometres east of its normal range (*Yann Muzika*).

There are various reasons why relying on magnetic fields alone would be risky – the angle of dip itself varies with latitude and may also become hard to discern in some regions, particularly towards the Equator. Another key feature of the avian magnetic compass is that it cannot truly differentiate between north and south – rather, it always points poleward from the Equator (Wiltschko & Wiltschko 1972). This means that if a bird moves across the Equator, its magnetic compass will 'flip', so a bird moving south out of the northern hemisphere will suddenly 'see' its magnetic compass pivot to point back towards the opposite pole. To counter this, transequatorial migrants must have some other mechanism to be aware that they have switched hemispheres, and thus adjust their orientation to control for the fact that their internal magnetic compass now points south.

Magnetic intensity itself varies, suggesting that the 'dip' pattern birds use to navigate may be very faint in some regions, making it difficult for birds to assess the precise point of maximum inclination. Another critical issue with the magnetic compass is that the orientation of the field varies from place to place (known as magnetic declination) meaning that in some areas the compass may become highly inaccurate. Across much of the globe this inaccuracy is relatively slight (between 0–20 degrees outside the true polar direction; Muheim *et al.* 2006), but at high latitudes the discrepancies between true and magnetic poleward orientation can be much more significant – up to 90 degrees in some cases. In the high Arctic, therefore, any bird naively using a magnetic compass to orientate could end up setting off in a direction up to 90 degrees outside the 'correct' orientation encoded in their genetic program (Alerstam *et al.* 2001).

Unsurprisingly, evidence suggests that migratory species breeding at high latitudes have evolved to rely much more heavily on other compasses – Arctic-breeding shorebirds, for example, follow routes that suggest they use a Sun compass almost exclusively, at least at high latitudes (Alerstam *et al.* 2001). For many other species, the favoured strategy appears to be to use a magnetic compass, which they then 'calibrate' using one or more other compasses at different stages of the migration. Strongest evidence for this comes from studies of Savannah Sparrows *Passerculus sandwichensis*, showing that despite breeding at high latitudes they do rely heavily on their magnetic compass, but prior to migration they correct for the declination effect by observing the 'true' north point using polarised light and use this to adjust the north point of their magnetic compass (Muheim *et al.* 2006). Having memorised this discrepancy, they can then apply it continuously as they begin to move (i.e. at night when the polarised light compass is no longer visible) to adjust their magnetic compass and find the true direction of north. This type of compass correction appears to be a common strategy – though some species might use other compasses as their reference category (solar or celestial cues). It provides an elegant way for birds to use their only compass that is always available – being visible day and night, and unaffected by weather conditions – despite that compass being 'flawed' in that it does not always point to true north. They do this by periodically working out how much they need to adjust their magnetic compass, by ground-truthing it against another compass during windows when weather conditions allow one of them to be used.

In light of this huge body of research, we can now go some way towards constructing a general picture of how the compass sense works in a typical juvenile migratory bird. It appears that to correctly develop the tools needed to robustly navigate on a first migration, a juvenile must go through two key steps:

- After hatching and prior to migration, a bird must develop a capacity to use at least one non-magnetic compass – either by memorising star maps (most important in nocturnal migrants) or developing a capacity to finely track the passage of the Sun and/or pattern of polarisation through the day (most important in diurnal migrants).
- Next, they must use these learned compass cues to calibrate their innate magnetic compass, ensuring they can use it to locate true north despite magnetic variability.

Clearly, failure to achieve both these steps during the early months of life could lead to a juvenile bird embarking on its migration without a robust method of navigation – placing it at risk of vagrancy. We would perhaps expect such a bird to wander almost randomly if it were unable to effectively orientate itself in its innate desired heading. Some vagrant occurrences do indeed fit a pattern of random movement, particularly in the case of vagrancy in very young juveniles that stray outside their normal distributions soon after the end of the breeding season (see page 55). Interestingly, however, many long-range vagrancy events fit much clearer patterns, with certain species appearing consistently as vagrants in particular areas and at particular times of year. This suggests that such vagrants are steadfastly moving in their chosen – but erroneous – direction, rather than randomly orientating from one flight to the next, as one might expect from a bird that has simply failed to develop a strong compass sense. In subsequent sections, we examine some of the most prominent hypotheses concerning how and why these erroneous orientations might arise. For now, however, we turn our attention to another key component of the navigatory process that may also play an important role in vagrancy – the ability to measure how far you have travelled.

▲ Purple Sandpiper *Calidris maritima* and Surfbird *Calidris virgata*, Playa del Borrego, San Blas, Nayarit, Mexico, 17 January 2015. This, the second record of Purple Sandpiper for Mexico and the first from the Pacific coast of North America, was presumably a misorientated individual, although it may have followed the Surfbirds south from the Arctic (*Steve Howell*).

The avian clock – deciding when to stop migrating

If we imagine that our inexperienced young juvenile is successfully able to orientate in the migratory direction given by its inherited genetic program, the only other barrier to successfully completing its first migration (assuming a simple straight-line scenario) is to judge when it has moved far enough to have reached its wintering grounds. This aspect of migration is perhaps less well understood than the compass component, but it is highly relevant to the question of vagrancy. Failure to stop on the first migration could lead individuals to overshoot their normal winter ranges – a phenomenon explored in more detail in the overshooting section on page 46. Perhaps more importantly from a vagrancy perspective, the mechanisms involved in sensing when to stop migrating may play an important role in determining how far an individual were to move if, for some reason, it set off in the 'wrong' direction. Many vagrants appear to move much further than their normal migration distances, suggesting some degree of failure in judging when to stop and thus continuing their migratory restlessness longer than necessary. Knowledge of the stopping mechanism can help us understand this intriguing component of avian vagrancy, although current research has yet to give a definitive answer as to how birds determine how far to migrate in their first year.

One possible cue that migrants may use to judge when to cease migration is time itself. Most animals possess an internal clock, and in birds the clock sense is known to be particularly advanced, with the capacity to keep track of time at high precision across daily and annual cycles (Åkesson *et al.* 2017). Key evidence concerning the role of the internal clock in determining migration distance came from experimental studies using captive Garden Warblers *Sylvia borin* that were kept for a year in light-sealed laboratories where day length was kept constant at 12 hours, and temperature and other environmental factors were controlled (Gwinner & Wiltschko 1978). Birds kept in these conditions therefore had no way of tracking the changing seasons, other than through their own internal circannual clock. Amazingly, these birds showed exactly the same patterns of migratory restlessness as birds exposed to natural light and weather – they became highly active and agitated at night during the precise spring and autumn periods where migration normally takes place. Even more remarkably, the direction of migratory orientation (assessed using Emlen funnels) shifted during the autumn, matching the shift in orientation expressed during the real migration as Garden Warblers pass through the eastern Mediterranean before turning south into East Africa.

▲ Garden Warbler *Sylvia borin*, Foula, Shetland, Scotland, 5 June 2008, a tired migrant resting on sea cliffs (*James Gilroy*).

This experiment, and others like it, revealed that the cues to start and stop migration appear to be hard-coded into the endogenous clocks of birds – in turn, we might expect that a vagrant would cease migration at a similar time to others of its species (and in turn at a similar distance from the breeding range). However, further experiments using another species, European Pied Flycatcher *Ficedula hypoleucos*, revealed an added wrinkle: here the expected changes in migratory orientation only happened in the laboratory-controlled birds when they were exposed to changes in magnetic fields, mimicking the magnetic environment they would have been moving through if performing real migration (Beck & Wiltshko 1982). This result strongly suggested that changes in the migratory program – in this case the direction of orientation – arise not only due to the internal clock of the bird, but also require some external confirmation that the bird has reached the target area to make the change. In this case, that information comes in the form of magnetic conditions, which change dramatically in field intensity and inclination at lower latitudes. Other cues could also be used, including changes in day length or other environmental features. Interestingly, magnetic cues are also apparently important in triggering switches between restlessness and fattening behaviour (Henshaw *et al.* 2006), indicating that birds may use a magnetic 'map' to determine when they have reached a stopover region, and thus to cease migration and fatten up before the next stage.

Crucially, these findings suggest that for migrants to move between the multiple phases of their migratory programs, they may need some cue to indicate that they have moved the correct distance and are ready to initiate the next phase. Magnetic cues appear to be favoured for this – and at some point, a final 'stopping' cue may indicate that the final phase is completed, and the migration can end, the destination reached.

A vagrant bird that has moved off in the wrong direction, or been displaced by weather, might therefore never encounter the correct cue to move to the next phase of migration. Such birds could become 'trapped' in a phase of the program, continuing along a single orientation potentially for much longer distances than would occur if they had set out along the correct heading. Migratory programs like those of Pied Flycatcher therefore could lead to individuals moving far longer than normal for their species, if their initial heading takes them along the wrong route, and so preventing them from ever encountering their genetically encoded 'waypoints' – which would indicate when to change course or stop.

The use of migratory waypoints – magnetic field characteristics or other landscape features – is also central to the more advanced navigational abilities that apparently develop in adult birds after their first migration, allowing them to migrate with much finer precision to target locations (Alerstam 1978). The development of this mental map of waypoints apparently happens through the course of the first year of life and can involve a suite of different cues.

The avian map sense and 'true' navigation

A juvenile bird that is equipped only with a clock and compass cannot perform 'true' navigation – that is, it has no way of knowing its actual position relative to its goal, or to other waypoints around it during the course of its journey. Rather, a navigator using a clock and compass will blindly follow its set of instructions (in this case instinctive movement directions and timings) in the hope that these will lead it to the right place. A true navigator, on the other hand, will be able to keep track of its progress relative to other waypoints, and would be aware of any movement that takes it off course.

It is well known that many birds do indeed possess remarkable abilities to perform 'true' navigation – indeed, from ancient Egypt through to World War II, humans have taken advantage of this uncanny ability, pinning messages to the legs of pigeons that could be relied upon to return to specific locations from almost any starting point (Blechman 2007). Charles Darwin was one of the first to hypothesise on exactly how birds might do this – he suggested that they might have a capacity to memorise landscape features, and thus retrace an outward journey with precision (Darwin 1873). Since then, hundreds of scientists have explored this phenomenon, and thousands of experiments have been conducted to try to uncover how these feats are achieved. Despite this effort, however, much about the mechanics of 'true' navigation in birds remains a mystery.

One of the seminal works in the science of bird migration involved an experiment by Perdeck (1958) in the Netherlands, where he and his colleagues captured more than 11,000 migrating Common Starlings *Sturnus vulgaris* over the course of several years and shipped them via aeroplane to Switzerland – about 650km perpendicular to their normal NE to SW migration route. The aim of the experiment was to see whether the birds were able to detect this displacement, which effectively mimicked the impact of a strong side-wind, and adjust their migratory direction to get back onto the correct track. A bird equipped only with a clock and compass would not necessarily be able to do this, as detecting displacement would require having some frame of reference (i.e. a map) to know that your position had shifted relative to your goal. Their results showed unequivocally that juvenile Starlings failed to detect and adjust for their displacement – they continued to migrate in a southwesterly direction from Switzerland, heading toward Spain. Adult Starlings, however, were able to adjust their orientation to the north-west, and many migrated back toward their 'intended' wintering sites in north-west Europe. This was the first study to establish the capacity for 'true' navigation in a wild migratory bird – in other words, the ability to accurately assess their position relative to a goal – and also to demonstrate that this capacity comes only with age.

◀ Masked Wagtail *Motacilla alba personata*, Camrose, Pembrokeshire, Wales, 30 November 2016. This, the first British and fourth European record of this Asian subspecies in the White Wagtail complex, was thousands of kilometres out of range (*Chris Batty*).

The development of a navigational map sense has been studied in most detail in homing pigeons, where studies have revealed a complex and multifaceted process that individual pigeons use to build up maps of their world with which to navigate. It appears that these maps are formed by birds memorising gradients in environmental features, with intersections between these gradients acting like spatial waypoints to help birds triangulate their position in space (Cherntsov *et al.* 2017). The actual cues they use to build these gradients are still subject to debate, although there is now some conclusive evidence that a key part of the gradient map is magnetic intensity. Pigeons appear to be highly sensitive to variations in the intensity of the magnetic field in space – and these variations, caused by features of Earth's geography, may be fairly stable over time and allow birds to build up a 'magnetic map' of areas they have explored in relatively fine detail. Magnetic maps have since been detected among adults of many other migratory species (Cherntsov *et al.* 2017), suggesting that the magnetic field is extremely important not only for the avian compass sense, but also for mapping out the world around them.

Intriguingly, there is considerable evidence to suggest that another important component of the mental maps birds generate involves their sense of smell. Experiments with pigeons have pointed strongly towards this – with pigeons experimentally deprived of the sense of smell being consistently disoriented when released at unfamiliar sites. More surprisingly, a recent experiment demonstrated a similar phenomenon in adult Grey Catbirds *Dumetella carolinensis* (Holland *et al.* 2009). Researchers trapped a number of these birds and subjected some of them to a zinc sulphate treatment that removed their sense of smell. After flying the birds 1,000km east from Illinois to Princeton, they radio-tracked the birds from a light aircraft and found that their untreated control birds were able to correct for displacement and return to their appropriate heading, whilst the smell-deprived birds could not (Holland *et al.* 2009). Another study recently found that both the visual and olfactory areas of the brain become more active at night during the migratory period, while they are most active during the day when birds are not migrating (Rastogi *et al.* 2011). This again suggests that smell may play an important role in migration, at least in experienced adults that have been able to map out a 'smell-scape' over the course of their previous movements. Disruption to the olfactory sense, for example through exposure to pathogens, or exposure to abnormal smells, could therefore plausibly play a role in driving vagrancy in some birds, especially adults.

▲ Whooper Swans *Cygnus cygnus*, Fair Isle, Shetland, Scotland, 5 October 2018. The crashing of waves on Fair Isle's cliffs may have been audible to this group of Whooper Swans arriving from Iceland long before the cliffs themselves came into sight (*Alexander Lees*).

A final feature that may form part of the avian map sense is sound – and more specifically, the spectrum of low sounds known as 'infrasound', which can be audible over hundreds if not thousands of kilometres. Researchers have hypothesised that birds could use stable geographic sources of infrasound as navigational cues – for example, deep sounds generated by waves along a rocky oceanic coastline, or wind-scraping sounds along rugged mountain ranges. The remarkable capacity for oceanic islands to attract vagrants could thus in part be explained by birds orientating towards the sound of waves breaking on the island shore. Direct evidence for navigation using sound has so far largely been limited to studies of pigeons, whose homing performance can be severely disrupted by infrasound disturbance, including the frequent disruption of pigeon races by sonic booms of aircraft (Hagstrum 2000). However, an experiment that removed the cochlea of homing pigeons did not produce any deficits in homing performance (Wallraff 1972). This suggests that while infrasound can clearly be part of the map sense, it is perhaps not a central component.

An important question in the context of vagrancy – if adult birds typically possess such a finely-tuned mental map of their world – is why some adult birds still end up as vagrants? One possibility is that some birds reach adulthood without managing to accurately internalise a working map during their developmental years. There are likely to be significant cognitive challenges involved in this complex feat, and it is perhaps inevitable that some individuals fail to build an accurate working map during their first migration, particularly if they undergo spells of disorientation due to poor weather, or other factors that might interrupt their ability to perceive key gradients. Birds migrating during solar flares, for example, might see considerable interruptions to the magnetic field, scrambling or confusing their ability to map out stable patterns of magnetic intensity. Another intriguing possibility is that exposure to disease might inhibit the development of the map sense – migratory birds are almost uniquely susceptible to exposure to new pathogens as they move around the world, and therefore may be highly prone to conditions that might inhibit their senses – particularly the ability to detect olfactory cues.

Another important facet of the map sense to recognise is that it is likely to be built around a combination of environmental gradients and 'beacons' that an individual has mapped out through direct experience. The map might therefore cease to be useful if weather displaces the bird outside the zone it has previously inhabited. Whilst adult birds are often able to compensate for wind drift, they can only do this when winds are relatively light – strong winds may still see them drifted for large distances (see page 36). If this drift pushes an adult bird into a new region outside its experience-based map, its ability to navigate may be seriously compromised. Once outside the realm of their internal map, such birds may be forced to revert to their innate clock-and-compass program of navigation – this would allow them to resume a normal migration heading after the drift event, but they would still be moving outside their normal migratory range. Alternatively, displacement outside the map-zone could lead individuals to make more significant navigatory errors, particularly if they attempt to orientate using the cues that form their internal map sense (e.g. olfactory cues or infrasound). For a bird displaced by winds to a new area, these cues might place a bird on a very different heading than it would have followed prior to displacement.

Another interesting wrinkle concerning the avian map sense relates to species that perform a loop migration, whereby the autumn migration route differs from that taken in the subsequent spring. Many species migrating between Europe and Africa perform looped routes – tracking studies of Common Cuckoos *Cuculus canorus*, for example, revealed that birds from central Europe take a very specific and circuitous route through western Africa during their southbound migration but follow a much more direct trans-Saharan route when returning in spring (Willemoies *et al.* 2014). For these species, the mental map developed during the autumn migration would be of little use in the following spring, as the return migration would not pass through the same regions – any waypoints or mapped gradients they have internalised would be irrelevant. The first spring migration must therefore be completed using the same error-prone clock and compass mechanism used in the first autumn. An interesting prediction from this is that spring vagrancy may be more prevalent in species that perform a loop migration, compared to equivalent species that retrace their autumn route the following spring – the latter would be able to navigate using their experience-based map sense and thus may be less prone to navigational errors, as well as more able to control for wind drift. To our knowledge, this prediction has yet to be empirically tested.

Finally, it is also worth noting that our broad generalisation of true navigational abilities being present in adult birds but not juveniles may in fact be far from clear-cut. There is compelling evidence that in some long-distance migrant species, juveniles do indeed show compensatory behaviour after being experimentally displaced, suggesting that they have some sense of where they have been moved to, relative to their previous position (Åkesson *et al.* 2005; Thorup & Rabøl 2007). Perhaps more compellingly, some of the migration routes undertaken by juveniles would appear simply impossible without some sense of where the individual is relative to its goal (i.e. a true map sense) – examples include species such as Marsh Warbler *Acrocephalus palustris*, which spread out over large areas of Eastern Africa during their southbound migration but later converge into narrowly defined routes, suggesting birds were aware they had reached some form of waypoint that triggers a complex and directed switch in orientation (Thorup & Rabøl 2001). For juvenile birds to perform such migrations alone, this implies that some form of innate map sense may be genetically encoded in the migratory program of these species, beyond the simple clock-and-compass program that is usually assumed.

◄ Marsh Warbler *Acrocepalus palustris*, Moscow, Russia, July 2007. The circuitous migratory route of this species across Africa involves multiple changes in orientation – the fact that inexperienced juveniles can navigate this successfully suggests they may be able to track their progress using an innate 'mental map', perhaps based on geomagnetic cues (*James Gilroy*).

VAGRANCY THROUGH COMPASS ERRORS

It is understandable that laypeople frequently assume that vagrant birds are usually 'blown off course' by adverse weather, but even the most cursory analysis reveals that a significant proportion of long-range vagrancy events cannot be explained by exogenous factors like winds alone. Vagrancy events are near constant, global and often entirely uncorrelated with weather patterns, indicating that there are invariably other factors at play – and indeed it is likely that most incidences of long-distance vagrancy are driven by factors that are endogenous to the birds themselves. Failures in the compass system are perhaps the most obvious and pervasive mechanism that could cause birds to stray from their normal ranges – even humans equipped with satellite GPS can still find navigation difficult and frequently get lost! It is therefore unsurprising that birds, especially inexperienced juveniles, commonly get their navigation wrong. And if embarking on a journey that spans continents, as many migratory birds do, even a small error in your initial bearing can lead to a very big error when the miles start to accumulate.

Many previous authors have considered the types of compass errors that might explain the patterns of vagrancy we observe, and several key candidates are often invoked – most prominently the associated phenomena of 'reverse migration' and 'mirror-image misorientation'. Whilst these two candidate mechanisms should feature prominently in any analysis of vagrancy causes, in reality they form part of a far broader spectrum of errors that could arise from the hierarchy of navigational tools that birds use. In this section, we will examine vagrancy scenarios arising from these errors, and consider their relative likelihood in light of recent research on the nature and mechanisms of avian navigation. To begin, we will focus on the mechanism that has arguably received most attention from scholars of vagrancy worldwide and is sometimes invoked as a blanket explanation for vagrancy among inexperienced migratory birds: reverse migration.

Reverse migration as a cause of vagrancy

When a migrating bird sets out on a migratory flight solo, it must make a decision about which direction to fly in, and subsequently ensure that it continues in that direction for the duration of the flight. Two steps are involved in this process – the first is to identify a reference point on the compass (i.e. north), and the second is to angle itself correctly with respect to that reference point, such that it is facing its genetically determined migratory direction (see page 9). In other words, a bird must 'think' something along the lines of "*OK, I know that north is that way, and I need to point myself 55 degrees to the right of north*". Mistaking south for north in step 1 – a simple error to make – would therefore lead our bird to migrate in the opposite direction to its innate heading, assuming it applied step 2 correctly with respect to this erroneous reference point. If the bird made this same error consistently throughout each migratory hop, the result would be a full reversal of its normal migration route.

► Red-breasted Flycatcher *Ficedula parva*, East Hills, Holkham NNR, Norfolk, England, 16 September 2009, a scarce passage migrant to the British Isles whose regular arrival in autumn many authors have attributed to reverse migration (*Alexander Lees*).

There is reasonable evidence that this form of reverse migration – i.e. individuals consistently mistaking north for south each time they check their migratory orientation – is a potentially significant cause of vagrancy, particularly among juveniles in their first migration. Patterns of relative abundance for vagrant species can often be explained reasonably well by models that assume reverse migration to be the principal causal mechanism of vagrancy – most classically, the occurrence of Siberian passerine migrants in north-west Europe each autumn (Thorup 1998). In the European context, a seminal example was put forward by Cottridge & Vinicombe (1996) involving the autumn occurrence patterns of two closely-related flycatcher species – Red-breasted Flycatcher *Ficedula parva* and Collared Flycatcher *Ficedula albicollis*. Both species breed in the temperate forests of central and eastern Europe, but they have very different migratory routes and wintering grounds. European populations of Red-breasted Flycatcher migrate south-eastwards in autumn to reach wintering grounds in southern Asia, whilst Collared Flycatchers from the same breeding regions migrate broadly due south to winter in East Africa. A reversal of the latter's migration route would take vagrants into eastern parts of Scandinavia and Russia but would be unlikely to carry them to north-west Europe (Figure 4). For the former, however, a reversed route would see individuals moving north-west towards the British Isles (Figure 4). As predicted under reverse migration theory, Collared Flycatcher is practically unknown in the British Isles in autumn, whereas Red-breasted Flycatcher is a frequent visitor with dozens or even hundreds recorded annually.

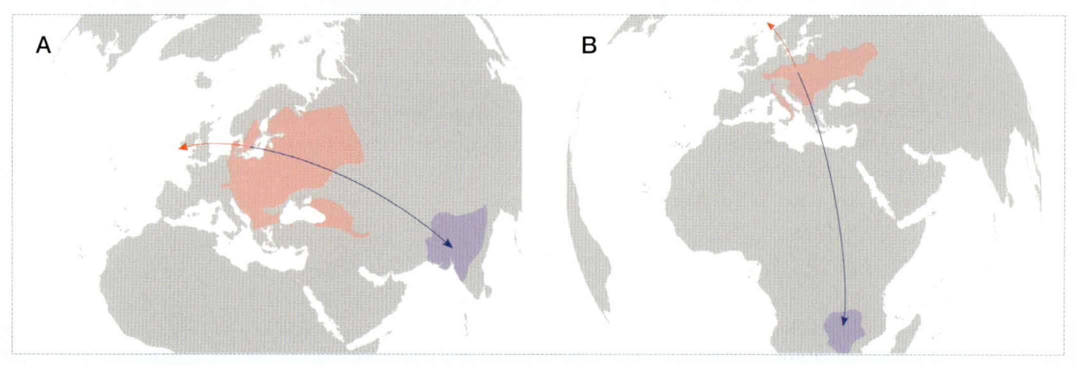

▲ **Figure 4.** Reverse migration routes (red) and normal migration routes (blue) for Red-breasted Flycatcher *Ficedula parva* (A) and Collared Flycatcher *Ficedula albicollis* (B). Only the former is a regular autumn vagrant to the British Isles.

Many other patterns appear to fit well with the reverse migration model. At vagrancy hotspots in Alaska, numerous frequently occurring autumn Siberian vagrants appear to be following direct reversals of their normal migratory routes from breeding areas in eastern Siberia and Kamchatka towards Indochina (Howell *et al.* 2014). The roll-call of autumn Old World vagrants to Japan and the Pribilof and Aleutian Islands of Alaska is long. It includes numerous species that would seem at face value extremely unlikely ever to reach the Pacific – including such species as Spotted Flycatcher *Muscicapa striata*, Wood Warbler *Phylloscopus sibilatrix* and Sedge Warbler *Acrocephalus schoenobaenus* which are largely restricted to western Eurasia as breeders, with wintering quarters exclusively in Africa. Despite the apparent remoteness of their ranges from the north-west Pacific, a reversal of the expected migration route from their westernmost breeding populations would indeed bring birds directly to the Bering Sea islands (Howell *et al.* 2014).

Given the apparent power of the reverse migration model to explain patterns of autumn vagrancy, it is interesting to consider just how such an error might arise. Clearly, reverse migration would involve mistaking north for south – effectively reading the compass upside down. As explained in the previous section, however, even naive juveniles should have between two and four different compass mechanisms at their disposal, each of which should be available to act as a fail-safe against erroneous readings from the others. If birds have multiple different sources of information at hand to help them find north and south – why is it that some birds apparently make this same consistent error, time and again, across all these compass sources?

One potential answer to this question may come from studies examining the way birds respond when their various different compass mechanisms each give conflicting indications of where north is. Experiments with caged birds show that exposing a young bird to conflicting signals – for example, a misaligned magnetic field alongside the correct star map for their location usually leads to birds focusing on one compass sense as a 'default' and ignoring the others (Muheim *et al.* 2006). Prior to the start of migration, it appears that this default

► Wood Warbler *Phylloscopus sibilatrix*, Gongendo Park, Satte, Saitama, Japan, 29 Oct 2016 – a major rarity in the Eastern Palearctic and candidate 'reverse migrant' (*Yann Muzika*).

is usually the star compass, but once birds start moving, they switch to trusting the magnetic compass above the others. A bird that has some fundamental problem with its magnetic compass could therefore consistently make compass errors throughout its migration, as it would inevitably get strongly conflicting indications from its different compass senses and would instinctively default to the erroneous magnetic sense every time it takes flight. Interestingly, there are further lines of evidence to suggest that the magnetic compass may play a particularly important role in causing reverse migration. As described in the previous section, the magnetic compass functions by birds 'viewing' the magnetic field around them using specialised structures in their retina (and/or magnetic particles in other sensory organs). This field 'flows' between the north and south magnetic poles, forming lines that are apparently visible in almost all conditions except complete darkness (Ritz *et al.* 2009). However, at any given moment, it may not be so obvious to the bird which 'end' of the magnetic line points towards the pole, and which points to the Equator. To determine this, a bird must evaluate the angle of the lines with respect to Earth's surface, and calculate the point where the lines 'dip' towards the horizon, rather than incline away from it (Ritz *et al.* 2009) – this being the poleward direction. Errors made in evaluating this dip angle could cause individuals to mistake north for south – and they may plausibly then do this consistently throughout migration (Muheim *et al.* 2006).

Another possible explanation for reverse migration relates to the way that the migratory program is encoded in young birds genetically. It is likely that this program must contain navigatory instructions for both the autumn migration and that of the subsequent spring. One possibility is that some individuals erroneously follow the migratory program for the wrong season, following the innate bearing that would be appropriate for their spring migration. These birds are therefore able to correctly read their various compasses, but they end up following the wrong innate heading for the season they are in (Pfeifer *et al.* 2007). Such an error would lead to reverse migration, provided that the innately encoded spring migration route of the species directly retraces that of the autumn. Many species follow a loop migration, however, whereby a different route is used in spring – such species might therefore follow a very different route than a reversal of their expected autumn migration under this hypothetical mechanism.

We can propose one further possible explanation for reverse migration, this time relating to another facet of avian migration that is quite commonplace, but seldom considered as a potential cause of long-distance vagrancy: the short-term reversal of orientation that birds often make in response to adverse weather. This phenomenon is often noted at migration hotspots worldwide, where in some conditions very large numbers of migrants appear to move *en masse* in the 'wrong' direction (Shamoun-Baranes & van Gasteren 2011). Usually, these events correspond with conditions that are unsuitable for onward migration; periods of very cold weather in spring, for example, often result in significant reverse flights of migratory birds in northern Europe (Lindström & Alerstam 1986). Radar studies have further confirmed the size and frequency of these events, with mass reversals often resulting in huge numbers of migratory birds temporarily moving 'backwards' along their normal route after encountering bad weather (Shamoun-Baranes & van Gasteren 2011).

These mass reverse flights are unlikely to lead to vagrancy by themselves – radio tracking suggests that such reverse flights are usually short-lived, with individuals moving tens to hundreds of kilometres backwards before stopping to rest and feed, prior to continuing in the correct direction once conditions improve (Nilsson & Sjöberg 2016). One intriguing possibility, however, is that the behavioural mechanism that underpins short-term reversal is also at play in driving vagrancy through sustained reverse migration. Several studies have shown that short-term reversal is strongly linked to the condition of the bird – birds in lean condition are far more likely to exhibit temporary reversed orientation than birds carrying fat reserves (Deutschlander & Mulhelm 2009). This suggests a strong innate response in migrants that run low on energy reserves, where a certain trigger causes them to switch their migratory orientation 180 degrees and 'back-track', helping them find better feeding conditions and avoid competition in coastal areas where large numbers of migrants may be grounded. Responses such as this are often referred to as 'reaction norms' – when a particular cue (weather conditions or body fat) reaches a certain level, a reaction is triggered – in this case the migratory orientation switches by 180 degrees. It is plausible that some individuals might become trapped in this reaction state, with their orientation system effectively stuck in reverse, even if the conditions that triggered the reaction return to normal. Such a scenario could hypothetically explain why some individuals consistently follow a reverse heading, despite having multiple compasses as fail-safes and multiple opportunities to detect their orientation error throughout the course of a migration.

Reverse migration in spring

Reverse migration is commonly invoked as a cause of vagrancy during autumn migration, but there is also compelling evidence that the same phenomenon can occur among individuals making their return migration in spring. Apparent examples of this phenomenon include the appearance of numerous Palearctic-Afrotropical migrants as vagrants in South Africa or even on subantarctic islands in the northern hemisphere spring, when birds should be returning northwards to their breeding grounds. A study of vagrant passerines on Marion and Prince Edward Islands in the far south of the Indian Ocean revealed remarkable records of many Eurasian species such as Barn Swallow *Hirundo rustica*, Common Whitethroat *Sylvia communis* and Willow Warbler *Phylloscopus trochilus*, with individuals often reaching this remote archipelago during the months of April and May (Oosthuizen *et al.* 2009). The trajectory of these birds is consistent with a reversal of the correct spring migration route from their wintering grounds. The frequent occurrence of Siberian vagrants on islands off northern Australia during the northern hemisphere spring might also fit this pattern (Clarke *et al.* 2016). Birders have recently uncovered a number of vagrancy hotspots, such as Ashmore Reef, Queensland, Australia, where it may be possible to find many Siberian species that normally winter in equatorial South-east Asia but have apparently reversed their northbound migration in spring to find themselves moving deep into the Southern Hemisphere. There is, however, also the possibility that some may have wintered further south in Australia and are being detected returning north, rather than heading south. The pattern of spring occurrence of several New World warblers (family Parulidae) and other Nearctic migrants in southern South America is also potentially an example of this phenomenon.

◄ Citrine Wagtail *Motacilla citreola*, Strandfontein WTP, Western Cape, South Africa, 27 April 2015. This, the fourth record for South Africa, may be a spring 'reverse migrant' that has wintered further north in Africa and erroneously migrated south on its return migration (*Niall Perrins*).

This pattern raises an interesting conundrum: whilst reverse migration appears to be the cause of these spring vagrancy events, it must also be the case that these individuals were able to orientate correctly southwards in their first migration – and hence reach their correct winter ranges – only to lose their capacity to tell north from south in the return stage. Why would this occur? One possible explanation again relates to the use of the magnetic compass and how this compass operates with respect to the Equator. As previously mentioned, that the avian magnetic compass does not function in the same way as a human compass – it indicates only the difference between polewards and Equator-wards, and does not actually identify north or south *per se*. When a bird crosses the Equator from north to south, the magnetic compass will become temporarily unreadable as the lines of magnetic field are extremely slight at very low latitudes. Moving further into the Southern Hemisphere, however, when the compass becomes readable again, it would be 'flipped' such that the lines of inclination now point towards the South Pole. To account for this change, the bird must be aware of the fact that it has crossed the Equator and be able to adjust for the fact that its magnetic inclination compass now points towards south rather than north. Any individual that then winters the opposite side of the Equator from its breeding range, but fails to mentally account for the fact that its magnetic compass has flipped, would then automatically mistake south for north in subsequent attempts to orientate.

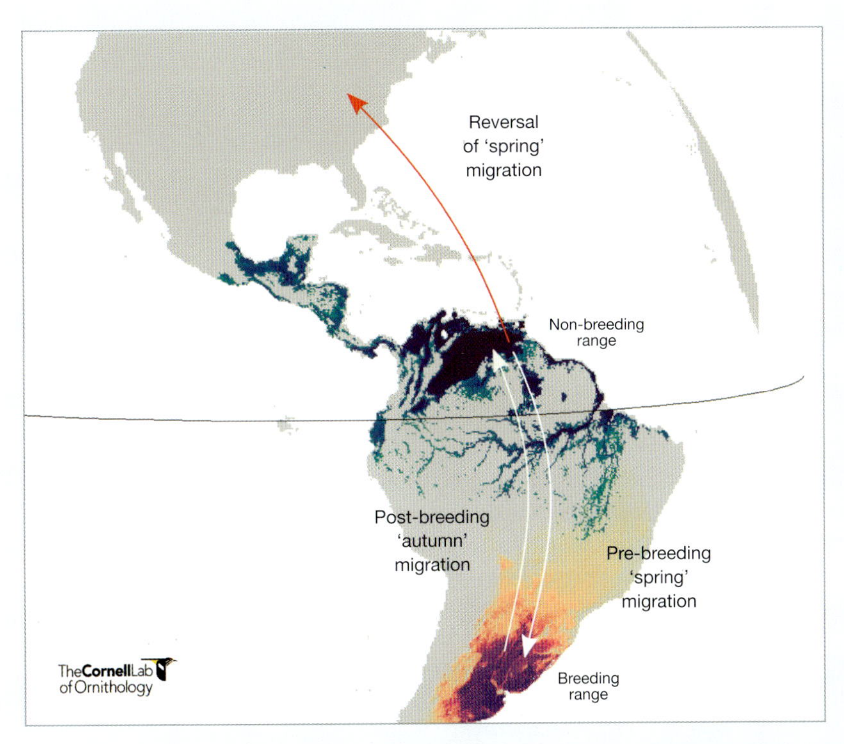

◄ **Figure 5.** Autumn vagrancy of Fork-tailed Flycatcher *Tyrannus savana* to North America by the migratory subspecies *savana* could be explained by reverse migration of the 'spring' leg of their austral migration (red arrow). This mechanism assumes that the birds in question navigate correctly on their first migration northwards but mistake north for south on the return journey – perhaps because they fail to compensate for the magnetic 'flip' that occurs when crossing the Equator. Abundance data from eBird Status and Trends provided by the Cornell Lab of Ornithology (Fink *et al.* 2020).

Clearly, birds could use other compass senses (Sun or star cues) to detect and adjust for this magnetic flip but, as previously described, these other mechanisms all require some degree of learning and are vulnerable to disruption by weather patterns. Many inexperienced juveniles might default to the magnetic compass during their first migration, and might therefore be vulnerable to navigating erroneously with a flipped magnetic compass after crossing the Equator. The same mechanism of spring reverse migration has been invoked as an explanation for the frequent occurrence of Fork-tailed Flycatcher *Tyrannus savana* and other South American austral migrants in North America (Howell *et al.* 2014). Occurrences of this species tend to peak in August–November, corresponding with 'spring' in the biological calendar of this austral migrant species. Reverse migration offers a convincing explanation for their occurrence, but again this relies on an assumption that these birds completed their first migration correctly, navigating from breeding areas in southern South America to spend their winter close to the Equator (Figure 5). In the austral spring (autumn in the Northern Hemisphere), birds that had crossed into the Northern Hemisphere would then need to correct for the fact that their magnetic compass now identifies north rather than south – any individuals that fail to recognise this change might perform a reverse spring migration to North America, if they were to rely solely on their magnetic compass.

◄ Fork-tailed Flycatcher *Tyrannus savana*, Pointe-Lebel, Manicouagan, Quebec, Canada, 14 October 2017. Appearing in Canada in October, this individual may have wintered successfully in northern South America, but instead of heading back south it has ventured north and ended up at least 7,000km away from the breeding areas of the migratory nominate subspecies in central Brazil (*Ian Davies*).

Mirror-image misorientation

As noted above, in order to orientate correctly using its internal compass, a bird must complete two steps – the first being to identify a reference point (north), and the second to rotate itself to the correct angle with respect to this reference point. Mirror-image misorientation can theoretically arise when a bird gets the first step correct (i.e. it identifies north and south) but then mistakes left for right when pivoting its body with respect to this compass reference point.

This mechanism was first proposed as a cause of vagrancy by De Sante during his studies of the annual appearance of migrants from eastern North America at vagrant hotspots in coastal California (de Sante 1973). De Sante observed that the relative abundance of some species, most notably eastern boreal migrants that normally migrate south-east along the Atlantic seaboard in autumn, could not be explained by weather patterns alone. In particular, the relative abundance of different eastern vagrant species in California was not straight-

◄ Blackpoll Warbler *Setophaga striata*, Cordell Bank pelagic, California, United States, 9 September 2008. This individual, which boarded a boat well offshore, seems to be on an orientation destined to take it to a watery Pacific grave (*Steve Howell*).

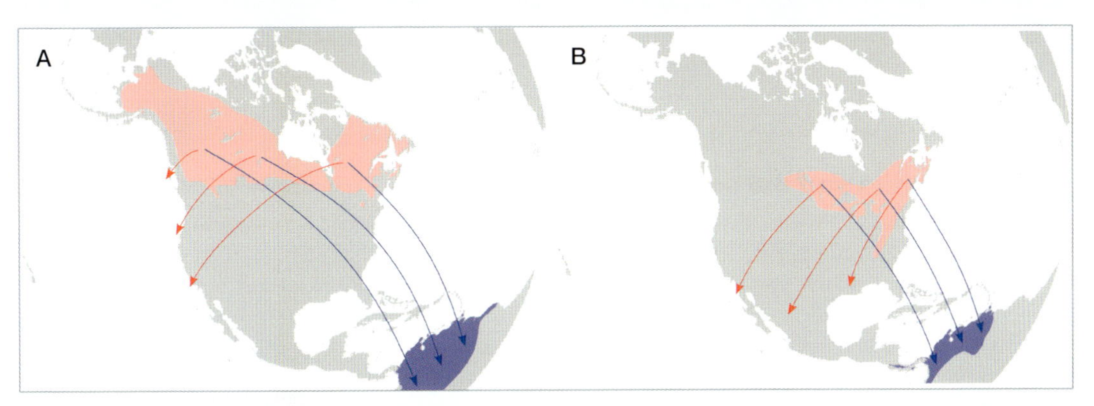

▲ **Figure 6.** Mirror-image routes (red) and normal migration routes (blue) of Blackpoll Warbler *Setophaga striata* (left) and Blackburnian Warbler *Setophaga fusca* (right). The higher relative frequency of the former as an autumn vagrant to California could be explained by mirror-image misorientation, although many other factors are also likely to be at play.

forward to explain in simple terms of population size, and it appeared to be related to migration routes. For example, one of the most frequent autumn vagrants to the region is Blackpoll Warbler *Setophaga striata*, a species that breeds abundantly in boreal forests across North America but migrates almost exclusively along the eastern North American coast in autumn. Populations at the western extreme of the breeding range must therefore initially follow an eastward orientation, a mirror image of which would bring birds directly to California (and other points along the Pacific coastline) (Figure 6). It is interesting to contrast this with another abundant boreal breeder – Blackburnian Warbler *Setophaga fusca* – a species with a similar migration but a breeding range that does not extend so far west across the North American continent. A mirror image of the autumn orientation for westernmost populations of Blackburnian Warbler would only take vagrant individuals as far west as New Mexico (Figure 6), with a low likelihood of bringing them to California. As predicted by the theory, Blackburnian Warbler is one of the rarer autumn vagrants in the state, outnumbered by Blackpoll Warblers more than 20 to 1 (although it is important to recognise that Blackpoll Warbler also has a far larger population size and breeds much closer to California than Blackburnian Warbler).

Further tentative support for the mirror-image orientation hypothesis came from tests carried out on vagrant Blackpoll Warblers trapped for ringing in California. Measurements of their overnight migratory orientation showed a mix of individuals showing both the 'normal' south-eastward orientation and its mirror image (De Sante 1973) – suggesting that at least some individuals were indeed fitting the expected pattern of misorientation. It is difficult to know how much to read into the apparent mixture of orientations shown by these tested birds, as the sample size was small. One possible explanation, however, is that individuals may not make the mirror image error consistently every time they attempt to orientate. In other words, their compass is not

► Tree Pipit *Anthus trivialis*, Hegurajima Island, Ishikawa, Japan, 23 April 2019. This vagrant to Japan might conceivably fit the pattern of mirror image misorientation (*Yann Muzika*).

inherently 'flipped', but when attempting to orientate ahead of any given flight, they may sometimes pivot the correct way, and at other times get it wrong. Such a mixture of mirror-image and correct orientations over the course of a migration could carry birds on a wide variety of possible route deviations, and hence give rise to a plethora of different vagrancy outcomes.

Compass errors and the axis of migration

A statistical analysis by Thorup (2004) found strong evidence that the frequency of vagrants within 'reverse migration shadows' in Northern Europe was far greater among species that migrate in an easterly or south-easterly direction in autumn (e.g. from Eastern Europe to South Asia) relative to species that migrate primarily north–south (e.g. from southern Europe to Africa). The seminal example of reverse migration described above, involving Red-breasted and Collared Flycatchers, also illustrates this pattern quite neatly – the west–east migrating Red-breasted Flycatcher is a common autumn vagrant within its reverse migration shadow in Western Europe, but the north–south migrating Collared Flycatcher remains extremely rare even in the zone covered by a reversal of its migration in Scandinavia (Figure 4).

What might explain this link between east–west migration and the likelihood of individuals making significant compass errors on their first migration? One possible explanation is that this simply reflects the relative difficulty of making orientation decisions when the desired direction deviates significantly from either north or south. If we recall our model of how a bird uses its compass to orientate, this involves locating north and then pivoting by a certain amount to ensure the bird is pointing in the correct direction. It is possible that the further a bird has to pivot from its north–south reference, the greater the likelihood it will become confused and/or unable to keep track of where north is. This seems particularly likely in the case of birds orientating using the magnetic compass, which as described above requires birds to evaluate the angle of dip in the lines of magnetic field visible in the sky above them. When a bird has to orientate itself horizontally with respect to these lines (which is the case when moving east or west), evaluating which way they dip may be more difficult than for when a bird's desired orientation is closer to the angle of the lines themselves.

Another possible explanation for the link between east–west migration and misorientation relates to the pattern of variation in magnetic fields across regions – and in particular the region of Siberia where many vagrants to Europe originate. Across Siberia there is a strong gradient in the pattern of magnetic declination – i.e. how different the apparent direction of magnetic north is from true north. A juvenile bird traversing the steepest part of this gradient from west to east would see marked shifts in the apparent direction of magnetic north from one day to the next. One possibility is that these frequent shifts could cause confusion and disorientation, increasing the chances that individuals will make a significant error like mistaking north for south. Furthermore, if the bird was attempting to calibrate its magnetic compass using other compass cues, such as the pattern of polarised light at sunset (as has been shown in some species), the angle of calibration would have to change significantly from day to day as the lines of the magnetic field shift. Potentially, this further increases the chances of confusion. Given that huge numbers of birds successfully complete these east–west migrations each year, it is clear that most individuals have the capacity to navigate these shifting fields in one way or another. Nevertheless, the apparent predisposition towards misorientation in some Siberian species could well be related to the significant variability in magnetic fields across the east–west gradient in that region.

Magnetic anomalies

The reverse migration and mirror-image hypotheses both involve large-scale errors made when individuals 'read' their internal compass – i.e. mistaking north for south, or left for right. As such, both hypotheses assume that the bird in question possesses some form of 'working' compass (i.e. a mechanism for consistently identifying north from south) but that they 'read' this compass incorrectly. Another possible explanation for navigational errors is that some individuals may lack the capacity to correctly discern any compass heading at all – in other words, they entirely lack a functioning compass. In the case of the magnetic compass, this could plausibly arise due to local-scale anomalies in the magnetic field associated with the properties of underlying bedrock.

As described in the previous section, three of the four avian compass mechanisms require some degree of learning during early life in order to be used effectively – a high-precision clock sense is needed to use the Sun or polarisation compasses, and the patterns of stars around the point of polar rotation must be learned in order to use the celestial compass. Particularly among long-distance migrants, it is easy to imagine a scenario where a late-fledged individual might have insufficient time – and in particular insufficient windows of clear weather – to

► Siberian Accentor *Prunella montanella*, Mossy Hill, Shetland, Scotland, 10 October 2016, the first British record of an unprecedented westward incursion by this species in 2016 (*Stuart Piner*).

properly develop the use of these compasses prior to embarking on migration. Such an individual would likely rely almost entirely on their magnetic compass to navigate, as this is generally assumed to function without any need for learning, or any set of weather conditions. It may not function correctly, however, in areas of significant magnetic anomaly (Winklhofer *et al.* 2013).

Experiments with homing pigeons and other species have repeatedly shown that they can become disorientated when passing over magnetic anomalies – and in particular zones with significantly higher or lower magnetic intensity than their surroundings (Wiltschko & Wiltschko 2005). Such anomalous zones tend to be quite concentrated in space – perhaps a few kilometres in diameter – and therefore a bird passing over such an area during migration should become disorientated only briefly, before subsequently re-finding its correct heading after moving away. A juvenile migratory bird that is hatched within such an anomaly, however, could plausibly develop a very different magnetic compass sense than birds fledging in surrounding areas. Importantly, experimental evidence from pigeons suggests that the magnetic sense is 'tuned' to a narrow band of magnetic intensity around that experienced by a bird in early life (Winklhofer *et al.* 2013). This means that a bird fledged within a magnetic anomaly zone may struggle to orientate correctly once it moves out into areas with more normal magnetic conditions, as its compass is tuned to the unusual pattern of intensity of its natal area. Experiments suggest that birds can re-tune their compass to new intensities, but this takes many days to achieve, and may not be possible if the differences are very stark (Winklhofer *et al.* 2013).

► Pechora Pipit *Anthus gustavi*, Gambell, Nome, Alaska, United States, 26 August 2004, a predictable species at vagrant traps in northern Europe and the Bering Sea region; it breeds at high northern latitudes overlapping areas with magnetic anomalies (*Brian Sullivan*).

It is possible to consider a set of 'perfect storm' circumstances for a juvenile bird originating from within such an anomaly. If such a bird fledged late in the season, it may not have had sufficient opportunities to properly learn its star map or Sun compass and would therefore need to rely solely on its magnetic compass. Upon commencing migration, however, the bird would move out of the anomalous zone and immediately lose its ability to correctly orientate using its one fully functional compass. Such a bird might become completely disorientated, and potentially follow a random heading on its subsequent migratory flights – potentially ending up almost anywhere within the plausible flight range of the species. Such a scenario may seem unlikely, but when we consider the tens of millions of migratory birds that fledge each year, it is likely that circumstances such as these happen quite frequently. Perhaps most tellingly, one of the regions known to have some of the strongest local magnetic anomalies worldwide is western Siberia (Alerstam 1990), a region supporting a suite of species that are well-known for long-distance vagrancy to both Europe and North America.

Limitations to the reverse migration and mirror-image hypotheses

Although the reverse migration and mirror-image hypotheses can apparently explain some broad patterns of vagrant occurrence around the world, many other patterns are hard to reconcile with either model. Consider the example of an archetypal Siberian vagrant to Europe, and one that is often seen as a classic example of a reverse migrant – Yellow-browed Warbler *Phylloscopus inornatus* (Cottridge & Vinicombe 1996, Thorup 2004). This species breeds across Siberia, reaching its western limit in the Ural mountain range, and it winters in South-east Asia. Numbers of vagrant individuals in Western Europe have increased markedly in recent years, and the species is now a relatively frequent sight in late autumn at migration hotspots along the Atlantic seaboard.

Interestingly, tests of the migratory orientation of vagrant Yellow-browed Warblers trapped on the Faroe Islands suggested that these birds were predominantly migrating along west or south-westerly headings, on a route that would likely see them either reach Iberia or perish in the Atlantic (Thorup *et al.* 2012). Thorup *et al.* (2012) hypothesised that these patterns could be explained by Yellow-browed Warblers following a 'great circle' route in reverse and failing to make the expected switch in orientation that would normally happen when birds reach the Pacific Coast – a reversal of the southward switch normally made would carry reverse-migrating birds northwards towards the pole (Figure 7). However, there is scant evidence that juveniles of any species are capable of navigating along great circles, as this is a complex feat requiring constant adjustments of orientation to maintain the correct course (Åkesson & Hedenström 2007). Rather, evidence suggests that migrants usually follow a single directional bearing for each leg of the journey (Åkesson & Hedenström 2007), resulting in a 'rhumb line' rather than a great circle route. Interestingly, reversal of a rhumb-line route does not carry Yellow-browed Warblers as far as the British Isles but rather tracks much further north towards the Arctic (Figure 7).

◄ Red-flanked Bluetail *Tarsiger cyanurus*, North Ronaldsay, Orkney, Scotland, 26 September 2008, now a routine autumn vagrant to Western Europe, which seems to be associated with its concurrent westward range expansion (*Alexander Lees*).

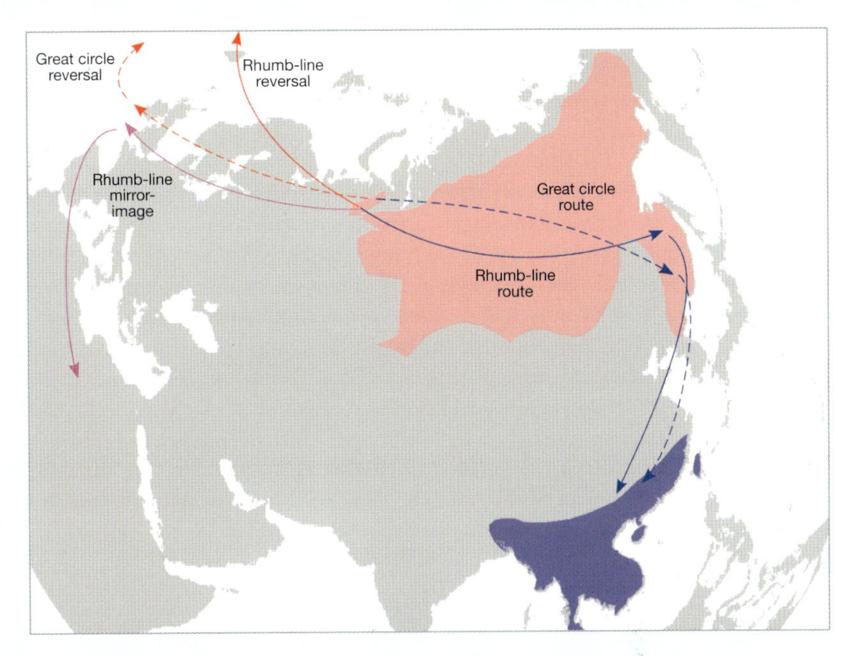

► **Figure 7.** Reversal and mirror-image migration trajectories of Yellow-browed Warblers *Phylloscopus inornatus* from the western range extent. Reverse migration (red) along a great circle route (dashed line) could explain the frequency of the species in western Europe, but there is scant evidence that any young birds are capable of great circle navigation. Rhumb-line navigation is more likely, but reversal of this would send birds north of the British Isles. Mirror-image misorientation of a rhumb-line heading is perhaps more plausible, but we would expect then to see more birds reorienting through the eastern Mediterranean in late autumn.

However, thousands of Yellow-browed Warblers are now recorded annually in the British Isles in autumn, with a clear seasonal pattern of initial arrivals in the north and a gradual filtering south-west (de Juana 2008). The species also appears in considerable numbers in Iberia during late autumn, some weeks after the peak of abundance in the British Isles and Scandinavia, again suggesting the birds are largely travelling on a south-west heading after arrival in northern Europe (de Juana 2008).

Mirror-image misorientation is somewhat more plausible as an explanation for vagrancy in this species, as mirrored routes would bring larger numbers of birds into northwestern Europe, in line with the normal pattern of observation (Figure 7). The south-west and westerly orientation of birds on the Faroe Islands observed by Thorup *et al.* (2012) would also fit well with the mirror-image hypothesis. However, a direct mirror route would then see birds reorienting back towards the eastern Mediterranean, reflecting the dog-leg nature of the normal route (Figure 7). In reality, there are far more records in the western Mediterranean than in the east, suggesting that a simple mirror mechanism does not perfectly explain the observed pattern either (de Juana 2008).

It is also worth noting that Yellow-browed Warbler's breeding range stretches eastwards almost as far as Kamchatka (Figure 7). Both reverse migration and mirror-image misorientation at the eastern end of the breeding range would see vagrants heading across the Bering Sea towards the Pribilof and Aleutian Islands in Alaska. There are indeed multiple autumn records of Yellow-browed Warblers from these vagrant hotspots, and numbers seem to be increasing (Howell *et al.* 2014). However, this species is still an extremely rare vagrant

► Yellow-browed Warbler *Phylloscopus inornatus*, Melrakkaslétta, Norðurland eystra, Iceland, 7 Oct 2014. This bird is an increasingly common feature of autumn, not just in Iceland but across western Europe (*Yann Kolbeinsson*).

in Alaska, despite there being multiple well-watched sites located firmly within the 'reverse migration shadow' at the eastern end of the range. If reverse migration was the primary mechanism driving the movement of thousands of Yellow-browed Warblers into Europe each autumn, why is the same thing not happening at the other end of the range?

One possibility is that Yellow-browed Warblers have established a new and self-sustaining migration route through Europe to unknown wintering areas in Western Africa (Gilroy & Lees 2003). Individuals using this route could therefore be 'correctly' following their internal compass, tracing a migratory program they inherited from their parents – they would thus not be vagrants at all but members of a viable subpopulation that was initially founded by vagrant individuals (perhaps under mirror-image misorientation). We have previously termed such cases 'pseudo-vagrants' (Gilroy & Lees 2003), as the individuals involved do not truly meet our definition of vagrants (birds outside their 'normal' range). In the case of Yellow-browed Warbler, the relative paucity of observations in possible African wintering areas, as well as the limited spring passage through Europe, perhaps suggests that this 'pseudo-vagrant' explanation may be unsatisfactory (de Juana 2008). However, wintering site fidelity has been confirmed for a Yellow-browed Warbler trapped and ringed in January 2018 near Tarifa in Andalucía, Spain, which was refound in November of the same year (Tonkin & Gonzalez 2019). Recent work with stable isotopes (deuterium-to-hydrogen ratios) suggests that birds passing through Scandinavia had their origins in the west of the species' range in the Ural Mountains (de Jong *et al.* 2019), but it is difficult to say whether these individuals are following a viable new migration route or are simply victims of a common (and as yet unidentified) navigation error.

Reverse migration and mirror-image misorientation both involve basic compass errors that probably arise commonly in nature, and there is little doubt that they are major causes of vagrancy in migratory species. However, they are likely just two of many factors that can cause navigation errors. Vagrants frequently appear in areas well outside their 'reverse migration shadows' (Cottridge & Vinicombe 2004), indicating that these simple models cannot offer a blanket explanation for all vagrancy. Indeed, for many species patterns of vagrancy do not correspond with reverse migration shadows at all – for example, Blyth's Pipit *Anthus godlewskii* and Richard's Pipit *Anthus richardi* are both regular vagrants to Western Europe, but this region lies completely outside the reverse shadow of their normal migration, which traverses broadly north–south across Central Asia. Indeed, reverse migration would predict both species to occur more frequently in the Bering Sea region, where neither has ever occured (Lees & VanderWerf 2011).

An element of randomness is inherent in vagrancy, as it is in any species trait. The migratory program that birds inherit genetically is likely to be a key source of this random variation – many vagrants may in fact therefore be reading their internal compasses correctly but following an errant heading. Such errors are likely to arise commonly due to mutations or transcription errors during the process by which migratory orientations are inherited. Such directional variability is likely to generate a universal degree of 'scatter' in migratory

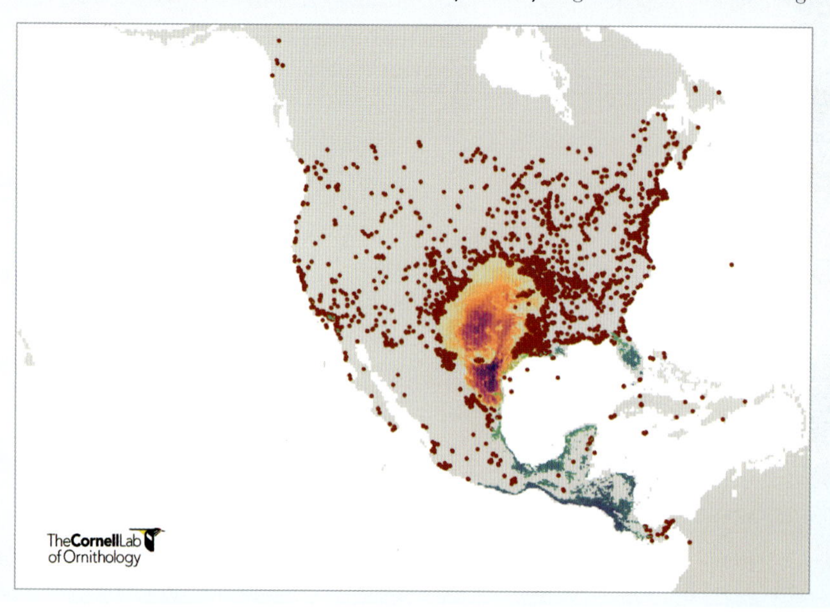

◄ **Figure 8.** Vagrant occurrences of Scissor-tailed Flycatcher *Tyrannus forficatus* (red dots) around the normal breeding (orange–purple) and winter (blue–green) ranges of the species. Vagrant records extracted from eBird (Sullivan *et al.* 2014) and abundance data from eBird Status and Trends provided by the Cornell Lab of Ornithology (Fink *et al.* 2020).

orientations, with the possibilities for vagrancy being limited only by the physiological capabilities of birds setting out in those directions. The scatter of Scissor-tailed Flycatcher *Tyrannus forficatus* occurrences across North America illustrates this random pattern very clearly (Figure 8). This species is perhaps an ideal model for understanding vagrancy – its central United States distribution means there is good observer coverage in all directions where vagrants might move, so patterns are unlikely to be biased by variation in detection rates. Correspondingly, vagrants have been found in all possible directions around the breeding range, with no particular bias towards reverse or mirror-image migration shadows (Figure 8, Gilroy & Lees 2003).

In reality, understanding the exact causes of vagrancy in any given species or region will always be difficult, due to the plethora of likely mechanisms at play, as well as the imperfect nature of the data at our disposal. Observer effort is hugely variable across most regions, particularly in Eurasia, potentially creating patterns in the data that are actually little more than artefacts of where people are looking (Gilroy & Lees 2003). Whilst simple forms of compass misorientation do offer compelling explanations for many vagrancy patterns, the reality is undoubtedly more complex and multifaceted than any simple models might suggest.

Vagrancy in social migrants

Thus far, we have focussed on the difficulties migrants face when navigating using their own sensory instincts, and the mechanisms by which these instincts can lead to vagrancy. It is important to recognise, however, that relatively few birds undertake their migrations entirely in isolation from other individuals. Even among nominally 'solitary' species (a category that includes many nocturnal passerine migrants), individuals usually have a tendency to associate with others when embarking on long flights. Some species migrate almost exclusively in flocks, with some even sticking to close-knit family groups while on the move. Socialising during migration provides a clear benefit in terms of safety in numbers, allowing individuals to reduce their exposure to predation risk during flights and stopovers (Lindström 1989). A potentially more important advantage, however, is that migrating in groups can reduce the likelihood of making navigational errors.

We have already explored how navigation is an acute and complex challenge for inexperienced birds. It stands to reason, therefore, that there may be advantages to individuals 'double-checking' their navigation decisions by comparing them against others around them. Migrating in groups, either closely or loosely, can provide migrants with this opportunity – effectively representing a fail-safe against making directional errors

▲ A pair of Tufted Ducks *Aythya fuligula* with a single drake Greater Scaup *Aythya marila* and a pair of Lesser Scaup *Aythya affinis*, Kuilima Wastewater Treatment Plant, Honolulu, Hawaii, United States, 19 February 2011, an eclectic small flock of diving ducks in the middle of the Pacific. On returning to breeding areas they may well travel together and this may lead to further vagrancy events given that Tufted Duck does not breed in sympatry with Lesser Scaup and vice versa (*Eric Vanderwerf*).

(Simons 2004). The vast synchronous departures that occur during peak migration may also help serve this purpose – the contact calls that fill the sky on such nights may in part serve to allow individuals to keep track of the flight direction of other individuals, providing a continuous error-checking service for inexperienced juveniles (Farnsworth 2005).

For species that migrate within single-species flocks, the social component of navigation may be even more direct. Evidence suggests that some species defer navigation decisions to a subset of experienced (usually older) individuals, with younger birds following behind and perhaps learning a trail of visual beacons along the way (Flack *et al.* 2012). Some flocking species also navigate collectively, with decisions apparently being made through some mechanism of group consensus building (Biro *et al.* 2006). Both Skylarks *Alauda arvensis* and White Storks *Ciconia ciconia* have been shown to make more accurate navigation decisions when flying in large groups, suggesting that more brains can mean better decisions – effectively suggesting that birds can 'crowd-source' a good migratory orientation (Rabøl & Noer 1973, Chernetsov *et al.* 2004).

Given that flocking improves navigation accuracy, we might predict that flock-migrating species should be less vulnerable to vagrancy than solitary migrants. However, heavy reliance on social navigation becomes extremely problematic for individuals that somehow become detached from their flock. For obligate social navigators such as cranes and geese, the cultural component of migration may be so strong that juveniles become completely disorientated and unable to navigate if they detach from their flocks (Mueller *et al.* 2013). Long-range vagrancy in immatures of these species is therefore quite common (Wolfson *et al.* 2020) and is counterintuitively even more frequent in rarer species that tend to have sparse and fragmented populations. An immature Siberian Crane *Leucogeranus leucogeranus* that arrived in Taiwan in December 2014 vividly illustrated this – having become detached from its family group during migration, the disorientated bird wandered far out of range, eventually being found in the car park of a subway station in Taipei. After being captured and released at a nearby marsh, the bird became a national celebrity and was even appointed its own bodyguard by the Taiwanese government, before eventually migrating to an unknown destination the following spring (Ramzy 2016).

A common outcome for lost juvenile cranes or geese is to join up with flocks of other species – sometimes entirely different genera – and migrate with them. These flocks can become 'carrier species' that often lead the stray individuals into long-range vagrancy. Each year, birders across the Holarctic avidly search through

▲ Pink-footed Goose *Anser brachyrhynchus* with Canada Geese *Branta canadensis*, Arthur J. Hendrickson Park, Nassau, New York, United States, 30 December 2016. The former is an increasing visitor to the north-east United States (*Jay McGowan*).

▲ Terek Sandpiper *Xenus cinereus*, migrating north with Common Ringed Plover *Charadrius hiaticula* and Dunlin *Calidris alpina*, Alkborough, Lincolnshire, England, 18 May 2020 (*Graham Catley*).

wintering flocks of Arctic-breeding geese and cranes in search of stray vagrants that have fallen foul of this mechanism. Some individuals will then continue to migrate back to the Arctic with their adopted species year after year, producing hybrid offspring in some cases (Ottenburghs *et al.* 2016). Species that migrate in family flocks such as geese may also be unusually prone to vagrancy from their second year onwards, when they finally leave the family group after completing their first migration as juveniles. Such birds must eventually leave the guidance of their parents and strike out on their own, at which point they may risk becoming lost or entrained in carrier groups of other migrating species.

Vagrancy arising through individuals following carrier flocks of other species is likely to be common in many flock-forming species. Arctic-breeding waders have some of the highest rates of global vagrancy among any species, driven in part by the extreme difficulty of compass navigation at high latitudes. The fact that they commonly migrate in flocks is likely to be another key factor, especially as it is common for waders in the Arctic to breed side-by-side with species that winter in completely different parts of the world. Many Nearctic-breeding wader species are surprisingly frequent vagrants to Europe, and they often appear in the company of Ringed Plovers *Charadrius hiaticula* of the *tundrae* subspecies, which breeds as far west as the Canadian Arctic. These westernmost populations migrate across the North Atlantic to winter in Europe and West Africa (Flegg 2004), and it seems likely that juveniles of Nearctic species regularly tag along with these flocks and follow them to the wrong side of the Atlantic.

Social navigation clearly plays an important role in explaining patterns of vagrancy in many species, both as a direct driver via carrier flocks, and as an indirect factor allowing inexperienced individuals to minimise their risk of vagrancy by error-checking their own orientation decisions. The true extent to which birds 'crowd-source' their navigation decisions is still unknown, particularly amongst nocturnal migrants, and much is still to be learned about the mechanisms that underpin the collective decision-making of birds during migration. Nevertheless, it is clear that social navigation can commonly lead to birds going astray during migration, and searching through flocks of migrants remains one of the most reliable strategies for birders intent on finding vagrants.

WIND DRIFT AND VAGRANCY

In the preceding sections we have explored how navigation errors can cause vagrancy in the absence of any external environmental influences. This 'endogenous' component is expected to generate a fairly constant baseline level of vagrancy among migratory birds in any given area and season, regardless of weather patterns or any other exogenous factors. However, any birder with an interest in vagrant-hunting knows that the chances of encountering rare birds can be a matter of feast or famine (and usually the latter), and that this variation can be largely attributed to the weather. The link between vagrancy and weather patterns is so strong that birders often become obsessive students of isometric charts and forecast models, trying to identify wind patterns that might bring flights of interesting birds in their direction. Such forecasting is a fickle business, however, and disappointment is usually the outcome of any attempt to predict the timing and location of arrivals of vagrant birds.

Whilst the association between winds and vagrant occurrence is often quite obvious, it can be difficult to say whether the wind itself is the fundamental driver of birds going astray in a given circumstance, or simply a secondary factor that acts to concentrate individuals with errant navigational instincts. In some cases, there can be little doubt that winds are the sole cause of vagrancy – particularly when large numbers are simultaneously displaced outside their normal ranges by storm events (Elkins 1979). Exploring the circumstances surrounding mass displacements can shed light on how birds respond to weather patterns and the conditions that might be expected to cause significant weather-driven vagrancy around the world.

◄ Hermit Thrush *Catharus guttatus*, Cape Clear, Cork, Ireland, 20 October 2006, the second record for Ireland of this Nearctic vagrant (*Chris Batty*).

To understand the role of the weather in driving vagrancy, a good starting point is to explore the ways in which migrating birds interact with winds from the bird's perspective. Most birds are utterly at home in the air. The sky is the main habitat for many species that have evolved complex behaviours to allow them to use winds to their advantage, helping them save energy, find food or escape predators. The ways that migrating birds interact with winds have been subject to intensive scientific study, and many insights have recently been gained from detailed examinations of the tracks of satellite-tagged birds, as well as the movements that can be inferred from radar images. These discoveries are gradually allowing us to build up a detailed understanding of the complex ways in which migratory birds are influenced by winds, and the circumstances in which wind patterns can switch from highly beneficial to potentially fatal.

The sky as a complex habitat

An important starting point in understanding the role of winds as a driver of vagrancy is to recognise that we, as humans, have very little intuitive grasp of the conditions that birds experience in the airspace above us. From our narrow perspective on the ground, we tend to think of the wind as a one-dimensional phenomenon – a movement of air that has a direction and strength, corresponding with the conditions we experience at ground level. Studying weather maps only serves to reinforce this perception, representing the wind as a one-dimensional flow across the planet, that can be easily described by simple sets of arrows or isobars. In reality, the windscape experienced by a bird is far more complex – a three-dimensional 'habitat' comprising an ever-changing maze of jets, boundaries and eddies, invisible to the eye, but keenly felt by any bird ascending through the atmosphere (Davy *et al.* 2017).

The ground-level winds that we experience represent the basement of a high-rise complex, with each 'floor' comprising a band of airflow that may differ in speed and direction from levels above and below it. To make matters more complicated, these floors are prone to shift unpredictably in space and time, creating a turbulent environment that must be perilous and confusing for inexperienced or exhausted birds (Dokter *et al.* 2011). This high-rise complex is referred to by meteorologists as the 'boundary layer' – the lower part of the troposphere where the air mass is heavily influenced by the shape of Earth's surface. It is within this layer – typically the first 2–3km above the ground – that the vast majority of bird flight and migration takes place. A brief snapshot of life in this environment can be gained during the take-off and landing phases of aeroplane flight, as we pass through what can feel like 'pockets' of turbulent air that buffet aircraft, sometimes causing surprising vertical lurches that can leave your stomach in your mouth. Given that these forces are strong enough to shake the metal hull of an aeroplane, it is easy to imagine just how disruptive the winds could be to an airborne bird.

► European Honey-buzzard *Pernis apivorus* Skaw, Unst, Shetland, Scotland, 23 May 2009. Wind drift, or perhaps a navigational failure, took this migrant far west over the Atlantic – an error that proved fatal, the bird probably having been killed and eaten by Great Skuas *Stercorarius skua* (*Rik Addison*).

We have previously discussed how challenging it must be for migratory birds to orient themselves in their desired migratory heading, given that it involves the mastery of multiple internal compass senses that each require different sensory inputs. Now, we must extend this thinking to acknowledge that birds must attempt to do this whilst also finding their way through the complex and potentially hostile aerial environment – a challenge that must make navigation even more difficult. Not only must birds constantly keep track of their various compass senses, they must do this whilst also seeking safe passage through the three-dimensional airspace, avoiding disruption and trying to locate bands of airflow that will help rather than hinder their onward passage.

Until recently, we knew very little about the behaviour of individual birds during the course of a single migratory flight. Recent advances in tracking have allowed researchers to follow individual birds in precise detail, revealing their routes, speeds and flight heights on a minute-by-minute basis. One of the first studies to track the flight of a passerine on migration was carried out by Bowlin *et al.* (2015), who attached tiny radio

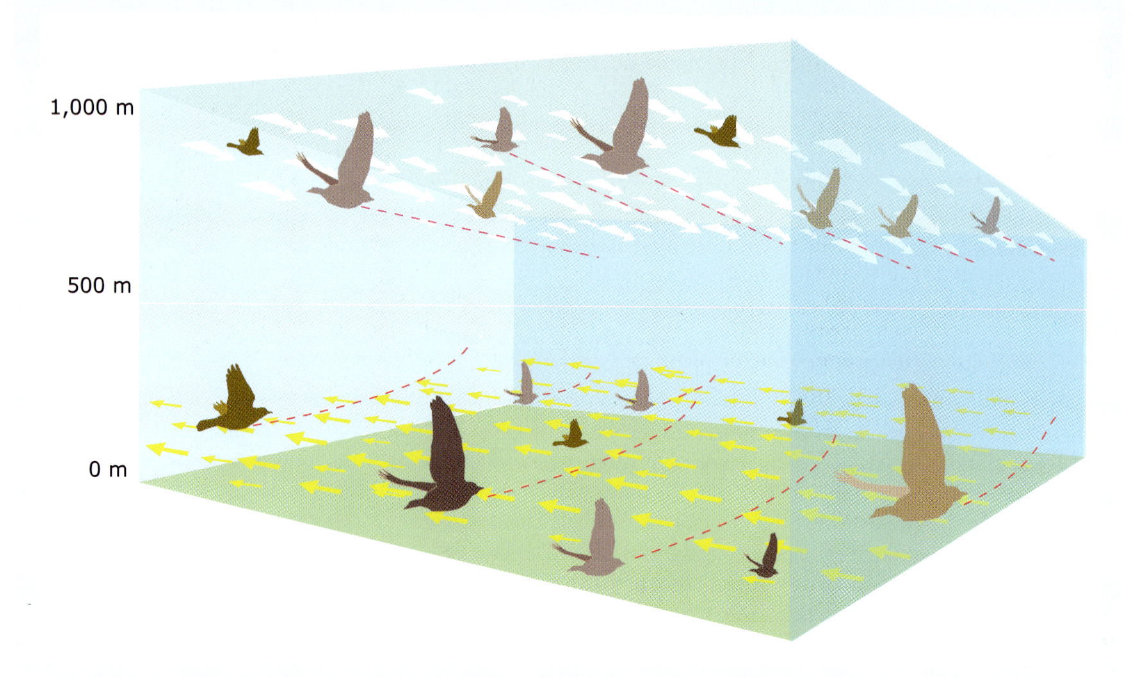

▲ **Figure 9.** Winds may simultaneously flow in multiple directions at different heights within the air column. In some circumstances, tailwinds (white arrows) may only be present at upper elevations, and birds migrating at ground level may be exposed to sidewinds (and thus drifted) unless they are able to gain enough elevation to reach the zone of optimal wind conditions above.

transmitters to nine Swainson's Thrushes *Catharus ustulatus* in Illinois; they were subsequently able to follow each of them for the duration of a single nocturnal migration flight, using a tracking antenna fitted to a vehicle on the ground. As is typical for nocturnal migrants, the thrushes all set off soon after dusk and rapidly gained height, with all birds showing a similar pattern of behaviour at the very start of their flights, climbing to an elevation of around 1km very soon after take-off (Figure 9).

After this initial ascent, however, the flight paths of the individual birds began to diverge, with each one exhibiting its own unique flight pattern. Some continued to climb throughout the night, others gradually descended, and some made continual ascents and descents of hundreds of metres as they continued on their journey. This variation surprised the researchers, as theoretical models of bird flight suggest that to maximise energy conservation, birds with flapping flight should seek to maintain a single elevation if possible. Bowlin and colleagues hypothesised that the thrushes were having to continually adjust their flight height in response to changing conditions within the atmosphere – perhaps moving up and down in search of tailwinds, trying to escape turbulent pockets of air, or moving above or below clouds to maintain a clear view of the stars above. It is easy to imagine circumstances whereby inexperienced migrants would become confused or disorientated within this ever-changing environment, potentially leading to frequent errors in navigation.

Similar patterns of complex vertical movement during migration have also been observed from radar stations. Intriguingly, radar signals often indicate that migrating birds form 'layered' patterns within the air column, with multiple narrow bands of intense migration at different levels above the ground. Dokter *et al.* (2013) showed that this happens frequently in Europe around high pressure systems in spring, where migrating birds appear to group into two distinct bands – one close to ground level and another at a considerable height (1–2km). They found that these patterns were almost always associated with a set of conditions where wind directions were unfavourable at ground level, but tailwinds could be found at greater altitudes within low-level jet streams about 1km above ground. This pattern of layered winds is a common feature of high pressure systems, and it is thus possible that birds have evolved to recognise barometric conditions where they can find ideal tailwinds at altitude, even when ground-level wind directions are unsuitable. This observation gives us an important insight into how winds might cause vagrancy. It is notable that of the two layers of migrating birds, only one was within the zone of optimal winds – the other, at ground level, concerns large numbers of birds flying in sidewinds that may cause them to drift (Dokter *et al.* 2013). Many birds are seemingly unable to escape this band by climbing to the better, higher zone, apparently becoming 'trapped' in sidewinds at ground level. These individuals have perhaps failed in their attempts to ascend into the jet stream, and instead find themselves vulnerable to drifting off course unless they are able to compensate by battling against the wind.

One obvious way that migratory birds can avoid drifting in sidewinds is to simply stop flying. Radar studies indicate that birds are indeed highly selective about when they depart, with little movement taking place during periods of strong adverse winds, followed by radar screens lighting up with thousands of departing birds when winds finally become suitable for onward passage (van Doren *et al.* 2018). Of course, departing when the wind is helpful does not necessarily guarantee an easy onward flight – birds do not have the luxury of a five-day weather forecast to help plan their movements. To some extent, they may be able to 'forecast' wind patterns if they have evolved to recognise weather patterns that are likely to mean conditions will be suitable for onward flight for at least the next few hours. In eastern North America, for example, it has long been known that flights of nocturnal passerine migrants in autumn often follow the passage of a cold front, these commonly being followed by light winds from the north that provide ideal tailwinds for southbound migrants.

The strategy of avoiding migration when winds are not ideal has one major drawback – migrants may end up waiting for days or even weeks for a window of perfect conditions, severely slowing their progress. Waiting around is a risky strategy for a migrant, as the bird might end up running out of food or being hit by colder conditions than it is physiologically capable of surviving. As the migration season progresses, migrants may become more and more likely to depart even if wind conditions are not suitable. This may be one of the reasons why vagrant occurrences tend to peak later in migration seasons, both in spring and autumn, when lagging individuals become increasingly willing to run the risk of wind drift. Juvenile birds in particular tend to migrate later in the year than adults and may therefore not have the luxury of waiting around for the perfect wind conditions before moving.

▲ Swainson's Thrush *Catharus ustulatus*, Cape May, New Jersey, United States, 30 September 2018. Radio transmitters fitted to this species revealed that they regularly make abrupt changes in flight altitude during migration, perhaps in search of zones with suitable tailwinds (*Michael O'Brien*).

How do birds respond to wind drift?

The extent to which sidewinds influence a bird's ultimate destination depends in part on the strength and direction of the wind relative to the bird's chosen direction, and also the bird's behavioural response to drift. In turn, this depends on whether birds are actually able to recognise the drift happening in the first place – and this may not be as simple as it first seems. As we discussed in the orientation section (page 9), most juvenile birds are likely to migrate using a simple 'clock-and-compass' approach, following a desired heading using an internal compass ('reading' the stars, Sun or magnetic field) for a given period of time. Such birds may struggle to detect wind drift during flight, as their only source of positional information comes from their innate compass sense – and the internal 'needle' of this will not change no matter how much the bird is drifted (Perdeck 1958). This may help explain why juvenile birds on their first autumn migration appear to be particularly prone to wind-driven vagrancy – in some cases they may simply be unaware that the drift is even happening.

One way that migratory species have solved this problem is by combining their innate compass sense with a set of 'beacons' – fixed features of the landscape below that are visible during flight and fall along the line of their desired heading; these are often referred to as 'leading lines' (Alerstam 1978). After deciding on a flight orientation initially using their internal compass cues (stars, Sun or magnetic field), birds may then be able to fix on visible features below them – perhaps mountain ranges, rivers or coastlines – and measure their progress towards these beacons during flight (Kullberg *et al.* 2007, see page 19). Unlike the internal compass, the position of these beacons will change whenever the bird is drifted sideways off its initial course. This will allow birds to 'see' drift as it happens, enabling them to compensate by changing their flight angle to maintain a net movement towards their desired heading. This may be one of the reasons why mass wind-drift events often correspond with low cloud or foggy conditions, leading to poor visibility, which prevents migrants from using their visual beacons to keep track of their course.

Whilst we might expect that birds will attempt to compensate for wind drift wherever possible, recent radar studies suggest that this might not always be the case, at least among nocturnal passerine migrants (Thorup *et al.* 2012). Horton *et al.* (2016) took advantage of recent advances in Doppler radar technology to measure not only the flight path of birds aloft, but also the direction they were facing, providing some remarkable new insights into how birds adjust their flight behaviour in response to drift. They found that large nocturnal flights often took place despite considerable sidewinds, and that in many cases birds showed little attempt to compensate for drift during the course of the night – rather, they simply continued to orientate themselves towards their desired heading and allowed the sidewinds to steadily drift them longitudinally. Interestingly, birds were far more likely to adjust their heading to account for wind drift when flying close to the Atlantic coast, relative to birds flying 200km inland.

There are two possible explanations for this, both of which are highly relevant to our understanding of wind-driven vagrancy. One relates to the use of 'leading lines' described above – it is possible that birds migrating along the coast are more aware of being drifted as they can use the coast as a leading line. Further inland, birds may lack such obvious landscape features upon which to fix, potentially making it more difficult to keep track of whether they are being drifted off course. Another possibility raised by Horton *et al.* (2016) is that birds may simply be less inclined to compensate for drift when flying further inland, because this compensation is energetically costly – the bird must pivot the axis of its body during flight to face further into the wind, such that the combination of its forward movement and the sidewind will cause it to move in the correct direction. This requires a much more forceful flight pattern, reducing the distance that the bird could fly before needing to refuel. Energy-conscious birds may therefore opt to maintain their normal axis of orientation and allow themselves to be drifted steadily off this course by the sidewind.

These observations indicate that vagrancy is perhaps most likely to arise when persistent sidewind conditions hit major inland migration routes, as birds on these flyways may be less inclined to battle against drift. This could in part explain a pattern that is very familiar to birders in Western Europe, where the presence of high pressure systems over Scandinavia or western Siberia in autumn can lead to a veritable glut of Eastern Palearctic vagrants arriving at coastal hotspots (Elkins 1988). These weather systems are usually associated with light easterly airflows and clear weather across Eurasia, potentially causing steady westward drift among migrants making overland movements south out of Siberia. If uncompensated, such drift could potentially accumulate over consecutive nights during stable anticyclonic conditions, gradually displacing large numbers of birds towards Western Europe (particularly those already following errant navigational instincts).

Wind drift over water

When migrating along coastlines, wind drift poses a more acute risk to landbird migrants – the danger of being pushed out to sea and thus into an area where landing to escape inclement conditions is impossible. Unless the species in question is physiologically adapted for long crossings over water (e.g. champion migrants like Blackpoll Warbler *Setophaga striata*), migrating landbirds are therefore more likely to pay the extra energetic cost of battling against wind drift when moving along coastlines, specifically to avoid being 'caught out' flying over water when they need to return to land and refuel (Horton *et al.* 2016). This is likely to be one of the key mechanisms that underpins the concentration effect of coastlines on bird migrations, leading to the often-spectacular visible landbird flights that occur wherever landmasses serve to funnel these coast-hugging migrants into narrow bottlenecks such as Cape May in New Jersey, United States and Falsterbo in Sweden.

► A fall of New World warblers (family Parulidae) drifted in strong offshore winds to Machias Seal Island, Charlotte, New Brunswick, Canada, on the night of 20 September 2017 (*Ralph Eldridge*).

The interaction between wind drift and water bodies is vital in understanding how weather patterns can give rise to extreme long-distance vagrancy in birds. A bird sensing that it is being drifted offshore by strong sidewinds will face a complex dilemma. The most risk-averse strategy will always be to head directly back towards land via an energy-intensive battle into the wind. The second option is to attempt to compensate for the wind drift and maintain the desired heading, again involving a potentially prolonged energy-intensive battle against the wind. The next option is to continue flying towards the desired heading and simply allow the offshore drift to happen, while the final option is to simply turn and fly in whatever direction the wind is blowing. It appears that the last response does indeed happen in many circumstances and is likely to be the cause of many well-established patterns of transoceanic vagrancy worldwide. Interestingly, radar studies suggest that as the length of time birds have spent over water increases, the likelihood of them trying to compensate for wind drift decreases (Horton *et al.* 2016). It seems likely that fatigue and depleting energy reserves mean that birds are less inclined to try to fight against the wind – and more inclined to allow themselves to be drifted with it. The strength of the wind is also likely to play a key role, as strong sidewinds may make it impossible for even freshly departed migrants to battle, leaving birds flying over water with little choice but to simply go wherever the wind takes them.

A vivid illustration of this response to drift came from a satellite tracking study of one of the world's champion transoceanic migrants – the Alaskan subspecies of Bar-tailed Godwit *Limosa lapponica baueri,* which performs a remarkable 11,000km overwater migration to New Zealand (Gill *et al.* 2008). This journey can take more than seven days of non-stop flight, and individuals must time their departure perfectly to maximise their likelihood of encountering good wind assistance. In most circumstances, godwits are able to compensate for even powerful sidewinds by altering their flight axis to ensure that they stick as closely as possible to their optimal route. However, the track of one individual, a female named 'E8', starkly illustrated how even these champion flyers can eventually reach a threshold where they must instead succumb to drift.

The bird in question encountered significant sidewinds during the second leg of its spring migration in 2007 (Figure 10). Having departed the Korean peninsula on 24 May with the backing of strong westerly tailwinds, E8 made rapid progress towards Alaska at the leading eastward edge of a cyclonic weather system (Conklin & Battley 2011). More than halfway through the journey, however, the depression gained momentum and merged with another system further north in the Pacific, battering E8 with strong side- and headwinds as she moved northward into the Bering Sea. After more than six days of continuous flight, she finally advanced to within just 110km of the Alaskan landmass. However, as the headwinds strengthened she eventually gave up her battle and turned 180 degrees to fly with the wind rather than against it. Remarkably, she then continued to fly due west for a further 30 hours, eventually making landfall on the Kamchatka Peninsula, Russia, some 1,200km west of her intended destination. Nine days later, in better weather, E8 set off again and this time flew successfully back to her Alaskan breeding site. However, the energy expended in battling the winds clearly took its toll, and she did not nest that year.

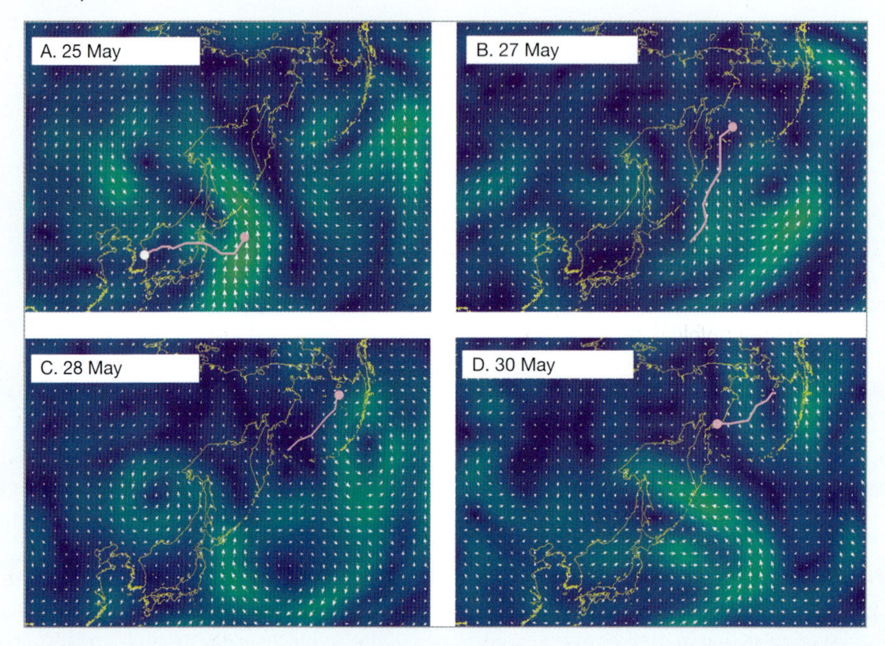

▲ **Figure 10.** The flight of 'E8', a satellite-tagged female Bar-tailed Godwit *Limosa lapponica*, in late May 2007. Having initially made rapid progress in good tailwind conditions, E8 (pink line and dot) was drifted northwards towards Kamchatka by strong sidewinds (A) and eventually became trapped in a battle against continuous headwinds lasting more than three days (B & C). Finally, having come within a stone's throw of reaching her Alaskan destination, she gave up the battle and turned to fly downwind back to Russia (D). (Bird flight data from Gill *et al.* 2008; wind data provided by the NOAA Physical Sciences Laboratory, Boulder, Colorado, United States, from their website at https://psl.noaa.gov/.)

This example gives a unique insight into the battles that migratory birds can face when making overwater crossings, and the speed with which they can be swept out of range once they make the decision to simply 'fly with the wind'. We understand very little about the precise thresholds at which birds decide to change their responses to lateral wind drift – it is likely that these vary hugely across species and individuals, with some species (particularly small-bodied birds) being far more prone to 'accepting' wind drift than others (Horton *et al.* 2018). In the next section, we explore some well-established patterns of transoceanic vagrancy and use these as examples to evaluate the key weather patterns associated with long-range vagrant displacement between continents.

Global transoceanic vagrancy patterns

Transoceanic wind-driven vagrancy occurs most frequently wherever coast-hugging migratory flyways pass through regions where prevailing winds are directed offshore. The eastern seaboard of North America is a prime example, where billions of birds annually funnel along the oceanic coastline, flying broadly perpendicular to the prevailing winds that flow principally west to east during autumn (La Sorte *et al.* 2014). In some circumstances, these winds can drift large numbers of migrating birds far out over the Atlantic Ocean, frequently carrying New World species as far as Western Europe. Like many weather-related vagrancy patterns, the meteorological

► American Bittern *Botaurus lentiginosus*, Lapa, Corvo, Azores, Portugal, 29 October 2013. This is the most frequent west-to-east transatlantic vagrant heron (*Richard Bonser*).

circumstances in which these drift events happen are complex and difficult to predict, involving a series of climatic events that must unfold in the right order, at the right times, and in the right places in order to take effect.

The first element that must fall into place involves the departure of large flights of southbound migrants. In eastern North America, huge departures often follow in the wake of eastward-moving cold fronts in autumn, when cool tailwinds and clear skies provide ideal conditions for overnight southbound departure (van Doren *et al.* 2016). In ideal conditions, vast numbers of landbirds ride these tailwinds directly to the Caribbean in a single overwater hop (DeLuca *et al.* 2015), while others hug the coastline in a series of southbound movements (La Sorte *et al.* 2016). In both cases, these migrants are highly vulnerable to being drifted further out across the Atlantic once they reach mid-latitude regions, where strong westerly winds are commonplace in autumn.

Migrants departing southwards in light tailwinds are most likely to get caught up in westerly airflows when they encounter northbound tropical storm systems – these frequently move polewards from the Caribbean along the Atlantic coastline in autumn, commonly tracking just offshore whenever high pressure sits over the continental landmass. The mechanism by which these systems drag migrants out into the Atlantic was first described in detail by Elkins (1979), who hypothesised that these storms entrain vast numbers of migrants when their arrival in American coastal waters coincides perfectly with overnight departures following a cold front. Feeding off energy provided by the warm waters of the Gulf Stream, these storms rapidly gather momentum, pulling the migrants into zones of strong offshore crosswinds around the trailing edge of the weather system. Having already completed many hours of continuous overwater flight, many entrained migrants may have insufficient energy reserves to keep battling against the sidewinds – much like godwit 'E8' described above. They may then reach a threshold where they simply abandon their desired heading and turn to fly with the wind. Once they reach this 'point of no return', the birds' fate will depend largely on the speed and trajectory of the weather system. Storms that cross the Atlantic within three days often carry multiple surviving migrants as far as Western Europe (Elkins 1979), though birds entrained in slower-moving systems are likely to perish unless they find shelter on ships or remote islands. Some tropical storm systems also follow a more northerly track up the eastern North Atlantic, delivering falls of southern Neotropical migrants as far north as Nova Scotia, Greenland or Iceland (McLaren *et al.* 2006).

In recent years, European birders have increasingly flocked to the mid-Atlantic outpost of the Azores in autumn in search of vagrants from the Americas. Interestingly, the exhaustive coverage of these islands – Corvo in particular – has revealed a somewhat different pattern of conditions associated with arrivals of vagrants. While tropical storm systems can bring large influxes of vagrants to the Azores, arrivals there show a far less clear association with these fast-moving systems. Some of the largest 'falls' of American vagrants on Corvo have actually occurred independently of storms, during periods of relatively light westerly winds, particularly when

▲ Northern Waterthrush *Parkesia noveboracensis*, Corvo, Azores, Portugal, 8 October 2019, the 16th record for the Azores and 30th for the Western Palearctic (*Paul French*).

these form a continuous vector across the eastern Atlantic (Alfrey *et al.* 2018). This strongly suggests that large numbers of migrants are routinely drifted far into the Atlantic during autumn, even when weather conditions are not particularly severe (Alfrey *et al.* 2018). It is perhaps even possible that many of these drifted vagrants may reorientate and return to their desired migration routes after a brief rest and refuel on these mid-Atlantic outposts.

Transoceanic vagrancy patterns have also been well studied in the North Pacific, again in a region where a major migratory flyway converges with a zone of prevailing offshore crosswinds. In this case, drift vagrants are frequently pushed eastwards towards North America from the vast migratory flyway that straddles the Pacific coast from the Russian Far East south to Japan, Korea and China. Here, ocean-crossing autumn storm systems are less frequent than in equivalent regions of the North Atlantic, with Pacific storms typically tracking northwards towards the Bering Sea rather than traversing the ocean basin. Long-range autumn drift of Asian vagrants across the Pacific does still occur, however, during conditions associated with a weather pattern known as the Pacific–North American teleconnection (Leathers & Palecki 1992). When in its positive phase, the teleconnection causes a shift in the position of the East Asian jet stream, pushing storm tracks in the West Pacific further south and west than normal, and placing them on a more direct trajectory towards North America. These patterns are positively associated with arrivals of Asian species in California, including Dusky Warbler *Phylloscopus fuscatus* and Red-throated Pipit *Anthus cervinus* (Howell *et al.* 2014), suggesting that these remarkable long-distance vagrants may be capable of making much of their Pacific crossing unaided.

North Pacific storm systems also frequently bring large numbers of Asian vagrants to the Aleutian and Pribilof Islands of Alaska. Here, wind-driven vagrancy patterns are well-established in both spring and autumn, with arrivals being dominated by species that migrate along the Pacific coastline and are thus more vulnerable to transoceanic drift than those migrating further inland (Howell *et al.* 2014). Some of the most detailed analysis yet conducted into the association between weather conditions and vagrancy was carried out by Hameed *et al.* (2009) in analysing the occurrence of vagrants on Attu Island in the Aleutians. Vagrancy to these islands is strongly linked to storm systems that pull westerly airflows from Kamchatka or Japan towards Alaska. The frequency of these storms is in turn linked to the El Niño oscillation, with strong El Niño events tending to generate stronger westerly airflows than normal, particularly in the northern part of the region, and hence producing bumper crops of drifted Asian vagrants (Hameed *et al.* 2009).

A particularly notable drift event occurred in the Bering Sea during the peak El Niño year of 1998, when the convergence of two storm systems in mid-May brought an unusually strong and consistent vector of winds stretching from the Aleutian Islands as far as the Sea of Okhotsk (Figure 11). On 17 May 1998, observers on Attu tallied some almost unbelievable totals of vagrant landbirds: some 180 Eyebrowed Thrushes *Turdus obscurus*, 225 Olive-backed Pipits *Anthus hodgsonii*, 193 Rustic Buntings *Emberiza rustica* and 366 Bramblings *Fringilla montifringilla*, with no fewer than 700 Wood Sandpipers *Tringa glareola* being found the next day (Hameed *et al.* 2009). Perplexed by this remarkable event, Hameed *et al.* (2009) developed a simulation model to reconstruct the most plausible scenarios that could have led to such huge numbers of vagrants being concentrated in such a small area. Given the speed of the weather system, their models suggested that the birds were likely to have opted to fly directly with the wind, rather than attempt to battle against it – and with the wind behind them, Hameed *et al.* estimated that it may have taken the quickest flyers only seven hours to complete the near-1,000km crossing from Kamchatka to Attu. Given that one of the most abundant vagrants in the fall, Eyebrowed Thrush, is a relatively scarce breeder on Kamchatka, Hameed *et al.* suggested that many of the vagrants in this event may have originated much further west, even beyond the Sea of Japan, as a migrant flying with the wind may have still been able to cover this much longer distance (up to 3,100km) in 24–38 hours (Hameed *et al.* 2009).

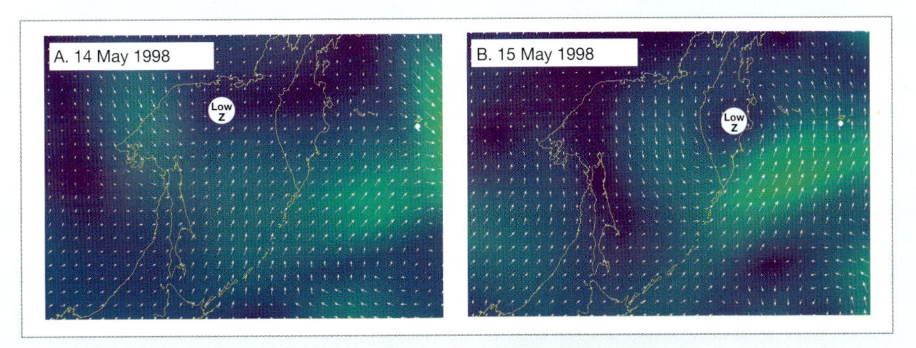

▲ **Figure 11.** Wind conditions leading to an unprecedented drift of Asian migrants to Attu Island, Alaska (white dot), in mid-May 1998. A cyclonic weather system ('Low Z') moved slowly across the Sea of Okhotsk, providing tailwinds that encouraged large numbers of migrants to depart north from Sakhalin and Kamchatka (A), before concentrating them in a strong easterly airflow that drifted large numbers into the Bering Sea (B). (Wind data provided by the NOAA Physical Sciences Laboratory, Boulder, Colorado, United States, from their website at https://psl.noaa.gov/.)

Jet streams and vagrancy

As noted previously, it is well established that migrating birds can take advantage of low-level jet stream winds during migration – typically those that form within the lower 2km of Earth's atmosphere (Dokter *et al.* 2013). In North America, for example, millions of migrants funnel annually through the central migratory flyway each spring to take advantage of the Great Plains low-level jet stream. This shallow northward conveyor-belt of air reaches speeds of up to 133km/hr and flows at altitudes of around 600–1,600m (Wainwright *et al.* 2016), allowing birds to migrate with minimum time and energy investment, as well as minimising the risk of predation along the way. Occasionally, radar observations have revealed migrant birds flying at much greater heights – in some instances up to 6,000m above ground level (Williams *et al.* 1977; Alerstam 1981). This has prompted some authors to speculate that migrants could sometimes become entrained in higher-level jet formations such as the Northern Hemisphere polar jet (Elkins 1979). In particular, this has been suggested as one way that North American vagrants may be brought to Europe, given that the polar jet typically flows eastwards across the North Atlantic at latitudes where transatlantic vagrancy is commonplace. However, this jet typically flows some 9–12km above ground (Hall *et al.* 2015), placing it substantially higher than the vast majority of observed migratory bird flight. Nevertheless, if birds were to become entrained in polar jets, they could theoretically reach ground speeds up to 180km/hr (Liechti & Schaller 1999), and thus be pulled very rapidly out of range. This could be a plausible explanation for vagrancy to Europe in some North American interior migrants such as Wilson's Phalarope *Phalaropus tricolor* and Franklin's Gull *Leucophaeus pipixcan*, species that are relatively rare along the eastern seaboard of North America yet appear with perplexing frequency across the Atlantic. For smaller-bodied species like passerines, temperature conditions within polar jets are likely to be prohibitively harsh to permit survival for long periods, suggesting that this is unlikely to be a widespread cause of long-range vagrancy for many species.

OVERSHOOTING

So far, we have considered the ways in which winds can drift migrating birds off course, but another common form of vagrancy involves individuals that continue to follow their 'correct' migratory bearing, but somehow overshoot their normal distribution to end up as vagrants. While overshooting is undoubtedly often weather-driven, it can also theoretically occur through a range of other mechanisms, including errant navigational cues or an errant internal clock that fails to provide a 'stop' signal at an appropriate time. Interestingly, overshooting may be one of the most important forms of vagrancy from an ecological perspective, particularly for species threatened by climate change. As the biosphere warms, zones of climatic suitability for each species are likely to shift poleward, and populations must move with them if they are to survive in the long term. These shifts could be facilitated by overshooting birds pushing the boundaries of their distributions into newly suitable climate zones. Overshooting vagrants can indeed sometimes locate mates and breed in new areas, for example, Iberian Chiffchaffs *Phylloscopus ibericus* nesting on the Gower Peninsula in south Wales in 2015 (Hunter 2018). Teasing apart the possible mechanisms that cause overshooting is difficult but essential if we are to understand this important form of vagrancy.

◀ Iberian Chiffchaff *Phylloscopus ibericus*, University of East Anglia, Norwich, England, 26 April 2007, the first record for Norfolk of a species that is being detected with increasing frequency in Britain (*James Gilroy*).

Overshooting in spring and autumn

In the Northern Hemisphere, birders typically think of overshooting as a spring phenomenon – in northern Europe, for example, vagrancy in springtime is often dominated by species that normally breed in the Mediterranean Basin and migrate to Africa. Warm southerly airflows can bring significant arrivals of species such as Alpine Swift *Tachymarptis melba* and Woodchat Shrike *Lanius senator* to areas well north of their normal distributions. In North America, birders often flock to see rare 'southern' migrants like Black-necked Stilt *Himantopus mexicanus* and Swainson's Warbler *Limnothlypis swainsonii,* which can appear well north of their normal ranges, particularly in late spring.

The mindset that sees overshooting as a spring phenomenon largely stems from bias in the ornithological literature towards case studies in the temperate zone of the Northern Hemisphere. In autumn, overshooting is seldom observed in these regions, primarily because there are few species whose normal autumn migratory routes overshoot into the well-birded parts of North America or Europe. In other parts of the world, however, autumn overshooting is a common and well-recognised form of vagrancy. In South Africa, for example, a major feature of the vagrant-hunter's calendar is the arrival of southbound migrants from Eurasia that normally winter in equatorial regions but can overshoot with a tailwind well to the south of their usual ranges in the cone

▲ Northern Shoveler *Spatula clypeata* (right) with Red Shoveler *Spatula platalea*, Lago Tonchi, Rió Chico, Santa Cruz, Argentina, 3 January 2012. This was the first record for Argentina, seemingly an individual that had overshot its wintering areas to reach the southern cone of South America (*Fabio Olmos*).

of Africa. Similar patterns are observed in Australia, where Siberian vagrants are anticipated at vagrant traps close to the Timor Sea, having overshot their Indonesian wintering grounds, and in southern South America, where boreal migrants may reach as far south as the Falklands. Weather-driven overshooting in autumn is also likely to be responsible for the arrival of European migrants such as Northern Lapwing *Vanellus vanellus* and Redwing *Turdus iliacus* in the northeastern United States. The normal migration of these species within Europe is on a broadly east–west axis, with huge numbers of individuals moving out of eastern Europe to winter along the Atlantic fringe. Individuals backed by strong tailwinds may find themselves pushed out over the Atlantic – and in such circumstances, if the tailwinds continue to push them westwards, they may ultimately travel all the way across the Atlantic (Howell *et al.* 2014). Overshooting is thus likely to occur commonly on both outbound and homebound migrations, with seasonal prevalence simply depending on where a given region lies in relation to the typical distributions of migratory birds.

How overshooting works

Most overshooting events tend to correspond, as expected, with tailwind weather conditions. Winds that are highly conducive to onward migration, if timed correctly, can sometimes trigger mass displacements of birds to areas well beyond their normal geographic ranges (Gunn & Crocker 1951). The mechanisms involved here are self-evident, as tailwinds can increase flight speeds and reduce energy expenditure, allowing migrating birds to travel much further than they would under other circumstances. However, given that migratory species are generally extremely adept at navigation, an important question is why such birds do not recognise that tailwinds are pushing them beyond their normal ranges, and simply stop flying?

In some cases, the answer may be obvious – landing may not be an option if the tailwind has pushed birds over the sea or some other inhospitable landing area like a mountain range or desert. However, overshooting remains a common phenomenon across large landmasses that lack significant inhospitable zones, including North America and Asia. Another explanation might be that overshooting occurs when tailwinds are strong enough to prevent entrained birds from landing quickly. In North America, vast numbers of spring migrants that take advantage of the Great Plains low-level jet stream to speed their northbound passage may be particularly vulnerable to tailwind drift (La Sorte *et al.* 2014). Jet stream travellers in particular might find it difficult to descend rapidly through the high-velocity airspace, potentially increasing the likelihood of short-distance overshooting. Because tailwinds offer such energy-saving benefits to migrating birds, it is also possible that

▲ Dark-faced Ground-Tyrant *Muscisaxicola maclovianus*, Parque Nacional da Lagoa do Peixe, Tavares, Rio Grande do Sul, Brazil, 8 May 2011, the first record for Brazil of this austral migrant, which had overshot its regular wintering range (*Paulo Ricardo Fenalti*).

some individuals temporarily ignore their navigatory instincts in the rush to take advantage of ideal flight conditions, leading them to miss key cues indicating they have passed their destination. Another possibility is that overshooting occurs when individuals simply make errors when judging how far they have come along their migratory path. Determining when to cease migration is a complex navigatory skill, involving careful evaluation of timing and/or geographic location – it is likely that errors in determining migration endpoints are relatively commonplace, particularly among inexperienced young individuals.

A final mechanism that may cause overshooting relates to the need to find other individuals of the same species in order to settle and cease moving. For many species, simply arriving in suitable habitat within the destination range may not be sufficient to trigger the decision to settle – it may also be necessary for individuals to locate other conspecifics, either as potential mates or flock members (Dale *et al.* 2006). Birds that find themselves close to the edge of their normal range following migration might find it difficult to locate other conspecifics – this could theoretically force them to continue moving ever further along their original migratory trajectory in search of others. This scenario is perhaps most likely to occur in conjunction with weather-driven overshooting – individuals that pass over the bulk of their breeding range may consequently find themselves having difficulty locating mates, and if they fail to compensate by back-tracking they may end up wandering well beyond their normal range. There are several lines of evidence supporting this hypothesis. One is that spring overshoots are often predominantly males (Newton 2008), which are perhaps the most likely individuals to be driven beyond their normal range in the continuing search for mates (although singing males are also easier to detect). Secondly, arrivals of overshooting vagrants often peak somewhat later than peak migration times in the normal range, suggesting that overshooting often occurs when most birds on the breeding grounds will be already paired up, driving remaining individuals to move beyond their normal range in the vain search for breeding opportunities. Interestingly, at least some overshooting vagrants are known to have been able to retrace their steps and return to their normal ranges to breed successfully. For example, a Common Rosefinch *Carpodacus erythrinus* ringed as a vagrant on Orkney, Scotland, in June 2013 was found breeding in central Germany the following year (Grantham 2014). It is possible that overshooting represents a relatively simple error to correct in navigational terms, requiring only a degree of backtracking along a course – successful returns may therefore be more frequent among overshoots than vagrants affected by sidewind drift, as the latter may be more difficult to detect and compensate.

EXTREME WEATHER AND IRRUPTIONS

Wind drift and overshooting are routine risks faced by migrating birds, and they conspire to push some individuals into vagrancy during every migration season. Extreme weather events occur only rarely, but these have the potential to cause major vagrancy events affecting both migratory and resident species alike. Many remarkable instances of global vagrancy have been linked to periods of extreme cold weather. Cold snaps in Western Europe, for example, can lead to westward 'escape movements' of birds towards the milder Atlantic coastal fringe, and when these movements are backed strong easterly winds, this can lead to transatlantic vagrancy to North America. Unusually cold winter weather in 1966 and 2010–11 led to influxes of Northern Lapwings *Vanellus vanellus* to Newfoundland, for example (Howell *et al.* 2014). Hard weather in eastern North America has also been associated with Killdeers *Charadrius vociferus* and American Robins *Turdus migratorius* moving in the opposite direction to Europe.

Periods of extreme hot weather also have the potential to drive escape movements, particularly in waterbirds if they trigger desiccation of wetland habitats. For example, influxes of Glossy Ibises *Plegadis falcinellus*, Black-winged Stilts *Himantopus himantopus* and Baillon's Crakes *Zapornia pusilla* to the British Isles in 2012 were suspected to be related to drought conditions in Iberia (Hudson *et al.* 2013). In the tropics, there is good evidence that birds can make significant movements to escape extreme rainfall events. White-ruffed Manakins *Corapipo altera* in Costa Rica, for example, are known to make erratic altitudinal movements in response to tropical rainstorms (Boyle *et al.* 2010). There are also records of long-range escape movements by high Andean species fleeing to lower elevations following severe cold, in species ranging from tyrant flycatchers to flamingos (Schulenberg *et al.* 2010).

► Killdeer *Charadrius vociferus*, Upper Lough Erne, Fermanagh, Northern Ireland, 3 March 2005, the 18th record for Ireland and a typical date for the species to occur in Europe (*Stuart Piner*).

Storms and seabirds

Weather conditions are often most extreme over the oceans, where winds unfettered by interrupting landmasses can reach enormous speeds. Any species of bird adapted to life at sea must therefore have evolved a strong capacity to cope with high winds. Indeed, seabirds tend to be highly adept at using winds to their advantage, making use of air currents to effortlessly glide in their chosen direction with precision and expertise way beyond any human sailor. However, wind conditions at sea can sometimes reach levels where even the most experienced avian navigator must find it impossible to avoid being drifted – particularly when tropical storms reach hurricane or typhoon severity. These systems may displace huge numbers of seabirds and push them outside their normal ranges – either through direct drift, or through birds sensing the storm and moving ahead of it to escape the worst of the winds (Bourne 1967).

▲ Masked Booby *Sula dactylatra*, Port Isabel Channel, Cameron, Texas, United States, 8 November 2018. Boobies are frequently displaced by tropical storms and hurricanes (*Ian Davies*).

In rare circumstances, fast-moving tropical storms make landfall and push significant numbers of disorientated seabirds considerable distances inland. Hardy birders in North America regularly head out in the immediate aftermath of hurricane conditions (sometimes even during the storms!) in the hope of finding storm-blown vagrants on inland freshwater bodies. In early September 1996, Hurricane Fran delivered an unprecedented inland displacement of seabirds as far as the Great Lakes region, including over 20 Black-capped Petrels *Pterodroma hasitata* reaching as far as 400km inland in the province of Ontario, Canada (Dobos 1997), and a Fea's Petrel *Pterodroma feae* reaching a reservoir in Mecklenburg County, Virginia, United States (Patteson *et al.* 2013). The deserts of the south-west United States, immediately north of the Gulf of California, are another hotspot for storm-driven seabirds, with inland wrecks sometimes involving large numbers of storm-petrels and even *Pterodroma* petrels such as Juan Fernandez Petrel *Pterodroma externa* and Cook's Petrel *Pterodroma cookii*.

In March 2007, an Atlantic storm centered offshore from the coast of southern Brazil became unusually active, developing and deepening rapidly to become the first-ever reported hurricane in the South Atlantic Ocean – Hurricane Catarina. The unprecedented nature of this storm meant it likely had a disproportionately strong effect on the seabird species within the region, with many potentially having had little evolutionary exposure to storms of such magnitude. As it made landfall on the Brazilian coast, the storm drove a massive inland displacement of petrels, particularly female Atlantic Petrels *Pterodroma incerta*. At least 354 individuals were found at 26 different locations up to 420km from the coast and up to 1,100m above sea level. Most of the birds affected were in heavy moult, and it is likely that the timing of the storm relative to the species' moult cycle played a major role in exacerbating its impacts (Bugoni *et al.* 2007). Another unusual tropical cyclone event struck the North Island of New Zealand on 9–10 April 1968, displacing 588 seabirds found dead or alive on beaches and inland (Kinsky 1968). This wreck included large numbers of albatrosses (110 Royal *Diomedea epomophora* and 26 Wandering Albatrosses *Diomedea exulans* in the Wellington area) in addition to rarer species such as Grey-faced Petrel *Pterodroma gouldi* and Black-winged Petrel *Pterodroma nigripennis*. Similar wrecks have also been recorded in South Africa associated with the passage of cold fronts and strong onshore winds (Ryan *et al.* 1989) and include large numbers of vagrants from the Southern Ocean such as Blue Petrel *Halobaena caerulea* and Kerguelen Petrel *Aphrodroma brevirostris*.

Irruptions and nomadism as a cause of vagrancy

Many otherwise resident species make occasional irruptive movements, particularly among species occupying extreme environments like the polar regions and deserts. It is tempting to think that irruptions typically occur in response to extreme weather events – cold in the case of polar regions, hot in the case of deserts – and there is indeed some evidence that this can be the case. However, most detailed studies carried out on irruptive species suggest that the key driver of these occasional movement events is food supply. A classic example is Snowy Owl *Bubo scandiacus*. Although birders often link the species' erratic appearances outside its normal Arctic range with periods of extreme cold, analysis of many years of occurrences across North America in fact suggests that food availability in the preceding summer – and in particular the abundance of small mammals – is the main driver of subsequent winter irruptions, with bumper summer feeding conditions typically leading to mass southward irruptions the following winter (Robillard *et al.* 2011).

Similarly, mass irruptive movements in many species of northern finches are driven by the availability of conifer seeds. Indeed, such irruptive species rather defy the concept of static species distributions, as in some years they may be rare or absent from large parts of their range and extralimital vagrancy may be routine. A recent study of irruptive patterns in Pine Siskin *Spinus pinus* (Strong *et al.* 2015) across North America shed light on the complexity underpinning these periodic movements. For Pine Siskins, irruptions can follow two directional axes: north–south and east–west. In both cases, movements correlate with sets of climatic conditions that cause seed production in a particular region to be low. Interestingly, Strong *et al.* found that conditions across the Pine Siskin range tended to form 'dipoles', where poor conditions in one part of the range are typically mirrored by good conditions in the opposite part of the range – be it north versus south, or west versus east. They suggest that irruptions work like a push–pull system, where movements are driven not only by the 'push' of falling local food supply, but also the 'pull' of knowing that on average, conditions in the opposite part of the range will be better. It is possible that irruptive behaviour evolves along these lines, with populations adapting to make rapid and huge movements between regions that tend to show opposing climatic patterns in any given season (Strong *et al.* 2015). Major vagrancy events are perhaps most likely to occur in those rare circumstances where this dipole breaks down – i.e. where, due to unusual meteorological circumstances, conditions for feeding

▲ Northern Hawk Owl *Surnia ulula*, Zwolle OV, Netherlands, 23 December 2013, the fourth record for the Netherlands of this boreal breeder, which occasionally irrupts southwards (*Gerjon Gelling*).

▲ Masked Woodswallow *Artamus personatus*, Cape Portland, Tasmania, Australia, 4 April 2018. Most woodswallows are nomadic and range widely in search of the best local climate conditions for insects, and these movements may take them from mainland Australia to Tasmania and New Zealand, where they are vagrants (*Peter Vaughan*).

are poor across the whole range. In these circumstances, populations are only exposed to the 'push', and may end up wandering far beyond their normal range in search of feeding opportunities.

Some tropical and subtropical species are also known to make nomadic movements in response to temporally unpredictable food resources, which can lead to vagrancy. For example, woody bamboos in South America flower fairly synchronously over large areas, providing a localised resource boom which appears then disappears, forcing bamboo-specialist birds like Temminck's Seedeater *Sporophila falcirostris* and Buffy-fronted Seedeater *Sporophila frontalis* to track this food source in a nomadic fashion (Areta *et al.* 2013). These influxes can sometimes lead to long-distance vagrancy events, particularly when large crashes follow a high peak.

Weather may play a more significant role in driving irruptions of species from arid areas and deserts. Nomadism in desert birds is well documented (Dean 2004), with deserts often showing dipole patterns similar to those described above, with rainfall creating suitable conditions in one region whilst others are unsuitable, resulting in population-level displacements in wandering desert species. These regular movements in search of suitable habitat conditions may be of the scale of hundreds or even thousands of kilometres, leading to irruptions of desert species like Trumpeter Finch *Bucanetes githagineus,* when multiple individuals may be discovered extralimitally in northern Europe. In Australia, patterns of rainfall are closely linked to irruptions of many species associated with arid areas – and the subsequent appearance of species like Flock Bronzewings *Phaps histrionica* (Pedler & Lynch 2016) and Princess Parrots *Polytelis alexandrae* (Pavey, *et al.* 2014) outside their normal ranges.

VAGRANCY AND NATURAL DISPERSAL

Throughout nature, the process of natural selection seldom favours organisms that lack the capacity to disperse to new areas. Even among sedentary plants, evolution has found ways for species to spread from one place to another, 'hitchhiking' on animals or using the wind to scatter their seeds and pollen. Almost all species have some mechanism for exploratory movement – and the longest of these movements are often vital for the long-term survival of populations, for a whole suite of reasons. In the absence of long-distance dispersal, populations will invariably become threatened by inbreeding, as local gene pools become more and more impoverished over time through the lack of mixing with individuals elsewhere. Dispersal can also be essential in allowing populations to persist when environmental conditions change, or when population densities within a local area grow to the point that competition for resources limits survival chances.

In many ways, vagrancy represents the endpoint of this tendency for dispersal. A significant proportion of vagrancy in birds may well be driven by individuals following urges to disperse in search of new territory, rather than being pushed by adverse weather or falling victim to faulty navigator apparatus (Newton 2008). Grinnell (1922) saw vagrants as representing "sensitive tentacles, by which the species keeps aware of the possibilities of a real expansion". Taking this viewpoint to its logical conclusion, perhaps we should see vagrancy as something

► American Purple Gallinule *Porphyrio martinica*, Parque Monsanto, Lisbon, Portugal, 9 November 2013, the first record for mainland Portugal. Rails are highly itinerant and adept at dispersing between distant wetland areas, which may lead to them moving between hemispheres on occasion (*Richard Bonser*).

► Cinereous Vulture *Aegypius monachus*, Nong Pla Lai paddies, Phetchaburi, Thailand, 2 January 2011, a vagrant to Thailand from Central Asia where the species may wander widely (*Wich'yanan Limparungpatthanakij*).

that is evolutionarily advantageous to species, allowing them to colonise new areas? Given that natural selection operates at the level of individuals and genes, however, a trait like vagrancy would need to result in relative benefits for the individuals themselves in order to be favoured by evolution. In most cases, vagrancy is unlikely to improve the wandering individual's prospects – the vast majority probably die without reproducing, unless they are able to make it back into their normal range. Consequently, we might expect any characteristics associated with long-distance vagrancy to carry little evolutionary advantage at the level of the individual.

Interestingly, however, detailed studies of how evolution shapes dispersal behaviour tend to paint a more complex picture. In general, evidence from wild populations suggests that selection does generally favour shorter dispersal distances, but interestingly there are often far more individuals moving long distances than researchers would expect if natural selection were heavily penalising vagrancy (Fraser *et al.* 2001). In fact, it appears that evolution tends to favour a mix of dispersal strategies within most populations, with the majority of individuals moving relatively little, but a significant proportion of 'explorers' wandering much further. This suggests that vagrancy can indeed be part of an evolved mechanism that causes some individuals to move much further than others in their population. But in what circumstances might this come about?

Exploratory vagrancy in resident species

The link between vagrancy and exploratory dispersal is perhaps most obvious when we consider the case of fully resident species. As a rule, sedentary birds tend to be far less prone to vagrancy than migratory species – a pattern largely explained by migrants being more exposed to drift by adverse weather, as well as being more prone to navigational errors. Even the most sedentary of bird species do still occur as long-distance vagrants, however, and it is likely that this is most often the result of extreme – but nonetheless 'deliberate' – exploratory wandering. This pattern was nicely illustrated in a 40-year ringing recovery study of the fate of 2,742 nestling Red-shouldered Hawks *Buteo lineatus elegans* in southern California, United States, from 1970 to 2009, a subspecies which is largely resident (Bloom *et al.* 2011). Of the 119 birds that were recovered, some 91.6 per cent moved less than 10km from their natal site, but a significant proportion (2.5 per cent) were recovered as vagrants, with some individuals moving up to 850km out of range and in all directions.

What might cause an individual of a resident species like Red-shouldered Hawk to leave the area in which it hatched and wander so far from its normal range? The vast majority of these movements are made by juveniles, and it seems likely that a major proximate cause is the necessity to find a territory. Resident species tend to be strongly territorial, and remaining in the same home range year-round means that survival is heavily dependent on their capacity to defend their local food resources from competitors. Adult resident birds will therefore only tolerate the presence of their young within the territory for a certain time after fledging – sooner or later, the young will be ungraciously ousted.

◄ Black Scrub-Robin *Cercotrichas podobe*, Samar, HaDarom, Israel, 22 March 2019. First recorded in Israel in 1981, this species seems now to be colonising the region. Intriguingly, studies suggest that the individuals involved in long-range dispersal events may possess 'personality traits' – including aggression – that increase their chances of being successful colonists (*Paul French*).

Birds solve this problem in various ways. In some strongly territorial species, young birds play a waiting game, becoming 'floaters' within their natal area and persisting by drifting from territory to territory, feeding whenever they can but never lingering in one place long enough to draw the ire of a territory holder (Newton & Rothery 2001). As a floater, their lives are not stable enough to allow them to rear young, and this strategy therefore comes at a cost. However, they are able to stay in constant touch with the distribution of other birds within their natal region and will therefore be ideally placed to step in when a territory becomes available. As soon as a territory holder perishes, the floater will be able to occupy the vacant territory. If it is lucky, it may take up the reins of a high-quality territory, and consequently its reproductively barren wait will have been worthwhile.

The alternative to becoming a local floater is to adopt a phase of much larger exploratory wandering during early life. It is quite easy to imagine how this scenario might give rise to long-distance vagrancy, particularly if juveniles search for specific habitat features that are necessary to trigger their instincts to stop and set up territory. If we imagine a desert species, for example, the itinerant wanderings of juveniles fledged close to the edge of a desert region might take them into other more vegetated ecosystems – and in the absence of suitable looking habitat, they may simply keep wandering ever further from their species' range. This logic implies that resident species with very specific habitat requirements may be more prone to vagrancy, as they may be searching for a particular set of habitat conditions – and the relative frequency of vagrancy among resident desert and montane species lends some weight to this (Lees & Gilroy 2009).

Another interesting wrinkle to the story of avian exploratory wandering relates to the link between exploratory behaviour and the 'personality' of individual birds. A fascinating insight into this was provided by Duckworth *et al.* (2007), who set out to determine whether long-range wandering in male Western Bluebirds *Sialia mexicana* was linked to the characteristics of individuals – and in particular how aggressive they were. They tested this experimentally by presenting bluebirds with a model Tree Swallow *Tachycineta bicolor* – a species with which bluebirds compete aggressively for nest sites – and measuring how dramatic the bluebirds' response was. They found huge variation in aggression amongst males, with some individuals merely engaging in a few fly-pasts of the models, whilst others relentlessly mobbed and attacked the apparent intruder. Intriguingly, they found that the highest levels of aggression were shown by bluebirds that had moved furthest from their place of hatching to set up their territories. Duckworth *et al.* hypothesised that aggression may be advantageous to long-distance wanderers, increasing their capacity to oust other individuals of the same or even different species from potentially suitable nesting sites.

Amazingly, Duckworth *et al.* also found that females may be able to control the levels of aggression in their male offspring – and hence whether they are likely to become long-range dispersers – by varying the timing of hatching of their male and female eggs. Many bird species can control the sex of their offspring through hormonal changes during egg development. In Western Bluebirds, males tend to be more aggressive if they are the earliest to hatch in a brood, and less aggressive if they hatch last. Duckworth's studies revealed that females tend to have far more early-hatching males (higher in aggression) in populations that are at high density – and therefore high levels of local competition for territories. They therefore appear to be able to evaluate whether it is advantageous for their male offspring to become wanderers, and so ensure that they hatch early and thus develop more aggressive personalities. This link between long-range exploratory wandering and aggressive personality has been detected in many other bird species, suggesting that bold and aggressive individuals may indeed be more prone to wandering and hence even vagrancy.

Exploratory vagrancy in migratory species

As noted above, vagrancy among migratory species may often be driven by adverse weather conditions or navigational failings, but a significant proportion of vagrancy events among migrants may also involve 'deliberate' exploratory wandering in search of new territories. In many migratory species, juveniles can make apparently random movements away from their natal sites during the first weeks after fledging, prior to commencing their proper directional migration. These random movements are thought to be an exploratory phase, where they may be searching out potential areas to return to and breed in the subsequent year (Brown & Taylor 2015). These exploratory movements may be very similar to those described above for resident species – and there may indeed be a proportion of individuals that wander much further than the rest of their population and become vagrants.

It has been postulated that early-autumn vagrancy in long-distance migrants is often the result of this random wandering – and that such vagrants may subsequently retrace their steps and migrate to their 'correct' winter range after initially wandering astray (Nisbet 1962). Examples of such movements may include the regular early autumn appearance of species like Barred Warbler *Sylvia nisoria* and Common Rosefinch *Carpodacus erythrinus* on the North Sea coasts of Scotland and England in anticyclonic conditions. Evidence for the reorientation of some individuals comes from ringing recoveries. For example, a Barred Warbler ringed on Shetland, Scotland, in September 1978 was recovered in the former Yugoslavia in February 1979 (Newton 2008).

Vagrancy among migrants during spring migration may also result from similar exploratory random movements. Unlike resident species, which have all winter to wander in search of good potential breeding territories, migrants have only narrow windows of time to explore their natal grounds. If juvenile birds fail to identify a future breeding territory in the autumn prior to their first migration, they may again begin exploratory wandering in the spring following their return migration. Many migrant species are highly faithful to their natal sites – with juveniles tending to come back and attempt to find mates very close to the area where they fledged (Paradis *et al.* 1998). If they return to find all the potential territories are full, they must again begin wandering – potentially ever further in search of suitable habitat and mates. It is often observed that vagrant birds in spring make shorter stopovers than in autumn (Nilsson *et al.* 2013), perhaps reflecting the more hurried nature of spring dispersal, where wandering individuals are constantly shifting location in the hope of locating a mate in time to breed.

◀ Bar-tailed Godwit *Limosa lapponica*, Icapuí, Ceará, Brazil, 10 November 2006. This, the first record for the Brazilian mainland, has been followed by the discovery of small flocks of up to 19 wintering in the state of Maranhão and Pará in the north-east of the country, presumably representing a 'pseudo-vagrant' population (*Ciro Albano*).

There is one other element of exploratory wandering that bears mention in the context of vagrancy. As described above, there are good reasons to suspect that evolution may favour some form of vagrancy capacity in any population, but we usually think of this in terms of exploration to locate new breeding sites. Among migratory birds, it may also be advantageous for populations to locate new non-breeding/wintering or stopover sites via exploratory wandering, for the same reasons that this is advantageous for breeders. Migratory species can travel vast distances in order to reach suitable areas for overwintering, but the quality or suitability of these areas will change over time. Performing exploratory dispersal around winter ranges may therefore be advantageous, allowing individuals to locate new areas of suitability, or avoid overcrowding in core winter sites (Cresswell 2014). The most adventurous of these wandering individuals may inevitably disperse beyond their normal winter range limits in search of suitable wintering areas, a mechanism that might be responsible for a significant proportion of vagrancy occurring during the non-breeding period.

Overall, it is likely that a significant component of vagrancy arises not because of extraneous factors like weather, or internal navigatory errors, but rather from the evolved tendency for some individuals to perform random exploratory movements. Particularly during early life stages, some degree of exploratory movement can be highly advantageous to individuals, and this evolved urge to disperse will inevitably drive a small proportion of individuals to move far further than the norm for their species, ultimately becoming extralimital vagrants.

HUMAN-DRIVEN VAGRANCY

Humans have been transporting wildlife around the world for millennia – sometimes deliberately, sometimes by accident – and often with catastrophic results for the natural world, particularly on oceanic islands where non-native species have caused some 90 per cent of all recent bird extinctions (Johnson & Stattersfield 1990). With the ever-increasing globalisation of trade, humans have created veritable transport highways that can entrain and carry wild species across and between continents. Active trade in wildlife – captive birds in particular – has also proliferated over the last century, with cagebird enthusiasts increasingly keeping more diverse sets of species, which have previously been difficult to sustain in captivity (e.g. small insectivores, referred to as 'softbills' by aviculturists). As a result, it is becoming increasingly difficult to tease apart instances of 'natural' vagrancy from those driven, at least in part, by human actions.

Thankfully, birds that escape from cages are often easy to recognise. Many carry telltale closed rings/bands, typically made of coloured metal or plastic and fitted shortly after hatching (unlike the open rings used for wild birds). Even if unringed, escapees often show other 'smoking gun' features indicative of a captive past. Many have damaged or aberrant plumage, particularly in the flight feathers and around the base of the bill where abrasion most often occurs. Cagebirds may have unusual states of moult or lack natural pigmentation, and they are often also tame or behave in unusual ways for the species in question. At the same time, many birds moved in trade may look and behave exactly like wild birds, and these individuals can severely muddy the waters of true natural vagrancy patterns (England 1974). National rarities committees often face a difficult task when deciding whether apparently wild vagrant birds might have actually escaped from captivity, as we invariably lack sufficient knowledge to properly evaluate the likelihood of either possibility. The true potential for natural vagrancy is almost impossible to estimate in many birds, especially for resident species, as natural variation within wild populations can occasionally lead to wholly unexpected natural vagrancy events (Gilroy *et al.* 2017). Species that appear completely outside their normal range therefore cannot always be assumed to be escapees – for example, Sulphur-bellied Warbler *Phylloscopus griseolus*, a short-distance migrant from South Asia, would seem a highly unlikely natural vagrant to Western Europe, but a bird of this species that appeared in Denmark in May 2016 could have been nothing other than a natural vagrant, given that the species (and indeed genus) is unknown in captivity. At the same time, we usually have little understanding of just how common bird species actually are in trade, not least because a considerable portion of the global cagebird market is illegal. Determining the origins of many apparent vagrant birds is therefore fraught with difficulty.

▲ Ross's Goose *Anser rossii* with Pink-footed Geese *Anser brachyrhynchus*, Stalmine, Lancashire, England, 7 February 2008. This species has yet to be officially accepted as a vagrant to the UK; although several individuals would appear to have good credentials for a wild origin there are also free-flying escapes roaming the country (*Chris Batty*).

Birders often look at the location and timing of records when trying to evaluate whether birds are likely to be escapes. Apparent vagrants that appear at unusual times of year, or away from traditional vagrant hotspots, are often viewed with scepticism. Similarly, unusual birds that turn up during migration periods at well-known vagrant traps tend to be given the benefit of the doubt. This can be problematic, however, not least because escaped birds can themselves travel large distances in search of conspecifics or suitable habitat, or simply following their inbuilt migratory program. Escapes are thus also prone to concentrate on the same peninsulas and islands that attract natural vagrants. For example, the Spurn area of Yorkshire, England – a renowned vagrant hotspot – has recorded at least 66 different species of escapes from across the avian tree of life (Roadhouse 2016). This 'escape migration' effect was further illustrated by the appearance of many ostensibly unlikely Asian vagrants along the North Sea coast in spring in the 1980s and 1990s, including multiple records of Japanese Grosbeak *Eophona personata* at locations and times highly suggestive of natural vagrancy. Tellingly, these spring occurrences entirely dried up following a ban on the importation of wild-caught birds to Europe in 2006 (European Communities Commission 2007). These records are thus likely to have been related to birds escaping from captivity around trade hubs in the Netherlands, Belgium and Germany and then reorienting and migrating north-west (Cottridge & Vinicombe 1996). The import ban has also toughened up regulations on close-ringing of captive birds in Europe, making more recent escapes easier to recognise – such as the ringed Godlewski's Bunting *Emberiza godlewskii* that arrived on the Isles of Scilly, England, in April 2019. Escapes are not always so obvious, and Milne & McAdams (2008) highlight the case of a metal-ringed Kentish Plover *Charadrius alexandrinus* in Cork, Ireland, in December–January 2008 which "...was eventually traced to Germany, where it had been hatched and reared in captivity, having originally been taken illegally from the wild. Following a raid on the dealer's premises by the authorities, this bird and numerous other waders were confiscated". Had these efforts not been made this record would have been assumed to pertain to a wild bird.

The worldwide volume of birds in trade has increased steadily over the last century in line with human population growth, although there have been declines in some regions recently due to tighter legislation following disease outbreaks, including the H5N1 'bird flu' epidemic of 2005. During the 1990s, BirdLife International (2008) estimated that 2–5 million birds per year were being traded annually worldwide, with almost one-third of the world's bird species being kept or traded as pets. The trade is particularly rife in parts of Asia – for example, Marshall *et al.* (2020) estimated that one-third of Java's 36 million households keep approximately 66–84 million cagebirds. There can be little doubt that a significant proportion of birds accepted as wild vagrants are in fact escapes from captivity – particularly for popular groups such as wildfowl and some brightly coloured seed-eating passerines. Likewise, a great many true vagrants worldwide have probably been dismissed as escapes – either by their observers or the relevant records committees (Veit 2004). Indeed, this is perhaps the most likely fate of many of the more extreme or unusual natural vagrancy events that have occurred (e.g. Lees & Mahood 2011). Uncertainty surrounding the origins of out-of-range individuals is inevitable and unavoidable, unless perhaps a universal system for cagebird tracking can be implemented at some point in the future.

◄ White-throated Bee-eater *Merops albicollis*, Dhakla, Western Sahara, 4 March 2017. This bird's plumage was in poor condition and hence there was significant debate about its provenance, but it does now seem to be widely recognised as a wild bird and hence the second for Western Sahara and the Western Palearctic (*Richard Moores*).

The accidental transport of birds by boats and other vehicles arguably presents an even greater headache for birders when trying to decide what vagrants count as 'wild' and therefore 'tickable'. Shipping, in particular, plays a major role in influencing the movements of birds across oceans and one that is hotly debated amongst birders and ornithologists. Birds, and migratory landbirds in particular, frequently alight on vessels at sea – sometimes in truly spectacular numbers. Whilst ships are slow-moving in avian terms, birds that remain on board for long periods can inevitably find themselves moving inadvertently out of their normal ranges, and there are a multitude of examples of wild birds being carried vast distances aboard ships – for example Snowy Sheathbills *Chionis albus* transported from the Antarctic to the British Isles (Jay 1993) and Taiwan (Lin *et al.* 2018).

The debate usually starts when such a bird makes landfall in a new area – should we count it as 'wild' and a 'natural vagrant', even though it has not made the journey under its own steam? This is a tricky question – and one that troubles legislators as well as birders, as 'wild' status typically determines whether individuals are protected under regional wildlife laws. Usually, organisms transported outside their native ranges by boats are considered non-native in their destination regions and may thus be targeted for culling to prevent them establishing invasive populations (Gilroy *et al.* 2017). Some of the most adept ship-born hitchhikers in the avian world are potentially significant pests. Perhaps the most prominent of these is House Crow *Corvus splendens*, a species that has taken advantage of global shipping to colonise 28 countries outside its native Indian Ocean range, as well as reaching an additional 23 countries as disjunct as Japan, Ireland, France, Hungary, Denmark and Chile (Ryall 2016). House Crows can have damaging impacts on local ecosystems, as well as being vectors of disease, and have consequently been targeted for removal in most of these locations (Ryall 2016). Another dramatic recent example concerns an apparent wild vagrant Nicobar Pigeon *Caloenas nicobarica* found in Kimberley, north-west Australia, in 2017. It was subsequently captured and confined to Adelaide Zoo by the Australian Quarantine and Inspection Services (Davis & Watson 2018) as it was viewed as being a potentially damaging invader. This latter case illustrates the legislative complexity concerning how wild vagrants are treated. Davis and Watson (2018) highlight the role of vagrants in allowing species to respond to climate change and argue that legislators need to develop a better framework to decide what constitutes 'natural colonisation' and what is an 'introduction', and to ideally let 'natural' vagrancy processes remain unmolested. They also argue that sites routinely visited by vagrants need better protection as stepping stones for colonisation, especially if the species in question are threatened.

When considering whether ship-assisted birds should be classed as 'wild', there is a huge grey area between the clear-cut examples of 'unnatural' port-to-port transport for largely sedentary species (such as House Crow) and the potentially much shorter (but no less 'unnatural') transport of other migratory birds that alight on vessels for a few days during their long overwater flights. At what point do we draw the line between unnatural vagrancy caused by ship assistance and natural transoceanic vagrancy? If a bird alights on a vessel for just a few hours during a transatlantic crossing, does that automatically invalidate it from consideration as a true vagrant? What about a bird that hops from one ship to another, perhaps spending only a short time on each, but using them as stepping stones to cross an ocean? Such movements are likely to be increasingly commonplace in the

► House Crow *Corvus splendens*, Cobh, County Cork, Ireland, 13 September 2010. This individual was placed on Category D of the national list as probably ship-assisted (*Stuart Piner*).

▲ **Figure 12.** Global relative density of maritime traffic (reprinted with permission from Halpern *et al.* 2008).

modern world, as global maritime traffic continues to increase at pace, with key shipping lanes forming 'highways' across the North Atlantic and North Pacific in particular (see Figure 12). Can we logically differentiate between birds that hop between vessels to disperse across oceans from those that hitch a ride on a single ship? Clearly, any decisions that birders or legislators make about where to draw the line between these scenarios must be fairly arbitrary.

Some national records committees consider ship-transported birds to still be 'wild' vagrants, as long as the species is deemed physiologically capable of making the crossing unaided (Dudley *et al.* 2006). However, in practice such assessments are extremely difficult to make – the true flight capacity of most bird species is unknown. Champion migrants like Blackpoll Warblers *Setophaga striata* are known to have physiological adaptations such as long wings and large fat deposits that enable long overwater flights, and these adaptations are usually lacking in more sedentary species. Nevertheless, short-distance migrants may still be capable of making long ocean crossings unaided, especially if the right weather conditions conspire to carry them. For example, molecular evidence suggests that thrushes (*Turdus* spp.) dispersed historically across the Atlantic from the Western Palearctic to the Neotropics, giving rise to the rich diversity of the genus in Latin America (Batista *et al.* 2020) and underlining the potential for remarkable transoceanic dispersal even in the absence of human involvement.

◄ Dunnock *Prunella modularis* with Royal Albatross *Diomedea epomophora*, Campbell Island, New Zealand, 18 November 2008. Vagrants of this introduced species have colonised subantarctic islands from New Zealand (*Steve Howell*).

Uncertainty about the overwater flight capacity of any given species is further compounded by the fact that traits associated with long-distance flight can evolve extremely rapidly when bird populations are exposed to new selection pressures. This means that historical movement patterns cannot necessarily tell us very much about a species' current vagrancy capability. A classic example concerns the House Finch *Haemorhous mexicanus* population that was established in New York State in the 1940s through the release of individuals from a sedentary population in California. The cold winter climates in this new range meant that the population was exposed to strong natural selection in favour of individuals that dispersed south in winter, and within just a few decades the population had effectively 'evolved' from being resident into a short-distance migrant (Able & Belthoff 1998). Similar rapid increases in the prevalence of dispersive wandering have been seen in other newly established populations, including the reintroduced Bearded Vulture *Gypaetus barbatus* population in Western Europe (McInerny & Stoddart 2019). Stark changes in migratory and dispersive capabilities can emerge within just a few generations in birds, highlighting the capacity for long-range wandering to emerge in even the most sedentary species.

Although boats are the primary mechanism of accidental human-mediated transport of rare birds, there are also numerous records of live birds successfully stowing away aboard aircraft. These include a Mourning Dove *Zenaida macroura* found in the cargo depot of Heathrow Airport, London, England, which had arrived on a plane from Chicago, United States (Anon 1998a), and a Barn Owl *Tyto alba*, which was ringed near Wendlebury, Oxfordshire, England, and recovered (long dead in a building) at Camp Bastion, Helmland Province, Afghanistan, having presumably arrived alive in the hold of a military aircraft (BTO 2021).

Indirect human impacts affecting vagrancy

A more subtle set of human impacts on vagrancy involve the changes we make to natural habitats, and how this can encourage and enable species to disperse outside their normal ranges. A prominent case concerns the spread of Barred Owl *Strix varia* into western North America (Livezey 2010), driven by widespread fire suppression, habitat fragmentation and tree planting. New populations founded by vagrant Barred Owls have had significant negative impacts on native populations of Spotted Owls *Strix occidentalis*, prompting the implementation of an experimental control programme by the U.S. Fish and Wildlife Service (Diller *et al.* 2016). This effort has sparked considerable debate about the ethics of controlling species expanding naturally in response to indirect human impacts (Carey *et al.* 2012). In this case, the new Barred Owl populations are founded by wandering individuals that would surely qualify as 'wild' vagrants, and therefore might arguably be afforded the same protections as any other wild birds in the area. However, the potential for new colonists to have negative impacts on native species raises questions about whether such range shifts are desirable.

A wide range of human-driven habitat changes has allowed birds to disperse and expand beyond their historic ranges in recent decades. Westward range expansions of many waterbirds in Europe, with vagrants at their growing fringe, have been associated with widespread recent wetland restoration and improved protection

► Masked Water-Tyrant *Fluvicola nengeta*, Ubatuba, São Paulo, Brazil, 21 May 2014. This species has undergone a rapid range expansion in Brazil with vagrants reaching remote areas in Amazonia, far from the main population expansion front (*Alexander Lees*).

▲ Great White Egrets *Ardea alba*, Blakeney Point, Norfolk, England, 14 October 2009. This migrating flock was the largest ever recorded in the UK at the time, but with the subsequent colonisation by the species it was quickly eclipsed by higher counts (*Alexander Lees*).

from persecution. In South America, rapid range expansion by some open-country species is also being driven by deforestation, with vagrants of species like Masked Water-Tyrant *Fluvicola nengeta* appearing in ever-increasing numbers beyond the growing fringe of the range. For many species, climate change is also playing a key role in facilitating range shifts, likely driven by higher survival rates among vagrant individuals that wander into areas that previously had less suitable climates. For example, Great White Egret *Ardea alba* is one of a suite of waterbird species that were major rarities in the UK as recently as the 1980s but have now established growing breeding populations. Human actions may also facilitate vagrancy by provisioning resources that vagrants may use, such as habitat stepping stones across otherwise inhospitable areas. For example, hummingbird feeders permit the survival and onward journey of wandering vagrants, particularly in late autumn and winter (Grieg *et al.* 2017).

Overall, there can be little doubt that human agency plays an increasingly major role in influencing patterns of avian vagrancy worldwide. As the impacts of humanity on the natural world continue to accumulate, it is becoming more and more difficult to identify what is 'natural' and what is not, as all aspects of our ecosystems are ultimately being impacted in some way by human activities. Going forward, birders and conservationists may need to increasingly embrace the 'unnatural', as more and more wild species adapt to take advantage of the opportunities afforded by human-made environments. Vagrants that cross barriers by piggybacking on human transport networks may have an important role to play in allowing species to respond to climate change and should perhaps therefore be viewed with a more open mind than many currently afford them.

CONSEQUENCES OF VAGRANCY FOR SPECIES AND ECOSYSTEMS

It is a common assumption that most vagrant birds are ultimately doomed, aside from the rare cases where individuals are able to reorientate and return to their normal ranges. In turn, it is also commonly assumed that vagrancy itself is a relatively inconsequential biological phenomenon. This is undoubtedly true for the majority of cases, as the most likely outcome of any given vagrancy event is that the individual will fail to find enough resources, and/or be exposed to inhospitable environmental conditions, and perish. However, there are many lines of evidence to suggest that vagrancy can, on rare occasions, dramatically alter the fate of populations, species or even whole ecosystems. Despite being infrequent, these events can be extremely important when viewed at the timescales over which ecological and evolutionary processes unfold. The most profound consequences of vagrancy relate to the establishment of new breeding sites, new migration routes and wintering locations. Each of these can occur through different mechanisms, and at different frequencies, and they each have their own unique importance. In this section, we will explore these phenomena in turn, as well as considering some additional and more unexpected consequences of vagrancy for wider ecosystems.

Establishing new breeding locations

Whilst most long-distance vagrants undoubtedly meet an untimely end, some individuals can survive for protracted periods in their new ranges – even decades in some cases – provided they are able to meet the basic requirements for survival. This is particularly obvious in long-lived species; multiple Black-browed Albatrosses *Thalassarche melanophris*, for example, have become famous for returning to Northern Gannet *Morus bassanus* colonies in the Atlantic year after year, each waiting in vain expectation for a potential mate to appear on their chosen ledge. Accounts of an individual that frequented a gannetry in the Faroe Islands for over 30 years in the 19th century suggest the bird reached such celebrity status in the archipelago that it gave rise to the Faroese name for albatross: *súlukongur*, or 'gannet king' (Dánjalsson & Ryggi 1951). By virtue of never breeding, and consequently escaping the physiological stresses of reproduction, it is possible that these lost individuals can survive even longer than their life expectancy in their native range (Erikstad *et al.* 1998).

This failure to reproduce – a likely consequence of long-distance vagrancy in most instances – is also the fundamental reason why vagrancy is often seen as having little relevance for the ecology and evolution of species. Without breeding, vagrant birds are 'dead-ends' in ecological terms, tragic in themselves but of no real consequence for the wider population. However, if they can somehow find a mate (ideally of the same species!) and enough resources to raise a brood, then a new breeding population can theoretically be established, extending the existing range of the species.

▲ Lesser Black-backed Gulls *Larus fuscus* with Ring-billed Gulls *Larus delawarensis* and American Herring Gulls *Larus* [*argentatus*] *smithsonianus*, Niagara Falls State Park, Niagara, New York, United States. Dozens of Lesser Black-backed Gulls may stage at this site (*Alexander Lees*).

The circumstances necessary for this to happen may occur quite frequently, particularly in the case of weather-driven vagrancy, where multiple individuals of a given species may be pushed out of range in the same weather system. Spring vagrancy, and particularly overshooting (see page 46), may be an important mechanism by which species colonise new areas – something that may be essential for species to persist in response to climate change (Davis & Watson 2018, see page 330). Whilst most shifts in species distributions involve small-scale movements at the expanding boundary of a population – and hence local dispersal rather than 'true' long-distance vagrancy – vagrants can play a critical role in 'leap-frogging' sea barriers or other inhospitable regions (Veit 2000). The colonisation of Greenland by Lesser Black-backed Gulls *Larus fuscus* in the late 20th century presents a clear example. Following the first confirmed breeding of a vagrant pair in 1990, the population had grown to well over 600 individuals just 10 years later (Boertmann 2008). This population is probably responsible for the increase in records on the western Atlantic seaboard of North America, and in time we can perhaps expect a new subspecies to emerge if gene flow with European populations becomes interrupted. Vagrant-founded range expansions such as this are likely to increase as further impacts of climate change unfold, with new temperature regimes providing opportunities for vagrants to play a key role in reshaping ecosystems as the planet warms.

Insights into the role vagrancy plays in shaping species distributions can also come from genetic reconstructions of the long-term history of populations that are well established. Many remarkable examples are 'hidden in plain sight', cases where a familiar widespread species owes its current distribution and status to a vagrancy event far back in their evolutionary history. One such example concerns Eurasian Wren *Troglodytes troglodytes* – one of the most abundant and familiar birds across much of Eurasia, and sister taxon to Winter Wren *Troglodytes hiemalis* and Pacific Wren *Troglodytes pacificus* in North America, with which it was formerly considered conspecific. The Eurasian Wren group has been subject to much study and provides several insights into the role of vagrancy in shaping species distributions. Wrens are, with the exception of the single Eurasian species, a family restricted to the New World. There are numerous exclusively New World families that have no representatives in the Old World, and it is unusual for a family with such clear origins in the New World to have a single representative in the Old World, and even more so one that it is so widespread. Eurasian Wren is undoubtedly of New World origin, and yet it has colonised vast swathes of the Old World, despite being an apparently weak flier. How did this occur?

Detailed reconstruction of the history of this species from DNA evidence revealed that the Eurasian Wren ancestor likely colonised the Old World more than one million years ago, when the North American and Asian continents were connected across the Bering Strait (Drovetski *et al.* 2004). Ancestral wrens are likely to have spread across the Beringean land bridge through a series of hops made by vagrant individuals wandering outside their historic ranges. The remarkable thing about the spread of Eurasian Wren across Eurasia is just

◄ Eurasian (Fair Isle) Wren *Troglodytes troglodytes fridariensis*, Fair Isle, Shetland, Scotland, 1 October 2018. This subspecies is endemic to Fair Isle and is one of six insular forms on Atlantic islands (*Alexander Lees*).

how comprehensive it was: the current distribution spans all wooded habitats across boreal and temperate zones, as well as Mediterranean and even arid climate zones in North Africa. Contrast this with the distributions of the sister taxa in North America, which are largely limited to boreal coniferous forests, perhaps largely because they are outcompeted by other wren species in the habitats that their sister species has been able to colonise in the Old World.

One can imagine the likely scenario that led to Eurasian Wren's burgeoning takeover of the Old World. Wrens radiated out of the New World tropics to fill a suite of niches across the biomes of North and South America, from humid forests to deserts. In the Old World, it appears that no ecologically similar species had evolved prior to the ancestral wren making the 'leap' across Beringia. As such, the 'wren' niche may have been vacant – by pure chance, no Old World species had evolved a similar set of morphological and ecological characteristics. When vagrant individuals made the leap across Beringia, they may have faced little competition from native species within their optimal habitat, enabling a rapid and prolific spread across the northern forests of the continent. Over the following millennia, they spread further by colonising other vacant niches in woodland and scrub habitats, well outside their traditional coniferous forest niche. Eurasian Wren's huge range therefore suggests a story of invasion that probably started with vagrant wanderers from the New World discovering fertile new ground to rapidly establish competitor-free populations. Similar processes to these have likely played out thousands of times over evolutionary history worldwide. Indeed, events such as this explain why so many avian families and even genera have distributions that span the globe. Cross-continental colonisation events, driven by vagrant individuals that are somehow able to find a mate and successfully breed, have undoubtedly played a huge role in shaping biogeographic patterns.

Nowhere is this more obvious than in the avifauna of remote offshore islands, which can only be colonised through long-distance vagrancy events. All ocean archipelagos support at least some landbirds, very often endemic species that have evolved in isolation for millennia, which cumulatively account for around one-fifth of all bird species (Johnson & Stattersfield 1990). Even these endemics must have as their origins a successful pair of historic vagrant colonists to thank for their position on the islands they inhabit. This is in stark contrast to continental systems, where low dispersal rates such as in the aforementioned tapaculos lead to high diversity. Without occasional epic overwater vagrancy events, oceanic islands would have far more impoverished avifaunas. Genetic studies that trace the origins of these species can often yield surprising results. For example, the 20 genera of endemic honeycreepers of Hawaii, which make up a highly diverse family dominated by nectarivores, are now thought to have evolved from a rosefinch of the Eurasian genus *Carpodacus* – a seemingly highly unlikely candidate to have colonised this mid-Pacific archipelago (Lerner *et al.* 2011). The mechanism by which such a colonisation and eventually speciation attempt might function in the middle of the Pacific Ocean was well illustrated by a flock of at least 13 Bramblings *Fringilla montifringilla,* which appeared on Kure Atoll

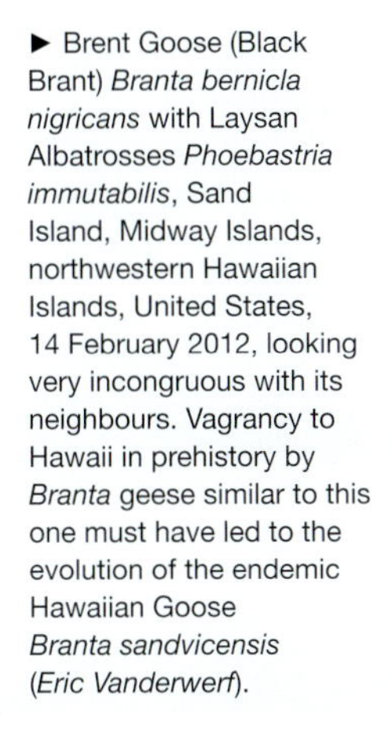

► Brent Goose (Black Brant) *Branta bernicla nigricans* with Laysan Albatrosses *Phoebastria immutabilis*, Sand Island, Midway Islands, northwestern Hawaiian Islands, United States, 14 February 2012, looking very incongruous with its neighbours. Vagrancy to Hawaii in prehistory by *Branta* geese similar to this one must have led to the evolution of the endemic Hawaiian Goose *Branta sandvicensis* (*Eric Vanderwerf*).

in the northwestern Hawaiian Islands in October 2014, with some staying till at least March 2015 (Pyle & Pyle 2017). This was the first record of the species from the Pacific region, and one could imagine that a similar mass arrival event of rosefinches must have kick-started the evolution of the honeycreepers in prehistory. Some examples of odd island species can be difficult to interpret until genetic studies can shine a light on deeper relationships. For example, recent genetic studies have cleared up a long-standing mystery of how two gallinaceous bird species with medium-sized bodies and short, rounded wings – unsuitable for overwater flights – came to inhabit landmasses isolated by major marine barriers. We now know that the ancestors of Madagascan Partridge *Margaroperdix madagarensis* in Madagascar and Snow Mountain Quail *Anurophasis monorthonyx* in New Guinea shared ancestorship with species related to the modern-day *Coturnix* quails, several of which are highly migratory (Hosner *et al.* 2017). These two island species subsequently evolved to become 'island giants', losing those traits associated with high dispersal and becoming more like 'normal' gallinaceous birds, which are often poor dispersers (although see page 82).

Very often, the communities of landbirds that establish on offshore islands are largely made up of subsets of the species that occur on their nearest neighbouring landmasses. The species within these subsets often share a predictable set of traits – not least, they tend to be species that are particularly prone to vagrancy. Often, island colonists are species that are relatively abundant in the neighbouring mainland, are often migratory (or at least partially migratory) and have large global ranges – a feature of species that are strong dispersers (Lees & Gilroy 2014). However, rates of vagrancy alone do not explain the likelihood that a given species will be adept at colonising offshore islands. Some species occur with remarkable frequency as vagrants on offshore islands, and yet have extremely low rates of success in colonising. Barn Swallow *Hirundo rustica*, for example, has been recorded on at least 44 oceanic island groups, but has successfully colonised only one. Others seem to be able to colonise almost every time they arrive on a new archipelago: the Barn Owl *Tyto alba* species complex, for example, has successfully colonised 12 of the 16 offshore island groups where it is known to have occurred (Lees & Gilroy 2014). The 'supertramp' theory (Diamond 1974) argues that alongside a propensity for vagrancy, a key feature of successful island colonists is the capacity to adapt to novel environments, as well as an ability to reproduce rapidly and thus quickly establish a self-sustaining population. This same principle likely applies far beyond island contexts, so species that will be most successful at colonising new areas in response to climate change, for example, are likely to be a relatively small set of tenacious 'supertramps' that are able to rapidly adapt and take advantage of emerging zones of climatic suitability, while a whole suite of other species may struggle to persist within their existing ranges. Ironically, despite being apparently poor island colonisers, Barn Swallows have successfully 'colonised' their winter range as breeders, and established a new migratory population in southern South America that migrates north to winter (Winkler *et al.* 2017).

Colonisation of new winter sites and migratory routes

In most species, vagrancy occurs most frequently during the period after the breeding season, when inexperienced juveniles are more vulnerable to making navigational errors on their first migration or fail to compensate for wind drift (see pag 40). Whilst autumn vagrancy is less likely to result in colonisation of new breeding areas than spring vagrancy, it can still have important impacts on the ecology and biogeography of species by enabling the colonisation of new routes to new wintering areas. However, the circumstances necessary for these routes to be colonised by populations are perhaps even more complex than those necessary for the colonisation of new breeding ranges.

A vagrant bird that is fortunate enough to 'discover' a route to a new location that is suitable for overwintering will only successfully colonise that site, in population terms, if it is able to pass its new migration route on to future generations. This can only happen if a number of circumstances are met: the individual must be able to return to an area where it can encounter a mate and successfully reproduce, and those progeny must be able to retrace the new migration route and also successfully overwinter at the new site. If this process is repeated over multiple generations, the new migration route may be successfully pioneered. However, it is clear that this process depends heavily on the capacity for migratory routes to be passed from one generation to the next. In some species, this can occur directly through juveniles following their parents (or other adults) on their first migration. In social migrants such as cranes and geese, cultural transmission of migration routes can facilitate the colonisation of new wintering areas. For example, Bewick's Swan *Cygnus columbianus* has a well-established wintering range in lowland eastern England, where up to 8,000 individuals migrate from their Russian breeding grounds to spend the winter. However, prior to 1938 this species wintered almost entirely in continental Europe

► Rufous-headed Robin *Larvivora ruficeps*, Phnom Penh, Cambodia, 18 November 2012, one of only a handful of records away from the breeding areas of this species. Its wintering area remains largely unknown, although it is now thought to include the highlands of peninsular Malaysia where there have been several recent records (*Rob Martin*).

– until an extreme cold snap during the winter of 1938–39 led to large numbers being displaced west into the British Isles in search of snow-free grazing habitat. After their 'discovery' of the low-lying grazing marshes of the East Anglian fens, flocks have returned there ever since (Cadbury 1975). However, the north-west European population of Bewick's Swan subsequently declined by 38 per cent between 1995 and 2010 and declines in the English wintering population are compounded by the birds 'short-stopping' on the Continent, perhaps facilitated by milder winters (Beekman *et al.* 2020).

Most migratory birds do not migrate in familial flocks, and therefore do not have a capacity for direct cultural transmission of new migration routes. These species must therefore rely on their inherited navigational abilities and migratory program in order to migrate. Thus, any new migratory route would need to be encoded genetically in order for successful long-term colonisation. Conclusive proof that migratory orientation can be genetically inherited was first found through genetic studies of Blackcaps *Sylvia atricapilla* breeding in Central Europe, where a migratory divide exists between individuals that migrate south-east around the Mediterranean to winter in East Africa, and others that migrate west to winter in western Europe (Helbig 1992). Experiments where males from one migratory route were paired with females from another showed that their offspring tended to migrate in a direction intermediate between the two, clear evidence that the direction of orientation, at least, can be directly inherited (Helbig 1992).

The Blackcap example reveals another significant potential barrier to the colonisation of new migratory routes through vagrancy – if just one individual successfully pioneers a new route and returns to breed somewhere close to its natal site, the likelihood is that it will pair with an individual that uses a 'normal' route for that species. If the inheritance of migratory routes follows the model shown by Blackcap, any offspring are likely to orientate towards some intermediate location, which may or may not ultimately lead to a successful wintering location. This highlights the complexity of changes in migratory routes, and in part explains why significant changes in migration routes are infrequently observed in most taxa.

Nevertheless, there are numerous examples of species apparently colonising new migration routes as a result of initial vagrancy events. As previously mentioned, the rapid rise in wintering Lesser Black-backed Gulls *Larus fuscus* overwintering in eastern North America probably indicates a new Nearctic migration route being pioneered among breeders from Greenland and Iceland (Hallgrimsson *et al.* 2011). Increasing numbers of several Siberian passerines as autumn migrants in Western Europe – particularly Yellow-browed Warblers *Phylloscopus inornatus* and Richard's Pipits *Anthus richardi* – may also represent successful pioneers of new migratory routes (see page 32), although this has yet to be proven conclusively (Gilroy & Lees 2003), and we highlight several other examples in the family accounts.

Impacts on other species and ecosystems

Vagrancy can also lead to more pervasive ecosystem impacts beyond the vagrant species themselves. Research into these effects is still in its infancy, but their importance is increasingly recognised. Rose & Polis (2000) described a set of five principal impacts that vagrants can have on ecosystems, particularly island systems. The first is predation – even if only brief, intensive predation by a vagrant can have important impacts on the dynamics of local species. For example, a Peregrine Falcon *Falco peregrinus* that reached remote Saint-Paul Island in the Indian Ocean predated at least 27 Macgillivray's Prions *Pachyptila macgillivrayi*, representing a serious threat to this endemic, whose global population numbers only 150 pairs (Jiguet *et al.* 2007). Similarly, a vagrant Short-eared Owl *Asio flammeus* on the 0.12km^2 Praia Islet, off Graciosa Island in the Azores, exclusively ate Madeiran Storm Petrels, killing one or two individuals every night from a population of only 200 breeding pairs (Bried 2003). Vagrant Peregrines on Midway Atoll killed ≥4 per cent of a newly translocated Laysan Duck *Anas laysanensis* population in 2006 and ≥2 per cent in 2008, and also succeeded in eating about 1 per cent of all Laysan Finches *Telespiza cantans* on Laysan Island during 2008–09 (Reynolds *et al.* 2015). In the latter cases, predation was facilitated by abandoned aircraft runways – open, 'Peregrine-friendly' hunting habitats – suggesting that restoration of these habitats could mitigate the problem. It is possible that global shipping has elevated the rates of vagrancy of raptors and owls, by providing stepping stones to distant oceanic islands.

◄ Peregrine Falcon *Falco peregrinus*, Midway, northwestern Hawaiian Islands, 12 December 2011, Vagrant Peregrines can have a disruptive impact on the avifaunas of remote offshore islands (*Cameron Rutt*).

Secondly, vagrants can carry parasites and pathogens to new environments, potentially leading to novel disease transmission. For example, three vagrant Silvereyes *Zosterops lateralis* of the Australian mainland subspecies *cornwalli,* captured on Heron Island, Australia, were all infected with avian haemosporidian blood parasites, which can cause avian malaria. In contrast, only 6.2 per cent of a sample of 195 of the resident endemic 'Capricorn' Silvereye *Zosterops lateralis chlorocephalus* were infected. Avian malaria had previously been thought to be absent on the island (Clark *et al.* 2014). Migrant passerines may host subadult *Ixodes* ticks for two to four days and whilst on active migration may routinely travel as far as 950km/day, facilitating the dispersal of ticks and tick-borne diseases over long distances, even transoceanic crossings. For example, of 11,324 spring migrants trapped on Appledore Island, Maine, United States, 1.2 per cent were infested with the tick *Ixodes dammini;* 65 per cent of these were taken from Common Yellowthroats *Geothlypis trichas.* Twenty per cent of these nymphal ticks were infected with *Borrelia burgdorferi,* which causes Lyme disease (Smith *et al.* 1996).

Vagrants can also potentially impact local species through competition for resources, and colonisation by vagrants may lead to the replacement of one species by another, even potentially driving global extinction (Gilpin & Diamond 1981). Such replacements may be facilitated by anthropogenic changes to habitats. For example, rapid declines of the endemic Cyprus Warbler *Sylvia melanothorax* are strongly correlated with land-cover change and with the duration of site occupancy by Sardinian Warblers *Sylvia melanocephala* (Pomeroy *et*

► Cyprus Warbler *Sylvia melanothorax*, Cape Greco, Ammochostos, Cyprus, 11 May 2019, a species in decline due to changes in land use and apparently accentuated by competition with colonising Sardinian Warblers *Sylvia melanocephala* (*Alexander Lees*).

al. 2016). Sardinian Warbler is a recent colonist of Cyprus, first confirmed breeding in 1993, and in concert with changes in vegetation associated with farming changes may eventually entirely replace the endemic Cyprus Warbler. Fourthly, although this seems less pertinent for most bird species, some vagrants may have direct impacts on habitats themselves, for example trampling, digging or burrowing, or by the transport and accumulation of material for nest-building.

Finally, and arguably most importantly, there are cascading ecosystem impacts on food webs when vagrants arrive transporting other species. Birds are the most important agents of dispersal for plant species to oceanic islands, and they can transport seeds internally in their gut – to be excreted on arrival – or externally via mud on their feet or stuck on their feathers. For example, Porter (1976) estimated that 378 'natural' species introductions could account for the 522 species of indigenous vascular plants occurring on the Galapagos archipelago. He estimated that of the species whose colonization was assisted by birds, 144 (64 per cent) arrived internally,

▲ Blue-winged Teal x Cinnamon Teal *Spatula discors* x *cyanoptera* with Blue-winged Teals *Spatula discors*, Montezuma NWR, Seneca, New York, United States, 6 November 2015. Hybrid waterfowl are relatively common, so it is impossible to know in this case if this hybrid is itself a vagrant from the American west or is the progeny of a vagrant Cinnamon Teal that paired with a Blue-winged Teal in the absence of conspecifics in the east. Vagrant waterfowl frequently hybridise with other congeneric species when 'marooned' in different biogeographic regions without access to members of their own species (*Jay McGowan*).

34 (15 per cent) in mud attached to birds, 28 (12.5 per cent) attached by sticky structures of seeds or fruits, and 19 (8.5 per cent) attached mechanically. An interesting contemporary example concerns the appearance of a single shrub of *Ochetophila trinervis* (family Rhamnaceae) on subantarctic Marion Island. An analysis by Kalwiji *et al.* (2019) concluded that the plant must have been naturally dispersed by birds more than 7,500km from the nearest location in the southern Andes. Endozoochory – dispersal by ingestion – is not restricted to seeds in birds, and can also involve animals. We now know that waterbirds can transport viable eggs of a range of aquatic invertebrates, including insect larvae (Green & Sánchez 2006), and even fish eggs (Lovas-Kiss *et al.* 2020). These complex impacts of vagrancy play out across millennial timescales; for example, Hawaiian Duck *Anas wyvilliana* is now thought to have arisen from hybridisation of vagrant Mallard *Anas platyrhynchos*-like ducks and Laysan Duck *Anas laysanensis*, which was an earlier colonist (Lavretsky *et al.* 2014). Vagrant Mallards thus not only gave rise to a subsequent duck speciation event but may well also have carried with them seeds and invertebrate eggs that would go on to become established and eventually also evolve into other new species.

▲ Buff-breasted Sandpiper *Calidris subruficollis*, Carrahane, Kerry, Ireland, 27 September 2011. Shorebirds are known to be efficient vectors of long-range plant dispersal by endozoochory; this Buff-breasted Sandpiper could have conceivably ingested and moved plant matter from North America to Ireland and could even have subsequently moved viable seeds from Ireland to South America if it was able to successfully reorientate (*Alexander Lees*).

FAMILY ACCOUNTS

In this, the major part of the book, we explore extralimital occurrences across the entire avian tree of life. Our family-by-family accounts are not intended to be an exhaustive account of vagrancy across all species within a family; indeed, for vagrancy-prone groups like waders, gulls and terns this would necessitate a book per family. Our review is intended as a broad overview of the patterns and processes that underpin extralimital movements and is global in scope. There are, however, many caveats, foremost that the literature on vagrancy is vast and diffuse and we have probably missed many important records, particularly from publications in languages other than Germanic or Romance ones, during our attempts at systematic searches. Moreover, it should be remembered that there are huge disparities in observer coverage between regions, and our understanding of avian vagrancy patterns is much more complete in some regions than others. This bias in space also creates another bias in taxonomy, as it means we know relatively little about extralimital vagrancy in bird families that are restricted to less well-covered regions, especially in the tropics (Lees *et al.* 2020). We include records up to the end of December 2020, and our taxonomy largely follows Clements *et al.* (2019).

STRUTHIONIDAE
Ostriches

Both species, Common Ostrich *Struthio camelus* and Somali Ostrich *Struthio molybdophanes*, can behave nomadically in response to shortages of food and water. Latitudinal north–south movements have been reported in Common Ostriches in the Sahel of West Africa in response to monsoon rains (Folch *et al.* 2020a). Occasional vagrancy to areas surrounding known ranges might be expected to occur as a result of these wanderings, but we are unaware of any that have been labelled as such.

RHEIDAE
Rheas

Unlike ostriches, rheas are strictly sedentary, although the range of Greater Rhea *Rhea americana* is expanding in Brazil as deforestation creates new suitable open arable habitats in southern Amazonia. We are not aware of any vagrancy records in this group.

TINAMIDAE
Tinamous

Tinamous are not known to perform any true migrations, and vagrancy is undocumented, but individuals of some species do perform 'escape movements' during cold snaps at high altitudes, as well as seasonal movements in and out of flooded forests (Cabot 1992). Forest-associated tinamous in the Amazon, such as Undulated Tinamou *Crypturellus undulatus,* have been observed crossing more than 500m of open water (Remsen & Parker 1983), which is a long flight for a tinamou! Occasional wandering is therefore possible, but long-range vagrancy seems unlikely in this family.

CASUARIIDAE
Cassowaries and Emu

Emu *Dromaius novaehollandiae*, like many Australian desert birds can be either sedentary or nomadic, depending on resource availability. When water is scarce, Emus may make long-distance north–south movements within Western Australia, and may even cross shallow bays to coastal islands (Folch *et al.* 2020b). Three adult Emus ringed near Cue in the Gascoyne Region of Western Australia in June 1969 were later recovered 440km WSW, 360km SW and 540km NW respectively within an eight-month period (Storr 1985), indicating the extent to which individuals are capable of moving. Cassowaries, by contrast, are likely to make far more limited movements as they live in more climatically stable forest regions, and we are aware of no records of vagrancy.

APTERYGIDAE
Kiwis

Kiwis are highly sedentary forest-associated species and year-round residents in small territories. We are unaware of any extralimital movements in this family.

ANHIMIDAE
Screamers

As is the case with many tropical waterbirds, understanding the nature of movements and hence vagrancy by screamers is hampered by seasonal shortfalls in observer effort, and a lack of knowledge about their regular ranges (Carboneras 1992). They are large, powerful birds, capable of soaring flight and some regularly form large flocks. Horned Screamer *Anhima cornuta* is usually regarded as quite sedentary, but

◄ Southern Screamer *Chauna torquata*, Tanqua, Rio Piracicaba, São Paulo, Brazil, 10 September 2013, a vagrant to the state, possibly driven by cold weather (*Robson Silva e Silva*).

there is an extralimital record from Uruguay in June 2016, around 1,100km south of the nearest known populations (Rabau & Löwe 2018), as well as a record further north in southern Brazil, in Rio Grande do Sul (Correia 2017). Southern Screamers *Chauna torquata* may be more prone to wandering still, particularly in response to cold fronts in central-southern South America. There is a wide scatter of vagrant records, for example to northwestern Mato Grosso state (Menq 2014) and Sao Paulo (Alves 2013) in Brazil, and occasional wanderers to the southeastern Peruvian Amazon (Servat & Pearson 1991).

ANSERANATIDAE
Magpie Goose

Magpie Goose *Anseranas semipalmata* is not truly migratory, but it is apparently quite nomadic and makes regular movements across the Torres Strait between Australian and New Guinean populations (Draffan *et al.* 1983). The status of vagrants is somewhat occluded by escapes and reintroductions (Nye *et al.* 2007). Records from Tasmania (Legge 1905) and south-west Australia (Nye *et al.* 2007) apparently pertain to vagrants, as does the single Indonesian record from Banda Island (Eaton *et al.* 2016).

ANATIDAE
Swans, Geese and Ducks

Waterfowl include among their ranks the full spectrum of avian migration syndromes, from highly sedentary tropical species to intercontinental migrants. Vagrants are frequently reported, yet some vagrant ducks, geese and swans are often viewed with a high degree of scepticism by birders, owing to their ubiquity in captivity and the frequency with which 'obvious' escapes appear in apparently wild settings, often without incriminating plastic leg-wear (e.g. England 1974). However, most records certainly do not deserve to be tarred with the 'wire-jumper' brush, as wildfowl are among the most highly dispersive of all bird species, with an incredible capacity for genuine long-range vagrancy. Although most species are associated with fresh water, they can also rest on the sea, which facilitates transoceanic movements. Indeed, of

► Green-winged Teal *Anas crecca carolinensis* with Laysan Duck *Anas laysanensis*, Midway Islands, northwestern Hawaiian Islands, United States, 25 December 2008 (*Eric Vanderwerf*).

◄ Lesser Scaup *Aythya affinis* with Common Scoter *Melanitta nigra*, Mývatn, Norðurland eystra, Iceland, 31 May 2020 (*Yann Kolbeinsson*)

the 100 or so bird species that have 'Holarctic' distributions – extending across the whole of Europe, Asia and North America, many are wildfowl. Engilis *et al.* (2004) lists 5,800 records of 36 species of migratory waterfowl from the Hawaiian Islands, some 3,680km from the nearest continental shore. Of these, 10 species were annual visitors, with Northern Shoveler *Spatula clypeata*, Northern Pintail *Anas acuta*, Lesser Scaup *Aythya affinis*, Common Teal *Anas crecca* and both American Wigeon *Mareca americana* and Eurasian Wigeon *Mareca penelope* accounting for 95 per cent of the records. There are also records of Falcated Duck *Mareca falcata*, Baikal Teal *Sibirionetta formosa* and Barrow's Goldeneye *Bucephala islandica* from the same archipelago – these are notable given that the status of all three species as wild vagrants to Europe has proven controversial (e.g. Barthel *et al.* 2018).

Despite the escape possibility that hangs over many waterfowl, proof of remarkable genuine vagrancy has come from a number of sources. Thanks to large scale ringing/banding programmes, and the less broadly welcomed efforts of sportsmen with shotguns, well over 30 ringing recoveries of Nearctic wildfowl have been amassed in Europe (Dennis 1990). From the Eastern Palearctic, genuine vagrancy to Europe has been supported by stable isotope analysis for Baikal Teal (Fox *et al.* 2007; Votier *et al.* 2009), while a candidate Marbled Duck *Marmaronetta angustirostris* was ruled unlikely with the same technique (Fox *et al.* 2010).

Assessing extralimital records of less migratory species, such as Ruddy Shelduck *Tadorna ferruginea* and White-headed Duck *Oxyura leucocephala*, is especially complicated by the high frequency of escapes. Although these species do not exhibit strong annual migration strategies, they are partially nomadic owing to their dependence on ephemeral waterbodies, so occasional long-range dispersal events might be expected. In the absence of stable isotope analysis (which itself cannot provide 'proof' of vagrancy, as it does not preclude rapid transit in the cagebird trade, or the possibility of captive birds being fed using imported food), the best way to uncover the status of these and other Mediterranean species is through statistical appraisal of records – as Jiguet *et al.* (2008) did for pelicans, examining all records and relating them to prevailing climatic conditions on a region-by-region basis (see page 175).

The occurrence of North American ducks in Europe is quite complicated, and difficult to tease apart due to the propensity for some individuals (often of unknown origin) to make long stays, sometimes wandering widely within their adopted range. Whilst clear seasonal peaks occur for most species, apparently 'new' birds are regularly detected throughout the year, clouding any clear pattern of arrival. There is little doubt that many do cross the Atlantic in fast-moving autumnal depressions in a similar manner to waders and land birds (McLaren *et al.* 2006), as reflected by annual multiple arrivals in western Britain and Ireland during autumn. It seems safe to assume that flocks of 15 Ring-necked Ducks *Aythya collaris* in Galway, Ireland, in October 2008, 13 American Wigeon in Kerry, Ireland, in October 1968 and 10 in Shetland, Scotland, in October 2000, and five Blue-winged Teals *Spatula discors* on the Outer Hebrides, Scotland, in September 2020 were all recent transatlantic arrivals. These records also demonstrate that it is not just individuals with 'faulty' navigatory apparatus involved, but whole flocks of exovagrants, presumably impacted by adverse weather whilst on southward passage. These systems may carry individuals well out of their normal latitudinal range, as is evidenced by a recent record of an American Wigeon from The Gambia (Cross *et al.* 2017).

Of the 30 recoveries of North American-ringed (banded) waterfowl recorded in Europe up to 1990, just two were ringed south of Canada – a Wood Duck *Aix sponsa* in North Carolina and a Blue-winged Teal in Maine (Dennis 1990). This strongly suggests a bias in transatlantic vagrancy towards Nearctic wildfowl at the northern edge of their range, although the results may also have been influenced by the intensity of ringing operations. Northernmost populations tend to have the longest migrations, and many may leapfrog southern populations to winter in Central America, the Caribbean and even South America (Bruun 1971). It is often assumed that Nearctic wildfowl also reach Europe as new arrivals in winter, perhaps as refugees from hard weather in the northernmost parts of their winter ranges (e.g. Dennis 1998). However, such arrivals are clouded by regular movements of wildfowl within Europe, inevitably involving vagrant individuals that have already made landfall undetected. Various species, including Lesser Scaup, leave their breeding grounds late in the year, usually at the onset of hard weather, but this almost always occurs before midwinter (Rogers 2005). Nevertheless, exceptionally hard weather within the winter period might trigger movements large enough to result in transatlantic arrivals, as was suggested by Cottridge & Vinicombe (1996) in the case of Bufflehead *Bucephala albeola*.

Many vagrant Nearctic waterfowl show a spring peak in occurrence in Europe, generally assumed to correspond to the northward movement of individuals that arrived in the Palearctic in previous autumns and have wintered further south in southern Europe or North Africa (or undetected somewhere else in the British Isles or Europe). Such birds presumably respond to increasing day length and migrate with 'carrier' flocks of other species, typically the closest ecological analogue within the wintering area (Bruun 1971; Cottridge & Vinicombe 1996; Dennis 1998). Some individuals undoubtedly attempt to recross the Atlantic, as is evidenced by the recovery of a Ring-necked Duck ringed at Slimbridge, Gloucestershire, in March 1977 and shot in Greenland in May of the same year (Grantham 2004). Others may attach themselves to flocks of related species and 'abmigrate' (Thompson 1923) with them to breeding areas in Scandinavia and western Russia, where some might even rear hybrid offspring (Cottridge & Vinicombe 1996). Genuine transatlantic vagrancy probably also occurs in spring, although it is likely to be a rare phenomenon (cf. Lees & Gilroy 2009) owing to the low frequency of suitable weather systems at this season (Elkins 1999). For instance, an unusual multiple occurrence of American ducks, waders and gulls in the British Isles in spring 2009 bore the signature of a fresh arrival, being associated with a fast-moving Atlantic storm system. The mix of species occurring as transatlantic and transpacific vagrants is very similar, with the same species recorded as vagrants in both Japan and Britain.

Records of many North American ducks are increasing in Europe and Asia, which may be a result of increased observer numbers and awareness, as well as increasing population sizes for some species

▲ Ring-necked Ducks *Aythya collaris*, Loch Phort Chorrúch, Inishmore, Galway, Ireland, 8 October 2008, a record-breaking flock arriving in the wake of a fast-moving Atlantic depression (*James Gilroy*).

◄ Steller's Eider *Polysticta stelleri* with Harlequin Ducks *Histrionicus histrionicus*, Borgarfjörður, Austurland, Iceland, 1 May 2012. The former is an almost annual vagrant to Iceland, with many individuals making protracted stays (*Yann Kolbeinsson*).

◄ Garganey *Spatula querquedula* with Mallards *Anas platyrhynchos*, Waller Park, Santa Barbara, California, United States, 9 December 2017. This 1st-year male bird was associating with dabbling ducks and was fairly tame but was accepted by the state rarities committee, highlighting the challenges of ascertaining provenance (*Tom Johnson*).

like Lesser Scaup. Similar positive trends have been noticed in North America with Eurasian waterfowl. For instance, Spear *et al.* (1988) were unable to trace a record of Garganey *Spatula querquedula* from North America or Hawaii prior to 1957 but catalogued 136 subsequent records up to 1985. Their analysis suggested that Garganey records in continental North America involved birds on northward spring migration, which had arrived in the New World in autumn or winter. This supposition was backed up by autumn records from the Gulf of Alaska, British Columbia and the Caribbean, all regions that regularly receive Palearctic vagrants.

A broad pattern of natural vagrancy to North America for other European species such as Common Shelduck *Tadorna tadorna* is becoming established, whereas records of Ruddy Shelduck are subject to the same burden of doubt that they face in northern Europe (Howell *et al.* 2014). Perhaps the exception is those recorded in west Greenland in mid-July 1892, although a captive origin for these birds is also possible as the timing closely mirrors extralimital movements of feral birds that we see now in northern Europe (C. Batty *in litt.*).

The true status of some vagrant ducks may also be clouded by identification challenges. Common Scoter *Melanitta nigra* was only recently first recorded in North America (away from Greenland) with recent records in California and Oregon (Chesser *et al.* 2017). Common Pochards *Aythya ferina* are extremely rare in the Lower 48 states of the United States, but some may pass undetected in large flocks of other diving ducks. Tufted Ducks are by contrast relatively regular visitors and might perhaps even breed in North America.

Vagrancy to temperate latitudes by tropical or subtropical waterfowl species tends to be viewed with considerable scepticism by rarities committees and birders in general. Boldly coloured tropical species are avicultural favourites and frequent escapees, yet nomadism in tropical waterbirds is also well documented. Extralimital records of other Afrotropical and Neotropical waterbirds – especially rails – to

► Paradise Shelduck *Tadorna variegata*, Lake Culburra, New South Wales, Australia, 1 January 2016, the first mainland Australian record of this New Zealand endemic (*Rohan Clarke*).

Europe and North America are usually treated as genuine vagrancy. The distributions of several tropical species are shared between continents, such as Fulvous Whistling-Duck *Dendrocygna bicolor* and Southern Pochard *Netta erythrophthalma*, suggesting a capacity for intercontinental movements.

Whistling-ducks *Dendrocygna* sp. are among the most frequent tropical waterfowl vagrants, with four species occurring as regular northbound vagrants in the United States; White-faced *Dendrocygna viduata*, Black-bellied *Dendrocygna autumnalis* and West Indian *Dendrocygna arborea* whistling-ducks (Howell *et al.* 2014) have predictable seasonal peaks. Elsewhere, Lesser Whistling-Duck *Dendrocygna javanica* has been recorded as a vagrant from sites as diverse as Israel, Oman (Haas 2012), Japan and Taiwan (Brazil 2009), whilst Plumed Whistling-Duck *Dendrocygna eytoni* is an occasional irruptive vagrant to New Zealand (Miskelly *et al.* 2013). Salim *et al.* (2020) reported a series of records of White-faced Whistling-Ducks *Dendrocygna viduata* from Iraq, which they regarded as of 'unknown origin' but maybe involved natural vagrancy. There is also a single record of Black-bellied Whistling-Duck *Dendrocygna autumnalis* from The Gambia in December 1998 (Borrow & Demey 2011), which could represent a case of transatlantic vagrancy given the capacity of these species for wandering. Records of Fulvous Whistling-Ducks in Europe are generally treated as escapes, but at least occasionally individuals or small groups do turn up in circumstances that may suggest genuine vagrancy.

► Black-bellied Whistling-Ducks *Dendrocygna autumnalis* with Ring-billed Gulls *Larus delawarensis*, Allan H. Treman State Marine Park, Tompkins, New York, United States, 10 July 2016. Although they may look rather incongruous and are commonly kept in captivity, whistling-ducks like these are also frequently recorded as genuine vagrants (*Alexander Lees*).

◀ Northern Shoveler *Spatula clypeata* with Pink-eared Duck *Malacorhynchus membranaceus* and Australian Shoveler *Spatula rhynchotis*, Western Treatment Plant, Wyndham, Victoria, Australia, 28 June 2013, the ninth record for Australia (*Rohan Clarke*).

Other tropical duck species have also crept into the Western Palearctic as vagrants, including records of Cape Teal *Anas capensis* and Southern Pochard from Israel (Perlman & Meyrav 2009) and Cotton Pygmy-goose *Nettapus coromandelianus* from Jordan (Haas 2012). Not all vagrancy in tropical wildfowl is controversial, as extralimital Masked Ducks *Nomonyx dominica* in North America are usually considered to be of wild origin, given that the species is rare in captivity, unlike White-cheeked Pintail *Anas bahamensis,* which is typically considered to be an escape away from Florida (Howell *et al.* 2014) – perhaps a risky assumption given that this species has also reached as far afield as the Falklands (Woods 2017).

Many Holarctic ducks winter at tropical latitudes, and while none routinely winter as far south as Australia, that continent has recorded vagrants of many Palearctic breeding species, including Garganey, Northern Shoveler *Spatula clypeata*, Eurasian Wigeon, Northern Pintail, Common Teal and Tufted Duck (Menkhorst *et al.* 2017). Northern Shoveler has even reached New Zealand and Argentina (Kinsky & Jones 1972, Pearman & Areta 2020). In addition to the aforementioned American Wigeon in The Gambia, transoceanic vagrancy of wildfowl to the tropics is also well documented, including two female Canvasbacks *Aythya valisineria* in November 2015 on Luzon in the Philippines (Robson *et al.* 2015) and multiple records

▼ Baikal Teal *Sibirionetta formosa*, with Eurasian Wigeon *Mareca penelope*, Eurasian Teal *Anas crecca*, Garganey *Spatula querquedula* and Northern Shoveler *Spatula clypeata*, Tanguar Haor, Sylhet, Bangladesh, 20 February 2011, one of only five winter records from Bangladesh (*Sayam Chowdhury*).

of Northern Pintail from the remote Brazilian island of Fernando de Noronha (Silva e Silva & Olmos 2006). These are, however, totally eclipsed by an old forgotten specimen of Australasian Shoveler *Spatula rhynchotis* from Entre Rios, Argentina, prior to 1897, which had been mislabelled as Red Shoveler *Spatula platalea*. If there is no evidence of confusion or fraud, then a transpacific dispersal event by an austral duck would represent one of the most unlikely instances of vagrancy known to date (Crozariol & Nacinovic 2017). Given that this species closely resembles hybrid Northern Shoveler x Blue-winged Teal – a hybrid that has even been recorded in Europe (Van Bemmelen *et al.* 2018) and would seem far more likely to occur in Argentina, it seems pertinent to examine the Argentinian specimen genetically to rule out this possibility.

Geese and swans are unusual amongst wildfowl in that they show extended parental care of offspring, and their migratory behaviour is culturally transmitted (i.e. learned) rather than inherited genetically. There have also been various cases where migratory behaviour has been taught by humans (Sutherland 1998). Transatlantic or transpacific vagrancy of geese is a relatively routine occurrence, occasionally as a result of wind drift, particularly when single-species flocks of family groups are involved, as in the example of flocks of White-fronted Geese *Anser albifrons* reaching Hawaii (Pyle & Pyle 2017). However, vagrancy may occur more commonly when a juvenile bird becomes detached from its parental group and attaches itself to a flock of another species. This may explain the occurrence of species like Snow Goose *Anser caerulescens*, Ross's Goose *Anser rossii* and Cackling Goose *Branta hutchinsii* arriving with flocks of Barnacle Geese *Branta leucopsis* or Pink-footed Geese *Anser brachyrhynchus* in Western Europe. Arriving with appropriate 'carrier species' (i.e. those originating from a geographically proximate area) is therefore often seen as an important indicator of wild status amongst vagrant geese.

As populations of many northern goose species have recovered from intensive historical hunting, vagrancy has also increased. In North America, this

► Pink-footed Goose *Anser brachyrhynchus* with Canada Geese *Branta canadensis*, Frederick Lake, Weld, Colorado, United States, 14 December 2018, the first state record (*Chris Wood*).

► Cackling Goose *Branta hutchinsii hutchinsii*, Longyearbyen, Svalbard, 6 June 2018. This individual, presumably a genuine vagrant, might have arrived on Svalbard with flocks of returning Barnacle Geese which winter in south-west Scotland (*Paul Dufour*).

▲ Coscoroba Swan *Coscoroba coscoroba* with Black-necked Stilts *Himantopus mexicanus melanurus*, Greater Yellowlegs *Tringa melanoleuca*, Lesser Yellowlegs *Tringa flavipes*, Great White Egrets *Ardea alba*, Snowy Egrets *Egretta thula* and White-faced Ibises *Plegadis chihi*, Dique do Furadinho, Cubatão, São Paulo, Brazil, 19 January 2015. This was the first record of Coscoroba Swan for São Paulo state (*Robson Silva e Silva*).

applies to records of Greylag Geese *Anser anser*, Taiga Bean Geese *Anser fabalis*, Tundra Bean Geese *Anser serrirostris*, Pink-footed Geese and Barnacle Geese, but Lesser White-fronted Goose *Anser erythropus* represents an exception, as this species has declined globally and is known from a single record on the Alaskan island of Attu in the Bering Sea in June 1994 (Howell *et al.* 2014). Although records of vagrant geese entrained with flocks of carrier species are generally viewed favourably, the possibility remains that some 'lost' individuals might wander for large distances in isolation from flocks, making the assessment of non-typical records difficult. Five species of Nearctic geese have reached the West Indies (Kirwan *et al.* 2019) whilst three

◀ Whooper Swan *Cygnus cygnus*, Al-Ansab Lagoons, Masqat, Oman, 5 January 2003. Oman lies around the southern range limit for vagrancy in this species (*Jens Eriksen*).

species, including Lesser White-fronted Goose, have reached Myanmar (Robson 2005) and White-fronted Goose has occurred south to Niger, Nigeria and Chad (Borrow & Demey 2014). Southern Hemisphere species are also migratory and prone to occasional long-range vagrancy; notable records include an Upland Goose *Chloephaga picta* at Lagoa do Peixe National Park in Rio Grande do Sul, southern Brazil (Bencke & de Souza 2013), and Ashy-headed Geese *Chloephaga poliocephala* on the Falklands (Woods 2017).

Vagrancy in *Cygnus* swans tends to be more geographically restrictive than that of geese. Still, there are exceptions, with some remarkable swan vagrancy including regular records of Whooper Swan *Cygnus cygnus* in western North America south to California (Howell *et al.* 2014), Trumpeter Swan *Cygnus buccinator* from Japan (Brazil 2009), Tundra Swan *Cygnus columbianus* in the West Indies and Black-necked Swan *Cygnus melancoryphus* from Antarctica (Silva *et al.* 1995).

MEGAPODIIDAE
Megapodes

Perhaps somewhat surprisingly for such rotund birds, megapodes are quite proficient dispersers (Jones & Birks 1992), leading to a very wide distribution including islands as remote as Palau and the northern Marianas. There are no contemporary records of significant vagrancy, but such events must happen periodically, although historical transport of megapodes between islands by people could theoretically explain their presence on some remote archipelagos.

CRACIDAE
Guans, Chachalacas and Curassows

Members of the family Cracidae are typically highly sedentary and poorly adapted for sustained flight. No species undertake any significant long-distance movements apart from Highland Guan *Penelopina nigra*, which is periodically reported at lower elevations, suggesting some form of altitudinal migrations or occasional vagrancy events (Brooks *et al.* 2006).

NUMIDIDAE
Guineafowl

All of the guineafowl, which include both forest and non-forest species, are sedentary and there have been no extralimital reports.

ODONTOPHORIDAE
New World Quails

Some short-distance altitudinal movements have been reported to occur in Spotted Wood-quail *Odontophorus guttatus* and in both Scaled Quail *Callipepla squamata* and Mountain Quail *Oreortyx pictus*, but there are no reports of significant vagrancy in this group.

PHASIANIDAE
Pheasants, Grouse and allies

Most people would not anticipate a substantial section on gallinaceous birds in a book on avian vagrancy. Indeed, most members of the family are heavy-bodied with short wings and tend to be terrestrial, all features associated with birds with poor dispersal capacity. Most members of the group fly only when hard-pressed to escape predators, or to ascend trees or rock faces to roost, and they have limited ability for sustained flight (Tobalske *et al.* 2003). However, a few members of the Phasianidae are partial migrants or altitudinal migrants, moving distances of tens of kilometres. One such is Satyr Tragopan *Tragopan satyra*, which has been recorded crossing multiple mountains and moving up and down slopes in a predictable fashion (Norbu *et al.* 2013). Bucking the trend of sedentarism are the Old World quails (*Coturnix* spp.), which routinely undertake long-distance transcontinental migrations, including long overwater movements, which have been recognised since at least biblical times. Common Quails *Coturnix coturnix* are widespread breeders in the Western Palearctic and most winter south of the Sahara. They have been found as vagrants on oceanic islands as isolated as the Seychelles and Iceland, with records from as far afield as Cameroon and Myanmar (McGowan *et al.* 2020). The intratropical migrant or nomadic Harlequin Quail *Coturnix delegorguei* has been recorded as a vagrant to Senegambia, Chad, Gabon and Oman (Jun, Nov, Dec) with similar incidences of widespread vagrancy known in Rain Quail *Coturnix coromandelica*, for example to Nepal and Sri Lanka, Japanese Quail *Coturnix japonica* to Cambodia and the Philippines, and Blue Quail *Synoicus adansoni* to South Africa (Perry 2006).

Also bucking this sedentary trend are several grouse species, for which wing shape and muscle composition suggest that they are better adapted to sustained flight and hence longer dispersal movements or migrations than other Galliformes (Tobalske *et al.* 2003). Normally thought of as being quite sedentary in Europe, there are big regional differences in patterns of movement and hence incidence of vagrancy. For instance, in the UK, the endemic subspecies *scotica* of Willow Grouse *Lagopus lagopus* (Red Grouse) is usually highly sedentary, rarely straying more than

▲ Common Quail *Coturnix coturnix*, Önundarfjörður, Vestfirðir, Iceland, 27 July 2008, the second record for Iceland of this species, the most migratory of all the Phasianidae (*Yann Kolbeinsson*).

► Willow Grouse *Lagopus lagopus*, Point Peninsula, Jefferson County, New York, United States, 26 April 2014. This long-distance vagrant was twitched by over 300 people (*Jay McGowan*).

1.5km from breeding areas (Cramp & Simmons 1980). However, even in Britain Red Grouse have occurred as vagrants, with recent records from Flamborough Head in March 2014 – around 30km from the nearest suitable habitat, and another on the Isle of May, Fife, in April 2015 which required a minimum 8km sea crossing. There are older records of Red Grouse from as far afield as the island of St Kilda, with singles in December 1959 and July 1970 requiring a sea-crossing of at least 65km (Harris & Murray 1989). In Europe, Willow Grouse move longer distances, increasing from regular movements of tens of kilometres in Scandinavia to routinely hundreds of kilometres in Russia (Cramp & Simmons 1980), driven by increasingly harsh winters further east.

In North America, where it is known as Willow Ptarmigan, Willow Grouse routinely make significant migrations in flocks numbering up to 2,000 individuals, particularly in Manitoba, Saskatchewan and Alberta (Hannon *et al.* 1998), with more than 1,000 observed on intertidal flats at Kotzebue Sound, Alaska, in September 1988 (Lehman 1989). This species is apparently found with some regularity as a migrant offshore, with Zimmerman *et al.* (2005) describing an encounter with a group of more than 100 in August 2003 in Kuskokwim Bay, Alaska, many of which landed on the ship and were even observed landing – and taking off from – the sea surface! Given this greater proclivity towards migration, it should not come as a surprise that there are quite a few extralimital records of Willow Grouse, even extending into southern Canada and North Dakota, Minnesota, Wisconsin and Maine in continental United States, for example. Two records that deserve particularly mention are the first acceptable record from New York – an individual present at Point Peninsula in Jefferson County in April 2014, which was at least 500km out of range – and an incredible record of one on Sable Island, Nova Scotia, in August 1966 (Tufts & McLaren 1986). The latter would have necessitated a minimum 150km sea crossing but is more likely to have travelled directly from Newfoundland, where the nearest population occurs, over 400km across open sea.

Willow Grouse co-occurs with Rock Ptarmigan *Lagopus muta* over much of its range, and although Rock Ptarmigan is seemingly less prone to extralimital vagrancy, there are regular patterns of movement established for populations at high latitudes (Bent 1932). On Svalbard such movements are in the range of 30–150km (Fuglei *et al.* 2017), and large flocks numbering into the thousands have been observed in spring and autumn on the northern coast of Labrador, a staging point for movements to and from breeding grounds in the Canadian Arctic Archipelago. Flocks have been observed in late September and early October, flying from Ellesmere Island toward Greenland (Montgomerie & Holder 2008). Clear vagrant records include a pre-1841 report from St Kilda in the Western Isles of Scotland (Harris & Murray 1989) and one shot at Elmsdale, Nova Scotia, in April 1922 (Piers 1923). The remaining grouse species move far less, usually at the scale of tens of kilometres in Blue Grouse *Dendragapus obscurus* (Cade & Hoffman 1993), Spruce Grouse *Canachites canadensis* (Herzog & Keppie 1980) and Greater Prairie-Chicken *Tympanuchus cupido* (Schroeder & Braun 1993), for example. Worthy of mention however are reported 'irruptive movements' in Western Capercaillie *Tetrao urogallus*, which may cover over 1,000km in Siberia, with reports such as "wandering birds even enter villages, where recorded crashing into buildings" (de Juana & Kirwan 2019).

PHOENICOPTERIDAE
Flamingos

Flamingos tend not to be truly migratory, but they are highly itinerant and may move vast distances between colonies to avoid severe weather. For example, in late April 2005 a group of flamingos alighted on the runway of Rio Branco International Airport, in the capital city of the state of Acre, in northwestern Brazil after a cold front swept through the central Andes and western Amazonian lowlands. One flamingo was injured and subsequently captured but died in captivity two days later. This individual proved to be a juvenile Puna Flamingo *Phoenicoparrus jamesi* and was the first record of this species for Brazil (Guilherme *et al.* 2005). There followed a second Brazilian record, of an adult bird 3,000km south-east on an oceanic beach in São José do Norte, in Rio Grande do Sul state, in October 2013 (Dias & Cardozo 2014). Puna Flamingos are essentially restricted to the high Andean plateaus of Peru, Chile, Bolivia and Argentina, so these extralimital records and others from coastal Peru (Vizcarra & Vicetti 2013) seem unexpected. However, taken together they fit into a broader pattern of vagrancy in the family, albeit one which is often confounded by escapes from captivity.

The Andes are the centre of global flamingo diversity, home to half of the world's six species. Puna Flamingo has the lowest incidence of vagrancy, followed by Andean Flamingo *Phoenicoparrus andinus*, which has also wandered to Amazonia where one was shot by fishermen curious as to the identify of a bird swimming in a river at the Amanã Sustainable Development Reserve in May 2007 (Bernadon & Valsecchi 2014). This species occurs with some regularity now on the coasts of both Peru (Høgsås *et al.* 2010) and Brazil (Ghizoni-Jr & Piacentini 2010). Finally, Chilean Flamingo *Phoenicopterus chilensis* has the largest distribution in the southern cone of South America and, breeding in both upland and lowland regions, also has the largest spread of vagrant records, north all the way to Colombia (Donegan *et al.* 2018), with records of island vagrants on both the Galápagos and the Falklands (Woods 2017).

American Flamingos *Phoenicopterus ruber* have a patchy distribution in the Galápagos, northern South America from Colombia to northern Brazil and throughout the Caribbean. Formerly more widespread, they used to breed commonly in the southern United States and Brazil south to Ceará (Sick 1997). Inter-colony movements are common between Cuba, the Bahamas and the Yucatán, Mexico, as evidenced by marked birds. Ringed birds have been recovered as vagrants in the southern United States, in Texas, Louisiana and Florida, with an increase in sightings in recent years (Galvez *et al.* 2016). American Flamingo was formerly 'lumped'

◄ Andean Flamingos *Phoenicoparrus andinus* with a Black-necked Stilt *Himantopus mexicanus*, Tijucas, Santa Catarina, Brazil, 25 May 2018. There may be a small, regular 'pseudo-vagrant' wintering population of the former in southern Brazil (*Ciro Albano*).

▲ Greater Flamingo *Phoenicopterus roseus* with Grey-headed Swamphens *Porphyrio poliocephalus*, Black-winged Stilt *Himantopus himantopus*, Common Greenshank *Tringa nebularia*, Asian Openbills *Anastomus oscitans* and Great White Egret *Ardea alba*, Bueng Boraphet Non-hunting Area, Nakhon Sawan, Thailand, 15 February 2016. This individual returned in several subsequent winters but has yet to be accepted as a wild bird and a first for Thailand (*Wich'yanan Limparungpatthanakij*).

with Greater Flamingo *Phoenicopterus roseus,* which has a wide Palearctic distribution and is well known for long-range inter-colony movements and vagrancy. For example, Greater Flamingo chicks ringed in the Camargue, Bouches-du-Rhône, France, have been resighted in Spain, Iran, Senegal and Russia (Johnson 1989), and birds ringed in Spain have even reached Guinea-Bissau (de Juana & Garcia 2015). Asian birds also move extensively, many moving south to southern India and Sri Lanka, with birds ringed at the large colony at Lake Uromieh, Iran, resighted from Libya to India. Against this background of intercontinental movements there are also reports, although surprisingly few, of clear extralimital vagrancy. Vagrant Greater Flamingos have, for example, reached the Maldives in the Indian Ocean (Ming 1999) and Lake Baikal in Russia (Mlíkovský 2009). Occasional displacement events are known to occur from the Camargue colony following strong winds or severe cold, leading to vagrancy to the French interior up to 500km from the coast (Johnson 1989). Interpreting patterns of vagrancy to northern Europe is complicated by small numbers of feral birds. A major source of these is a mixed-species colony of Greater and Chilean Flamingos on the German/Dutch border at Zwillbrocker Venn (van den Berg & Bosman 1999), which may periodically include wild Greater Flamingos. However, there are records of unquestionable wild origin. These include a flock of six immatures at Fanel, Lac de Neuchâtel, in Switzerland from September to November 1998, one of which had been ringed in the Camargue (Sharrock & Davies 2000). Subsequently, one of a group of five at Baie de Somme, Picardie, northern France in late June 2010, had been ringed as a nestling in the Camargue in 2005 and had been resighted in Portugal and in the Camargue before moving north to Picardie (Slaterus 2010).

The large colonies of Greater Flamingos in Spain and France regularly attract Lesser Flamingos *Phoenicopterus minor,* which have bred successfully in

▼ Lesser Flamingo *Phoeniconaias minor* with Greater Flamingo *Phoenicopterus roseus*, Ein Evrona Salt Pools, Eilat, Arava, Israel, 20 March 2006. An increasingly regular vagrant outside Africa (*Chris Batty*).

both France and Spain (de Juana & Garcia 2015). Some known escapes have been recorded in northern Europe, but genuine vagrancy is supported by the increasing frequency of such observations and an amazing ringing recovery of one ringed as a chick in Kenya in 1962 subsequently found dead in Western Sahara in 1997 (Childress & Hughes 2007). Such intercontinental movements are comparable to those found in Greater Flamingos, and analyses of nuclear DNA have indicated a lack of population genetic structure and asymmetric gene flow – which hint at regular interchange of Lesser Flamingos between colonies in Kenya and Gujarat, India (Parasharya *et al.* 2015). This result also supports a natural origin for birds recorded along the coastlines of the Indian Ocean and Arabian Peninsula.

PODICIPEDIDAE
Grebes

Grebes include a mix of sedentary and at least partially migratory species and many have very large ranges, including five species that occur right across Eurasia from Ireland to Japan. Only one species, the North American Pied-billed Grebe *Podilymbus podiceps,* is a vagrant to Eurasia, and it exhibits a pattern of vagrancy that mirrors many North American wildfowl. One celebrated occurrence included a successful incidence of extralimital breeding, when a male Pied-billed Grebe paired to a female Little Grebe *Tachybaptus ruficollis* in Cornwall, England, in 1994 producing three intergeneric hybrid young from two clutches (Ogilvie 1994).

Grebe species that breed at high latitudes are frequently recorded as vagrants far south of their normal range. Slavonian Grebe *Podiceps auritus* has, for instance, been recorded as a vagrant south to Israel, Kuwait, India and Thailand, and Red-necked Grebe *Podiceps grisegena* has straggled to Bangladesh (Pender 2010). All of the Eurasian species, plus Pied-billed Grebe, have occurred as vagrants to the Azores archipelago, but the Old World species Great Crested Grebe *Podiceps cristatus* and Little Grebe have yet to be recorded in the Western Hemisphere. Two western North American species – Western Grebe *Aechmophorus occidentalis* and Clark's Grebe *Aechmophorus clarkii* – are regular and extreme vagrants respectively to the eastern half of the continent. A final species, Least Grebe *Tachybaptus dominicus,* is predominantly Neotropical but has a breeding toehold in the southern United States, in Texas, and has occurred as a vagrant to Arizona, Louisiana, Arkansas and Florida. Incredibly, four grebe species have made it to Hawaii – Pied-billed Grebe (which has bred), Slavonian Grebe, Red-necked Grebe and Black-necked Grebe *Podiceps nigricollis.* Grebes, unlike divers, are also widely distributed in the tropics and in the Southern Hemisphere. Several species in the Old World and New World are present at both high and low latitudes, with highly migratory populations

◄ Pied-billed Grebe *Podilymbus podiceps*, RSPB Leighton Moss, Lancashire, England, 1 May 2015. Vagrants of this species in Europe often make protracted stays spanning multiple years (*Stuart Piner*).

► Clark's Grebe *Aechmophorus clarkii*, Oswego Harbor, Oswego, New York, United States, 4 March 2017. This species is much less frequent as a vagrant to eastern North America than its congener Western Grebe (*Jay McGowan*).

► Black-necked Grebe *Podiceps nigricollis*, Padma River, Aricha, Dhaka, central Bangladesh, 5 January 2012. This bird was the second record for Bangladesh (*Sayam Chowdhury*).

where winters are very cold, and also highly nomadic populations in the tropics where waterbodies may be ephemeral.

In the southern cone of South America, several species are Austral migrants, moving north towards the tropics in winter. For example, Silvery Grebe *Podiceps occipitalis* has occurred as a vagrant to Brazil (Bornschein *et al.* 2004) and Bolivia (Tobias & Seddon 2007), with small flocks recorded on multiple occasions; for example, in Brazil there were 10 in Curitiba, Parana, in October 2010 (Forone 2004) and eight at the Represa do Guarapiranga, São Paulo, in September 2018 (Schunck 2018). Great Grebe *Podiceps major* has also been recorded making some unusual movements in the same region. For example, it has occurred as a vagrant north to the Brazilian state of Espirito Santo (Simon *et al.* 2005), six were reported with Magellanic Penguins *Spheniscus magellanicus* in November 1994, 352km from the Argentinian coast (Montalti *et al.* 1999) and the species has reached the Falklands on multiple occasions (Woods 2017). There are also two old specimen records from Spain of this species from the turn of the 20th century, earning it a place on the Spanish rarities holding 'category D'. Genuine vagrancy would appear unlikely, and it would seem more likely that the records are fraudulent.

Hoary-headed Grebe *Poliocephalus poliocephalus* is somewhat nomadic in Australia and was first recorded in New Zealand on the Snares Islands in the 1970s, this event being followed by colonisation of the mainland (Barlow 1976), but this persistent breeding was short-lived, and the species has remained a vagrant to the archipelago in the last 20 years (Miskelly *et al.* 2017). This rush of extralimital records in the mid-1970s included multiple individuals on Macquarie Island, midway between Australia and mainland Antarctica (Copson & Brothers 2008).

COLUMBIDAE
Pigeons and Doves

Most pigeons and doves are sedentary, but a few species are at least partially migratory, and the family seems to be highly prone to vagrancy. This has resulted in their colonisation and speciation across many oceanic island groups. In the Western Hemisphere, migration is most prevalent in the genus *Zenaida*, with the Mourning Dove *Zenaida macroura* of North America being a prime example. Northern populations of this species are wholly migratory and transatlantic vagrancy to Europe and Greenland in both spring and autumn is well documented (Haas 2012). At the other end of the Americas, Eared Dove *Zenaida auriculata* is also a partial migrant, with southern populations moving north in the austral winter. Vagrants are regular on the Falklands (Woods 2017) and have even reached north to Costa Rica (Garrigues *et al.* 2016) and – ship-assisted – east to South Georgia (Clarke *et al.* 2012). Further north, a flock of four Eared Doves was recorded on the Brazilian island of Trindade, over 1,000km from the mainland (Port & Fisch 2013). Vagrancy in small groups like this likely permits colonisation and eventual speciation in wandering pigeons, and the Brazilian island of Fernando do Noronha has its own endemic subspecies of Eared Dove, which presumably arose after such a dispersal event. Historic vagrancy must have given

◄ Mourning Dove *Zenaida macroura*, Inishbofin, Galway, Ireland, 4 November 2007, the first record for Ireland (*Anthony McGeehan*).

◄ White-winged Doves *Zenaida asiatica*, Fenway and Victory Gardens, Suffolk, Massachusetts, United States, 20 February 2017. This species is expanding its range northwards in North America, likely benefitting from increasing urbanisation and backyard bird feeders (*Marshall Iliff*).

► Zenaida Dove *Zenaida aurita*, West Kendall Agricultural Area, Miami-Dade, Florida, United States, 31 July 2018 (*Tom Johnson*).

rise to other better-differentiated endemics such as Galápagos Dove *Zenaida galapagoensis* and Socorro Dove *Zenaida graysoni*. Zenaida Doves *Zenaida aurita* are essentially restricted to the Caribbean but have occurred as a vagrant fewer than a dozen times to Florida (Howell *et al.* 2014). More surprising was a record of a Zenaida Dove in April 2017 on São Miguel on the Azores, presumably ship-assisted and the first record for the Western Palearctic (Ławicki & van den Berg 2017d). White-winged Doves *Zenaida asiatica* are also partial migrants, with northernmost populations in the southern United States migratory – vagrants have been recorded across much of North America, north to Alaska, Hudson Bay and Labrador, and east to Bermuda (Schwertner *et al.* 2002, eBird). In this context, the first South American record, from Colombia in 2009, about 700km out of range, was not unexpected (Strewe 2015).

In the southern cone of South America, several species are migratory, including Chilean Pigeon *Patagioenas araucana,* Picazuro Pigeon *Patagioenas picazuro* and Black-winged Ground Dove *Metriopelia melanoptera,* all of which have occurred as vagrants to the Falklands (Woods 2017). In North America, Band-tailed Pigeon *Patagioenas fasciata* is a partial migrant, with most from the north Pacific coast breeding range migrating south in winter, coupled with some apparent nomadism, and interior populations also moving south in winter. Despite this ostensibly western North American range, Band-tailed Pigeons are regular as vagrants throughout much of North America, north-east to Hudson Bay, east to Nova Scotia and with records scattered down the United States east coast to Florida, where one shot vagrant had been ringed the previous year in Oregon (Stevenson & Anderson 1994).

The tropical quail-doves in the genus *Geotrygon* might strike most people as unlikely vagrants, but many species may undertake at least partial migrations following peaks in fruit abundance, such as those made by the Ruddy Quail-Dove *Geotrygon montana* in Amazonia (Stouffer & Bierregaard 1993). In this context, records of vagrant Ruddy Quail-Doves from Texas and Florida in the United States seem less exceptional (Howell *et al.* 2014). The insular Key West Quail-Dove *Geotrygon chrysia* is also a vagrant to Florida, where it formerly bred. Elsewhere, understanding what constitutes a vagrant, migrant or even rare resident population of quail-doves is hampered by their secretive nature in their tropical forest haunts. One candidate for vagrancy might be the single documented record of Sapphire Quail-Dove *Geotrygon saphirina* from western Amazonian Brazil, although the species may yet prove to be regular there (Kirwan *et al.* 2015).

Across Eurasia, pigeon migration and hence vagrancy is most prevalent in the *Streptopelia* turtle doves. European Turtle Doves *Streptopelia turtur* migrate on a broad front from Eurasia to subsaharan African wintering grounds and there is a corresponding wide scatter of vagrant records. Overshooting birds have been recorded several times from South Africa (Sinclair *et al.* 2002) but also on the Seychelles (Skerrett *et al.* 2006) and more amazingly at least five times on the Prince Edward Islands, at least 5,000km out of range, and entailing an over-water crossing (or ship assistance) of over 1,700km (Oosthuizen *et al.* 2009). This species is a fairly regular vagrant to the Atlantic islands of Iceland, the Azores and Cape Verde but has also occurred as a transatlantic vagrant on three occasions – in Florida and Massachusetts in the United States and Saint Pierre and Miquelon in Canada (Howell *et al.* 2014). Surprisingly, there are no records from the Caribbean region, but the record from Florida in early spring might well have involved a transatlantic crossing at lower latitudes the previous autumn and subsequent northward movement. Rufous Turtle Dove *Streptopelia orientalis* is sedentary over much of its range, but two subspecies – *meena* and nominate *orientalis* – are highly migratory and both have occurred as vagrants to Western Europe. Rufous Turtle Dove has also occurred as a vagrant to North America with some regularity, principally in Alaska and the Bering Sea region, but also south as far as California, where there are two records (Howell *et al.* 2014). A flock of six Rufous Turtle Doves came aboard a cargo ship in the Sea of Japan in May 2016 and one individual remained onboard for a week until the ship approached the California coastline (eBird). Other members of the genus are less migratory but have still been recorded as vagrants with some regularity – Red Collared Dove *Streptopelia tranquebarica* has, for example, been recorded from the Arabian Peninsula, Christmas Island, Australia (James & McAllan 2014), Korea, Japan and the Russian Far East (Brazil 2009).

◀ Rufous Turtle Dove *Streptopelia orientalis meena* with Collared Dove *Streptopelia decaocto*, Stromness, Orkney, Scotland, 6 December 2002 (*Chris Batty*).

◀ European Turtle Dove *Streptopelia turtur*, Húsavík, Norðurland eystra, Iceland, 23 October 2019. This is an annual vagrant to Iceland (*Yann Kolbeinsson*).

Several of the large pigeons of the genus *Columba* are at least partial migrants, and vast flights of Common Wood Pigeons *Columba palumbus* are a feature of late autumn migration in Europe. Common Wood Pigeons are vagrants to Iceland, where they have also recently bred, as well as to other disparate northern archipelagos such as Greenland, Svalbard and Jan Mayen, as well as south to Mauritania. One photographed near Romaine, in Quebec, Canada, in May 2019 (Pyle 2020) was the first for North America. Given the low diversity of pigeons in eastern North America following the extinction of the Passenger Pigeon *Ectopistes migratorius* and the success of Common Wood Pigeon in the human-modified landscapes that characterise so much of the Northern Hemisphere, colonisation of North America via Iceland does not seem beyond the realms of possibility. Interestingly, Passenger Pigeon has recently been shown to be unrelated to the *Zenaida* doves (it was previously thought to be their closest relative), but is in fact the sister taxon of all other New World pigeons in the genus *Patagioenas,* and may have colonised North America from Asia (Johnson *et al.* 2010).

The eternally optimistic rarity finder in western Europe might check itinerant flocks of Common Wood Pigeons and Stock Doves *Columba oenas* for Yellow-eyed Dove *Columba eversmanni*. This species has been recorded reliably once in the Western Palearctic, near Orenburg, Russia, in May 1881 (Arkhipov *et al.* 2010). It is a highly migratory species which breeds in western Asia east of the Caspian Sea and winters in Pakistan, north-west India and Afghanistan, and it might be expected again as a vagrant even as far as western Europe.

The fruit-doves in the genera *Treron* and *Ptilinopus* are largely sedentary, but some species do undertake at least partial migrations and even those that do not seem prone to long-range dispersal events. Their penchant for overwater movements has helped the group to occupy a very large range in the Old World tropics and colonise many remote island groups. Notable records of vagrancy in this group include Bruce's Green Pigeon *Treron waalia* at Luxor in Egypt in January 2011 (Veen 2011), Orange-breasted Green Pigeon *Treron bicinctus* on Taiwan (Brazil 2009) and Orange-bellied Fruit-Dove *Ptilinopus iozonus* on Boigu Island in the Torres Strait, Australia in November 2004 (Clarke 2007). Surpassing these even is the sole New Zealand record of Rose-crowned Fruit-Dove *Ptilinopus regina*, a juvenile that landed on a ship anchored 75km off the south Taranaki coast in August 2019. This individual was taken ashore and controversially euthanised due to biosecurity concerns (Miskelly 2020).

This tendency towards being non-migratory, but still maintaining a high rate of dispersal has helped pigeons and doves reach oceanic islands and speciate *in situ*, such that the Azores, Canaries, Caribbean and Oceania all have their own endemic pigeons (with the notable exception of Hawaii). Many island pigeons later evolved to be flightless, such as the celebrated

▲ Common Ground-Dove *Columbina passerina*, Jones Beach State Park, Nassau, New York, United States, 16 November 2014, the second state record of this largely sedentary species whose normal range extends no nearer than South Carolina (*Jay McGowan*).

Dodo *Raphus cucullatus* and Rodrigues Solitaire *Pezophaps solitaria,* whose ancestors were presumably vagrant Nicobar Pigeons *Caloenas nicobarica* or a very closely-related extinct relative (Shapiro *et al.* 2002). Nicobar Pigeon has itself recently reached northern Australia as a vagrant (Davis & Watson 2018). Future endemic forms are likely to again emerge with the occasional appearance of vagrants, especially if we can preserve the ecological integrity of island archipelagos.

MESITORNITHIDAE
Mesites

There is no evidence for movements in either the Subdesert Mesite *Monias benschi* or the White-breasted Mesite *Mesitornis variegatus,* which appear to be strictly sedentary. Periodic absence of Brown Mesite *Mesitornis unicolor* from regular sites may hint at movements and possibly altitudinal migration (Evans *et al.* 2020), which may lead to vagrancy, but no such events have been recorded yet.

PTEROCLIDAE
Sandgrouse

Migratory behaviour in sandgrouse is poorly known. Most species are assumed to be sedentary or nomadic, but both Black-bellied Sandgrouse *Pterocles orientalis* and Pin-tailed Sandgrouse *Pterocles alchata* are migratory in Central Asia, fleeing the region's very severe winters. Pin-tailed Sandgrouse have been recorded as vagrants from the United Arab Emirates, Egypt (Waschkies *et al.* 2005) and Oman, and also noteworthy is a record from Fuerteventura (Canary Islands) in December 1998 (de la Puente *et al.* 1999). Lichtenstein's Sandgrouse *Pterocles lichtensteinii* has been reported as a vagrant from Lebanon and Iraq (Sawan 2020), although this crepuscular species may be overlooked as a breeding species there. Pallas's Sandgrouse *Syrrhaptes paradoxus* is an extremely enigmatic vagrant, appearing thousands of kilometres out of range in irruption years. Exceptional numbers arrived in western Europe in eight years between 1859 and 1908, leading to extralimital breeding in six western European countries. There have been relatively few subsequent records in Europe, bar very minor incursions in 1964 and 1969. Irrupting birds may also move east, with major incursions into northern China in 1860 and Manchuria in 1912/13 and 1922/23, and birds have reached India (Prater 1926) and Japan (Brazil 2018). These irruptions have been attributed to food shortages – particularly of the seed-bearing orache plant *Agriophyllum squarrosum,* triggered by either prolonged drought (Newton 2008) or heavy snowfalls (Dementiev & Gladkov 1969). These incursions also apparently involved an adult Chestnut-bellied Sandgrouse *Pterocles exustus* reported to have been shot with a flock of Pallas's Sandgrouse in Hungary in August 1863 (Gorman 1996) – a remarkable occurrence if genuine, given that the ranges of the two species do not overlap..

OTIDIDAE
Bustards

Out-of-range bustards are among the most sought-after vagrants. Many species of bustards exhibit migratory tendencies; some are genuinely nomadic and a few are fully migratory, at least in some parts of their range. Migratory behaviour, and hence vagrancy potential, is concentrated among bustards that inhabit a band of Eurasia stretching from France to eastern Russia and China where three species – Great Bustard

▲ Great Bustard *Otis tarda*, east of Ma'ale Gamla, HaZafon, Israel, 25 January 2008, the seventh record for Israel (*Yoav Perlman*).

Otis tarda, Little Bustard *Tetrax tetrax* and MacQueen's Bustard *Chlamydotis macqueenii* – are exposed to long, cold winters. These frequently push birds out of colder grassland regions, potentially leading to random escape movements in any direction. Populations of Great and Little Bustards in the north and east of their range (e.g. eastern Europe) frequently undertake long-distance hard-weather movements, and it is generally assumed that these mobile populations are the most likely source of vagrant influxes (Brown & Grice 2005). The destruction and fragmentation of habitat has driven strong population declines in both species, particularly in eastern Europe, making such vagrancy events considerably rarer.

The collapse of populations of Little Bustard in central and northern France – which winter in Iberia (Villers *et al.* 2010) – have also probably driven the decline of this species as a vagrant to Britain. With fewer birds migrating north in spring, there are fewer to overshoot, and with a reduced breeding population there are fewer juveniles to disperse in autumn. This, coupled with the spectre of climate change (e.g. the continuing trend towards milder winters), suggests that Little and Great Bustards look set to become even rarer extralimitally. Great Bustards in the east may move farther than western ones, with three females being satellite-tracked making a 2,000km migration between breeding grounds in northern Mongolia and wintering grounds in Shaanxi province, China (Kessler *et al.* 2013), and in this context their regular vagrancy to Japan is not surprising (Brazil 2018).

MacQueen's Bustard is globally rarer still; the last western European records include singles in Switzerland in 2008, Belgium in 2003, Sweden in 1974 and England in 1962. Recent investment in satellite tracking across a vast swathe of Asia from western Kazakhstan to the Gobi Desert in China has shown that birds leave in large numbers and take different migration routes (Combreau *et al.* 2011). In some cases these routes circumvented high mountain ranges but at other times birds crossed them, attesting to a considerable migration prowess, to winter in disjunct areas from the Arabian Peninsula to Pakistan. All five British records fall in the period October–November and four of them fall in a 19-day period in October, suggesting that some sort of extreme post-juvenile dispersal might be the most likely driving mechanism. However, two of the four aged individuals were adults, suggesting a failure to 'turn off' nomadic responses rather than an orientation error. Future records would be exceptional until the global population recovers from habitat loss and unsustainable hunting pressure. Elsewhere in Asia, considerable uncertainty surrounds potential movements of both Bengal Florican *Houbaropsis bengalensis* and Lesser Florican *Sypheotides indicus*, confounded by the difficulty of finding them after the breeding season, but records of the latter species from Bangladesh have been classed as vagrants. Several African species are also nomadic, especially in the savanna and Sahelian zone, but there are no clear-cut cases of extralimital vagrancy at present (Collar 1996).

MUSOPHAGIDAE
Turacos

Turacos are highly sedentary forest-associated species for which there is no evidence of any long-distance movements. Plantain-eaters and go-away-birds, which are found in more seasonal habitats, are suspected of making short-distance movements in relation to fluctuating food and water supplies (Turner 1997) but there is no evidence of substantive vagrancy in this group.

CUCULIDAE
Cuckoos

The cuckoos include species that perform among the longest migrations of any landbirds, and transoceanic migrations are routine for species in both the Old and New Worlds. Migration is most prevalent in species that breed at high latitudes, but many low-latitude species also undertake complex intratropical migrations. In the Western hemisphere, cuckoos of the genus *Coccyzus* are perhaps the most frequent vagrants. Two species that breed in North America – Yellow-billed Cuckoo *Coccyzus americanus* and Black-billed Cuckoo *Coccyzus erythropthalmus* – have ecological traits that predispose them to long-distance vagrancy; these traits include long migration span, large population size and a transoceanic migration strategy (McLaren *et al.* 2006). Yellow-billed Cuckoo is one of the most frequent transatlantic landbird vagrants, principally to peninsulas and archipelagos on the western seaboard of Europe, with over 60 records from Britain and 50 records from the Azores, as well as eight records from as far east as Italy (Lewington *et al.* 1991). Individuals may routinely lose half their body weight during ocean crossings that form part of their regular migration, and over half of European records involve dead or dying individuals that have clearly exceeded their viable flight capacity (Cramp 1985). Beyond Europe, this species has also overshot its wintering areas to occur as a vagrant south to Chile (Torés *et al.* 2013) and the Falklands (Woods 2017), but most impressive is a record from Tristan da Cunha in the southern Atlantic Ocean, some 3,200km from the South American mainland (Bond & Glass 2016). Black-billed Cuckoo is rarer as a vagrant to Europe, probably because this species takes a more westerly migration route, with fewer individuals routinely crossing the Atlantic, but it has still penetrated the European continent east to Germany and Italy (Lewington *et al.* 1991).

▲ Black-billed Cuckoo *Coccyzus erythropthalmus*, Dale of Walls, Shetland, Scotland, 18 September 2017, the second record for Shetland (*Rebecca Nason*).

In addition to these boreal migrant *Coccyzus* cuckoos there are also austral and intratropical migrant *Coccyzus* in South America with a track record of vagrancy. Most notable is Dark-billed Cuckoo *Coccyzus melacoryphus,* which has reached

Grenada in the West Indies (Kirwan *et al.* 2019), Clipperton Atoll in the Pacific Ocean more than 1,000km from the Mexican coast, Chile (Jaramillo *et al.* 2003) and the Falklands (Woods 2017). The first North American record, from the Lower Rio Grande Valley in February 1986 was initially rejected (Howell *et al.* 2014) but the species was recently added to the United States list after a second individual appeared in Florida in February 2019 (Pyle *et al.* 2019). Pearly-breasted Cuckoo *Coccyzus euleri* does not breed as far south as Dark-billed and the nature of its movements are poorly known, but it has been recorded as a vagrant to Anguilla in the West Indies (Kirwan *et al.* 2019) and is probably underreported because of its similarity to Yellow-billed Cuckoo. Like Dark-billed it is a potential vagrant to North America and conceivably a transatlantic vagrant to, for example, Cape Verde. Mangrove Cuckoo *Coccyzus minor* does not make such long-distance movements but has occurred as a vagrant to the Gulf of Mexico coast (Sibley 2000).

In the Old World it is members of the genus *Cuculus* that undertake the most impressive migrations. Common Cuckoo *Cuculus canorus* has the longest migration span and the most extensive distribution – extending right across Eurasia, almost all of which winter in Africa. Common Cuckoos are among the more regular Palearctic vagrants to North America, mostly to islands in the Bering Sea region, but there have even been records of birds holding territory on the Alaskan mainland (Howell *et al.* 2014). Elsewhere in the Western Hemisphere there are single records from California and Massachusetts in the United States (Howell *et al.* 2014), two records from Barbados in the Caribbean and an amazing recent record from Fernando de Noronha in Brazil in February 2018 (Whittaker *et al.* 2019). In the Old World there are many records from remote islands, such as Greenland, Iceland and Borneo, as well as oceanic archipelagos such as the Prince Edward Islands and Palau (Erritzøe *et al.* 2012). Understanding vagrancy in *Cuculus* cuckoos is to an extent confounded by identification difficulties, with several very similar species inviting confusion. Oriental Cuckoo *Cuculus optatus* is easily mistaken for Common Cuckoo and is probably a more regular visitor to the Western Palearctic than the few records from western Russia (Erritzøe *et al.* 2012) and, recently, Finland suggest. For the same reason, the absence of records of Common Cuckoo from Australasia may reflect vagrants being overlooked for Oriental Cuckoos, which are regular there. Oriental Cuckoo has also occurred as a vagrant to Alaska and also widely in South-east Asia and Australasia.

Another potential vagrant to the Western Palearctic and the Nearctic is Lesser Cuckoo *Cuculus poliocephalus*, which undertakes a long transoceanic migration from India to east African wintering grounds. Vagrants of this species have been recorded from Thailand, Uzbekistan and Turkmenistan (Erritzøe *et al.* 2012), but European observers would do well to pay close attention to any late-season cuckoos. African Cuckoo *Cuculus gularis* makes long intratropical migrations and occurs north to southern Chad; it might conceivably occur as a vagrant to Morocco or Egypt. Other African cuckoo species make longer intratropical migrations, notably Diederik Cuckoo *Chrysococcyx caprius*, which has occurred as a vagrant twice in the Western Palearctic – in Israel and Cyprus respectively (Haas 2012). Another cuckoo that should be regarded as a potential vagrant across a broad swathe of the Northern Hemisphere is Jacobin Cuckoo *Clamator jacobinus*, which like Lesser Cuckoo appears to move between the Indian subcontinent and Africa in

► Great Spotted Cuckoo *Clamator glandarius*, Penally, Pembrokeshire, Wales, 13 March 2014, a less than annual spring overshoot to the British Isles (*Stuart Piner*).

▲ Jacobin Cuckoo *Clamator jacobinus*, Thai Mueang Health Garden, Phangnga, Thailand, 10 March 2018. This long-distance migrant is a potential candidate for natural vagrancy to Europe (*Wich'yanan Limparungpatthanakij*).

◄ Common Hawk-Cuckoo *Hierococcyx varius*, Masirah, Oman, 14 January 2010, the second record for Oman and one of few away from the Indian subcontinent (*Jens Eriksen*).

addition to intra tropical migrations by other African populations. Jacobin Cuckoo has been recorded as a vagrant east to Japan's Ryukyu Islands and the Philippines, in addition to scraping into the Western Palearctic with a single record from Chad (Haas 2012). More controversial is a record from Finland in September 1976, which has not been accepted by the national rarities committee on the grounds of escape risk (Lees & Mahood 2011) but should perhaps be given more credence as a wild bird. Other members of the genus *Clamator* also have a strong track-record of vagrancy, with Great Spotted Cuckoo *Clamator glandarius* recorded from countries across Europe east to Russia, Armenia and Turkmenistan (Erritzøe *et al.* 2012); juveniles are more frequently reported than adults, although the latter do overshoot too, especially in spring. Similarly, the highly migratory Chestnut-winged Cuckoo *Clamator coromandus* has occurred as a vagrant several times to Japan, all the way west to Palau in Micronesia (Erritzøe *et al.* 2012), the Cocos (Keeling) Islands of Australia in December 2018 to February 2019 and tantalisingly close to the Western Palearctic at Ayn Hamran, Oman, in December 2019 (Lehikoinen & Forsman 2020).

There are a number of highly migratory cuckoos in the Australo-Pacific region, which move between temperate regions of Australia and New Zealand to winter in the tropics. The suite of species making such movements includes Channel-billed Cuckoo *Scythrops novaehollandiae*, Horsfield's Bronze Cuckoo *Chrysococcyx basalis* and Pallid Cuckoo *Cacomantis pallidus,* which have been recorded widely as vagrants to subantarctic islands and New Zealand. Meriting special mention in this group is Long-tailed Koel *Eudynamys taitensis,* which breeds on New Zealand and migrates north and east to winter on islands dotted around the South Pacific, one of the most remarkable migrations performed by any bird species. Migrants routinely undertake 3,000km overwater crossings, and vagrants have occurred west to Queensland in Australia, and north to Palau (Erritzøe *et al.* 2012). Asian Koel *Eudynamys scolopaceus* has also been widely reported as a vagrant from as far afield as Kuwait (Al-Sirhan & Al Hajji 2010), Turkmenistan (Rustamov *et al.* 2016), Japan (Brazil 2018) and western Australia (Menkhorst *et al.* 2017).

Many other cuckoo species do not undertake spectacular migrations and are consequently less prone to vagrancy, although even nominally resident or short-distance migrants do make odd long-range movements. Such vagrancy events are well documented in Groove-billed Ani *Crotophaga sulcirostris,* which has occurred widely as a vagrant in the United States, principally in autumn (Mlodinow & Karlson 1999). Meanwhile, Smooth-billed Ani *Crotophaga ani,* which apparently originally colonised Florida by way of hurricane-displaced vagrants (Erritzøe *et al.* 2012) is considerably rarer, but has occurred as a vagrant as far north as Ohio (McLean *et al.* 1995). Migratory behaviour in Greater Ani *Crotophaga major* is poorly understood, but the species is probably an intratropical migrant whose movements are tied to seasonal rains (Erritzøe *et al.* 2012). Vagrants have reached the Virgin Islands and Barbados in the Caribbean (Kirwan *et al.* 2019) and there is a record from Miami-Dade County, Florida, in December 2010, which although rejected by the state rarities committee – given the apparent presence of Greater Anis in the bird trade in the state (Greenlaw 2013) – should, however, be expected as a vagrant to the United States.

► Groove-billed Ani *Crotophaga sulcirostris*, Braithwaite, Plaquemines, Louisiana, United States, 22 December 2017. This species has been widely recorded as a regular vagrant across the southern United States (*Matt Brady*).

PODARGIDAE
Frogmouths

There is no evidence for migration or vagrancy in frogmouths. If migratory behaviour were to be detected it might also be found as partial migration in temperate zone populations of species such as Tawny Frogmouth *Podargus strigoides.* However, the documentation of torpor (a state of decreased physiological activity) in that species suggests that the group may rely on other mechanisms to deal with seasonality (Körtner *et al.* 2000). As such, vagrancy seems unlikely.

CAPRIMULGIDAE
Nightjars

Of all the nightbirds in the order Caprimulgiformes, the nightjars are by far the most species-rich and the most migratory – factors that conspire to make them prone to vagrancy. Yet, as a group they are also very difficult to detect outside of their breeding range, so vagrants are likely to be heavily under-recorded. Common Nighthawk *Chordeiles minor* undertakes among the longest migrations of all nightjars, moving all the way from Canada to Brazil and has appeared as a vagrant to western Europe on several occasions, with even more impressive records from Malta in October 2018 (Ławicki & van den Berg 2018c) and Tristan da Cunha (Ryan 1989), suggesting that vagrancy to southern Africa might even be a possibility. Lesser Nighthawks *Chordeiles acutipennis* are less migratory, but several populations do apparently undertake long-distance movements and there are odd records of extralimital vagrants from the northeastern United States, Canada and Bermuda. The latter hints that vagrancy to the Western Palearctic is not out of the question. In the Palearctic, the most highly migratory species is European Nightjar *Caprimulgus europaeus,* an Palearctic-African migrant which has occurred as a vagrant to many of the Atlantic islands including Iceland, the Faroes, Azores, Madeira, the Canary Islands and even Ascension Island, and in the Indian Ocean to the Seychelles and Socotra (Cleere 1998). Given the record from Ascension, this species would seem a likely candidate for vagrancy to the New World, most likely in the Caribbean or northern South America.

In the Eastern Palearctic the mantle of the migratory high-latitude nightjar is taken by Jungle (Grey) Nightjar *Caprimulgus indicus,* for which there is a wide scatter of vagrants. These include records west to Bhutan and Kamorta Island, central Nicobars (Dalvi *et al.* 2017), south to north-west New Guinea and to remote Ashmore Reef (Carter 2009) and Christmas Island (James & McAllan 2014), Australia, east to Palau and once even to the United States on the Aleutian island of Buldir in May 1977 (Howell *et al.* 2014). This species would appear to be a very likely candidate for vagrancy to the Western Palearctic.

Beyond these long-distance migrants there are also numerous records of vagrancy in shorter-distance migrants; these include multiple extralimital records of two near-mythical species in northern Europe – the restricted-range Red-necked Nightjar *Caprimulgus ruficollis* and the enigmatic Egyptian Nightjar *Caprimulgus aegyptius.* The former species, which breeds in Iberia and North Africa and winters in tropical West Africa has reached the Britain, Denmark and France and has occurred west to the Canary Islands and east to Malta and Israel (Melling 2009). On paper, Egyptian Nightjar seems an unlikely vagrant to northern Europe, but it has a long migration span (especially populations from

◀ Common Nighthawk *Chordeiles minor*, Galgorm, Antrim, Northern Ireland, 8 October 2019, the first record for Northern Ireland (*Josh Jones*).

Arabia and Kazakhstan) and there is a wide scatter of vagrants including from Britain, Denmark, Germany, Sweden, Cyprus, the Canary Islands and, most recently, Georgia. Most extralimital records come from Malta and southern Italy, which are geographically most proximate to breeding areas and migration routes (Schärer & Cavaillès 2019). Recent records of the partially migratory Sykes's Nightjar *Caprimulgus mahrattensis* from Oman (Burgas & Ollé 2017) and Bangladesh (Alam *et al.* 2019) also hint at the possibility of vagrancy to the Western Palearctic. Elsewhere, some tropical species are partial migrants, leading to some limited extralimital records; for example, observations of Mozambique Nightjar *Caprimulgus fossii* in the northern Ivory Coast and southern Ghana, and Rufous-cheeked Nightjar *Caprimulgus rufigena* in the Central African Republic and western Sudan (Cleere 1998). Pennant-winged Nightjar *Caprimulgus vexillarius* makes longer trans-equatorial movements and vagrants have been recorded a long way out of range from The Gambia, south-east Chad, southern Somalia and South Africa (Cleere 1998). Savanna Nightjar *Caprimulgus affinis* is thought to be largely sedentary but there is at least one documented record from Christmas Island, Australia (James & McAllan 2014), which may even have involved a bird from the migratory northern subspecies and not one of the more geographically proximate taxa.

In the New World, shorter-distance migrants like Eastern Whip-poor-will *Antrostomus vociferus* have been recorded in western North America (in northern California and Oregon) and Mexican Whip-poor-will *Antrostomus arizonae* has been recorded as far north as northern Utah and northern California. Movements made by Neotropical nightjars are generally poorly understood, as many involve partial migration at southern or northern range edges, such as that of Rufous Nightjar *Antrostomus rufus,* but extralimital records might still be expected. Similarly, the partially migratory Buff-collared Nightjar *Antrostomus ridgwayi* has been recorded as a vagrant to both California and Texas. Some populations of Band-winged Nightjar *Systellura longirostris* are among the most migratory of South American nightjars and there is a single record of this species from the Falklands in April 2004 (Woods 2017).

NYCTIBIIDAE
Potoos

Potoos are widely reported in the ornithological literature as being highly sedentary (Cohn-Haft 1999) but there is ample evidence that southern Brazilian populations of Common Potoo *Nyctibius griseus* is a partial migrant in the southern part of its range (Belton 1984), moving north in the austral winter. Migrants are regularly detected in unusual circumstances in gardens and urban areas where the species does not normally occur – habits that also betray migration in other poorly-known nightbirds. It is unknown how far these individuals might migrate, or where to. In this context, a record of a bird identified as a Common Potoo ship-assisted from the Panama Canal to Florida is rather intriguing (Cohn-Haft 1999). The first Common Potoo for Chile (Barros *et al.* 2016) was photographed on 9 September 2015 at Iquique, a date that might suggest a major overshoot by a returning migrant. Northern Potoos *Nyctibius jamaicensis* have also been reported as vagrants from two tiny Caribbean islands west of Puerto Rico – Mona and Desecheo – requiring a sea crossing (or ship assistance again) of at least 100km (Kirwan *et al.* 2019).

STEATORNITHIDAE
Oilbird

The movements of Oilbirds *Steatornis caripensis* are something of an enigma within a mystery. This species is known regularly to make long-distance movements of hundreds of kilometres between cave systems (Roca 1994), but the nature of these, and of various nominally extralimital records, is confounded by a lack of knowledge of exactly where Oilbirds might be regularly found, as there may be various undiscovered colonies.

Potential out-of-range vagrants include records from locations as diverse as western Amazonian Brazil (Whittaker *et al.* 2004) and Amazonian Peru (Tello *et al.* 2013), in Costa Rica (Arévalo & Sánchez 1999) and even in northern Chile (Barros & Schmitt 2013). Overwater movements include records of apparent vagrants on Tobago (Thomas & Kirwan 2019) crossing at least 30km of open sea and suggest that the potential for longer-range vagrancy in this species may be very significant.

AEGOTHELIDAE
Owlet-nightjars

All species of owlet-nightjars are considered sedentary and there are no documented instances of vagrancy. However, the existence, or former existence of endemic owlet-nightjars in the Moluccas, New Caledonia and New Zealand indicates that such long-distance dispersal events happen at least occasionally over evolutionary timescales.

APODIDAE
Swifts

Swifts are consummate fliers and all species occurring outside the tropics are highly migratory, with movements driven by the seasonal availability of their aerial invertebrate food. Annual swift migrations are among the longest of all landbirds – for instance, Common Swifts *Apus apus* breeding in China migrate to South Africa in the non-breeding season, where they mix with Common Swifts from Europe. Some, perhaps most, swift species remain airborne for most of their non-breeding period; for example, Common Swifts remain airborne for >99 per cent of the time during their 10-month non-breeding period, sleeping on the wing by gaining height and shutting down one half of the brain at a time (Hedenström *et al.* 2016). Migration speed can be very fast: Alpine Swifts *Tachymarptis melba* have been known to cover 1,620km in three days (Bourliere 1950). This combination of being able to feed and sleep on the wing with efficient flight means that swifts have an extremely high capacity for long-distance movements, easily traversing biomes that are hostile to many landbirds, including deserts and open oceans. As a result, transoceanic vagrancy events may be easier for swifts than for most species.

Common Swifts have one of the broadest spreads of vagrant records of any landbird. Individuals of this species have been recorded at various locations on the northeastern Atlantic seaboard from Greenland (Ławicki & van den Berg 2015b) and Newfoundland (Hoell *et al.* 2014) to Bermuda (Howell *et al.* 2014), Puerto Rico (Ławicki & van den Berg 2016a) and the West Indies (Kirwan *et al.* 2019). Common Swifts have also reached the Pacific coast of North America, with two records from the Pribilofs (Howell *et al.* 2014) and more unexpectedly a record from Riverside County, California, in 2013 (Rottenborn *et al.* 2016). One of the Pribilof specimens and the California bird were identified as the *pekinensis* subspecies from Asia. A low-latitude bias for western Atlantic seaboard records suggests that birds cross the ocean assisted by trade winds at tropical latitudes, an assumption helped by the incredible record of a Common Swift photographed at sea 280km off the coast of Surinam in July 2012 (de Boer *et al.* 2014). Elsewhere, Common Swifts have made it to the Prince Edward Islands, almost halfway from South Africa to Antarctica in the Southern Ocean (Burger *et al.* 1980), as well as to Thailand, the first for South-east Asia (Pierce *et al.* 2015). Future records from Australia's vagrant traps would appear likely.

Alpine Swifts have occurred as transatlantic vagrants to the West Indies an incredible seven times (Kirwan *et al.* 2019) and there is also a South American sight record from French Guiana in June 2002 (Ottema 2004). This species is also a regular vagrant to northern Europe; such overshooting birds tend to be associated with spells of warm southerly airflows (Moss 1995). Two migratory Asian species – Pacific Swift *Apus pacificus* and White-throated Needletail *Hirundapus*

► Pacific Swift *Apus pacificus* with Common Swifts *Apus apus* and Pied Avocet *Recurvirostra avosetta*, Trimley Marshes, Suffolk, England, 15 June 2013. This, the seventh British record uncharacteristically stayed put for two days, allowing many people the chance to see it (*Josh Jones*).

► White-throated Needletail *Hirundapus caudacutus*, Tarbert, Harris, Outer Hebrides, Scotland, 26 June 2013. This was the first in Britain for 22 years but was unfortunately killed when it collided with a small community wind turbine (*Stuart Piner*).

caudacutus – have been recorded as vagrants widely across the Palearctic and even the Nearctic. White-throated Needletail has reached the Aleutian Islands on four occasions (Howell *et al.* 2014), and there are widely spread records from the Indian Ocean on the Seychelles, Christmas Island and the Cocos (Keeling) Islands, as well as to Fiji in the Pacific and even the Australian island of Macquarie in the South Tasman Sea (Chantler & Driessens 2000). There are records from a number of northern European countries as well as Malta and Spain. Records of Pacific Swifts follow a similar pattern, although there are more records of this species from the Bering Sea region, with a single mainland North American record from the Yukon in September 2010 (Howell *et al.* 2014). There have also been records from several western European countries, the United Arab Emirates and Macquarie Island. One English site, Spurn Point, Yorkshire, has hosted Pacific Swift on five occasions (Roadhouse 2016), in addition to five other swift species, highlighting the importance of geographic bottlenecks for detecting vagrant swifts. House Swift *Apus nipalensis* of southern Asia is usually considered non-migratory but seems to have a penchant for wandering; there are records of flocks from northern Australia and incredibly a recent moribund individual from British Columbia, Canada (Szabo *et al.* 2017).

Vagrancy in some swifts is probably under-recorded due to identification difficulties – Pallid Swift *Apus pallidus*, for example, is probably overlooked in areas where Common Swift occurs – the concentration of northern European records early and late in the season may reflect the fact that any swift at these times receives extra scrutiny. The relative rarity of the Asian swifts in Europe, in comparison to the regularity of many Siberian breeding species, may suggest that swifts are less susceptible to wind drift or misorientation than other species. The clustered nature of records of these species in western Europe probably means that very few individual birds may be involved, with re-sightings of the same individuals occurring in multiple years. However, the Alaskan records include multiple records of flocks of Pacific Swifts, with 40 on Attu on 18 September 2004 (Howell *et al.* 2014), suggesting mass displacement by storm systems.

Chimney Swift *Chaetura pelagica* is somewhat exceptional in both being the only transatlantic vagrant swift to have occurred in Europe, and a species that occurs in vagrant flocks with some regularity, particularly associated with late October/early November storm events. Hurricane Wilma in 2005 displaced record numbers of Chimney Swifts to the northeastern United States and Europe late in the season; for example, 130 were recorded on the Azores and 18 in Britain and Ireland (Elkins 2008). Those that made landfall were the lucky ones; this storm event is inferred to have killed half of the Chimney Swifts in Quebec, Canada (Dionne *et al.* 2008), highlighting how destructive large storm events can be for migratory bird populations. Understanding vagrancy in *Chaetura* swifts is also complicated by identification difficulties – although there are records of both Chimney and Vaux's Swifts *Chaetura vauxi* in western and eastern North America respectively. Most of the species diversity in this genus is found in South America where identification challenges are even more acute, confounding the status of vagrant individuals. However, there are records of vagrancy in some species, most notably of the highly migratory Ashy-tailed Swift *Chaetura meridionalis* on the Falklands (Woods 2017) and Short-tailed Swift *Chaetura brachyura* to several islands in the northern islands in the West Indies, where they don't breed (Kirwan *et al.* 2019). Ashy-tailed Swift, which migrates from central-southern South America to winter in northern South America, is perhaps even a candidate for transatlantic vagrancy to Cape Verde or the African mainland.

Vagrancy in South American swifts is not confined to the genus *Chaetura*, as White-collared Swift *Streptoprocne zonaris* has also been recorded as a vagrant, with nine records of 10 birds as far north as Ontario, Canada, 2,400km north of its regular range (Howell *et al.* 2014), and one record on the Falklands, 1,800km south of its regular range (Woods 2017). This species could conceivably appear anywhere in the Americas. The Antillean Palm-Swift *Tachornis phoenicobia* has occurred as a vagrant to Florida and Puerto Rico (Howell *et al.* 2014), which would seem surprising, given that palm-swifts tend to be very sedentary, but again occasional long-distance dispersal events might be expected in a species that has obviously island-hopped in its evolutionary history.

Intratropical movements by Afrotropical swifts are particularly poorly known. Vagrancy has been reported for Mottled Swifts *Tachymarptis aequatorialis*, with a sight record from Senegal (Devisse 1992) and the southern Cape Province, South Africa (Chantler & Driessens 2000). Extratropical vagrancy in this species is not unthinkable – there is already a convincing but unaccepted report of this species from Britain at the aforementioned swift hotspot of the Spurn peninsula (Roadhouse 2016). Two ostensible Afrotropical swifts – Little Swift *Apus affinis* and White-rumped Swift *Apus caffer* – have colonised Spain in the 20th century, highlighting the role of vagrancy in aiding range extension; this is a not dissimilar situation to the historic colonisation and speciation of Plain Swift *Apus unicolor* and other island endemics. Little and White-rumped Swifts are now increasingly regular vagrants in northern Europe, but Plain Swift has yet to be reported in Europe, although since the species is known to winter south as far as the Ivory Coast (Norton *et al.* 2018), overshooting to Iberia or beyond seems likely.

◄ Antillean Palm-Swift *Tachornis phoenicobia*, Grassy Key, Monroe, Florida, United States, 17 July 2019, the second record for North America (*Tom Johnson*).

HEMIPROCNIDAE
Treeswifts

Treeswifts are all apparently non-migratory and there are no reports of vagrant individuals. However, the extensive occupancy of disjunct island archipelagos by Grey-rumped Treeswift *Hemiprocne longipennis* does, for example, indicate historic overwater dispersal events of nominally vagrant individuals.

TROCHILIDAE
Hummingbirds

That hummingbirds occur as vagrants at all may come as a surprise given their physiological limitations – their tiny size and high metabolic rates, which necessitate regular feeding. Yet, vagrancy in the group is well documented and, in some cases, prolific – even among species that do not undertake pronounced annual migrations. Vagrancy is most frequently documented among temperate zone species in North America and southern South America. That is not to say that tropical hummingbirds are poor dispersers; we know from dispersal challenge experiments (Moore *et al.* 2008) and direct observation of gap-crossing experiments (Lees & Peres 2009) that hummingbirds are among the most vagile of lowland tropical birds. A lack of extralimital records may therefore reflect either a lack of long-distance movements in most species, or a combination of identification difficulties and low observer coverage in the tropics.

In southern South America there are several highly migratory hummingbird species, notably Green-backed Firecrown *Sephanoides sephaniodes,* which migrates north towards the tropics in the austral winter. This species has been recorded offshore when crossing the Corcovado Gulf in Chilean Patagonia (Segre *et al.* 2018), a flight that requires regular overwater crossings of up to 70km. This is far from the limit of the species' flight capacity, however, as Green-backed Firecrowns have colonised the Juan Fernandez Islands, 667km off the coast of Chile at least twice. An early colonisation event led to the evolution of Juan Fernandez Firecrown *Sephanoides fernandensis*, and a later one by the same species means that the islands now have two resident hummingbird species (Roy *et al.* 1998). Given Green-backed Firecrown's status as the southernmost breeding hummingbird, its at least partial migratory status and history of long-

► Mexican Violetear *Colibri thalassinus*, Clarksville, Howard, Maryland, United States, 25 October 2011. It has been hypothesised that autumn vagrants of these species in eastern North America may be 'reverse migrants' from northernmost migratory populations in Mexico *(Tom Johnson)*.

distance overwater crossing, it should not come as a surprise that there are several extralimital records. These include at least 14 records from the Falklands – necessitating an overwater flight of 500km (Woods 2017) and several records from eastern Argentina (Marinero *et al.* 2012). This species would seem a likely future vagrant to southern Peru, Brazil and Uruguay at least.

Elsewhere in South America, vagrancy in hummingbirds is poorly understood but has been reported in some migratory species such as the Ruby-topaz Hummingbird *Chrysolampis mosquitus*, which has strayed as far north as central Panama (Braun & Wolf 1987) as well as reaching Grenada in the West Indies (Kirwan *et al.* 2019). It has also been recorded well south of its regular South American range in Misiones province, Argentina, in April 2001 (Pugnali & Pearman 2001) and west as far as Acre in Brazil (Guilherme & Dantas 2008) and Peru in May 2019 (Alferez & Delgado 2019). As such, this species seems a plausible vagrant to North America. Arguably also deserving to be on the radar of North American rarity-finders is White-necked Jacobin *Florisuga mellivora*. Although movements of this hummingbird are even more poorly understood than the preceding species, there are records from the Grenadines in the West Indies (Kirwan *et al.* 2019) and the Atlantic Forest of Brazil in Espirito Santo (Ruschi & Simon 2012) and São Paulo (Silva 2018). Even more incredibly, the first for Argentina frequented a feeder on the edge of Buenos Aires in February–March 2016 (Ferroni *et al.* 2017). This observation of an ostensibly Amazonian hummingbird over 2,000km out of range hints that regular movements in this species may be far more dramatic than currently known.

Hummingbird vagrancy is best documented and most prevalent in North America, where several species including Ruby-throated Hummingbird *Archilochus colubris*, Black-chinned Hummingbird *Archilochus alexandri* and Calliope Hummingbird *Stellula calliope* are fully migratory. The last is regarded as the world's smallest long-distance avian migrant, and despite its diminutive size and western North American distribution it has been widely recorded in the eastern United States and southern Canada. Even more impressive is a record of an adult female on Herschel Island on the shores of the Beaufort Sea in Yukon Territory, Canada (eBird). Vagrancy in the Nearctic species is almost exclusively an autumn phenomenon; spring vagrancy is much less prevalent and some of the handful of east coast spring records of western hummingbirds may represent vagrant birds moving north from wintering areas in the east (M. Iliff in litt.). Among the most celebrated of North American vagrant hummingbirds are a suite of species that are essentially confined to Central America and occur regularly no closer than central Mexico. These include the recently split Mexican Violetear *Colibri thalassinus*, a species known to be a partial migrant which has been widely recorded in eastern North America – all the way north to Ontario, Quebec and Maine. Howell *et al.* (2014) inferred that the eastern bias in records in this species may reflect biases in misorientation and these Mexican species tend to occur as vagrants in late summer, but some into late autumn. The increasing frequency with which this species is recorded (and, indeed, many vagrant hummingbird species) probably reflects an increase in observer density, competence and familiarity, and especially the high rates of provisioning of feeders and ornamental planting.

◀ Berylline Hummingbird *Amazilia beryllina*, Cochise, Arizona, United States, 21 August 2019. This species is one of the more regular vagrant hummingbirds to the southern United States and has even bred there (*Tom Johnson*).

► Plain-capped Starthroat *Heliomaster constantii*, Portal, Cochise, Arizona, United States, 19 August 2015, one of several Neotropical hummingbird species that regularly stray to the southern limits of the United States in late summer (*Tom Johnson*).

Green-breasted Mango *Anthracothorax prevostii* shares a broadly similar distribution to Mexican Violetear, albeit penetrating into northern South America, and has been recorded in the United States in Texas, North Carolina, Wisconsin, Georgia and Louisiana. Students of avian vagrancy in the North American birding community would have probably not been terribly shocked with the first United States record of Amethyst-throated Mountain-gem *Lampornis amethystinus* in western Texas in October 2016, but this was incredibly preceded by one in Quebec, Canada, in July of the same year. Other species that breed closer to the United States southern border – including Plain-capped Starthroat *Heliomaster constantii,* Cinnamon Hummingbird *Amazilia rutila* and Berylline Hummingbird *Amazilia beryllina* – have been recorded as vagrants to states on the southern border, probably only involving movements of a few hundred kilometres (some have bred extralimitally in the United States). More surprising was a recent record of Berylline Hummingbird much further out of range in Michigan in 2014. Also exceptional are extralimital records of Xantus's Hummingbird *Hylocharis xantusii*, an endemic of the southern half of the Baja California peninsula, Mexico, which occurs regularly as close as 400km to the United States border. This species evidently makes some limited northward movement accompanying winter rains, so while the two records in southern California are not truly unexpected (one is a sight record only), the Xantus's Hummingbird that appeared just north of Vancouver, British Columbia, in 1997 was utterly remarkable (Toochin 1997).

A suite of other Neotropical hummingbirds maintains a toehold in the United States on the country's southern border in south-east Arizona, western Texas or both. Several of these species have also occurred widely as vagrants, with Blue-throated Mountain-gem *Lampornis clemenciae* (formerly Blue-throated Hummingbird) having occurred west to California, north to Colorado and east to Louisiana, and Rivoli's Hummingbird *Eugenes fulgens* (formerly Magnificent Hummingbird) occurring even more widely – west to northern California and east to Georgia and Virginia. Broad-billed Hummingbird *Cynanthus latirostris* shares a broadly similar range with these two species and has been recorded across a wide swathe of North America, up the west coast to northern Oregon, north to three Canadian provinces – Ontario, Quebec and New Brunswick – and south-east to southern Florida. This big variation in spread of vagrancy seems likely to reflect both overall population sizes and the nature of the poorly known partial migrations in these species; understanding the migration spans and bearings taken of these movements will be crucial to understand the spatio-temporal nature of vagrancy patterns.

Two species of hummingbirds are also vagrants to the United States from the Caribbean. Both are extremely rare, and the Cuban Emerald *Chlorostilbon ricordii* has yet to make it onto the official list, as none of the 14 or so reports from Florida are documented by specimens or photo/video/sound recordings (Howell *et al.* 2014). Bahama Woodstar *Nesophlox evelynae* has an even more restricted-range distribution as a Bahamian endemic and is known from a half dozen

◄ Lucifer Hummingbird *Calothorax lucifer*, Ash Canyon Bird Sanctuary, Cochise, Arizona, United States, 24 July 2005 (*Jay McGowan*).

or so better-documented records. Most of the latter are from southern Florida as one might expect, but incredibly in April 2013 one appeared in south-east Pennsylvania (Weidensaul & Lockerman 2013). Considering that this species is basically resident and makes only small inter-island movements, its appearance at a time of year not known for hummingbird vagrancy in a random inland location is all the more amazing (M. Iliff *in litt.*).

Most of the common hummingbirds in the United States and Canada are found in the western half of the continent, with just a single species – Ruby-throated Hummingbird *Archilochus colubris* – regular over much of the east. Recent increases in detection of this species during its regular southward migration period (August to October) on the west coast surely represents increased observer attention, a better understanding of identification criteria, and advances in digital photography since it is very hard to separate from its congener, Black-chinned Hummingbird *Archilochus alexandri* (M. Iliff *in litt.*). Future records of this species from northern South America would seem likely. Several western hummingbird species are now turning up with increasing frequency in eastern North America in late autumn and early winter (e.g. Mitra & Bochnik 2001). Members of the genus *Selasphorus* dominate this pattern, with Rufous Hummingbird *Selasphorus rufus* most abundant but both Broad-tailed Hummingbird *Selasphorus platycercus* and Allen's Hummingbird *Selasphorus sasin* recorded with increasing frequency. Underpinning this increase may be both greater observer awareness and the provision of late-season feeders after Ruby-throated Hummingbirds have departed, but the magnitude of the increase suggests a genuine change in behaviour and raises the spectre of the founding of a new nominally pseudo-vagrant wintering population for Rufous Hummingbird on the Gulf Coast at least (Hill *et al.* 1998).

Although they are now an exclusively New World lineage, hummingbirds that resembled modern forms formerly occurred in Europe, where there is an excellent fossil record of stem-group hummingbirds from the early Oligocene of Germany, around 32 million years ago (Mayr 2004). European rarity-finders have long speculated about the possibility of modern occurrence of hummingbirds along the Atlantic fringe and the obvious candidate for transatlantic vagrancy would be the eastern North American Ruby-throated Hummingbird. This species occurs almost annually on Bermuda, an oceanic crossing of over 1,000km and there are records north to Hudson Bay, west to Baja California and east to the Dominican Republic. Zenzal & Moore (2016) investigated the stopover ecology of Ruby-throated Hummingbirds along the northern coast of the Gulf of Mexico and calculated potential flight ranges based on wingspan, wing area, fat-free mass and fuel load. These calculations gave an overall estimated flight range of 2,261km (±1,152km) and an upper limit of 4,960km, which would seemingly put the Azores well within the physiological capacity for flight of a vagrant individual. Should a Ruby-throated Hummingbird make it to Europe, it will, however, not be the first recent Old World hummingbird, as Rufous Hummingbird has occurred on multiple occasions in spring on the Chukotka Peninsula in the Russian Far East (Brazil 2009, Sorokin 2017).

► Allen's Hummingbird *Selasphorus sasin*, Leola, Lancaster, Pennsylvania, United States, 21 December 2009, the first state record; the provision of feeders in late autumn and early winter in the east not only makes vagrants detectable but also ensures their survival in regions and at times when natural food sources will have vanished (*Tom Johnson*).

► Black-chinned Hummingbird *Archilochus alexandri*, Boston, Massachusetts, United States, 21 November 2015. Field separation of this species from Ruby-throated Hummingbird *Archilochus colubris* is a challenge that is increasingly being conquered (*Marshall Iliff*).

► Calliope Hummingbird *Selasphorus calliope*, Apalachicola, Franklin, Florida, United States, 16 February 2013. This species is regarded as the smallest long-distance migrant bird and has been relatively rarely but widely recorded as a vagrant across eastern North America (*Chris Wood*).

OPISTHOCOMIDAE
Hoatzin

Hoatzins *Opisthocomus hoazin* are weak fliers and are regarded as highly sedentary. It is thus something of a surprise to note a well-documented extralimital record from Yacambú National Park, Venezuela, in July 2013 (Buitrón-Jurado 2014). A single individual was photographed in emergent vegetation at a reservoir in an area of premontane cloud forest at 1,448m and considered by the observer to be a vagrant rather than an escape/released bird. Although only 46km out of range, the individual was over 1,000m outside of the species' usual altitudinal habitat envelope! As such, this is one of the least-expected incidences of short-range avian vagrancy.

SAROTHRURIDAE
Flufftails

Flufftails are African and Madagascan rail-like birds which are very difficult to detect and are hence rather poorly known. Several species are suspected to perform migrations, or at least erratic nomadic movements, with individuals of several species reported as being attracted to lights in cities during nocturnal movements. Some vagrancy has been reported, albeit with the caveat that some records may later prove to be migrants or even residents in these areas. These include, for example, records of Buff-spotted Flufftail *Sarothrura elegans* from Somalia, northern Kenya, northern Botswana and Namibia (Taylor & Kirwan 2020). Their itinerant movements may even take flufftails over significant waterbodies, with a Streaky-breasted Flufftail *Sarothrura boehmi* taken at sea (in June) 150km off the coast of Guinea (Urban *et al.* 1986). White-winged Flufftail *Sarothrura ayresi* is among the most enigmatic of the group, and periodic long-distance dispersal is suspected to occur between Ethiopia and South Africa (Colyn *et al.* 2019), although the species has now been confirmed to breed in both countries. Nevertheless, records from Zambia and Zimbabwe might reflect dispersing individuals or vagrants. If this species does routinely undergo epic trans-continental migrations, then vagrancy to the Palearctic would appear to be possible (and an old undocumented report to a United Kingdom internet birding forum of a flufftail species at an Egyptian Red Sea resort remains intriguing).

RALLIDAE
Rails, Gallinules and Coots

Unlikely as it may seem for such apparently weak fliers, many members of this family are long-distance migrants and have a remarkable capacity for extreme vagrancy. Corncrake *Crex crex*, for example, has been recorded as widely as eastern North America, the West Indies, the Brazilian island of Fernando de Noronha, Ascension Island, the Seychelles, Sri Lanka, the Andaman and Nicobar Islands, and even Australia and New Zealand (Bräunlich & Rank 2001, Burgos & Olmos 2013), many thousands of kilometres from its normal migration routes between Palearctic breeding and African wintering grounds. Despite their apparent ungainliness, rails and gallinules are remarkably strong fliers when on active migration, moving in a fast and direct course by continued flapping, exclusively at night (Bent 1926).

The dispersal prowess of rallids is illustrated by their widespread presence on remote archipelagos – few other avian groups have shown such consistency in colonising even the remotest island settings. However,

▲ Corncrake *Crex crex*, Cedar Beach, Suffolk, New York, United States, 8 November 2017. Formerly more regular as a vagrant to North America, this was the first New York record since 1963 (*Jay McGowan*).

the history of island rails is also rather tragic. After establishing insular populations, many rails rapidly evolve flightlessness, and their subsequent vulnerability to humans and their companion fauna (pigs, dogs, cats and rats) has made them highly prone to extinction. Although the true number of extinct island rails and gallinules is unknown, palaeontologists estimate that thousands of rail species may have been lost on islands in the Pacific alone (Steadman 1995).

Many temperate zone rail species are highly migratory. For example, Sora *Porzana carolina* may move from northern Canada all the way to northern South America. Migrants may be prone to wind drift, and falls of thousands of Soras occurred on Bermuda in the 19th century (Bent 1926), where considerable numbers still occur on passage (Amos 1991). This strongly suggests that high numbers of this diminutive crake still use this transoceanic migration route. Small numbers have made landfall in the Western Palearctic, where there is a scattering of records between Iceland and Morocco, including 20 records from Britain; vagrants that survived the crossing established territories and sang in Sweden in 1966 and 1987. Soras have overshot their breeding grounds to reach Rio de Janeiro in Brazil, the only Brazilian record, in January 2015 (Camacho & Accors 2016), and they have even occurred three times on the south-eastern Hawaiian islands (Pyle & Pyle 2017).

Another frequent New World vagrant is Purple Gallinule *Porphyrio martinica* – a largely Neotropical species with migratory populations breeding in the southeastern United States and wintering in the Caribbean and northern South America. Vagrants are frequently recorded across North America, and Farnsworth *et al.* (2015) found that these records were associated with warm summer weather. Perhaps unsurprisingly, there are numerous autumn records from Europe involving individuals (often emaciated) being drifted across the Atlantic by storms. Bizarrely, however, Purple Gallinule occurs more frequently as a vagrant to South Africa than to Europe and is also recorded with some regularity on mid-Atlantic islands, including Tristan da Cunha, Ascension and St Helena, as well as the Falklands (Prater 2012, Woods 2017). Unlike the European occurrences, these Southern Hemisphere records typically occur in the northern spring (February–August) and are usually associated with strong westerly winds between southern South America and South Africa. Quite why vagrancy is more prevalent in the Southern Hemisphere for this species is unclear, and it is the only species of any taxonomic group to show this pattern.

The only other west-to-east transatlantic vagrant is American Coot *Fulica americana,* with 35 records from the Azores and 17 from Britain and Ireland. The highly migratory Yellow Rail *Coturnicops novebora-*

◀ Purple Gallinule *Porphyrio martinica*, Waiehu Stream, Maui, Hawaii, United States, 26 February 2018. This species is a relatively frequent intercontinental vagrant (*Eric VanderWerf*).

censis, which has reached Bermuda and the Bahamas (Kirwan *et al.* 2019), would also seem a candidate to cross the Atlantic at some point. Vagrancy in the other direction in temperate zone breeding species is also limited. Besides the numerous instances of Corncrake vagrancy, there are a few records of Eurasian Coot *Fulica atra* from north-east Canada (Howell *et al.* 2014) and two records of Spotted Crake *Porzana porzana* from the West Indies, on St Martin and Guadeloupe (Kirwan *et al.* 2019).

Vagrancy in most bird families is more prevalent among temperate zone species than tropical ones, but this rule seems to be largely broken among the Rallidae. Several Neotropical species have occurred as vagrants to the United States, including single records of Paint-billed Crake *Mustelirallus erythrops* in Texas and Spotted Rail *Pardirallus maculatus* in Texas and Pennsylvania, an Azure Gallinule *Porphyrio flavirostris* in New York and a Rufous-necked Wood-Rail *Aramides axillaris* in New Mexico (Howell *et al.* 2014, Williams & Daw 2016). Several of these records had a rough ride with rarities committees, given their seemingly one-off nature, but with increasing observer coverage they can be now seen to form part of a wider pattern of extreme dispersal among tropical rallids. A Chilean record of Spotted Rail was killed by a cat on Robinson Crusoe Island, in the Juan Fernández archipelago 600km west of the mainland in September 1906 (Parkes *et al.* 1978), whilst Paint-billed Crake has reached the Galapagos in recent decades and has even started breeding (Taylor 1998). We know little about the 'normal' movements of many tropical species; a recent, highly unexpected record of the Brazilian endemic Little Wood-Rail *Aramides mangle* in French Guiana (Ingels *et al.* 2011) was followed by the revelation that this species is actually migratory (Marcondes *et al.* 2014) and hence more likely to occur as a vagrant. Until 2012, the sole Brazilian record of Black Rail *Laterallus jamaicensis* was a single individual that collided with the window of the ornithology department of the Goeldi Museum in Belém, Pará, in October 1994. It was assumed to be a vagrant from North America until birds were discovered breeding at multiple coastal sites in nearby Pará and Maranhão (Lees *et al.* 2014).

Afrotropical rails are also regular vagrants to the temperate zone and, in the case of Allen's Gallinule *Porphyrio alleni*, also transoceanic vagrants. An immature Allen's Gallinule reached the Brazilian island of Fernando de Noronha in February 2018 (Whittaker *et al.* 2019), whilst another landed on a boat 240km off the coast of Amapa, Brazil, in January 2017. There are also multiple records from the equatorial Atlantic islands of Ascension and St Helena (Olson 1971, Prater 2012) as well as in the subantarctic on South Georgia (Prince & Croxall 1983) and to Rodrigues in the Indian Ocean. Most Western Palearctic records of this species are from North Africa and southern Europe, but there are now three from Britain, all of which have occurred during mini incursions of this species into the Western Palearctic. Allen's Gallinules are 'rains migrants' (Brooke 1968), with their breeding and movements regulated by the local seasonality of rainfall in the Sahel. Dispersive movements in this species tend to occur at the onset of droughts in the period December–March, with individuals dispersing widely in order to locate ephemeral wetlands within this unpredictable environment. Under certain synoptic conditions, these movements can evidently

► Allen's Gallinule *Porphyrio alleni*, El Rocio, Huelva, Spain, 17 January 2007. Allen's Gallinules are the most frequent Afrotropical vagrant to Europe (*Chris Batty*).

lead to vagrancy on a massive scale. Movements to Europe have been associated with disturbed weather after prolonged spells of anticyclonic conditions, with favourable southerly winds blowing from northwest Africa (Hudson 1974). Such conditions have also brought other rarer Afrotropical rallids north, including Striped Crakes *Amaurornis marginalis* to Kuwait, Malta, Italy, Algeria and Spain; African Crakes *Crex egregia* to the Canary Islands, Cape Verde, Spain and Israel; and Lesser Moorhens *Paragallinula angulata* to Spain (four), Madeira, the Canary Islands and Cape Verde (Haas 2012). Remarkably, Lesser Moorhen has also straggled out to the mid-Atlantic, to the Brazilian archipelago of São Pedro and São Paulo (Bencke *et al.* 2005), whilst Striped Crake and African Swamphen *Porphyrio madagascariensis* have reached St Helena (Prater 2012).

In Asia, several tropical crakes have wandered widely, with Red-legged Crake *Rallina fasciata* recorded as a vagrant north to Taiwan (Brazil 2009), whilst the first Ruddy-breasted Crake *Zapornia fusca* for Oman and the Middle East was recorded in November 2014 (Olsson 2015). Both of these species have reached Australia (Menkhorst *et al.* 2019), as has Watercock *Gallicrex cinerea*, which has been widely recorded as a vagrant – from Japan and Kamchatka (Brazil 2009) west to Oman, Socotra and the UAE (Menzie 2014). Vagrancy at southern temperate latitudes is also pronounced. There are records of the poorly known Austral Rail *Rallus antarcticus* from the Falklands, and Red-fronted Coot *Fulica rufifrons* and Red-gartered Coot *Fulica armillata* have also been recorded from those islands (Woods 2017); Horned Coot *Fulica cornuta* has been recorded in Peru (Chalco

► Watercock *Gallicrex cinerea*, Christmas Island, Australia, 8 February 2011 (*Rohan Clarke*).

▲ Band-bellied Crake *Zapornia paykullii*, Chinese Garden, Singapore, 6 March 2014, the first record for Singapore (*James Eaton*).

◄ Red-legged Crake *Rallina fasciata*, Whim Creek, Roebourne, Western Australia, Australia, 2 June 2007, the second record for Australia following a specimen record from Broome, Western Australia, in July 1958 (*Rohan Clarke*).

2013). In addition to the aforementioned Corncrake, New Zealand has been visited by Australian Crake *Porzana fluminea*, Common Moorhen *Gallinula chloropus*, Dusky Moorhen *Gallinula tenebrosa* and Black-tailed Nativehen *Tribonyx ventralis*; the last species is prone to irruptions and has reached north to Christmas Island and seems a likely future vagrant to Indonesia (James & McAllan 2014).

It seems likely that members of this family possess an unusually strong tendency to disperse from their natal sites. Such a tendency could be related to their reliance on wetland habitats, which are very often ephemeral and prone to disappearance through vegetative succession and desiccation. Populations not capable of colonising new marshes would thus be doomed to eventual local extinction, generating a strong evolutionary pressure for frequent dispersal. As a result, even fully 'resident' species may also undergo unpredictable movements in search of new wetlands (Remsen & Parker 1990) especially if triggered by droughts. Such weather phenomena have, for example, been implicated in an incursion of Baillon's Crakes *Zapornia pusilla* to Britain in the summer of 2012, when at least seven birds were singing at three sites (Hudson *et al.* 2013) and an unprecedented northward displacement of Western Swamphens *Porphyrio porphyrio* into northern Europe, including the first British record in July 2016, in Suffolk then relocating to Lincolnshire (Harvey 2018). Previous records of swamphens in Britain have all been considered to be escapes from captivity, and this is one of few rallids for which escapes are a significant problem in determining vagrancy. There are also very few naturalised populations of rallids, a notable exception being Grey-headed Swamphen *Porphyrio poliocephalus,* which is now expanding in Florida, probably the source of a remarkable record from Bermuda in October–November 2016 (Pranty 2012). Bermuda has previously hosted an African Swamphen in October 2009 (Norton *et al.* 2010), which seems likely to be a genuine vagrant, underscoring the difficulty faced by rarities committees in ascertaining provenance of exceptional vagrants.

HELIORNITHIDAE
Finfoots and Sungrebe

This tiny family of three species are among the most sought-after of tropical birds. Formerly considered to be highly sedentary – a pervasive assumption about many tropical species – some have been recorded making long-distance dispersive movements. This behaviour is best documented in Sungrebe *Heliornis fulica,* for which there are multiple records of extralimital vagrancy to the south and east of its distribution in Brazil, including the southernmost global record in Rio Grande do Sul in November 2015 (Longo 2015). More impressive is a photo-documented record from Bonaire in November 2010 (Rozemeijer 2011), which crossed at least 90km of the Caribbean Sea. The most celebrated vagrant Sungrebe is the first United States record present at Bosque del Apache National Wildlife Refuge, New Mexico, in November 2008 (Williams *et al.* 2009). Masked Finfoot *Heliopais personatus* of the Asian tropics is thought to be at least a partial migrant in Cambodia and peninsular Malaysia (Bertram *et al.* 2020) with vagrants reaching Java (1984) and Sarawak (February–May 2004) (van Balen *et al.* 2013). We are unaware of any vagrant records of African Finfoot *Podica senegalensis,* although we infer that similar long-range dispersive movements are likely to occur, which might even take them to the Western Palearctic, on the Cape Verde islands or in Mauritania/Western Sahara.

ARAMIDAE
Limpkin

The sole member of the family Aramidae, Limpkin *Aramus guarauna* is nomadic, like many predominantly tropical or subtropical waterbirds, which allows it to respond to the dynamics of temporary wetland areas. There are strong seasonal biases in records of vagrant Limpkins in the eastern United States away from their breeding range, with these records all occurring between April and November. Vagrants have been recorded north and west as far as Ohio and Illinois. Limpkins have recently bred in both Georgia (Dobbs *et al.* 2019) and Louisiana (Marzolf *et al.* 2019), assisted by the concurrent establishment of the invasive Giant Apple Snail *Pomacea maculata*; evidence of the importance of such wandering vagrants in establishing new populations when the opportunity arises. Elsewhere, Limpkins are vagrants to some Caribbean islands away from breeding areas including Aruba (Mlodinow 2006) and the Cayman Islands (Kirwan *et al.* 2019). More exceptional was a first recent record from Chile at Huentelauquén Sur in September 2016 (Piñones & Bravo 2017) and an incredible record of a single photographed on the Rio Tel in southern Argentina in March 2018, at least 800km south-west of its normal breeding range.

PSOPHIIDAE
Trumpeters

Trumpeters are essentially rainforest understorey 'cranes' of the Amazon Basin and tend to run rather than fly if spooked. There are no extralimital records, and large Amazonian rivers are evidently effective barriers to dispersal with unique lineages having evolved in each interfluvial region (Ribas *et al.* 2011). Significant vagrancy is therefore not expected to occur.

GRUIDAE
Cranes

Several of the cranes of the Northern Hemisphere perform some of the most epic migrations of all large-bodied birds, crossing high mountain ranges and broad deserts and undertaking at least short sea crossings. In common with other waterbirds, including geese, swans and storks, the migratory routes of cranes are culturally transmitted rather than genetically inherited, so juveniles that become separated from the parents are vulnerable to getting lost (Chernetsov *et al.* 2004). Arguably the most spectacular instances of crane vagrancy concern the occurrence of transatlantic vagrancy of Sandhill Cranes *Antigone antigone* to Europe. Vagrancy is perhaps most likely to be the result of wind displacement (Riddiford 1983); that first-summer birds were involved in two of the three British occurrences corroborates a wealth of evidence that young soaring migrants are less able to compensate for wind drift (e.g. Hake *et al.* 2003).

Cottridge & Vinicombe (1996) constructed a reverse great circle route for the passage of the 1991 Sandhill Crane, which appeared first on Shetland in the United Kingdom, based on extrapolation back from its final recorded destination in the Netherlands. This route crossed a potential staging area in Greenland

▲ Sandhill Crane *Antigone canadensis* with Hooded Cranes *Grus monacha*, Izumi Crane Observation Center, Kagoshima, Japan, 11 February 2016. This species is now wintering with increasing frequency in Japan (*Paul French*).

▲ Demoiselle Crane *Anthropoides virgo* with Hooded Cranes *Grus monacha*, Izumi Crane Observation Center, Kagoshima, Japan, 14 January 2020 (*Yann Muzika*).

from a 'jump-off' point in the Canadian Arctic. This route would indeed seem fairly likely, as there are now two records of Sandhill Crane from Greenland (Haas 2012). The reverse migration hypothesis is highly unlikely, however, given that cranes do not have an inbuilt migratory orientation, thus making a genetically controlled reverse migration impossible (Tacha *et al.* 1992; Chernetsov *et al.* 2004). Wind drift, or simply the adoption of a random direction by an inexperienced and isolated individual, would seem a more likely proximate cause. Large birds with a high wing-loading are relatively unlikely transatlantic vagrants, given their poor adaptation to sustained periods of flapping flight over water (e.g. Tacha *et al.* 1992; Bildstein *et al.* 2009), so a northern route via various island stepping stones seems most likely. That none of the European records involved juveniles is perhaps surprising; it may be that such young inexperienced individuals are less capable of surviving long journeys once separated from adults. First-year birds, in contrast, may be physiologically stronger, but still face the difficulty of migrating without direct assistance from adults, perhaps rendering them most likely to stray long distances from their normal range.

Sandhill Cranes are not restricted as breeders to North America, and there is an expanding population in north-east Siberia. These Siberian breeders undertake the longest migration of any crane, crossing the Bering Sea and continuing south to northern Mexico. The expansion in Siberia has been accompanied with an increase in vagrants further south in Asia in China, Japan and South Korea where vagrants have been recorded associating with flocks of five other crane species (Gao *et al.* 2019). A high proportion of all vagrant crane records have involved individuals becoming entrained with flocks of other crane species. In some instances this may have even been the mechanism of vagrancy, for example Common Cranes *Grus grus* following Sandhill Cranes from Siberia and ending up being detected on Sandhill staging areas on the North American Great Plains (Howell *et al.* 2014). On other occasions, it seems that wandering vagrant cranes have eventually joined up with flocks of other crane species due to their gregarious nature. Examples include many of the continental European records of Sandhill Cranes travelling with Common Cranes (de Juana & Garcia 2015), a Common Crane associating with Black Crowned-Crane *Balearica pavonina* in Chad (Demey 2017), and a Wattled Crane *Grus carunculatus* found with Grey Crowned-Crane *Balearica regulorum* in eastern Uganda (Kasozi & Byaruhanga 2009). Once entrained with cranes outside of their normal distribution, they may accompany the flocks for years; it is thought that most of the United States records of Common Crane relate to just a few returning individuals, for example. A similar pattern is evident in Europe, with a Sandhill Crane tracked moving with Common Cranes from wintering areas in Spain through central Europe to Scandinavia in two consecutive years (de Juana & Garcia 2015).

Most tropical crane species are non-migratory or move relatively short distances between breeding and non-breeding areas, although some vagrancy has been noted in such species. In addition to the aforementioned record of Wattled Crane in Uganda, this species has also been recorded as a vagrant from Guinea-Bissau (Hazevoet 1997), and Blue Cranes *Grus paradisea* have been reported as vagrants from Botswana, Lesotho, Swaziland and Zimbabwe. Understanding the pattern of natural vagrancy in cranes is also hampered by a relatively high frequency of escapes from captivity. This is especially the case for Demoiselle Crane *Anthropoides virgo,* which is a long-distance migrant but also commonly kept in captivity. Several western European countries have accepted records of this species, e.g. Spain (de Juana & Garcia 2015), which have been found associating with Common Cranes. The German rarities committee recently added this species to the national list based on an individual shot on Heligoland in 1837 (Krüger 2018), and the species has occurred as a vagrant to Kamchatka in Siberia, Korea, Japan, Taiwan and the Philippines. British records have thus far failed to impress the relevant records committees. One that generated considerable attention at Spurn, Yorkshire, in September 1993 was touted as potentially a wild bird but subsequently became hand-tame after being relocated at a site in Essex (Roadhouse 2016).

A better candidate for a national first was the Demoiselle Crane first found with Sandhill Cranes in California in September 2001, which overwintered and was subsequently seen on northbound migration with the Sandhills in British Columbia and then Alaska the following May; it was ultimately not accepted as a first for the United States or Canada (Howell *et al.* 2014). A subsequent record of a Hooded Crane associating with Sandhill Cranes in multiple states on the Great Plains in two consecutive winters was also not accepted by the national records committee (Pranty *et al.* 2014), although this species has a track record of some vagrancy including westward to Bhutan (Oatman 2015). Association with Sandhill Cranes is obviously a pointer towards natural vagrancy, but escaped cranes also seek out heterospecifics and escaped individuals of both Grey Crowned Cranes and Sarus Cranes *Antigone antigone* have been found associating with Sandhill Cranes in Florida. Cranes remain a difficult species for rarity committees!

CHIONIDAE
Sheathbills

◀ Snowy Sheathbill *Chionis albus*, Parque Nacional da Lagoa do Peixe, Tavares, Rio Grande do Sul, Brazil, 25 April 2015. This species is seemingly a fairly regular vagrant to Brazil (*Paulo Ricardo Fenalti*).

Of the two members of the Chionidae, Black-faced Sheathbills *Chionis minor* are sedentary and have never been recorded away from their subantarctic island homes in the southern Indian Ocean, whilst Snowy Sheathbills *Chionis albus* are largely migratory, with most leaving subantarctic breeding areas and migrating to southern South America. Most winter around the Falklands, in Tierra del Fuego and southern Patagonia but are regular north to Uruguay. Vagrants have been recorded up the Brazilian coast from Rio Grande do Sul, which has multiple records, north to Rio de Janeiro (Fonseca *et al.* 2000); the Arquipélago de Abrolhos in Bahia (Intera-minense *et al.* 1996); and, amazingly, to Recife, Pernambuco, in July 1993 (Telino Júnior *et al.* 2001), which is only eight degrees south of the Equator.

Ship assistance is suspected to have been responsible for many, if not all, of the records beyond South American temperate latitudes, as the species is not averse to settling on ships even on its regular migration route from Antarctica to South America (de Brum *et al.* 2018) and is a fairly regular visitor to South Africa's Cape Region. Ship assistance was suspected for one that appeared on St Helena in May 1968 (Loveridge 1969), whose discovery followed the arrival of a Russian tanker, which had passed through the Antarctic. The most celebrated of such stowaways concerned an individual that arrived in Plymouth on the English south coast in September 1982 aboard the Royal Fleet Auxilliary vessel *Pearleaf*, returning from the Falklands conflict. The bird was apparently not held captive during the stay but was probably frequently fed and was forcibly ejected on arrival in the United Kingdom, where Chief Officer R.S. Jay decided "that its unsavoury habits were not conducive to good order and discipline, and I personally shooed it ashore, although it took some time to get the message" (Jay 1993). This is not the only northern European record of this species, however: one was shot near Carlingford lighthouse, in County Down, Northern Ireland, in December 1892 (Ussher & Warren 1900). More recently, in June 2017 a juvenile Snowy Sheathbill was discovered near the International Port of Kaohsiung in Taiwan, weighing just 292g – it had lost almost half its expected bodyweight and died in care (Lin *et al.* 2018). It seems that Snowy Sheathbills could conceivably be recorded as stowaways to almost anywhere.

PLUVIANELLIDAE
Magellanic Plover

Magellanic Plovers *Pluvianellus socialis* are partial migrants, with some birds moving north or towards the coast in the austral winter. The species is regarded as a vagrant or rare winter visitor to the Beagle Channel region of Argentina (Couve & Vidal 2003) and there is a hypothetical record of a vagrant on the Falklands (Woods 2017).

BURHINIDAE
Stone-curlews, Thick-knees

Most species of thick-knees are rather sedentary, with the exception of the Eurasian Stone-curlew *Burhinus oedicnemus,* which is a long-distance migrant from Eurasia to northern and central Africa. Consequently, most instances of long-distance vagrancy in this group involve this species. These records include considerable overwater movements to the Seychelles (Safford & Hawkins 2013) and Atlantic islands from Príncipe to the Azores and Iceland (Snow & Perrins 1998), suggesting that transatlantic vagrancy is not impossible. In the remaining species, vagrancy tends to be over considerably shorter distances, such as the single record of Senegal Thick-knee *Burhinus senegalensis* from Israel in July 2015 (Harrison 2016) and July–August 2020; this species breeds as close as the Nile Delta, Egypt. Also, the sole record of Double-striped Thick-knee *Burhinus bistriatus* from the United States was in Texas in December 1961 (Howell *et al.* 2014), this species being regular a few hundred kilometres south of the United States–Mexico border. Double-striped Thick-knees have a track record of wandering in the Caribbean region and have even bred extralimitally on Great Inagua in the Bahamas (Kirwan *et al.* 2019). Elsewhere, Water Thick-knee *Burhinus vermiculatus* is considered a vagrant to Senegal (Borrow & Demey 2001), whilst Great Thick-knee *Esacus recurvirostris* is a regular vagrant to the Arabian Peninsula.

PLUVIANIDAE
Egyptian Plover

In response to monsoon rains Egyptian Plovers *Pluvianus aegyptius* undertake intratropical migrations, which may regularly be of the order of 600–800km (Maclean & Kirwan 2020). The species historically bred north to the Nile Delta, where it was last documented in Egypt in 1937 (Snow & Perrins 1998). This extinct population may have supplied records of vagrants to Jordan and Israel in the 19th century (although these are poorly documented and have not been accepted by all authorities, Shirihai 1996). Similar questions remain over 19th-century records from Tenerife on the Canary Islands (Bannerman 1919) and in Spain (de Juana & Garcia 2015), but a record from Libya as recently as August 1969 is apparently regarded as genuine (Snow & Perrins 1998). More recent records from Poland, France and Belgium have all been suspected or proven to be escapes, although one in south-east Spain in May–June 2008 did generate some interest among birders; it was, however, ultimately rejected by the national rarities committee (Dies *et al.* 2010).

RECURVIROSTRIDAE
Stilts and Avocets

The stilts and avocets are a globally distributed family of elegant waders that exhibit extensive variation in migratory behaviour both within and between species. A tendency towards vagrancy is predominantly associated with north temperate zone species. Pied Avocets *Recurvirostra avosetta* have a huge breeding range in Eurasia, where they are mostly migratory, especially in the east, and in Africa where they are partially nomadic. Vagrants have reached several Atlantic islands including Iceland, the Faroes, Azores and Madeira as well as the Seychelles and Madagascar, and parts of South-east Asia including Thailand, Vietnam, Borneo and the Philippines (Pierce & Kirwan 2020). Vagrancy in American Avocet *Recurvirostra americana* has been recorded primarily to the south of the species' range, with birds reaching as far as Colombia (Verhelst *et al.* 2018) and throughout the Caribbean region (e.g. Kirwan *et al.* 2019). Spring overshoots have reached as far north as Alaska (Gibson & Kessel 1992) and there is an amazing record of one collected in western Greenland in November 1937 (Boertmann 1994), which hints at the possibly of vagrancy to the Western Palearctic, with the Azores perhaps the most likely location. The single record from Hawaii (Pyle & Pyle 2017) indicates that long overwater flights are not an insurmountable barrier for this species. Red-necked Avocet *Recurvirostra novaehollandiae,* which is essentially restricted to Australia, has also occurred as a vagrant north to Irian Jaya (Bostock 2000) and south and east to Tasmania and New Zealand.

Among the stilts, Black-winged Stilt *Himantopus himantopus* from the Old World has been most frequently recorded as a vagrant and is a regular spring overshoot in northern Europe. There, a combination of climate change – with droughts leading to greater dispersal distances (Figuerola 2007) – and habitat creation are leading to frequent extralimital breeding, which probably represents the early stages of colonisation. Black-winged Stilts have wandered north to Scandinavia and Iceland and one on Fair Isle in October 2019 was the first for Shetland and the 393rd species for the island. Vagrants have reached several other Atlantic islands including the Azores and Madeira and there is one record of transatlantic

► American Avocet *Recurvirostra americana*, Montezuma NWR, Seneca, New York, United States, 17 May 2018, a vagrant to northeastern North America (*Jay McGowan*).

vagrancy: two on Guadeloupe in August 2014 (Kirwan *et al.* 2019). Vagrancy to the eastern United States seems likely in the fullness of time, but the species has already made it on to the national list courtesy of three spring records from Alaskan islands (Howell *et al.* 2014), in addition to records from the Hawaiian archipelago (Pyle & Pyle 2017). Black-necked Stilt *Himantopus mexicanus* has not been recorded outside of the Americas with the geographic exception of Hawaii, where there is an endemic breeding subspecies, *knudseni*. Overshooting birds have been recorded north and east to the Atlantic Canadian provinces and the southernmost records concern three sightings from the Falklands (Woods 2017).

IBIDORHYNCHIDAE
Ibisbill

Some Ibisbills *Ibidorhyncha struthersii* are altitudinal migrants, but apart from some movement into northern Myanmar in the winter, most birds remain fairly close to their breeding grounds all year and no vagrancy has been reported to our knowledge.

HAEMATOPODIDAE
Oystercatchers

Most oystercatchers are sedentary, with the exception of Eurasian Oystercatcher *Haematopus ostralegus*, which migrates over relatively short distances, with some populations undertaking long-distance overwater movements such as between Scotland and Iceland. Oystercatchers are noisy and easy to detect, which has meant that we have a good idea of their history of extralimital occurrences. Vagrancy has been recorded in all four subspecies of European Oystercatcher. For example, *longipes*, which winters coastally from East Africa to India, has been recorded from various islands in the Indian Ocean, including several from

◄ Eurasian Oystercatcher *Haematopus ostralegus*, probably of subspecies *longipes* ('Siberian Oystercatcher'), Ribeira Grande, São Miguel, Azores, Portugal, 18 October 2013, potentially the first European record of this form (*Josh Jones*).

the Seychelles, but also remarkably from the Azores in 2012–13 (van den Berg & Haas 2013a, Hockey *et al.* 2020). The nominate subspecies of the Western Palearctic has wandered widely, having occurred on Atlantic islands from remote Ascension though the Cape Verdes, Azores and Madeira, north all the way to Spitsbergen with a few having straggled across the Atlantic to Greenland and Newfoundland. Other oystercatchers are less well travelled, although notable records include Blackish Oystercatcher *Haematopus ater* on the Juan Fernández Islands, 800km off the Chilean coast, and in south-west Ecuador (Félix & Haase 2016); and African Oystercatcher *Haematopus moquini* from Angola (Simmons *et al.* 2009) and more incredibly The Gambia, a record recently confirmed by DNA analysis of a historic specimen (Senfield *et al.* 2019). In the Americas, both American Oystercatcher *Haematopus palliatus* and Black Oystercatcher *Haematopus bachmani* are migratory in the north of their ranges and such individuals are perhaps responsible for records of Black Oystercatcher in the Pribilof Islands and even north-east Siberia (Brazil 2009), and the scattering of inland records of American Oystercatchers in North America. Elsewhere, Pied Oystercatcher *Haematopus longirostris* and Sooty Oystercatcher *Haematopus fuliginosus* of Australasia have both wandered to Indonesia (Mason 1997, Gilfedder *et al.* 2011), and South Island Oystercatcher *Haematopus finschi* from New Zealand has reached Australia and New Caledonia (Barré & Bachy 2003).

CHARADRIIDAE
Plovers and Lapwings

The plovers range from the extremely sedentary island endemic St Helena Plover *Charadrius sanctaehelenae* to intercontinental migrants like Common Ringed Plover *Charadrius hiaticula*. Among them, the 'tundra plovers' in the genus *Pluvialis* are among the most vagrancy-prone of all birds. It is possible, for instance, to encounter migrant or wintering Grey Plovers *Pluvialis squatarola* in almost any suitable coastal habitat on Earth. Ascribing individuals as vagrants is usually only done for the most remote oceanic islands or inland locations, or in the gap in its otherwise almost circumpolar breeding distribution in Greenland and Spitsbergen. The remaining three tundra plovers, the three 'golden plovers' each have discrete high-latitude breeding ranges; European Golden Plover *Pluvialis apricaria* is rather the odd one out of the three as it is not a trans-equatorial migrant and consequently exhibits a narrower spread of vagrancy. However, this species' wanderings have stretched west to Alaska twice, to the eastern seaboard of North America with some regularity (Howell *et al.* 2014), and east all the way to India (Abhinav & Dhadwal 2014), China

► Caspian Plover *Charadrius asiaticus* with White-rumped Sandpiper *Calidris fuscicollis*, Reservoir area, Corvo, Azores, Portugal, 19 October 2012, the first record of the former for the Azores and the westernmost record in the Western Palearctic (*Vincent Legrand*).

(Liu *et al.* 2013) and Japan (Miyajima *et al.* 2012, Nakamura *et al.* 2013). The spread of the remaining two species, American Golden Plover *Pluvialis dominica* and Pacific Golden Plover *Pluvialis fulva* is even more impressive. The former is a routine transatlantic vagrant to north-west Europe, with up to 39 in a single year in Britain (White & Kehoe 2016) and beyond Europe has occurred in 10 western African countries (Schmaljohann & Thoma 2005), Turkey (Kirwan 1994), Israel, Oman (Grieve *et al.* 2005), India, Hong Kong and Japan (Brazil 2009), south all the way to Australia and New Zealand (Menkhorst *et al.* 2017). Pacific Golden Plover is a vagrant to the eastern United States, most of Europe, central and western Africa, including Angola (Mills 2015), and inland to Uganda and Zimbabwe. Intriguingly, Jukema & Piersma (2002) argued that the species might have been a regular visitor and candidate pseudo-vagrant to the northern Netherlands near the Zuiderzee before the area was reclaimed. Pacific Golden Plovers winter across East Africa, Asia and the Pacific islands, with a small number wintering in southern California; a lost European wintering population does not seem beyond the realms of possibility.

► Semipalmated Plover *Charadrius semipalmatus*, Demco Beach, Broome, Western Australia, Australia, 16 November 2010 (*Adrian Boyle*).

Some smaller *Charadrius* plovers are also ultra long-distance migrants. Common Ringed Plover *Charadrius hiaticula* in the Old World and Semipalmated Plover *Charadrius semipalmatus* in the New World have both occurred as vagrants to their opposing hemispheres, but both are likely under-recorded given difficulties of field identification. Where both species are vagrants, e.g. in Australasia or on remote islands in both the Atlantic and the Pacific, records may be more frequent as any 'ringed plover' is worth checking.

Vagrancy in most plovers, indeed in most shorebirds, tends to be biased towards early to mid-autumn, when more naive juveniles are performing their first migration and stronger storm systems are in play. Spring vagrancy tends to be rarer and might often involve reorientating individuals. An exception to this rule is the pattern of Killdeer *Charadrius vociferous* vagrancy to Europe, which tends to be concentrated in late autumn and winter (Cottridge & Vinicombe 1996) and may sometimes be associated with cold-weather movements. Another group of globetrotting *Charadrius* plovers with a long history of vagrancy is the sand plovers and their allies, a group that predominantly breeds in the Eastern Palearctic but whose wintering areas extend from East Africa to Australasia. Both Greater Sand Plover *Charadrius leschenaultii* and Lesser Sand Plover *Charadrius mongolus* broadly overlap in the breeding and wintering areas and share a similar pattern of vagrancy to Europe, and southern and western Africa. More impressively, Lesser Sand Plover has reached North America with some regularity, with records in the United States, Canada and Hawaii (Howell *et al.* 2014), and even Argentina in South America (Le Nevé & Manzione 2011). Greater Sand Plover has reached the United States twice (Howell *et al.* 2014) and a controversial sand plover photographed in southern Brazil may also be this species (Franz *et al.* 2018). Two other large *Charadrius* plovers are similar to the sand plovers in breeding in central and eastern Asia but differ in wintering in very disjunct areas: Caspian Plover *Charadrius asiaticus* in eastern and southern Africa and Oriental Plover *Charadrius veredus* in northern Australasia. Consequently, both are rarer as extralimital vagrants and yet those vagrants that have been recorded are widely scattered, perhaps owing to their capacity for long-haul flights without stopping between breeding and wintering grounds, which means that misorientated individuals can travel vast distances. Oriental Plover has been recorded from Greenland (Boertmann 1994), the Netherlands, Norway, Finland (Haas 2012), Kazakhstan, the Seychelles, Japan (Brazil 2009) and New Zealand, while Caspian Plover has wandered to many western European countries, out into the

▲ Common Ringed Plover *Charadrius hiaticula*, Semipalmated Plover *Charadrius semipalmatus* and Least Sandpiper *Calidris minutilla*, Point Reyes, Marin, California, United States. This was the second state record of the first species (*Steve Howell*).

Atlantic on the Azores, to West Africa, the Seychelles, India, Sri Lanka and even northern Australia.

The remaining plovers include many southern temperate and tropical species that do not typically migrate far, and consequently there is far less inter-continental vagrancy. Most extralimital records in these species are hundreds, rather than thousands of kilometres from their regular range. Examples include the largely non-migratory Collared Plover *Charadrius collaris* in Texas, United States (Howell *et al.* 2014), and the West Indies (Kirwan *et al.* 2019), the austral migrant Rufous-chested Plover *Charadrius modestus* overshooting to Peru (Schmitt & Pariset 2005), south-east Brazil in São Paulo (Cestari 2008) and Rio de Janeiro (Simpson & Simpson 2011a), and the regular appearance of Kentish Plover *Charadrius alexandrinus* north of its regular range in Europe. Some records have been regarded as of uncertain provenance. For example, an old record from India of the Australasian Black-fronted Dotterel *Elseyornis melanops* has not been met with widespread acceptance, but the species was recently recorded for the first time in Indonesia (Ericsson 2015), indicating that long-range vagrancy in this species is not impossible. Even the range-restricted Wrybill *Anarhynchus frontalis* has been known to wander 800km from New Zealand to the Chatham Islands (Bell & Bell 2000). More unusual in this context is Kittlitz's Plover *Charadrius pecuarius*, which is generally considered non-migratory but has been recorded widely as a vagrant in North Africa, the Middle East and southern Europe (Spain, Greece and Cyprus) with an exceptional specimen record from Norway (Snow & Perrins 1998). This species is kept in captivity, but these extralimital records are widely treated as genuine and a repeat occurrence in northern Europe seems likely in the fullness of time. The genuine nature of these long-distance events has been reinforced by molecular studies; it is now known that prehistoric vagrancy in Kittlitz's Plovers to Madagascar and St Helena – oceanic islands on opposite sides of Africa – led to the evolution of Madagascar Plover *Charadrius thoracicus* and St Helena Plover respectively (Dos Remedios *et al.* 2018). Vagrant Killdeers have bred in recent years on the Azores (van den Berg & Haas 2010), and colonisation of Europe by this adaptable species would not seem an ecological impossibility.

The lapwings in the subfamily Vanellinae include a suite of largely sedentary species, many of which are found in the tropics and for which vagrancy is either unknown or relatively discrete. There are, however, a few species that penetrate further north into temperate latitudes where they have evolved strongly migratory behaviour. Arguably the most celebrated

▲ Three-banded Plover *Charadrius tricollaris*, Ma'ayan Tzvi Fish Ponds, Haifa, Israel, 24 April 2020, the first record for Israel (*Yoav Perlman*).

is Northern Lapwing *Vanellus vanellus,* with its broad Palearctic breeding distribution stretching from Ireland to Japan. This species typically moves from northern temperate to southern temperate latitudes in winter. Transatlantic vagrancy to the northeastern United States and eastern Canada in this species is particularly associated with cold weather 'escape movements' aided by strong easterly winds (Howell *et al.* 2014). Elsewhere, the species has occurred widely as a vagrant in the High Arctic on Spitsbergen and Jan Mayen and in Alaska (Howell *et al.* 2014); widely in the Caribbean (Kirwan *et al.* 2019); south in Africa to Senegambia (Borrow & Demey 2001) and Kenya (Jackson 1997); and south-east to the Philippines (Jensen *et al.* 2015). Grey-headed Lapwing *Vanellus cinereus* is an even longer-distance migrant, albeit much more range-restricted, and has been widely recorded as a vagrant from locations as widely spaced as Oman (Harrison & Grieve 2012), the Philippines (Allen *et al.* 2006) and Australia (Clarke *et al.* 2008). This species had been predicted as a vagrant to the Western Palearctic and the first record, from Turkey, came in March 2018, followed by a second individual tracked through Norway, Sweden and the Netherlands in May–June 2019. Several species of lapwings are kept in captivity, but unless escapes are known to be at large, this species should be treated as a likely vagrant across the Palearctic.

Even more enigmatic is Sociable Lapwing *Vanellus gregarius,* a Critically Endangered species that breeds on the steppes of southern Russia and Kazakhstan and migrates to disjunct wintering areas in the Indian subcontinent, Middle East and North-east Africa. Surprisingly given its relative rarity, vagrancy is routine – the species has been recorded in most European countries, albeit with declining frequency, and as far afield as China (Brazil 2009), Sri Lanka and the Maldives (Rasmussen & Anderton 2005), the Seychelles (Skerrett 2003), Cameroon (Messemaker 2004) and the Canary Islands. The species is an annual visitor in small numbers to Iberia (de Juana 2011) which may represent a small disjunct pseudo-vagrant wintering population, or perhaps indicate the magnetic draw of westbound Northern Lapwing flocks that may entice wandering Sociable Lapwings away from their regular wintering grounds. Breeding further south than Sociable Lapwing, but overlapping in some wintering areas, White-tailed Lapwing *Vanellus leucurus* has been widely recorded as a vagrant to western Europe, most of which have occurred in a series of influxes, e.g. in 1975, 1984 and 2000, probably associated with drought events. This species has recently bred extralimitally in Europe in Romania (Kiss *et al.* 2001). It has also occurred east to China (Ding & Ma 2012).

Vagrancy in other species of lapwings is more modest and often confounded by uncertainty concerning the risk of escapes from captivity. This is illustrated well by Southern Lapwing *Vanellus chilensis,* for which a record from the United States from Florida and Maryland in June 2006 has been mooted as genuine (Howell *et al.* 2014) but not accepted by the national committee, given the occurrence of known escapes in the past. However, this species is undergoing a rapid population expansion following deforestation in Central America, has recently been recorded for the first time in Mexico (Martin 1997) and is expanding its range in the West Indies (Kirwan *et al.* 2019). Vagrancy

◀ White-tailed Lapwing *Vanellus leucurus*, Seaforth LWT, Lancashire, England, 28 May 2010. This, the seventh record for Britain, visited Essex/Greater London, Gloucestershire, Lancashire and north Merseyside (*Stuart Piner*).

to the Juan Fernández and Desventurada Islands off Chile (Jaramillo 2003) and the Falklands (Woods 2017) indicates that substantial overwater movements are not an issue for dispersing individuals. Understanding the provenance of extralimital Spur-winged Lapwings *Vanellus spinosus* in northern Europe has also proved problematic; the individuals breeding in southeastern Europe are migratory and perhaps the source of most records from elsewhere in southern Europe, from Spain to Romania. However, records from northern Europe have all either been known escapes or regarded as being of suspect origin, despite the strong potential for natural vagrancy. A similar question mark hangs over the Red-wattled Lapwing *Vanellus indicus* that toured western Europe from Croatia through Germany to France in May–June 2019. Extralimital movements in the remaining, largely tropical species is rare but includes a single old record of Black-headed Lapwing *Vanellus tectus* from the Israel/Jordan border in 1869 (Haas 2012) and an amazing history of wandering to oceanic islands by Blacksmith Lapwing *Vanellus armatus*. This species has been recorded from St Helena (Rowlands *et al.* 1998), Prince Edward Island (Cooper & Underhill 2002) and the Crozets (Stahl *et al.* 1984), which perhaps opens the possibility of transatlantic vagrancy to South America.

PEDIONOMIDAE
Plains-wanderer

Despite its name, Plains-wanderer *Pedionomus torquatus* is considered to be sedentary, based on data from tracking and ringing studies (Baker-Gabb *et al.* 1990), although this species can be very difficult to detect.

THINOCORIDAE
Seedsnipes

Of the four species of seedsnipes, Least Seedsnipe *Thinocorus rumicivorus* shows the strongest migratory tendencies, withdrawing from the southern limits of its distribution during the harsh austral winter. Some of these partial migrants presumably overshoot their regular wintering areas, which lie as far north as Buenos Aires in Argentina and in southern Uruguay, and may reach southern Brazil between March and August. Vagrants probably occur annually in Rio Grande do Sul, with one making it as far north as São Paulo (Castro *et al.* 2012). This species occurs as a regular vagrant or scarce migrant to the Falklands in both spring and autumn (Woods 2017). There is even a record from Antarctica on the South Shetland Islands in December 1996, which was presumably an over-shooting spring migrant (Favero & Silva 1999). Some White-bellied Seedsnipe *Attagis malouinus* undertake altitudinal migrations and such movements may have underpinned vagrancy in this species on several occasions to the Falklands (Woods 2017).

ROSTRATULIDAE
Painted-snipes

Of the three species of painted-snipes, only Greater Painted-snipe *Rostratula benghalensis* is thought to undertake any migratory activity, and this is restricted to partial migrations or rains migrations in some populations. This species has been widely recorded as a vagrant, mostly at subtropical latitudes, with records from, for example, Morocco (Thévenot *et al.* 2003), Iran (Scott 2008), the Seychelles (Skerrett 2017), the Andaman Islands (Gokulakrishnan *et al.* 2018) and north to Ussuriland in Russia (Brazil 2009). This hints at the possibility of vagrancy to Europe. Australian Painted-snipe *Rostratula australis* is largely resident but has occurred as a vagrant to Tasmania and Western Australia (del Hoyo *et al.* 2020a) and the similarly sedentary South American Painted-snipe *Nycticryphes semicollaris* has been recently documented in a number of areas remote from known breeding areas in Brazil, although these may yet prove to be range extensions rather than vagrants. However, a single bird recorded in November 2006 on the Falklands was an unquestionable vagrant at least 900km out of range (Woods 2017).

JACANIDAE
Jacanas

Most jacanas do not make predictable annual to-and-fro migrations, but most species may make local movements that may extend for hundreds of kilometres to occupy seasonal wetlands. Vagrancy to the southern United States by Northern Jacanas *Jacana spinosa* is a good example, and the species even temporarily colonised Texas between 1967–78, occurring subsequently as a vagrant to Texas and Arizona (Howell *et al.* 2014). Wattled Jacana *Jacana jacana* has wandered to Chile, necessitating a movement over the Andes (Jaramillo *et al.* 2003). Reported instances of vagrancy in other species tend to be over short distances, such as Bronze-winged Jacana *Metopidius indicus* from Bhutan, which lies adjacent to source populations of this species (Spierenburg 2005). The exception to the jacanas' non-migratory rule is Pheasant-tailed Jacana *Hydrophasianus chirurgus,* whose northern populations are entirely migratory although the destination of these migrants is unknown. Pheasant-tailed Jacanas have been recorded as vagrants both to the west – in Iran (Abbasi *et al.* 2019), the UAE (Smiles 2014), Oman and Socotra (Porter & Aspinall 2010), for example – and to the east, in southern Japan, South Korea (Brazil 2009) and Bali, Indonesia (Mason 2011). Vagrancy to the south-east of the Western Palearctic seems a likely future possibility.

▲ Pheasant-tailed Jacana *Hydrophasianus chirurgus*, The Wave Muscat, Masqat, Oman, 1 July 2017, a very scarce visitor to northern Oman (*Jens Eriksen*).

SCOLOPACIDAE
Sandpipers and allies

▲ Grey-tailed Tattler *Tringa brevipes* (centre) with Wandering Tattlers *Tringa incana*, Midway, northwestern Hawaiian Islands, 26 February 2012, the eighth record for Hawaii (*Cameron Rutt*).

These shorebirds make up a genuinely global group that performs some of the longest and most spectacular migrations of all birds. Consequentially a high proportion of shorebird species, particularly those breeding at high northern latitudes, tick the two key boxes for vagrancy likelihood: large population sizes and long annual migrations. Their broad association with wetland habitats also predisposes them to occur at well-watched locations, increasing the probability that any given stray individual will be detected by birders. Almost all species, with the exception of a few insular woodcocks *Scolopax* spp., highland snipes *Gallinago* spp. and the island endemic Tuamotu Sandpiper *Prosobonia parvirostris* have thus been recorded with some regularity as vagrants. Sanderling *Calidris alba* is, for example, one of the northernmost breeding birds in the world and although this species has quite high fidelity to both stopover sites and wintering locations, it is conceivably a visitor to sandy ocean beaches almost anywhere on the planet, including the remotest oceanic islands. Nominal vagrants have occurred inland in the middle of all continents except Antarctica. Sanderlings breed throughout the High Arctic, but many other shorebird species have more restricted breeding and wintering ranges. For example, White-rumped Sandpiper *Calidris fuscicollis* only breeds in Canada and Alaska and migrates south through the Americas to winter principally in southeastern South America. This species is a regular vagrant to western Europe, principally in autumn, sometimes occurring in flocks displaced by fast-moving transatlantic depressions. Beyond Europe this species is nowhere regular as a vagrant but there are records from such scattered places as the Ivory Coast (Demey & Fishpool 1991), Mozambique (Allport 2020), Turkey (Browne 1997), the United Arab Emirates (Campbell & O'Mahoney 2013), Azerbaijan (Himmel 2019), China (Wu *et al.* 2015), Japan (Senzaki 2014), Australia and New Zealand (Menkhorst *et al.* 2017) and Hawaii (Pyle & Pyle 2017). White-rumped Sandpiper is the only regular shorebird vagrant to Antarctica (Korczak-Abshire *et al.* 2011a) although both Upland Sandpiper *Bartramia longicauda* and Hudsonian Godwit *Limosa haemastica* have occurred in this region at least twice (Juáres *et al.* 2010, Petersen *et al.* 2015).

White-rumped Sandpiper's occurrence pattern of regular vagrancy to north-west Europe and occasional records from other corners of the planet is shared by other Nearctic waders such as Semipalmated

◀ White-rumped Sandpiper *Calidris fuscicollis* with Curlew Sandpipers *Calidris ferruginea*, Macaneta, Maputo, Mozambique, 22 September 2018. This was the 28th record for southern Africa (*Gary Allport*).

Sandpiper *Calidris pusilla*, Baird's Sandpiper *Calidris bairdii* and Lesser Yellowlegs *Tringa flavipes* (Cramp *et al.* 1983). All these species often take a transoceanic 'short cut' across the western North Atlantic, flying from staging areas in northeastern North America to destinations in the Caribbean or northern South America in a single hop. This route involves a non-stop flight of at least 2,000km over water, requiring individuals to accumulate significant fat reserves prior to departure. This means that in addition to being vulnerable to drift in east-moving low pressure systems, birds following this transoceanic route are also likely to be carrying large fuel reserves, increasing their likelihood of surviving a transatlantic crossing. Semipalmated Sandpipers on autumn migration are known to carry fat reserves sufficient for continuous flights of at least 3,200km (Dunn *et al.* 1988), whilst White-rumped Sandpipers may be capable of flying up to 4,200km (McNeil & Cadieux 1972). By comparison, most exclusively Eurasian waders like Black-tailed Godwit *Limosa limosa*, Common Redshank *Tringa totanus* and Eurasian Curlew *Numenius arquata* do not undertake transatlantic flights and remain very rare in eastern North America (Howell *et al.* 2014). Britain has hosted 54 species in the family Scolopacidae, of which 18 are almost exclusively North American breeders, versus 14 of 59 species in Japan, 12 of 44 in New Zealand and five from 36 in South Africa.

◀ Eurasian Curlew *Numenius arquata* (right) with Far Eastern Curlew *Numenius madagascariensis*, Point Douro, Harvey, Western Australia, Australia, 4 March 2016, the third record for Australia (*Rohan Clarke*).

The migratory strategy of Hudsonian Godwit ties in neatly with other regular transatlantic vagrants, with almost the entire population making a long transoceanic autumn crossing from a staging area around Hudson Bay to northern South America in a single hop (Morrison 1984). The fat reserves carried by individuals making this journey should easily allow them to reach western Europe, and the route taken would put individuals firmly in the path of east bound Atlantic weather systems. Despite these factors, records are extremely rare in Europe; the fact that there is only a single record from the Azores underlines the species' low propensity for eastward vagrancy, given that those islands are not far removed from the actual migration route. The relatively small world population of Hudsonian Godwits (70,000 individuals; Delaney & Scott 2006) might account partially for their rarity as eastbound vagrants, although as discussed above, small population sizes do not necessarily preclude regular vagrancy.

Bizarrely, whilst Hudsonian Godwits apparently almost never stray to the eastern side of the Atlantic, they occur with remarkable regularity in New Zealand, where records are almost annual and sometimes involve small parties (Higgins & Davies 1996). Obviously, wind drift from the normal transatlantic migration route could not possibly account for these hugely extralimital records. The only plausible explanation is that Hudsonian Godwits from the Alaskan breeding population occasionally get caught up with local flocks of Bar-tailed Godwits *Limosa lapponica*, eventually following them on their epic transpacific migration to New Zealand (Gill *et al.* 2008). This is perhaps the most obvious and dramatic example of vagrancy caused by association with 'carrier-species', a mechanism that may well be important in driving other patterns of wader vagrancy. The lack of suitable carrier species to bring Hudsonian Godwits eastwards from their breeding grounds across the Atlantic therefore provides a further partial explanation for their relative rarity in Europe. Nevertheless, the apparent lack of wind-drifted individuals in autumn remains surprising given their primary migration route. One possibility is that Hudsonian Godwits respond differently to wind drift than our more regular wader vagrants. Given their large size and strong flight capacity, it is possible that godwits may be less prone to adopting the 'adaptive drift' strategy, preferring to battle into crosswinds rather than allowing themselves to be drifted long distances. The fact that godwits undertake such huge transoceanic migrations over the Pacific could support this hypothesis: a non-stop flight of 11,000km will inevitably involve passing through multiple weather systems with significant potential for crosswind drift. As such, godwits might have evolved a stronger capacity to compensate for drift in order to prevent themselves being displaced huge distances off course during their epic migrations.

In turn, the relative rarity of other abundant North American waders in Europe, such as Western Sandpiper *Calidris mauri*, can be explained by their near absence from the Atlantic seaboard transoceanic migration route (Butler *et al.* 1996). Whilst the broad east–west dichotomy of west–east transatlantic wader vagrancy can be explained by responses to wind drift, there are various specific cases that do not fit neatly

► Hudsonian Godwit *Limosa haemastica*, Inishmore, Galway, Ireland, 15 September 2015, the first record for Ireland; this individual was previously present at Inishdawros, Galway, in July 2015 (*Alexander Lees*).

◀ Lesser Yellowlegs *Tringa flavipes*, with Australasian Swamphens *Porphyrio melanotus*, Boyters Lane, Jerseyville, Kempsey, New South Wales, Australia, 19 January 2009. This individual returned to winter at the same site in three consecutive seasons (October 2007 to March 2008, October 2008 to March 2009 and September 2009 to January 2010) (*Adrian Boyle*).

into any expected pattern. Of particular interest are a number of North American species that occur extralimitally much more frequently than we might otherwise expect given their population sizes and migratory routes. Perhaps most perplexing is the case of Buff-breasted Sandpiper *Calidris subruficollis*; its principal autumn migration route passes through the interior of North America, to a relatively restricted wintering area in southeastern South America, and the species occurs only rarely on the northeastern seaboard of North America, the undoubted source area for most of our North American waders. However, the species is one of the most frequent of all North American waders in the Western Palearctic, sometimes occurring in flocks of over 20 individuals, despite having a relatively tiny and declining world population (Lanctot *et al.* 2002; Morrison *et al.* 2006). Beyond Europe this species is a truly global wanderer with records from 10 sub-Saharan African countries, including multiples from The Gambia (two), Ghana (four), Namibia (four), Senegal (nine), South Africa (12) and an amazing six records from the Seychelles (Donald *et al.* 2019). Further east they have reached India (Rajeevan & Thomas 2011), Indonesia (Hjerppe *et al.* 2015), Japan,

▼ Nordmann's Greenshank *Tringa guttifer* (second bird from the right) with Oriental Plovers *Charadrius veredus*, Bar-tailed Godwit *Limosa lapponica*, Little Whimbrel *Numenius minutus*, Great Knot *Calidris tenuirostris*, Common Greenshanks *Tringa nebularia* and White-winged Black Tern *Chlidonias leucopterus*, Eighty Mile Beach, Broome, Western Australia, Australia, 2 December 2006. This was the first confirmed record of Nordmann's Greenshank for Australia, although all subsequent records could pertain to the same returning individual (*Adrian Boyle*).

▲ Ruff *Calidris pugnax* with Southern Lapwing *Vanellus chilensis* and Lesser Yellowlegs *Tringa flavipes*, Campos de Quatipuru, Pará, Brazil, 5 October 2013, the second documented record for Brazil. There may be a small breeding population of Ruffs in the Canadian Arctic, which gives rise to wintering records further south in the Americas (*Alexander Lees*).

Korea, Taiwan (Brazil 2009), New Guinea (Pratt *et al.* 2015) and Australia and New Zealand (Miskelly *et al.* 2015).

Similarly, Wilson's Phalarope *Phalaropus tricolor* has a distinctly western breeding range in North America and a narrow central migration route passing through Mexico to wintering areas on the Pacific coast of South America. A high proportion of the world population makes a late summer stopover at Mono Lake in California to build up fat reserves, prior to making the journey to Ecuador in a single hop (Colwell & Jehl 1994). As a result, this species is relatively rare along the eastern seaboard of North America and is seemingly unlikely to get caught in eastbound Atlantic depressions. Nevertheless, it is a regular vagrant to Europe, being particularly abundant in the 1980s and 1990s when up to 18 occurred annually in the British Isles in addition to appearances in South Africa, Japan, Australia and New Zealand. The frequency of these two species in Europe is difficult to explain, but could potentially be explained by wind drift in the upper atmosphere – might Wilson's Phalaropes and Buff-breasted Sandpipers migrate at heights sufficient to bring them into contact with low-level jet stream winds?

Logically, it might be expected for Europe to get more vagrant shorebirds from the east than the west, given that vagrants from the east have few significant barriers to cross before reaching Europe, unlike North American species, which have the whole Atlantic ahead of them. Certainly, this is the pattern we find for other strongly migratory groups like the passerines, amongst which eastern vagrants greatly outnumber those from the west. The Eastern Palearctic is home to a wide range of wader species that have breeding distributions and migration routes similar to regularly visiting Siberian passerine vagrants. For example, Red-necked Stint *Calidris ruficollis*, Long-toed Stint *Calidris subminuta*, Sharp-tailed Sandpiper *Calidris acuminata* and Great Knot *Calidris tenuirostris* are all arguably well placed to occur as vagrants to the Western Palearctic, and yet they remain extremely rare or even unrecorded in most countries. By comparison, several are commoner in eastern North America than they are in Europe.

Asian waders are unlikely to be drifted towards Europe from the east: almost the entire journey will be over land, and consequently there will always be opportunities to settle and wait out unfavourable weather before continuing on the desired heading. As such, wind-drifted migrants are inevitably less likely to reach Europe from very far to the east. In many ways, this conclusion is relatively obvious – indeed, most previous authors on the subject have agreed that vagrancy from the far east is unlikely to be caused by wind drift as the distances are too great (Elkins 1988; Cottridge & Vinicombe 1996). Additionally, the rarity of species like Broad-billed Sandpiper *Calidris falcinellus* in the British Isles, despite being relatively common on the near-continent, underlines the apparently strong capacity for eastern waders to resist wind drift. As with passerines, it appears likely

◄ Pectoral Sandpiper *Calidris melanotos*, St Martin's, Isles of Scilly, England, 25 September 2008, an exceptionally confiding individual showing little innate fear of humans (*James Gilroy*).

that most vagrant waders arriving at our shores from the Eastern Palearctic are likely to be birds with deviant migratory orientations, or alternatively birds that simply 'follow the wrong flock', being brought along the wrong migration route by another carrier species. This is especially the case for species which may be expanding their ranges towards the Siberian migratory divide, as is apparently the case for Pectoral Sandpiper *Calidris melanotos,* and it may lead to the evolution of novel migration routes (Lees & Gilroy 2004, Hjort 2005).

There is another possibility, that some European records of Siberian shorebirds may have transited through North America. For example, most adult Sharp-tailed Sandpipers follow a direct southward route through coastal east Asia to reach wintering sites in Australasia, but juveniles adopt a much more circuitous and surprising strategy. After fledging, most juveniles move east or even north-east from breeding grounds to visit productive feeding sites in coastal north-east Russia, with a high proportion moving as far as Alaska (Mlodinow 2001). After accumulating large fat reserves, they then turn south-west to make a direct crossing of the Pacific, presumably making stopovers on Pacific islands such as Hawaii before eventually rejoining the adults on the austral wintering grounds (Tomkovich 1982; Mlodinow 2001). This migratory pattern also explains the relative frequency with which vagrant juvenile Sharp-tailed Sandpipers occur across North America, given the initial east-

◄ Common Greenshank *Tringa nebularia* (centre with white back) with Long-billed Dowitchers *Limnodromus scolopaceus* and Greater Yellowlegs *Tringa melanoleuca*, Edwin B. Forsythe NWR, Atlantic, New Jersey, United States, 23 October 2017. This was the first state record and one of only a few from eastern North America (*Tom Johnson*).

► Marsh Sandpiper *Tringa stagnatilis* and Willet *Tringa semipalmata*, Estero Punta Banda, Baja California, Mexico, 12 October 2011, the first for Mexico (*Steve Howell*).

► Great Snipe *Gallinago media*, Kilnsea, Yorkshire, England, 16 September 2013. This extremely confiding individual was later found dead, the suspected victim of a domestic cat (*Graham Catley*).

ward heading of birds leaving Siberia (Mlodinow 2001), and the corresponding rarity of juveniles as vagrants in Europe relative to adults. It further hints at the possibility that some British records, particularly juveniles occurring on western coasts, might have travelled all the way across North America and the Atlantic to reach the British Isles. The same may be true for some Red-necked Stints – this species and Sharp-tailed Sandpiper are the only shorebirds that occur regularly as vagrants both in the British Isles and along both coasts of North America (Mlodinow 2001). Despite their relative rarity in Europe several of these Siberian species have still been recorded widely elsewhere. For instance, Sharp-tailed Sandpiper has been recorded from Bolivia (Knowlton 2016), Oman (Eriksen & Victor 2013) and Mozambique (Allport 2018); Red-necked Stint from Peru (Schulenberg *et al.* 2007), Israel (Granit & Smith 2004) and Somalia (Ash & Miskell 1998); and both Great Knot (Cohen & Winter 2003) and Asian Dowitcher *Limnodromus semipalmatus* have reached South Africa (Deighton 2005).

Beyond the globe-trotting High Arctic species,

there are still some impressive feats of vagrancy on record for species usually considered to be shorter-distance migrants, such as the snipes and woodcocks. The notion of 'shorter distance' is, however, being challenged by satellite telemetry work. We now know, for example, that Eurasian Woodcocks *Scolopax rusticola* from Russia may migrate over 10,000km to winter in Spain (Arizaga *et al.* 2015). This species has occurred several times as a vagrant to eastern North America, Greenland and Spitsbergen (Howell *et al.* 2014) in addition to Cape Verde (Demey *et al.* 2012), Jordan (Andrew *et al.* 1999) and Bangladesh (Thompson & Johnson 2003). American Woodcock *Scolopax minor* has a more restricted range, and perhaps a shorter migration span, but has occurred at least once as a transatlantic vagrant to France (Haas 2012).

TURNICIDAE
Buttonquails

Buttonquails form an enigmatic family of species whose distribution and movements are poorly understood owing largely to their skulking nature. They may seem like unlikely migrants, but their long, pointed wings make them morphologically well adapted for long flights. Some species are partial migrants, especially at higher latitudes, and some may even be nomadic. Migrant Red-backed Buttonquails *Turnix maculosus* are known to cross the Torres Strait between Australia and New Guinea. Vagrancy has been reported for several species of buttonquail, although this comes with a caveat about whether the species might be a regular passage migrant, or even a scarce breeder in several of these areas. For instance, northern populations of Yellow-legged Buttonquail *Turnix tanki* are partially migratory, and the species has been considered a vagrant around Lake Baikal (Mlíkovský 2009) and in Bangladesh (Thompson *et al.* 2014) whilst Common Buttonquail *Turnix sylvaticus* is regarded as a vagrant to Saudi Arabia (Babbington & Roberts 2014) and Oman (Eriksen & Victor 2013).

▲ Yellow-legged Buttonquai *Turnix tanki* with Shore Lark *Eremophila alpestris*, Er La Pass, Qinghai, China, 1 July 2014, a migrant grounded at an altitude of 4,490m on passage (*James Eaton*).

DROMADIDAE
Crab-Plover

Migratory movements of Crab-Plover *Dromas ardeola* are rather poorly known but probably include impressive transoceanic movements across the Indian Ocean, as well as significant overland movements, with grounded migrants having been reported in the interior of Saudi Arabia. Vagrants have occurred east to peninsular Malaysia (Robson 2005), south to South Africa and north as far as Jordan, Israel and once to Turkey (Bouwman 1987, Kirwan *et al.* 2010).

GLAREOLIDAE
Coursers and Pratincoles

Most coursers are non-migratory, although like many desert birds they can be highly nomadic according to rainfall and hence conditions for foraging. An exception is the nominate race of Cream-coloured Courser *Cursorius cursor*, whose northern populations apparently cross the Sahara Desert in winter (Cramp & Reynolds 1972) and have occurred as an overshooting winter vagrant south to Kenya (Stevenson & Fanshawe 2002). This species has been recorded widely as a vagrant in Europe, north to Norway, Sweden and Finland; flocks in excess of 30 birds have been reported from Malta (Lewington *et al.* 1991).

Of the eight species of pratincoles, three species breeding in northern temperate latitudes are highly migratory and have been widely recorded as vagrants. The nominate race of Collared Pratincole *Glareola pratincola*, which moves from the Palearctic to the savannas of the Sahel, is arguably the most prodigious vagrant. Collared Pratincoles are relatively routine vagrants to northern Europe but have occurred as far afield in the Old World as the Chagos Islands (Carr 2015), Mongolia (Otgonbayar *et al.* 2017), China (Brazil 2009) and most recently Australia in February 2019 (eBird). Collared Pratincoles are also exceptional

► Cream-coloured Courser *Cursorius cursor*, St Mary's, Isles of Scilly, England, 30 September 2004. This bird was taken into care and died in captivity after spending a whole month on the islands (*Chris Batty*).

◀ Collared Pratincole *Glareola pratincola*, Caucaia, Ceará, Brazil, 21 April 2015, an amazing find in a soybean field in north-east Brazil and the first confirmed record from South America after a pratincole species photographed in the mid-Atlantic on Atol das Rocas, Brazil, in March 1990 (*Ciro Albano*).

in having crossed the Atlantic on multiple occasions: there are at least two records from the West Indies – from Guadeloupe and Barbados (Kirwan *et al.* 2019) – and a recent record of one found in a soybean field in the state of Ceará in north-east Brazil in April 2015 (Albano 2015). Oriental Pratincole *Glareola maldivarum* has also occurred as a vagrant twice to the New World, both times to the Bering Sea region in Alaska (Howell *et al.* 2014). This species migrates from southern Siberia to Australia and has been recorded with increasing frequency as a vagrant west as far as Europe, probably as identification awareness has improved. Vagrants have also reached Turkey (Kirwan *et al.* 2018), Egypt (Baha el Din & Baha el Din 1996) and Bhutan (Tobgay 2017), and records from Madagascar, Mauritius and the Seychelles (Safford *et al.* 2013) suggest that the species is overlooked as a vagrant to sub-Saharan Africa. Finally, Black-winged Pratincole *Glareola nordmanni*, which has a very long migration span of up to 10,000km from Eurasia to southern Africa, has been recorded with some regularity across the Western Palearctic (Snow & Perrins 1998). Interestingly, this species has been recorded more frequently in the Netherlands than Collared Pratincole, which breeds much closer in southern Europe, although the latter has a much smaller migration span (van den Berg & Bosman 1999). Elsewhere, as a vagrant there are relatively

◀ Black-winged Pratincole *Glareola nordmanni*, with Common Shelduck *Tadorna tadorna* and Eurasian Oystercatcher *Haematopus ostralegus* Titchwell Marsh RSPB Reserve, Norfolk, England, 5 June 2009. This exceptional nature reserve has hosted three species of vagrant pratincoles (*Alexander Lees*).

few records away from the Seychelles, which may lie close to its normal migration route (Safford *et al.* 2013). The species is very rare on passage in the Middle East and may entirely overfly the region in a 'long-jump' migration; this capacity for very long non-stop flights was recently evidenced by the appearance of a first-winter Black-winged Pratincole in a Subantarctic Fur Seal *Arctocephalus tropicalis* colony on Amsterdam Island in the southern Indian Ocean – 4,200km from the nearest South African coastline (Bigonneau *et al.* 2020). Given the aerial abilities of this group, it is perhaps surprisingly that there are not more extreme extralimital vagrant records of these three species.

The remaining pratincoles do not undertake such epic migrations, although northwestern populations of Little Pratincole *Glareola lactea* are migratory and the species is a vagrant to Yemen, the Arabian Gulf region and Oman (Porter & Aspinall 2010). Future records from the eastern limits of the Western Palearctic seem a possibility. Madagascar Pratincole *Glareola ocularis* is an endemic breeder on Madagascar and migrates to East Africa; vagrants have been recorded from a wide scatter of Indian Ocean islands including the Seychelles, Mauritius, Réunion and the Comoros (Safford *et al.* 2013).

STERCORARIIDAE
Skuas

The three small skuas, or jaegers to North American birders, Pomarine Skua *Stercorarius pomarinus*, Arctic Skua *Stercorarius parasiticus* and Long-tailed Skua *Stercorarius longicaudus,* are all high-latitude breeders and disperse throughout the world's oceans in the non-breeding season. Geolocator data from Long-tailed Skuas tagged in Greenland found that they may travel annually 43,900–54,200km, reaching locations between 10,500–13,700km from their breeding sites (Gilg *et al.* 2013). All three small skuas can be encountered almost anywhere on the high seas, and can scarcely be described as 'vagrants' anywhere in the marine realm. All three are also notable for routinely making overland migrations to and from their Arctic breeding areas, with birds regularly recorded in large numbers at landlocked waterbodies such as the North American Great Lakes (Sherony 1999) and the Aegean, Black and Caspian Seas (Arkhipov & Blair 2007). The relative rarity of detection by human observers of these inland jaegers on their trans-continental migrations is probably because they fly very high in most weathers – for example, a Long-tailed Skua collided with a passenger aircraft in-bound to Auckland Airport, North Island, New Zealand, in November 2010, at a staggering altitude of 4,084m (Galbraith *et al.* 2013). Whilst records of jaegers in the interior of North America and Asia are unsurprising, there are also a number of records from the interior of Africa and South America, far from known migration routes. These include a Long-tailed Skua photographed over the Napo River in Amazonian Peru in September 2008 (Freile *et al.* 2017); two specimen records of Pomarine Skua from the Brazilian Amazonian state of Pará in 1960 and 1984 (Lees *et al.* 2014); and a specimen of Arctic Skua from the Brazilian Amazonian state of Roraima taken in 1964 (Naka *et al.* 2006). Unusual African skua records include Long-tailed Skuas in Uganda (Hayman & Thorns 2015), Botswana (Tyler *et al.* 2008) and the Kalahari Desert in South Africa (Spearpoint 1981). These presumably represent storm-driven individuals, as was certainly the case for an Arctic Skua collected in Kruger National Park, South Africa, in January 1976 in the wake of Cyclone Danae; six Sooty Terns *Onychoprion fuscatus* were also displaced there by the same storm (Joubert 1977).

Species in the larger former '*Catharacta*' group of skuas do not undertake routine long overland migrations and have less extensive pelagic distributions than the smaller species. The most restricted-range species is Chilean Skua *Stercorarius chilensis,* which is typically confined to the southern temperate waters of South America, but vagrants have been recorded north to Ecuador (Freile *et al.* 2017) as well as Espírito Santo (Tavares *et al.* 2012) and Bahia in eastern Brazil (Silva e Silva *et al.* 2002). The most migratory of the big skuas is South Polar Skua *Stercorarius maccormicki,* which moves from Antarctica north into the North Atlantic, North Pacific and Indian Oceans (Weimerskirch *et al.* 2015). The species is considered a vagrant anywhere near European coasts but is likely to be regular off the continental shelf (de Juana & Garcia 2015), and there are records of vagrants all the way up the Red Sea to Israel (Shirihai 1999); wrecked birds have occurred in the United States far from the coast in Oklahoma,

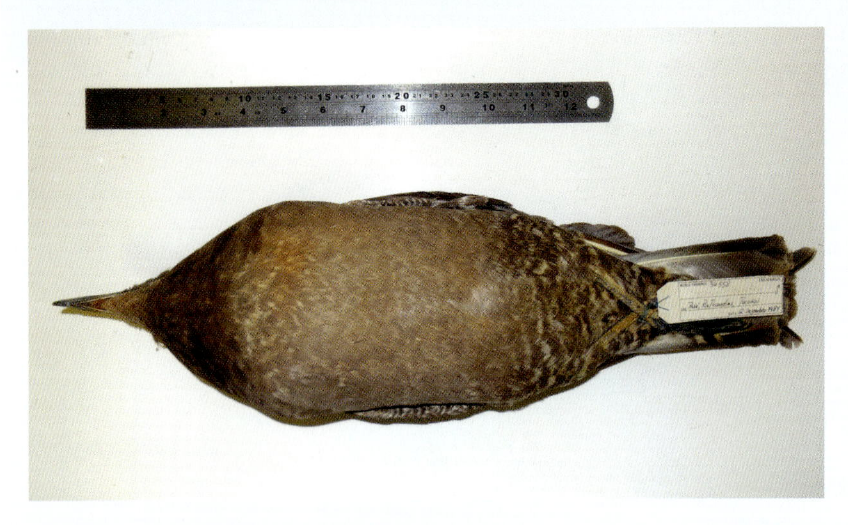

◄ Pomarine Skua *Stercorarius pomarinus* specimen (MPEG #36558) collected along the Tocantins River at the Tucuruí Dam, Pará, Brazil, on 12 December 1984. This individual (originally misidentified as an Arctic Skua *Stercorarius parasiticus* – see Lees *et al.* 2014) was found 400km from the coast at this Amazonian mega-dam. Two Atlantic Petrels *Pterodroma incerta* had been found at the same site in September of the same year (*Alexander Lees © Museu Paraense Emílio Goeldi*).

Tennessee and Georgia (eBird). There have been several putative records of Brown Skua *Stercorarius antarcticus* from the Northern Hemisphere, and some wrecked 'southern skuas' from Britain were initially suspected of being this species although this was later rendered doubtful (Votier *et al.* 2007), and several old records from the Indian Ocean have been reidentified (Praveen *et al.* 2013), so it seems likely that all these will prove to be South Polar Skuas (Newell 2008). Great Skuas *Stercorarius skua* have crossed the Equator to penetrate as far south as northern Brazil (Furness 1978) and have also wandered as far as Turkey (Kirwan *et al.* 2014) and the Black Sea in Ukraine (Redinov *et al.* 2014).

ALCIDAE
Auks

With their relatively short wings, auks seem unlikely candidates for long-range vagrancy, yet several species are famous for appearing tens of thousands of kilometres out of range. The champion alcid vagrant is Long-billed Murrelet *Brachyramphus perdix*, a western Pacific species which has been recorded widely across North America and three times in Europe (Switzerland, England and Romania). Its sister species, Marbled Murrelet *Brachyramphus marmoratus*, breeds along much of the eastern Pacific coastline, yet is unknown as a vagrant to the interior of North America. Comparison of their respective life histories reveals Long-billed to be a longer-distance migrant and hence potentially more predisposed to vagrancy (Konyukhov & Kitaysky 1995; Mlodinow 1997). Mlodinow (1997) suggested that vagrancy was most likely to involve birds that had become disoriented during normal post-breeding dispersal/migration, after finding a relationship between Long-billed Murrelet occurrences and storms within the regular winter range. Intriguingly, vagrant records were also correlated with periods when jet stream circulation tracks (at *c.* 4.5km altitude) stretched from the Gulf of Alaska and North Pacific into the interior of Alaska.

The idea that Long-billed Murrelets might get caught in the jet stream could go some way to explaining their occasional appearance so extremely far from home. But why would murrelets be flying in low-level jet streams in the first place? There is no evidence that migration in any alcid involves flight at high altitude, although jet stream activity might also correlate with strong winds at lower levels. Maumary & Knaus (2000) suggested that juvenile Long-billed Murrelets might 'vagrate' on their first flights, flying in the opposite direction to the sea from their nest, a hypothesis supported by the discovery of grounded fledglings far inland (Ralph *et al.* 1995). However, the likelihood of an inexperienced individual surviving on such a misorientated heading for any long distance is very slim, given the inevitably limited foraging opportunities they would encounter. The fact that Ancient Murrelet *Synthliboramphus antiquus* has also occurred as a regular extralimital vagrant further weakens this hypothesis, as this species leaves the nest within one

► Long-billed Murrelet *Brachyramphus perdix*, Dawlish, Devon, England, November 2006, the first for Britain and second for the Western Palearctic (*Rebecca Nason*).

day of hatching, with young being cared for by adults whilst still flightless at sea (Seally 1976). So immediate post-fledging vagrancy is impossible in Ancient and, by proxy, unlikely in Long-billed. The pattern of occurrences of Ancient Murrelet mirrors that of Long-billed, which has also been recorded right across the North American continent, in addition to reaching Europe – a single bird which spent three breeding seasons between 1990 and 1992 on Lundy Island in Devon, England (Haas 2012). The Lundy Ancient Murrelet occurred following an excellent winter for the species on the Pacific coast and after four inland records in the United States in the same period (Washington, Washington/Oregon, Michigan and Idaho. This suggests that the Lundy individual might have crossed the entire North American continent during a population-level displacement, rather than representing an individual with 'faulty' migratory instincts. That two Long-billed Murrelets occurred in Europe in November 2006 is also suggestive of some weather-linked event, rather than the movements of 'freak' individuals. Maumary & Knaus (2000) concluded that the shortest route for a Long-billed Murrelet to reach western Europe would involve a straight westward transit from the Sea of Okhotsk through Yakutia, the Taimyr Peninsula, Kara Sea, Barents Sea, Finland and the Baltic Sea, a distance of around 7,000km. This route is presumably more likely than the 10,000km through continental Siberia, and much more so than the 17,000km journey across North America and then the Atlantic Ocean following the same latitude.

The unlikely nature of vagrancy of Pacific seabirds to Europe and the apparent increase in their rate of occurrence prompted Vinicombe (2007) to speculate that such individuals may have crossed the North Pole to reach Europe, with vagrancy being encouraged by the shrinking of the polar ice cap, removing a significant barrier to dispersal at polar latitudes (e.g. Reid *et al.* 2007). Vinicombe (2007) suggested that seabirds taking advantage of new feeding grounds afforded by a minimal ice sheet in late summer might then find themselves on the 'wrong' side of the pole at the onset of the freeze, resulting in being pushed south into Europe. However attractive this hypothesis may seem, we should perhaps be cautious in its application. Neither Long-billed nor Ancient Murrelets are true Arctic alcids; they do not occur in the Beaufort Sea or even in the northern Bering Sea, and the spread of records on the continental United States does not show a northern bias (notwithstanding poor observer coverage in the far north). Furthermore, the supposed 'increased' rate of occurrence of such species (which have been documented since 1860) could also be related to increased observer coverage and competence (and even the 'digital revolution'). Nevertheless, the 'polar overshoot' theory fits well for Atlantic Ocean vagrancy in Parakeet Auklet *Aethia psittacula* (one record from Sweden) and Crested Auklet *Aethia cristatella* (records from Iceland and Greenland), and both Tufted Puffin *Fratercula cirrhata* and Horned Puffin *Fratercula corniculata*, all of which are unrecorded across most of the United States. Sealy & Carter (2012) suggested that such vagrants probably travelled from the Chukchi Sea east through the Canadian Arctic archipelago to the Atlantic Ocean. It is possible that diminishing Arctic ice might facilitate this, although the auklet records occurred long before any significant contraction in sea-ice extent (e.g. Overpeck *et al.* 1997) and the number of records of such Pacific alcids does not seem to have increased over the last two centuries (Sealy & Carter 2012).

Amazingly, Burnham *et al.* (2020) documented both species of Pacific puffins at an Atlantic Puffin *Fratercula arctica* colony on Dalrymple Rock (76.47°N, 70.22°W) in north-west Greenland – a single Horned Puffin in 2002–06 and 2013–19 (the first for the North Atlantic) and a Tufted Puffin in 2019. However, this is not the first time that all three Puffins have been recorded at the same colony – all three were present in 2008 at Talan Island (59.3°N, 149.1°E) in the Sea of Okhotsk (Kharitonov 1999), and the single vagrant Atlantic Puffin recorded there was the first for this ocean basin. Tufted Puffins have also been recorded in the Davis Strait off south-west Greenland in 2009, preceding a record from Kent, England, the same year, as well as records from Maine, United States, in the 1831–32 winter and in 2014, Sweden in 1994, and a returning individual on Svalbard in the summers of 2019 and 2020, which adds credence to the polar-transit hypothesis (Burnham *et al.* 2020). The other Pacific alcids do not have such an exceptional pattern of vagrancy and most have not been recorded far south of the regular range, e.g. Cassin's Auklet *Ptychoramphus aleuticus*. A few notable records include records of Kittlitz's Murrelet *Brachyramphus brevirostris* from Japan (Brazil 2009) and San Diego, California (Unitt 2004), and four species have made it to Hawaii – Ancient Murrelet, Parakeet Auklet, Rhinoceros Auklet *Cerorhinca monocerata* and Horned Puffin (Pyle & Pyle 2017).

In the Atlantic, Little Auk (Dovekie) *Alle alle* and Atlantic Puffin are the alcids most prone to long-range movements. Little Auks are occasionally wrecked by winter storms in both the eastern and western Atlantic and have occurred well inland in North America, e.g. in Manitoba, Ontario, Minnesota, Wisconsin and Michigan (Montevecchi & Stenhouse 2002), as well as in Europe. Little Auk is the only alcid that has occurred south to the West Indies (Kirwan *et al.* 2019) and in the Western Palearctic south to Morocco (Bergier *et al.* 2005). Little Auks do also breed in the Bering Sea region; where these birds winter is unknown, but there is a single record from British Columbia (Halpin & Willie 2014). Atlantic Puffins are more rarely wrecked inland but have reached as far inland as Ohio in the United States (Peterjohn 1989), as well as the aforementioned Pacific record, and one was recently found dead on a beach in Israel

▲ Brunnich's Guillemot *Uria lomvia*, Lerwick, Shetland, Scotland, 1 December 2005. Many individuals of this species have been found dead or dying, so this bird which spent 20 days in and around the harbour in Lerwick attracted many birders (*Chris Batty*).

(Fayet & Becciu 2018). Perhap rather surprisingly, there are just four records of this species for Florida, versus five records of Long-billed Murrelet (Kratter & Brothers 2010). The other auks are typically very rare outside of their regular range. Brunnich's Guillemot (Thick-billed Murre) *Uria lomvia*, which mostly winters in 'low' Arctic waters, is a highly sought-after species in western Europe and in the temperate zone in the Pacific. It is a vagrant south to California and is a regular vagrant on the eastern seaboard of the United States south as far as Florida (Gaston & Hipfner 2000). The occasional appearance of individuals in north-west Europe might perhaps relate to birds with 'faulty' navigatory apparatus or storm-driven individuals, perhaps accompanying flights of Little Auks from the same region. Pelagic observations of this species within 960km of Ireland (Cusa 1949) and occasional multiple occurrences around the Scottish Northern Isles and in the North Sea suggest that the former might lie close to the regular winter range of this species. It is therefore surprising that records are not more frequent in the British Isles following winter storms, although it is possible that the species is particularly adept at avoiding storm drift by avoiding flight during high winds. Whilst the vast majority of vagrancy records involve birds appearing at coastal sites in winter, there are two remarkable records of adult Brunnich's Guillemots at inland sites in Europe: in Germany in August 1987 and Belgium in August 2006 (Van Bemmelen & Wielstra 2008). The causes of these unusual events are unclear, although there are also multiple inland records from Scandinavia and the Great Lakes region of North America, suggesting that occasional overland movement is not exceptional in this species (Van Bemmelen & Wielstra 2008).

LARIDAE
Gulls, Terns and Skimmers

Thanks to their gregarious nature and preference for open habitats, flocks of gulls and terns tend to be relatively easy to observe, and therefore often fall under birders' scrutiny. Vagrant individuals are detected on a regular basis by armies of avid gull-watchers scrutinising landfills, harbours, reservoirs and other sites where larids aggregate. Thus, unlike the case of many more cryptic groups (e.g. passerines), it is likely that we detect a relatively high proportion of vagrant gulls, at least across Europe and North America. Many of the rarest vagrants are often tracked between sites across the country, e.g. the British Isles' first Glaucous-winged Gull *Larus glaucescens*, an exceptional vagrant from the North Pacific, which was detected in Gloucestershire, Carmarthenshire and Greater London (Sanders 2010), highlighting the quality of coverage.

► Herring Gull *Larus argentatus vegae* ('Vega Gull') with American Herring Gulls *Larus [argentatus] smithsonianus*, Tullytown Landfill, Bucks, Pennsylvania, United States, 15 December 2012. This was the first state record of this taxon, which is increasingly being identified extralimitally in Europe and North America thanks to improved understanding of identification criteria (*Tom Johnson*).

◀ Ivory Gull *Pagophila eburnea*, Pismo Beach, San Luis Obispo, California, United States, 6 November 2010, the second record for California (*Brian Sullivan*).

Whilst some gull species are strongly migratory, many others show limited migratory tendencies, reflecting their ability to adapt to a wide range of environmental conditions and exploit varied food resources. However, even the most sedentary species usually show patterns of dispersal away from breeding sites during the non-breeding season, and inevitably some individuals venture further than others. As a result, vagrancy occurs frequently in both migratory and non-migratory gull species, with rates depending also on geographic range size and population size. For example, Yellow-footed Gull *Larus livens*, which is essentially sedentary around the Gulf of California in Mexico, has occurred as a vagrant only north to Nevada, Arizona and Utah (Howell & Dunn 2017), up to 1,000km out of range, as well as regularly occurring in southern California's Salton Sea region. By contrast, the pelagic Swallow-tailed Gull *Creagrus furcatus,* restricted as a breeder to the Galapagos Islands and Malpelo Island, has a large at-sea distribution in the southern Pacific between Costa Rica (Young *et al.* 2010) and Chile. It has occurred off the Pacific coast of North America on three occasions: twice in California (Howell *et al.* 2014) and once near Seattle, Washington State, in August–September 2018 (eBird), which was at least 5,000km out of range.

Harsh weather probably plays a significant role in driving long-distance movements of more sedentary species, particularly amongst those inhabiting northern latitudes and continental regions (Elkins 1988). Vagrancy by two specialists of the high Arctic, Ivory Gull *Pagophila eburnea* and Ross's Gull *Rhodostethia rosea,* represent prime examples. Both species occur almost exclusively within the northern polar region throughout the year, the former being strongly tied to the edge of the polar ice whilst the latter presumably spends the winter in the near-perpetual darkness of offshore Arctic waters, perhaps associated with polynyas (areas of open water surrounded by sea ice), although its exact wintering areas remain a mystery (Divoky 1976; Densley 1999, Maftei 2014). Broadly, there is a reasonable correlation between European occurrences and periods of cold weather with a strong northerly component to the wind; such occurrences fit with the idea of dispersive movement to escape harsh conditions within the normal range (Elkins 1988). This is intriguing, as Ross's Gull occurs with greater frequency in western Europe than it does in eastern North America, perhaps suggestive of an unknown wintering area somewhere in the North Atlantic or the Norwegian Sea, or alternatively just wandering birds from the Davis Strait or the Labrador Sea (Maftei 2014). An influx of over 100 to Hokkaido in January 2001 also hints at a significant wintering population somewhere in the North Pacific (Malling Olsen & Larsson 2003). However, the correlation with winter storms is not particularly strong, and records of vagrancy in both species have occurred at all times of year and in all weather conditions, suggesting that other factors are also likely to be at play. It seems likely that some individuals possess stronger dispersive instincts than are typical of the population, leading them to move significantly beyond the normal wintering range. Such dispersers might have difficulty finding habitats that fit their perceptual image of suitability, resulting in a protracted search that may carry them ever further from the normal range. This includes records of Ivory Gulls south to Portugal (van den Berg 2014) and Switzerland (Malling Olsen & Larsson 2003), and Ross's Gulls south to Macedonia, Spain and Italy. Both Ross's and Ivory Gulls have also made it to California (McCaskie 2007, Weintraub & San Miguel 1999).

► Ross's Gull *Rhodostethia rosea*, Tupper Lake, Franklin, New York, United States, 29 January 2017, an amazing discovery inland in the Adirondack Mountains (*Ian Davies*).

Transoceanic vagrancy in gulls happens across all the ocean basins. Vagrant gulls frequently reach the eastern Atlantic and western Pacific from North America and, like most Nearctic vagrants, their occurrence is strongly linked to weather patterns. Species that perform significant southward migrations along the eastern North American seaboard (e.g. Laughing Gull *Leucophaeus atricilla* and Bonaparte's Gull *Chroicocephalus philadelphia*) are prone to eastward drift across the Atlantic in fast-moving depression systems during migration periods (Elkins 1988). Other less strongly migratory gull species (e.g. Ring-billed Gull *Larus delawarensis* and American Herring Gull *Larus* [*argentatus*] *smithsonianus*) appear to be relatively unaffected by these autumnal weather systems, with transatlantic movements peaking in winter and early spring, suggesting that movement may be caused by harsh weather in the normal range or depletion of food resources within core areas, or even birds following ships for extended periods. It has also been suggested that the early spring peak in Ring-billed Gull occurrence in the British Isles might reflect a northward movement of individuals that have been displaced across the Atlantic in more southerly regions during the preceding autumn and winter (Cottridge & Vinicombe 1996).

The importance of wind drift as a mechanism of vagrancy in Nearctic gull species was illustrated dramatically in late October 2005, when Hurricane Wilma (the most intense hurricane on record in the Atlantic basin) moved eastwards across the Atlantic after wreaking havoc in the Caribbean and southern United States, dragging a series of extensive frontal systems with it. Wilma's arrival in southwestern Britain brought an unprecedented influx of Laughing Gulls – at least 32 individuals were detected within the

▼ Laughing Gull *Leucophaeus atricilla* with Great Crested Terns *Thalasseus bergii*, Manulu Lagoon, Kiribati (Christmas Island) (*Eric Vanderwerf*).

first few days after the weather system's passage, mostly in Cornwall and Devon (Fraser & Rogers 2007), and the total movement was likely to have involved at least 60 birds (Ahmad 2005). Along with the Laughing Gulls came a handful of Franklin's Gulls *Leucophaeus pipixcan*, brought by the same weather system. This provided the first conclusive evidence of direct weather-driven vagrancy to Europe in this intriguing species, whose occurrence here had previously been the subject of much conjecture (Hoogendorn & Streinhaus 1990; Cottridge & Vinicombe 1996).

Breeding in the western North American prairies and migrating through the central United States to winter along the Pacific coast of South America, Franklin's Gull was long considered an unlikely candidate for 'direct' transatlantic vagrancy. The events of October–November 2005 demonstrated beyond doubt that direct transatlantic vagrancy does occur, despite the species' rarity along the eastern seaboard of North America. It is possible that the unusual weather conditions associated with Wilma's presence may have led to drift or displacement of individuals eastwards from their normal migration route, eventually leading them to be entrained in the Atlantic system that carried so many Laughing Gulls across to Europe.

Prior to the events of 2005, most Franklin's Gull records in the British Isles occurred outside the autumn period and involved birds in their second year or older, suggesting that the vagrancy caused by Hurricane Wilma may have been a one-off event. It has previously been postulated that Franklin's Gulls occurring in the British Isles arrive as a result of individuals crossing into the Atlantic at the southern extremity of their South American wintering range, before moving north up the Atlantic to arrive in Europe in the following summer (Hoogendorn & Steinhaus 1990). Regular records from southern Brazil, South Georgia (Reid 1998), and southern and western Africa certainly support this theory (Lewington *et al.* 1991; Cottridge & Vinicombe 1996). That said, Franklin's Gull is perhaps the most widely recorded of all western hemisphere bird species. It has been recorded as far afield as Antarctica (Maftei 2013), New Zealand (Onley & Schweigman 2004), the Philippines (Robson *et al.* 2015), China (Holt 2005), the Seychelles (Daly 2017), India (Holt *et al.* 2013), Kazakhstan (Wassink *et al.* 2013), Israel (Smith 2004), Libya (Bourass *et al.* 2013) and Ethiopia (Campbell & Berhe 2020). The long migration span of Franklin's Gull might perhaps predispose the species to long-distance vagrancy, but the only partially migratory Laughing Gull, with a considerably smaller migration span, has been recorded from an almost equally wide scatter of locations, across the Americas, e.g. even inland to Bolivia (Brady *et al.* 2019), across the Western Palearctic and Africa, and from Fiji (Dutson & Watling 2007), Malaysia (Tebb *et al.* 2003), Japan (Brazil 2009), Australia and New Zealand (Mislelly *et al.* 2019).

Transatlantic movements by Nearctic or Palearctic gulls, whether they occur in the Northern or Southern Hemisphere, evidently occur frequently

▼ Franklin's Gull *Leucophaeus pipixcan* with Hartlaub's Gulls *Chroicocephalus hartlaubii* and a Cape Shoveler *Spatula smithii*, Strandfontein WTP, Western Cape, South Africa, 24 August 2014 (*Niall Perrins*).

▲ Little Gull *Hydrocoloeus minutus*, with Bonaparte's Gulls *Chroicocephalus philadelphia*, Ring-billed Gulls *Larus delawarensis* and an American Herring Gull *Larus* [*argentatus*] *smithsonianus*, Long Point SP, Cayuga, New York, United States, 8 May 2016. This species is a scarce passage migrant through North America (*Alexander Lees*).

as a result of wind drift, harsh weather movements or extreme dispersal. It is interesting to speculate on what happens to these individuals, and in particular whether some individuals are able to return to their normal ranges. There are some clear instances of vagrant Nearctic gulls becoming 'stuck' on this side of the Atlantic, including adult Laughing Gulls that reappeared for multiple consecutive summers in gull colonies in Norfolk (Rogers 2000) and Tyneside (Rogers *et al.* 1988). It was postulated that the former individual also spent at least five summers touring gull colonies across western Europe, apparently performing an ever-widening search for willing suitors (Ottens 2007). Ring-billed Gulls have also been recorded visiting Common Gull *Larus canus* breeding colonies around the British Isles, successfully hybridising on some occasions (Kehoe 1992). The same Yellow-legged Gulls *Larus michahellis* are suspected of returning to the same site in Newfoundland in consecutive winters (Howell *et al.* 2014). Such cases appear to be relatively rare, however, at least in relation to the number of Nearctic vagrants that are annually displaced across the Atlantic during the non-breeding period.

Given that many gull species are generalists, they are likely to be quite capable of surviving in locations geographically removed from the normal niches and one would expect rates of survival amongst vagrants to be relatively high. As such, the low rate of return amongst individuals displaced from during the breeding season suggests that many may indeed reorientate to their natural ranges. This is supported by an example of a Ring-billed Gull captured in Norway that returned to north-east Canada and then back to Norway in successive winters, whilst another Norwegian-ringed individual was shot on Iceland, presumably en route back to North America (Cottridge & Vinicombe 2001). The events associated with Hurricane Wilma in 2005 were also subtly suggestive of a high rate of return; it was widely anticipated that after the major autumn influx, the following spring would see a bumper crop of northbound individuals passing through the British Isles. In the event, Laughing Gull numbers were slightly higher than average, with around a dozen spring 'passage' individuals detected (mostly in northern Scotland), but very few lingered and there was little evidence of individuals roaming this side of the Atlantic in search of breeding sites (Fraser 2007). Whilst some inevitably perished after their transatlantic journey, it seems likely that many of the displaced individuals were able to navigate back across the Atlantic after replenishing energy reserves in Europe.

Range expansions may also be driving change in status of some species. It seems likely that the colonisation of Greenland by Lesser Black-backed Gulls *Larus fuscus* (Boertmann 2008) has probably underpinned the rapid increase in North America, as there are still very few North American ringing recoveries of Lesser

Black-backed Gulls marked in Europe (Hallgrimsson *et al.* 2011). These Greenland Lesser Black-backed Gulls are already 'fairly common' winter visitors to the West Indies (Kirwan *et al.* 2019) and have occurred south to north-east Brazil (de Almeida *et al.* 2013). Little Gull *Hydrocoloeus minutus* seems to have a foothold as a breeding bird in North America, with a regular passage through the eastern half of the continent, although the location of the putative Nearctic colonies of this species, presumed to be somewhere in Arctic Canada, remain unknown (Ewins & Weseloh 1999). Vagrants in the western hemisphere have reached as far as Colombia (Blokpoel *et al.* 1984) and French Guiana, where a juvenile was photographed 290km north-east of Cayenne in September 2011 (Rufray *et al.* 2019).

Audouin's Gull *Larus audouinii* is also an interesting case. For most of the 20th century this species was considered a global rarity with a population estimated to number below 1,000 pairs (Witt 1977) and with few extralimital occurrences. However, a marked population increase has seen this vagrant appearing with increasing frequency in northern Europe. Although not strongly migratory, large numbers of Audouin's Gulls move westwards from the Mediterranean to winter along the Atlantic Coast of West Africa. Vagrants to northern Europe are perhaps most likely to originate in this area, perhaps tagging on with other migratory gull species bound for northern Europe, e.g. Lesser Black-backed Gull (Walker 2004). This rapidly increasing population has also sent vagrants out into the Atlantic, first to the Azores in 2005 (Gordinho & Martins 2009), followed by Trinidad in Dec 2016–Aug 2017 (Lallsingh 2018), with probably the same bird relocated on the South American mainland, 835km away in Suriname in March 2018 (Kasius *et al.* 2019). Future records from North America seem likely, and elsewhere this species has now been recorded in the eastern Atlantic as far south as Ghana (Kelly *et al.* 2014) and Guinea (Ławicki and van den Berg 2017a). The rapid population expansion of Audouin's Gulls was facilitated by fisheries discards and now supplemented by landfill waste, factors that have contributed to increases in many other gull species and led many formerly coastal species to move inland in the non-breeding season. Such shifts in resources may have drawn species like Glaucous-winged Gull, Slaty-backed Gull *Larus schistisagus* and Black-tailed Gull *Larus crassirostris* eastwards in North America. This has led to subsequent European records of the first two, and the last-named has now occurred as close to Europe as Bermuda (Hopkin 2011) and Newfoundland (Howell *et al.* 2014). In the east, Slaty-backed Gull has wandered south to Indonesia and Australia (Iqbal & Albayquni 2016) and Black-tailed Gull to Thailand (Robertson 1994) and Australia (Menkhorst *et al.* 2017).

It is not only north temperate zone gulls that wander extensively. Belcher's Gull *Larus belcheri* is an ostensibly unlikely candidate for long-range vagrancy, with a restricted range along the eastern Pacific coast in Chile and Peru, but it has occurred in Costa Rica (Bonilla *et al.* 2014), California, Florida and Panama (Howell *et al.* 2014), perhaps reaching Florida via the Panama Canal. Kelp Gull *Larus dominicanus* is also being recorded with increasing frequency in both

▲ Audouin's Gull *Ichthyaetus audouinii*, Anderby Creek, Lincolnshire, England, 18 August 2008 (*Graham Catley*).

► Slaty-backed Gull *Larus schistisagus*, Van Cleef Lake, Seneca, New York, United States, 17 February 2018. Slaty-backed Gulls are now recorded with increasing frequency in North America and Europe; whether this solely reflects an increase in observer awareness of their identification criteria or also a change in status remains unclear (*Jay McGowan*).

North America and Europe. This species has bred extralimitally in Louisiana, United States (Howell *et al.* 2014), and in Morocco, where nominal 'Cape Gulls' *L. d. vetula* established a mixed colony with a likewise extralimital group of Great Black-backed Gulls *Larus marinus* (Bergier *et al.* 2009, Jönsson 2011). Grey-hooded Gulls *Chroicocephalus cirrocephalus* also wander extensively and have occurred twice in the United States, in Florida and New York (Howell *et al.* 2014), in addition to Barbados (Kirwan *et al.* 2019) and at least seven countries in the southern half of the Western Palearctic; however, many may relate to escapes (Malling Olsen & Larsson 2003). Further east, Silver Gull *Chroicocephalus novaehollandiae* has recently been recorded for the first time beyond Australasia, with two records from Bali (Gardner & Gilfedder 2011). There would seem to be few limits to larid vagrancy, and it would be possible to write an entire book on their history of vagrancy. Other remarkable records include Black-legged Kittiwakes *Rissa tridactyla* in Pakistan (Manzoor *et al.* 2013) and Bangladesh (Robson *et al.* 2016), Red-legged Kittiwake *Rissa brevirostris* in California (Hamilton & Willick 1996), Grey Gull *Leucophaeus modestus* in Brazil (Chupil *et al.* 2019) and a Common Gull of the 'short-billed' form *brachyrhynchus* on the Azores (Alfrey & Ahmad 2007).

Unlike most gulls, terns tend to have relatively narrow ecological requirements, invariably specialising on live prey caught on the wing. Outside the tropics, virtually all tern species are strongly migratory, and unsurprisingly they too have a significant propensity for long-distance vagrancy. Rare terns are

► Kelp Gull *Larus dominicanus* with Lesser Black-backed Gulls *Larus fuscus*, Akhfenir, Laâyoune-Sakia El Hamra, Morocco, 3 April 2017. This species seems to be expanding northwards and has now bred in Morocco (*Paul French*).

usually drawn to large concentrations of commoner species, be that at breeding colonies or roost sites, which makes detection and subsequent relocation of vagrants surprisingly predictable. Arguably the most spectacular record of tern vagrancy concerns the sole European record of Aleutian Tern *Onychoprion aleuticus*, an individual that visited the Farne Islands, England, in May 1979 (Dixey *et al.* 1981) and stands out as an example of the near-limitless potential for long-distance vagrancy in migratory birds. At the time of occurrence, the wintering range of this species was unknown, but more recent discoveries have shown it to lie in the central-west Pacific and south at least occasionally as far as New South Wales, Australia (Hill & Bishop 1999, Goldstein *et al.* 2019). The most tenable explanation for the appearance in Europe would be via a northward crossing of the polar region, either in autumn as a nominal 'reversed migrant' (Cottridge & Vinicombe 2001) or as a spring overshoot. Either scenario is plausible given the orientation of its normal migration route from the Bering Sea region towards the Philippines. Nevertheless, given the global rarity of this species (an estimated 30,000 individuals worldwide; Delaney & Scott 2006), this record remains a remarkable event that seems unlikely to be repeated. Incredibly, the Farne Islands have attracted three terns in the genus *Onychoprion*; in addition to Aleutian Tern, both Bridled Tern *Onychoprion anaethetus* and Sooty Tern *Onychoprion fuscatus* have occurred in the huge tern colony there. Both these tropical species have been widely recorded as storm-driven vagrants elsewhere, including at inland locations worldwide.

A third largely tropical tern, Lesser Crested Tern *Thalasseus bengalensis*, has also been recorded on the fortunate Farne Islands, an individual which hybridised with Sandwich Terns *Thalasseus sandvicensis* in several summers between 1984 and 1997 (Brown & Grice 2010). Lesser Crested Terns maintain a

◀ Aleutian Tern *Onychoprion aleuticus*, Farquhar Inlet, Old Bar, New South Wales, Australia, 20 December 2017. Aleutian Terns have returned to this site since 2017 and were perhaps historically overlooked in Australia (*Rohan Clarke*).

▼ Bridled Tern *Onychoprion anaethetus*, Inner Farne, Farne Islands, Northumberland, England, 1 July 2013, one of three species of *Onychoprion* tern to have visited this island (*Chris Batty*).

▲ Elegant Tern *Thalasseus elegans* with Common Terns *Sterna hirundo*, Shinnecock County Park, Suffolk, New York, United States, 5 July 2013, the first state record (*Jay McGowan*).

small breeding population in the Mediterranean off Libya, which is probably the source of both northern European records and occasional attempts by this species at colonising Spain (de Juana & Garcia 2015). Identification of this species has been complicated by the apparent 'colonisation' of the region by small numbers of Elegant Terns, which have bred in both Spain and France with Sandwich Terns, with at least one pure pair in Spain (Dufour *et al.* 2017). This incredibly high rate of vagrancy to Europe for a Pacific seabird is all the more interesting as molecular analysis indicated that interspecific gene flow has not occurred regularly in the past, suggesting this must be a new phenomenon (Dufour *et al.* 2017). Elegant Tern is being increasingly recorded as a vagrant in eastern North America (Shoch & Howell 2013) where hybridisation has occurred with Sandwich Tern in Florida, and there is an even more extraordinary recent record of one from the Niagara River border of the United States and Canada in November 2013 (Roy & Pawlicki 2014). Both adult Elegant Terns and one of the hybrid young from France have been found wintering in South Africa (O'Connell 2019), which might suggest an initial colonisation route from the South Atlantic, with individuals tracking north to Europe with Sandwich Terns. Given that 90–99 per cent of the world's Elegant Terns are thought to nest on Isla Rasa in the Gulf of California, Mexico, the spread of records of this species seems all the more incredible. Greater knowledge of the field identification of 'yellow-billed' terns is leading to the detection of more extralimital records, including for the rarest of them all – Chinese Crested Tern *Thalasseus bernsteini,* which was recently recorded for the first time in Japan, near Irabu Island, Okinawa Prefecture, in October 2018 (Dinets 2018).

Even some pelagic tropical terns considered largely resident have been recorded extensively as vagrants. For example, White Tern *Gygis alba* has been recorded from Bermuda (Wingate & Watson 1974), the Bahamas (White *et al.* 2014) and mainland India (Jayson *et al.* 2013), and tideline corpses have been found as far south as New Zealand (Powlesland & Pickard 1992). Most impressively, Tropical Storm Nakri wrecked a White Tern 110km inland in Chungbuk province, South Korea, in August 2014, the first national record (Jeong *et al.* 2014). This species would seem like a good candidate to occur as a vagrant to the Atlantic islands of the Western Palearctic, or at sea in the Gulf Stream off the eastern United States. Similarly, there is only one acceptable record of Brown Noddy *Anous stolidus* from the Western Palearctic, from Schleswig-Holstein, Germany in 1917 (Haas 2012), although one was photographed 352km west of Flores in the Azores in August 2018 (Leitão 2011). Members of the small tern genus *Sternula* have a cosmopolitan distribution and are frequent long-distance vagrants. Arguably, this is true of the Nearctic Least Tern *Sternula antillarum* in the most spectacular fashion. This species has occurred as a vagrant east once to Europe – a returning individual present at a Little Tern *Sternula albifrons* colony in East Sussex, England, between 1983 and 1992 (Yates 2010) – and also west to Japan (Brazil 2018). Incredibly, both Least and Little Terns have bred on Hawaii (Pyle & Pyle 2017), an archipelago that has attracted 19 species of terns and noddies, of which only seven have bred.

It should not come as a surprise that highly pelagic tern species like Arctic Tern *Sterna paradisaea*, which routinely cover more than 90,000km during the non-breeding season (Fijn *et al.* 2013), are recorded widely as vagrants. That the *Chlidonias* marsh terns are also regular transoceanic vagrants comes as more of a surprise, given their association with freshwater habitats. Whiskered Tern *Chlidonias hybrida* has occurred in the western hemisphere on at least three occasions, twice in North America, in New Jersey and Delaware (Howell *et al.* 2014), and once in South America, incredibly in landlocked Paraguay (Clay *et al.* 2017). White-winged Black Terns *Chlidonias leucopterus* are more frequent in the New World, with most records coming from eastern North America and concerning adults. Howell *et al.* (2014) suggested that these may have involved many returning birds 'stuck' in the New World and that they may have arrived first in South America or the Caribbean and moved north with *Sterna* terns. There are several records from eastern South America from Brazil (Aldabe *et al.* 2010) and Argentina supporting this hypothesis, including three adults together at Mana, in French Guiana, in March–April 2013 (Rufray *et al.* 2019). White-winged Black Terns are regular vagrants to western Europe and influxes typically correspond with periods of easterly or southerly winds during key migration periods (Cottridge & Vinicombe 1996). However, the link to wind conditions may not be as clear-cut as might be expected. For example, classic late spring easterly drift conditions in 2007 and 2008 brought record influxes of White-winged Black Tern to the Netherlands and Denmark (numbering in the thousands), but virtually no individuals moved the extra few hundred kilometres to reach the British Isles during these periods. Black Terns *Chlidonias niger* of the Palearctic and Nearctic breeding subspecies have been recorded as transatlantic vagrants in both directions and these records include a record of one ringed in Berlin, Germany, in 1984 and subsequently recovered at Macau in Rio Grande do Norte in Brazil in September 1986 (Sagot-Martin *et al.* 2020).

◄ White-winged Black Tern *Chlidonias leucopterus*, Nessmuk Lake, Tioga, Pennsylvania, United States, 10 August 2017. It is thought that many records of this species from North America relate to the same individuals resighted in multiple years (*Ian Davies*).

◄ Whiskered Tern *Chlidonias hybrida* with Laughing Gull *Leucophaeus atricilla*, Cape May, New Jersey, United States, 13 September 2014, the third record for North America (*Jay McGowan*).

Skimmers are also occasionally recorded as vagrants. African Skimmers *Rynchops flavirostris* have occurred once outside the continent, in Oman in 2015 (Harrison 2015), where Indian Skimmer *Rynchops albicollis* has also occurred as a vagrant. African Skimmers are fairly regular in southern Egypt (Sueur & Siblet 2010) suggesting that vagrancy to southern Europe or elsewhere in the Middle East is not entirely fanciful. Black Skimmers *Rynchops niger* are regularly displaced north of their regular North American range by hurricanes and there is a single documented Old World record from South Africa in October 2012 (Wood and Wood 2012), with another (conceivably even the same bird) reported two months earlier off County Mayo in Ireland (although not submitted to the national rarities committee). Alongside the Aleutian Tern in the United Kingdom, another case for most unlikely tern vagrancy concerns the multiple extralimital records of Large-billed Tern *Phaetusa simplex,* which include three records from the United States (Howell *et al.* 2014), three from the West Indies (Kirwan *et al.* 2019) and one from Bermuda (Wingate 1973), which opens up the possibility of future trans-atlantic vagrancy.

► African Skimmers *Rynchops flavirostris*, Khawr Taqah, Dhofar, Oman, 6 February 2015, the first record of this species outside Africa (*Jens Eriksen*).

RHYNOCHETIDAE
Kagu

Kagus *Rhynochetos jubatus* are flightless and generally very sedentary.

EURYPYGIDAE
Sunbittern

Sunbitterns *Eurypyga helias* are thought to be largely resident, although some birds may have to undertake local movements in response to wetland desiccation, moving temporarily to permanent watercourses. There is currently some debate over documented records of the race *meridionalis* in eastern Ecuador; these could represent vagrancy, or perhaps more likely range extension or issues with field identification of this subspecies meriting re-evaluation (Freile *et al.* 2013).

PHAETHONTIDAE
Tropicbirds

The three species of tropicbirds are impressive wanderers, dispersing widely in the non-breeding season. These movements are not regarded as predictable migrations in the strictest sense, and their nature also depends on the latitude of their breeding grounds. White-tailed Tropicbirds *Phaethon lepturus*, for instance, are only present seasonally around Bermuda but may be present year-round further south in the Caribbean (Lee & Walsh-McGee 1998). Tropicbirds seem to be particularly prone to hurricane displacement, and White-tailed Tropicbirds have been widely recorded inland in the eastern United States and southern Canada after storms. Red-billed Tropicbird *Phaethon aethereus* remains considerably rarer as a storm-driven vagrant inland in North America, and rarer in general in the Gulf Stream region, but both species have also been recorded as storm-wrecked birds inland in New South Wales, Australia (Morris 1979). Red-billed Tropicbird is occurring with increasing frequency in the Western Palearctic and apparently colonising the Canary Islands and the Azores (e.g. Furness & Monteiro 1995), with a corresponding increase in vagrancy elsewhere; vagrants have been recorded north as far as the British Isles (Blamire 2004). White-tailed Tropicbird is also appearing with increasing frequency in the Macaronesian region (Monticelli & Aalto 2011, Haas 2012) and has been recorded as a vagrant to South Africa (Sinclair *et al.* 2011) and Angola (Lambert 2011). Several tropicbird records in north-west Europe have involved tideline corpses, including Red-billed in the Netherlands (Bruinzeel 1986), and both Red-billed (Knox *et al.* 1994) and White-tailed (BOU 2015) in Britain but have not been admitted to national lists given that the specimens may have drifted over long distances or died aboard ships and been washed overboard. The stomach of a tideline corpse of a Red-billed Tropicbird at Landguard in Suffolk, England, in February 1993 contained the bones of flying fish (probably *Hirundichthys* sp.) suggesting that its last meal was far from the North Sea, in warmer waters (Knox *et al.* 1994).

Red-tailed Tropicbird *Phaethon rubricauda*, which breeds in the Indian Ocean and western and central Pacific, is the most enigmatic of the tropicbirds. It is perhaps the most oceanic of all the Phaethontiformes, and ringing recoveries have demonstrated some huge movements. For example, one ringed in Western Australia was recovered nearly 6,000km away on Réunion island three years later (Le Corre *et al.* 2003). Red-tailed Tropicbird is now known to be regular over deep waters off the California coast, with occasional vagrants onshore (Howell *et al.* 2014) and there is a record of one found dead in a forest on Vancouver Island, British Columbia, Canada, in June 1992; this was probably a storm-wrecked bird that was predated by one of the local Bald Eagles *Haliaeetus leucocephalus* (Whittington 1992). This species is extremely rare in the Atlantic, where there are records

◄ White-tailed Tropicbird *Phaethon lepturus*, Corvo, Azores 25 October 2011, the third record for the Western Palearctic (*Richard Bonser*).

from the Arquipélago dos Abrolhos, Brazil (Couto *et al.* 2001), and multiple records, presumed to relate to the same adult, on Inaccessible Island, Tristan da Cunha, in 2011–15 (Bond *et al.* 2015). Given the vast distances this species wanders, a record from the northern Atlantic does not seem beyond the realms of possibility. More broadly, the increases in tropicbird sightings at north temperate latitudes may reflect increasing observer effort, or even the consequences of warming sea temperatures, but the primary driver may be increasing population sizes of many tropical seabirds after millennia of persecution.

GAVIIDAE
Divers

There are only five species of divers, or loons, all of which are migratory, and in early 2008 all five were found wintering along a relatively small stretch of the coast around Cornwall, England. Three of these, Great Northern *Gavia immer*, Black-throated *Gavia arctica* and Red-throated *Gavia stellata*, regularly winter in southern England, whilst White-billed Diver *Gavia adamsii* is rare (though no longer considered a vagrant to Britain) and Pacific Diver *Gavia pacifica* is now an annual vagrant, after being first recorded in Britain as recently as 2007 (Maher 2010). The rise in detection of White-billed Divers around the British Isles and elsewhere in western Europe was arguably catalysed by the discovery by Folvik & Mjøs (1995) that there was a regular northward passage of this high Arctic species off south-west Norway (around the same latitude as Shetland) in late April and May. They extrapolated a figure for an annual passage of *c.* 100 individuals. Less than a decade later a regular staging area was discovered off the island of Lewis in the Outer Hebrides, Scotland (Rogers 2003), and double-figure counts from Aberdeenshire in spring are now routine. Evidence is now surfacing that many of these birds probably winter over the Dogger Bank in the central North Sea and some may even perform a loop migration back through the Baltic where there is also a regular spring passage (Bellebaum *et al.* 2010). As such, the predictable occurrence of small numbers on the Atlantic fringe suggests that White-billed Diver is probably a pseudo-vagrant (*sensu* Gilroy & Lees 2003), with a small population wintering each year in pelagic waters that are poorly inventoried by ornithologists. Recent boat-based and aerial surveys of the Dogger Bank in the North Sea estimated a wintering population of 67 individuals (Forewind 2014). White-billed Divers remain very rare inland in the United Kingdom, although two of the three inland records from England were from the same stretch of the River Witham in Lincolnshire. Elsewhere, however, White-billed Divers are found with some regularity away from the coast at large inland lakes, with records spread across Europe and the United States, south as far as south Texas, central Bulgaria and South Korea.

► White-billed Diver *Gavia adamsii*, Inishbofin, Galway, Ireland, 19 May 2013. Formerly regarded as a vagrant to the British Isles, this species is now understood to be a rare but regular winter visitor (*Anthony McGeehan*).

If the pieces of the puzzle surrounding White-billed Divers wintering in Europe are beginning to fit together, the same cannot be said yet for Pacific Diver. Although previously touted as a potential vagrant (Birch & Lee 1995), the occurrence of the first three Pacific Divers for the Western Palearctic in the same winter took everyone by surprise. Pacific Divers probably breed regularly as close as western Greenland (Bent 1919; Burnham & Mattox 1984), but winter almost exclusively in the Pacific region and are consequently rare in eastern North America, although records are widespread there and the species has even reached Hawaii (Pyle & Pyle 2017). Considering the cryptic similarity between this species and Black-throated Diver, it is possible that this species may long have been a regular vagrant to European coastal waters, and it took a first inland record to bring the species to the forefront of birders' attention. It was initially suggested that the Pacific Divers' sudden appearance was a genuine freak event caused by severe weather (there was a notable 'wreck' of inland divers during the 2007/2008 winter). However, two of the Pacific Divers returned to the same wintering grounds the following year, and there have been over 20 subsequent records from across Europe, east to Sweden, south as far as Spain and even inland in Switzerland (Ammitzboell *et al.* 2017). It has been suggested that global climate change may even have played a role in their increasing appearance (Vinicombe 2007a).

The remaining three divers have occurred as vagrants south of their usual winter range with some frequency. In Eurasia, both Black-throated Divers and Red-throated Divers winter with some regularity in the Mediterranean, the Black Sea and the Caspian Sea. As such, multiple records of both species from both India and Iran are perhaps not surprising, although the sight of a Black-throated Diver flying high over the Potasali River in Assam, India, in the shadow of the Himalayas must have been extremely incongruous (Steijn & Vries 2009, Khaleghizadeh *et al.* 2011), much like the recent first record from Kuwait. Red-throated Divers breeding in Alaska routinely winter south to northern Mexico and northern China across a migratory divide (McCloskey *et al.* 2018), and some presumably press on further south, too, as vagrants. Black-throated Divers remain extremely rare in North America away from the Bering Sea region, with records dotted along the western seaboard south to Baja California (Howell *et al.* 2001) but also including records well inland in Washington, Colorado and Ohio. Finally, Great Northern Divers are the most widespread of all divers and routinely winter south as far as Colima in Mexico on the coast and on large inland lakes, with nominal vagrants south all the way to Oaxaca, Mexico, and on Cuba (Garrido & Kirkconnell 1993, Villagómez *et al.* 2017).

▲ Black-throated Diver *Gavia arctica*, San Simeon Creek mouth, San Luis Obispo, California, United States, 13 January 2012, the ninth record for California (*Brian Sullivan*).

SPHENISCIDAE
Penguins

Despite their inability to fly, vagrancy in penguins is surprisingly common as they disperse far and wide in the non-breeding season. This pattern of extralimital dispersal by swimming is also frequently seen in other pelagic non-avian taxa such as seals (Shaughnessy *et al.* 2012) and turtles (e.g. McAlpine *et al.* 2004), which also frequently appear thousands of kilometres out of range. Such occurrences might perhaps be driven by individuals becoming entrained in ocean currents that carry them out of range, or perhaps by severe storm events, and in some cases we know they are driven by food shortages. However, many species also undertake regular seasonal migrations covering thousands of kilometres, and occasional long-range vagrancy events may reflect navigation errors – or just the endpoint of a continuum of natural dispersal.

Arguably the most celebrated vagrant penguin was the immature Emperor Penguin *Aptenodytes forsteri*, the world's largest extant penguin species, which came ashore on Peka Peka Beach, on the Kapiti coast of New Zealand in June 2011. It was the second national record and became the most twitched bird in the country's history. After ingesting over 2kg of sand – they normally eat ice to drink – it was taken into care and spent 72 days in rehabilitation before being released at sea north-east of Campbell Island in September 2011 (Miskelly *et al.* 2012). Satellite-tagged Emperor Penguins from East Antarctica have been shown to forage 1,200km north of the Antarctic pack ice, so this bird was perhaps only a modest 1,500km out of range; another vagrant, an adult-plumaged bird, reached Punta Arenas, Chile, in May 2016 and was also taken into care (Barros *et al.* 2017).

Penguins generally return to breed at the colonies in which they were born. Such natal philopatry is common in the smallest of the group, Little Penguin *Eudyptula minor*, for example (Marchant & Higgins 1990). Ringing and genetic studies have shown that dispersal occurs infrequently and may routinely extend for several hundred kilometres, with birds sometimes stopping to breed at non-natal colonies (Dann *et al.* 1992, Priddel *et al.* 2008). This dispersal is apparently extensive enough to prevent genetic divergence between Australian and New Zealand colonies (Peucker *et al.* 2009). In this context, the fact that this 1.5kg species has occurred as a vagrant to South Africa, and at least five times to Chile (Matus & Jaramillo 2008), swimming a distance of over 9,000km to the other side of the Pacific comes as only slightly less of a shock.

As extraordinary as it may seem, such long-range longitudinal vagrancy events are routine in penguins in the Southern Ocean. Other examples include: three records of Snares Penguins *Eudyptes robustus* from the Falklands, a movement of *c.*8,000km from the Snares Islands, New Zealand (Demongin

► Moseley's (Northern) Rockhopper Penguin *Eudyptes moseleyi* with Southern Rockhopper Penguins *Eudyptes chrysocome*, Diamond Cove, Falklands, 26 November 2009. Vagrant Moseley's Rockhopper Penguins have hybridised with Southern Rockhopper Penguins on this archipelago (*Alan Henry*).

et al. 2010); and vagrancy of Moseley's (Northern) Rockhopper Penguins *Eudyptes moseleyi,* which nest on Tristan da Cunha, Gough, Amsterdam and St Paul Islands, to South Africa, Australia, New Zealand and the Falklands (Matias *et al.* 2009). Arguably most extraordinary are movements by Royal Penguins *Eudyptes schlegeli,* which breed only on Macquarie Island and nearby islets 1,200km south-west of New Zealand. Royal Penguins have been recorded on Marion Island in the Indian Ocean, in Tasmania and south-east Australia, New Zealand, Antarctica, South Georgia, the Falklands and Chile, effectively spanning most of the Southern Ocean (Matus & Jaramillo 2008, Dehnhard *et al.* 2012). Regular long-distance vagrancy events also frequently lead to hybridisation, for example between Snares Penguins and Erect-crested Penguin *Eudyptes sclateri* on the Snares (Morrison & Sagar 2014) and Southern Rockhopper *Eudyptes chrysocome* x Macaroni Penguins *Eudyptes chrysolophus* on Heard and Marion Islands (Woehler & Gilbert 1990). Vagrancy in penguins may thus be a disruptive evolutionary force, which acts to slow down further penguin speciation and diversification despite their natal philopatry.

Significant latitudinal penguin vagrancy within the Southern Hemisphere is also well documented. This includes southward vagrancy by subantarctic species like Macaroni Penguin into Antarctica (Gorman *et al.* 2010) and also northward vagrancy into temperate or even tropical latitudes by some more northern species. In particular, three species in the genus *Spheniscus* deserve special mention. African Penguins *Spheniscus demersus* sometimes become beached out of range in KwaZulu-Natal and north as far as the Limpopo River mouth in Mozambique (25°13'S). These records typically pertain to first-year birds between June and October, during or immediately after the run of sardines *Sardinops sagax* (Wilkinson *et al.* 1999). On the west coast, birds have been recorded much further north, as far as Sette Cama, Gabon, at 2°32'S (Malbrant & Maclatchy 1958) – 2,500km north of the species' breeding range.

On the other side of the Atlantic, Magellanic Penguins *Spheniscus magellanicus* move northward for the winter following the seasonal pulse of anchovy *Engraulis anchoita* spawning activity and are regular in tropical waters as far north as Rio de Janeiro in the Atlantic and southern Peru in the Pacific (Zavalaga & Paredes 2009). In occasional irruption years associated with low prey abundance, such as the austral winter of 2008, penguins are found much further north, including off north-east Brazil (García-Borboroglu *et al.* 2010). One bird in that irruption made it all the way to the state of Ceará at 2°52'S, 41°16'W – almost to the Equator. Others have moved east to reach Tristan da Cunha, South Africa and even Marion Island in the Prince Edward Island group in February 2006 (Oosthuizen *et al.* 2009), at which point vagrancy to the Western Palearctic on the Cape Verde Islands seems within the realms of possibility. On the Pacific coast, Humboldt Penguins *Spheniscus humboldti* are restricted to the coasts of Peru and Chile affected by the Peruvian Current (Hays 1986) and have been recorded as natural vagrants north to Colombia (Morales Sanchez 1988) and, even more surprisingly, a juvenile made its way across the Equator to El Salvador in June 2007 (Jones & Komar

◄ Macaroni Penguin *Eudyptes chrysolophus* with Adelie Penguin *Pygoscelis adeliae*, Avian Island, Antarctica, 14 January 2013 (*Cameron Rutt*).

► Apparent Macaroni Penguin *Eudyptes chrysolophus* with Royal Penguins *Eudyptes schlegeli*, Sandy Bay, Macquarie Island, Australia, 15 November 2014 (*Fabio Olmos*).

2008). More controversially, there are also a series of Northern Hemisphere records, some well documented with photos, from the west coast of North America in Alaska, British Columbia and Washington state. Van Buren & Boersma (2007) traced five records, concluding that natural vagrancy was 'unlikely' and ship assistance – perhaps better termed abduction or 'penguin-napping' – was more likely, perhaps aboard tuna vessels that fished the west coast of the Americas from northern Chile to California. A subsequent sighting in August 2011 at Willoughby Rock, Washington State, prompted Scordino & Akmajian (2012) to re-open the case for natural vagrancy, given the legal barriers to transportation now in place and the time elapsed from the sightings in the 1970s and 1980s, such that it was unlikely that those 'kidnapped' birds were still at large. Other seabirds from the Humboldt Current have also made it to the northwest seaboard of North America, including Swallow-tailed Gull *Creagrus furcatus* and Nazca Booby *Sula granti*; vagrancy events may be associated with the periodic climate-induced food shortages that affect their regular range (Howell *et al.* 2014). Given the continuation of North American reports, these latter records may need re-evaluation as potentially genuine vagrants, as far-fetched as that might seem.

DIOMEDEIDAE
Albatrosses

Albatross diversity peaks in the Southern Hemisphere, where most species have broadly similar dispersal patterns. After breeding (and even during breeding), individuals of most species disperse widely over the Southern Ocean. A few are quite range-restricted year-round, such that records of Salvin's Albatross *Thalassarche salvini* off Uruguay (Jiménez 2013) and Buller's Albatross *Thalassarche bulleri* off Argentinian southern Patagonia (Tamini & Chavez 2014) are considered vagrants. Identification difficulties confound knowledge of the at-sea distribution of many species but use of satellite tracking has led to a rapid reappraisal of the status of many species. The enigmatic Amsterdam Albatross *Diomedea* [*exulans*] *amsterdamensis* has, for example, been shown to range from waters off the east coast of South Africa all the way to Western Australia (Rivalan *et al.* 2010). Whilst they are ubiquitous in the Southern Ocean, albatrosses are considerably rarer in the Northern Hemisphere, where the only breeding species are confined to the North Pacific. Trans-hemispherical vagrancy is rare but known for both southern albatrosses arriving in the Northern Hemisphere and vice versa. This rarity is likely a consequence of their morphology and difficulty in flying in totally calm weather, such that the equato-

◀ Sooty Albatross *Phoebetria fusca*, Durban pelagic, Kwazulu-Natal, South Africa, 4 June 2016, a major rarity in South African waters (*Niall Perrins*).

rial doldrums probably present a considerable barrier to this family. Two species are responsible for most of the North Atlantic records, Black-browed Albatross *Thalassarche melanophris* and Yellow-nosed Albatross *Thalassarche chlororhynchos*. Yellow-nosed is considerably more regular in the western Atlantic than in the east, and Black-browed vice versa. This is likely to be related to the fact that Yellow-nosed Albatross routinely occurs in tropical waters off eastern Brazil, whereas Black-browed shows a preference for colder waters. Individual Black-browed Albatrosses have made extended stays in the northeast Atlantic – one fraternised with Northern Gannets *Morus bassanus* on the Faroe Islands for 34 years until it was shot in 1894. Subsequently, several more individuals have made long stays, predominantly at gannetries around Scotland, with the exception of one that has returned to the German island of Sylt since 2014; the latter's amorous advances towards Mute Swans *Cygnus olor* have been rebuffed and on one occasion it narrowly missed falling prey to a pair of White-tailed Eagles *Haliaeetus albicilla*. This same individual once grounded itself in a local aviary, apparently attempting to make advances towards a Red-crowned Crane *Grus japonensis* and necessitating rescue; a lovesick albatross will try it on with anything, seemingly! The species remains very rare in the western Atlantic (Patteson *et al.* 1999) although it may prove to be more regular in more northern colder waters south of Greenland (Coffey 2012).

▲ Black-browed Albatross *Thalassarche melanophris*, with Mute Swans *Cygnus olor* and Brent Geese *Branta bernicla*, Rantumbecken, Sylt, Schleswig-Holstein, Germany, 27 April 2018. This individual has returned to the North Sea, and particularly this German lake, since 2014 (*Josh Jones*).

One of the most remarkable of all seabird occurrences involved the appearance of a young Yellow-nosed Albatross in Somerset, south-west England in 2007, the first British record (Rowlands *et al.* 2010). After being rehabilitated by a local care group, this individual disappeared, only to be found at a small fishing lake in Lincolnshire, before then reappearing in the Baltic Sea and then eventually disappearing into the interior of Sweden. These events were particularly remarkable because the entire journey occurred in relatively clement weather conditions. This bizarre behaviour was certainly not weather-related, and it is perhaps more likely that the individual suffered from some kind of cognitive abnormality that caused it to deviate so far from its normal range, habitat and behaviour.

Records of other albatrosses in the Western Palearctic are vanishingly rare but include a record of a Shy Albatross *Thalassarche cauta* in the Red Sea off Egypt and Israel in 1981 (Haas 2012) and two records of Tristan Albatross *Diomedea [exulans] dabbenena*; one collected in Sicily in 1981 (Haas 2012) and another photographed at sea off Norway in May 2019 (Ławicki & van den Berg 2019b). There are old specimen records from Europe, but considerable doubt surrounds their provenance, and many may have been transported aboard boats against their will (Soldaat *et al.* 2009).

Although many vagrant Atlantic albatrosses probably do come to a lonely end, there remains the possibility that one day a pair will meet at a remote gannetry and breed. Not all records in the Northern Hemisphere pertain to singles, and one Black-browed Albatross was collected from a reported group of 10 off Martinique in the West Indies in November 1956 (Kirwan *et al.* 2019). Such an event might even create the nucleus for the recolonisation of the North Atlantic by this family, which was apparently lost around 405,000 years ago when a breeding colony of Short-tailed Albatrosses *Phoebastria albatrus* on Bermuda was submerged by rising sea levels (Olson & Hearty 2003).

Finding a vagrant albatross is not an impossible dream for land-locked birders. In circumstances reminiscent of the aforementioned touring Yellow-nosed Albatross, another individual appeared on Lake Ontario in southern Canada; it was taken into care but eventually died in captivity (Martin & Di Labio 2011). Black-browed Albatross has occurred twice inland in England, in Cambridgeshire and Derbyshire, and Laysan Albatross *Phoebastria immutabilis* has occurred twice on the Salton Sea, California (Patten *et al.* 2003), and twice in Yuma, Arizona (Monson & Phillips 1981). Laysan Albatrosses are restricted as breeding birds mostly to the northwestern Hawaiian islands, but have more recently recolonised islands off Baja California in Mexico and outlying Japanese islands. Despite being essentially restricted to the North Pacific, there is a single record from the Lachlan Banks, 30km off Cape Kidnappers, New Zealand, in December 1995 (Medway 2000) and, even more remarkably, one photographed off Cape Agulhas, South Africa, in April 1983 (Harrison 1983).

There is only one tropical breeding species, Waved Albatross *Phoebastria irrorata* of Española in the Galapagos Islands, which forages in the cold waters of the Humboldt Current. This species has occurred twice extralimitally off Costa Rica (Obando-Calderón *et al.* 2014). In contrast to the Atlantic, the North Pacific has three species of breeding albatrosses that wander widely, and the

► Salvin's Albatross *Thalassarche salvini* with Black-footed Albatrosses *Phoebastria nigripes*, 39km south-west of Pillar Point, San Mateo County, California, United States, 26 July 2014, the first record for California and the second for North America (*Alvaro Jaramillo*).

region has hosted five additional vagrant species. This is in part due to fairly intensive pelagic birding by North American observers, and incredibly includes three species in the 'Shy Albatross' complex: multiples of Shy Albatross *Thalassarche cauta*; a single Salvin's Albatross *Thalassarche salvini* in the western Aleutians, off California and on Midway Atoll; and a presumed returning Chatham Albatross *Thalassarche eremita* off California in 2000–01 (Howell *et al.* 2014). There are two records of Gibson's Albatross *Diomedea [exulans] gibsoni*, one of which made landfall in Sonoma County, California, in July 1967 and roosted overnight on a coastal cliff. Another two Wandering Albatrosses of undetermined (sub)species were seen in November 1970 near the Senkaku Islands, Japan (Brazil 2018), and one was captured in the Bay of Panama in August 1937 (Murphy 1938). Finally, there is a single, incredible, Northern Hemisphere record of Light-mantled Albatross *Phoebetria palpebrata,* which was found associating with Black-footed Albatrosses *Phoebastria nigripes* on a pelagic trip off Marin County, California, in July 1994 (Howell *et al.* 2014).

OCEANITIDAE
Southern Storm-Petrels

Among the austral storm-petrels, the most widely travelled is the transequatorial migrant Wilson's Storm-Petrel *Oceanites oceanicus,* which is regular in all the world's oceans. Previously thought of as a vagrant in north-west Europe, it is now known to be regular in deep Atlantic waters and even enters the Mediterranean with some regularity (Bonaccorsi 2003), and the northern Red Sea at least occasionally (Shirihai 1999). There are also now several records from the North Sea region (Kitching 2002). The sole member of this group to also breed in the Northern Hemisphere, White-faced Storm-Petrel *Pelagodroma marina* has a broad distribution across the world's oceans but is always highly sought-after by birders. This species remains extremely rare in north-west Europe, where ocean conditions may not be to its liking, but is regular in the western Atlantic where hurricane-wrecked birds have been recorded inland as far as Ontario. Its status in much of the temperate South Atlantic is poorly known, but it seems it may still persist as a breeder on St Helena Island (Bolton *et al.* 2009), whose small colony might be the source of vagrants to Brazil (Lima *et al.* 2002). Uncertainty also surrounds the distribution of Black-bellied Storm-Petrel *Fregetta tropica,* which is now being recorded with increasing frequency in both the western (LeGrand *et al.* 2004, Howell *et al.* 2014) and eastern sides of the North Atlantic (Correia-Fagundes & Romano 2011, Gutiérrez *et al.* 2013), which might suggest that small numbers of this species regularly cross the Equator, supported by a recent record of one photographed at sea off French Guiana in July 2018. Even more enigmatic is New Zealand Storm-Petrel *Fregetta maoriana*, a recent record of which off Fiji may have been a vagrant, or perhaps form part of the regular range of this species (Flood & Wilson 2017).

◀ White-faced Storm-Petrel *Pelagodroma marina*, Durban pelagic, Kwazulu-Natal, South Africa, 4 June 2016, a major rarity in South African waters (*Niall Perrins*).

► Black-bellied Storm-Petrel *Fregetta tropica*, Banco de la Concepción, Lanzarote, Spain, 10 September 2018, the sixth record for Spain and seventh for the Western Palearctic (*Paul Dufour*).

HYDROBATIDAE
Northern Storm-Petrels

Our knowledge of the at-sea distribution of many storm-petrels remains very poor. The identification of many species is fraught with difficulty (and mired in scepticism) such that we are probably overlooking many species as vagrants. Swinhoe's Storm-Petrel *Oceanodroma monorhis* is a highly enigmatic vagrant seabird, supposedly occurring with regularity no closer to Europe than the northern Indian Ocean and sometimes the Red Sea (Bailey *et al.* 1968); yet, this species is now detected frequently in the Atlantic Ocean from Cape Verde to Norway, and in the Mediterranean Sea. The capture of several individuals with vascularised brood patches during routine petrel-ringing operations seems to suggest that the species might even breed in Atlantic waters. That the 'Atlantic' Swinhoe's Storm-Petrels are genetically inseparable from those breeding in the north-west Pacific Ocean (Bretagnolle *et al.* 1991; Cubitt *et al.* 1992; Flood 2009) indicates that these birds do not represent an ancient relict population, but are recent arrivals. The closest we have come to the discovery of a breeding site are regular records from Selvagem Grande Island in the Savage Islands archipelago, including a pair duetting in a burrow (Silva *et al.* 2016). The discovery of birds off the east coast of the United States (O'Brian 1999, Howell *et al.* 2014) indicates that birds are not confined to the eastern Atlantic; incredibly, one was caught in a mist-net set for gulls and terns in French Guiana in June 2017, the first for South America (Flood *et al.* 2017). How they reached the Atlantic is also a mystery, perhaps taking a route from the Indian Ocean round the tip of South Africa, via the Agulhas Current and then into the Atlantic (Morrison 1998). However, multiple records from the Mediterranean, including the first European record from the island of Benidorm (King & Minguez 1994) suggest that overland movement from the Red Sea to the Mediterranean remains a possibility. Nevertheless, until a breeding colony is discovered on an Atlantic island somewhere, the mystery of Swinhoe's Storm-Petrel looks set to rumble on.

Storm-petrels are well known for occasionally being wrecked by storms and deposited inland, occasionally appearing in large numbers at inland water bodies. A major 'wreck' of Leach's Storm-Petrels *Hydrobates leucorhous* in the late autumn of 1952 resulted in over 7,000 casualties in Britain and Ireland, and some birds were carried as far as Switzerland (Snow & Perrins 1998). Both the aforementioned species and Madeiran (Band-rumped) Storm-Petrel *Oceanodroma castro* have been recorded at inland locations, with some regularity along the eastern seaboard of North America and inland to the Great Lakes. Madeiran Storm-Petrel remains very rare extralimitally in Europe, but wrecked individuals have appeared in midwinter in both Finland (Anon 1993) and

◀ Swinhoe's Storm-petrel *Oceanodroma monorhis*, South Haven, Fair Isle, Shetland, Scotland, 13 August 2013. This individual was one of two attracted to routine European Storm-Petrel *Hydrobates pelagicus* tape-luring sessions on the island in 2013 (*Chris Batty*).

▼ Leach's Storm-Petrel *Oceanodroma leucorhoa* hunted by Merlin *Falco columbarius*, Myers Point, Tompkins, New York, United States, 30 October 2012, one of many seabirds displaced far inland by Hurricane Sandy (*Jay McGowan*).

Switzerland (Maumary & Baudraz 2000), and the first for Britain was recorded only fairly recently from a pelagic off the Isles of Scilly (Flood 2012). It seems that this species might be more regular along the eastern Atlantic shelf edge than is currently appreciated. Perhaps the most spectacular inland displacement of storm-petrels involved the wreck of 500–1,000 Least Storm-Petrels *Oceanodroma microsoma* at the Salton Sea in California on 10 September 1976 in the wake of Tropical Storm Kathleen (Patten & Minnich 1997).

California has an enviable reputation for storm-petrels and indeed seabirds in general and is responsible for the sole North American records of several species. These include two Tristram's Storm-Petrels *Oceanodroma tristrami* and the only Northern hemisphere record of the enigmatic Ringed Storm-Petrel *Oceanodroma hornbyi* (Howell *et al.* 2014). Until recently, it also had a monopoly on records of Wedge-rumped Storm-Petrels *Oceanodroma tethys,* with four of a total of 13 recorded in 2015 associated with a strong El Niño event and unusually high water temperatures. Tropical Storm Newton brought 13 individuals inland to southern Arizona in September 2016 (Rosenberg *et al.* 2019), a displacement that also brought Black *Oceanodroma melania* and Least Storm-Petrels to the landlocked state. A Markham's Storm-Petrel *Oceanodroma markhami*, recorded 543km west-south-west of San Nicolas Island, California, further underlines the potential for new discoveries for seabirders off California (Pyle 1993).

PROCELLARIIDAE
Shearwaters and Petrels

The possibilities for vagrant petrels and shearwaters seem almost limitless, with birds seemingly routinely appearing in ocean basins or indeed hemispheres where they are not normally anticipated. For example, Monterey Bay, California has received visits from vagrant Grey-faced Petrel *Pterodroma gouldi* from New Zealand and the south Pacific, Streaked Shearwater *Calonectris leucomelas* from the western Pacific, and Great Shearwater *Ardenna gravis* from the Atlantic (Howell *et al.* 2014), among others. How these individuals have come to be tens of thousands of kilometres out of range has long intrigued ornithologists. Although storms may temporarily displace highly pelagic seabirds within sight of land or even leave them wrecked upon it, weather systems rarely cross from one hemisphere to another and are therefore unlikely to be the driver of ultra long-distance vagrancy. It seems likely that for perhaps innate reasons, some individuals end up dispersing between ocean basins of their own volition.

The historic pattern of long-range vagrancy is occluded by problems of provenance of early specimens, many of which may have been fraudulently passed off as vagrants (Bourne 1967). However, increasingly we are realising that some early records ruled 'impossible' in a vagrancy context, may actually be quite plausible. Despite this apparent proclivity to wandering and the lack of physical barriers to dispersal, different petrel species tend to show strong preferences for specific oceanic conditions (Murphy 1936). For example, of a suite of species breeding around the Macaronesian islands of the Azores, Canaries and Cape Verde, some are regular migrants to north-west Europe whilst others are among the rarest of vagrants. These differences reflect the oceanic niches and life histories of these species. Bulwer's Petrels *Bulweria bulwerii*, for instance, feed mainly at night on bioluminescent prey species that migrate to warm surface waters in the dark (Zonfrillo 1986), potentially limiting their capacity to survive in cooler waters. After breeding they migrate to deep oceanic waters of the tropical Atlantic and rarely venture anywhere near the coast (Dias *et al.* 2015). This species was formerly on the British and Irish list, but all records were recently judged to be unacceptable (Harrop 2008, Carmody & Hobbs 2015). No sooner had this happened than a Bulwer's Petrel was found on 20 July 2015 on a small reservoir at Kressbachsee,

▲ Streaked Shearwater *Calonectris leucomelas*, centre with Pink-footed Shearwater *Ardenna creatopus* and Pomarine Skua *Stercorarius pomarinus*, Monterey Bay, Monterey, United States, 30 September 2006, the 14th record for California (*Brian Sullivan*).

◄ Apparent Jouanin's Petrel *Bulweria fallax*, off Santa Cruz County, California, United States, 12 September 2015. This record is still under review, but subsequently an adult was captured at Santa Barbara, California, in June 2016 and became the first accepted record for California and North America (*Fabio Olmos*).

Baden-Württemberg, Germany (Hachenberg *et al.* 2017), so perhaps this species will eventually reinstate itself on both British and Irish lists, although it is still likely to be genuinely extremely rare in the cold waters of the North Atlantic. This species has also occurred as a vagrant to both the Atlantic and once to the Pacific coast of North America – another remarkable trophy for Monterey Bay (Howell *et al.* 2014).

By contrast, the frequency with which Fea's Petrel *Pterodroma feae* occurs off north-west Europe suggests that such individuals are not vagrants, but that the region represents the periphery of the globally-rare species' normal distribution. Recent studies using data loggers have revolutionised our understanding of the at-sea distribution of many rare and enigmatic *Pterodroma* petrels and we now know, for example, that both Fea's Petrel and Zino's Petrel *Pterodroma madeira* (Ramos *et al.* 2017) winter south to Brazil and routinely range into western European and North American waters. This is also true of Bermuda Petrel *Pterodroma cahow* and Trindade Petrel *Pterodroma arminjoniana*, the latter on a routine migration to the north-west Atlantic where it was formerly considered a vagrant. Trindade Petrel is one of several species that have been recorded as a storm-driven vagrant to the interior of North America on several occasions. Even more exciting, and perhaps the candidate for the best 'yard-bird' of all time, was the first documented record for North America of Juan Fernandez Petrel *Pterodroma externa* photographed flying over Tucson, Arizona, in the wake of Tropical Storm Newton in September 2016 (Rosenberg *et al.* 2019). Arguably stranger still is the record of three Atlantic Petrels *Pterodroma incerta* collected 400km away from the coast at the newly inaugurated Tucuruí Dam in the Brazilian Amazonian state of Pará in September 1984 (Teixeira *et al.* 1986). Context for this incredible record was provided in March 2004, when Hurricane Catarina, the first-ever hurricane recorded in the South Atlantic, wrecked at least 129 Atlantic Petrels inland in Rio Grande do Sul State in southern Brazil, including a flock of 50 on a reservoir 190km inland (Bugoni *et al.* 2007).

As populations of several *Pterodroma* petrels rebound following intensive conservation efforts such as removing invasive mammal species, we may see dispersing individuals founding new colonies. A Bermuda Petrel which returned to the same site in the Azores between 2002 and 2006 might perhaps be a precursor to colonisation (Haas 2012). The same cannot, however, be said of spectacular out-of-range records such as the Soft-plumaged Petrel *Pterodroma mollis* from the South Atlantic photographed in a fjord in Arctic Norway, or both Soft-plumaged and Atlantic Petrels found in the Red Sea off Israel and Jordan (Haas 2012). Recent first records of White-necked Petrel *Pterodroma cervicalis* from Baja California, Mexico (Dunn 2015), and the Russian Far East (Korobov & Glushchenko 2014) might also portend a first record from the United States or Canada.

A few petrels are relatively sedentary, with both Snow Petrel *Pagodroma nivea* and Antarctic Petrel *Thalassoica antarctica* having limited ranges associated with pack ice. Snow Petrel has, however, occurred as a vagrant north to Macquarie Island, Australia (Scofield & Wiltshire 2004), the Falklands (Woods 2017) and the Kerguelen Islands (Shirihai 2007). Antarctic Petrel has occurred in those same regions and straggled even further north to South Africa (Chadwick 1991) and Australia. In the Northern Hemisphere, Northern Fulmar *Fulmarus glacialis* is limited to relatively high latitudes, foraging in cold

▲ Soft-plumaged Petrel *Pterodroma mollis*, Nesseby, Troms og Finnmark, Norway, 6 June 2009, the first for Europe and the third for the Western Palearctic. Seen from land on a calm summer day, this bird must rank as one of the most surprising vagrant discoveries of all time (*Graham Catley*).

► Trindade Petrel *Pterodroma arminjoniana*, Hudson Canyon, 200km south-east of Long Island, New York, United States, 20 August 2018, the third state record but probably regular over deep Gulf Stream waters (*Jay McGowan*).

► Bermuda Petrel *Pterodroma cahow*, 190km south-east of Nantucket, Massachusetts, United States, 21 September 2019. As with Trindade Petrel this species may be more regular in northeastern North America than the few sightings suggest (*Tom Johnson*).

▲ Antarctic Petrel *Thalassoica antarctica*, approximately 333km north-west of Macquarie Island, Australia, 4 November 2011, the third accepted record of a live bird in Australian waters (*Adrian Boyle*).

waters, so two recent records from the Southern Hemisphere come as a major surprise. The first was photographed at sea near the Snares Islands, New Zealand, in February 2014 (Miskelly *et al.* 2017) and was followed by one photographed in February 2017 off Valparaíso, Chile (Marin *et al.* 2017). Conversely there are three specimen records of Southern Fulmar *Fulmarus glacialoides* from Maxaranguape, in Rio Grande do Norte in northeastern Brazil, found dead on a beach in June 1863 together with a Cape Petrel *Daption capense*, suggesting a major northerly historical wreck of Antarctic seabirds (Sagot-Martin *et al.* 2020).

Most records of vagrant petrels and shearwaters involve single individuals, but there are occasionally major displacements, ecological events that might even be important for colonisation and speciation over evolutionary timescales. Scattered records of Great Shearwaters *Ardenna gravis* in the Pacific from Chile to California and Alaska have typically involved single birds, but there was a major displacement of at least 50 Great Shearwaters into Australian and New Zealand waters in March–April 2011. Roger & Hull (2016) speculated that this might have been triggered by an oil spill in March 2011 around their breeding colonies off Nightingale Island, Tristan da Cunha, which might have forced birds further south than normal and caused them to be displaced by Antarctic weather systems.

Like *Pterodroma* petrels, shearwaters are also sometimes wrecked inland. Striking examples include entirely incongruous records of Streaked Shearwaters in Wyoming, United States (Howell *et al.* 2014), Wedge-tailed Shearwaters on the Salton Sea, California (Patten *et al.* 2003), and one over flooded paddyfields near Phong Nha Ke Bang National Park, Quang Binh province, Vietnam (Le *et al.* 2017), and a Barolo Shearwater *Puffinus baroli* on a lake in Cheshire, England (Conlin & Williams 2018). In a Western Palearctic context, the best place for vagrant shearwaters is the Red Sea coast of Israel, Jordan and Egypt, which has records of presumably storm-displaced Tropical Shearwater *Puffinus bailloni*, Wedge-tailed Shearwater *Ardenna pacifica,* Flesh-footed Shearwater *Ardenna carneipes* and Streaked Shearwater, all of which are regular in the Indian Ocean (Haas 2012). A relative dearth of extreme vagrant shearwaters elsewhere in the Western Palearctic might just reflect a lack of investment in pelagic trips, and the difficulty of convincing rarities committees that extraordinary claims of seabirds seen from land were genuine (Bourne 1992a). However, in the summer of 2020 there were records of a Flesh-footed Shearwater from the Azores, Short-tailed Shearwaters *Ardenna tenuirostris* from Ireland and France, and a White-chinned Petrel *Procellaria aequinoctialis* in Scotland, which does suggest a linked displacement of Southern

► White-chinned Petrel *Procellaria aequinoctialis* with Western Gull *Larus occidentalis*, off Half Moon Bay, California, 15 September 2015, the fourth record for California (*Fabio Olmos*).

▼ Tahiti Petrel *Pseudobulweria rostrata*, Durban pelagic, Kwazulu-Natal, South Africa, 11 November 2018, the first African record of this species (*Niall Perrins*).

Hemisphere seabirds in that year. Better observer effort on boats in the western Atlantic has produced some incredible records of vagrant seabirds, including the sole Buller's Shearwater *Ardenna bulleri* for the Atlantic, seen off New Jersey in October 1984 (Boyle 2011). More recently, a Tahiti Petrel *Pseudobulweria rostrata* was in the Gulf Stream off North Carolina in May 2018; this was followed by a second record photographed off South Africa in November 2018 (Allan & Perrins 2019) and then one photographed in February 2019 on a pelagic off Mirbat, Dhofar, Oman.

The small black-and-white *Puffinus* shearwaters are among the most difficult to get to grips with in a vagrancy context, given difficulties of field identification compounded by the fact that vagrants often associate with large flocks of other species in the same genus. For example, Manx Shearwaters *Puffinus puffinus* have been recorded with increasing frequency in the north-east Pacific since the mid-1970s between Alaska and Canada, and it has been speculated that they may now be breeding somewhere in the Pacific Northwest (Mlodinow 2004). Records further south from Mexico (Miguel & McGrath 2004), Ecuador (Freile *et al.* 2017), Peru (CRAP 2016) and Chile (Howell 2007) suggest that these birds may have entered the Pacific from wintering areas in the South Atlantic. Elsewhere this species has wandered as far afield as Israel (Harrison 2016), Australia (Palliser *et al.* 2013) and New Zealand (Powlesland & Pickard 1992). The very similar Newell's Shearwater *Puffinus newelli* has wandered to California from Hawaii (Howell *et al.* 2014) whilst Yelkouan Shearwater *Puffinus yelkouan* has recently been accepted as a vagrant to Britain with several other records pending (Darlaston & Langman 2016). Perhaps the most famous vagrant in this group is Bryan's Shearwater *Puffinus bryani*, which was first described as a new species to science in 2011 based on a vagrant collected in 1963 on Midway Atoll (Pyle *et al.* 2011). A second prospecting bird was recorded on Midway between 1990–92 and the species' breeding grounds were subsequently discovered on the Bonin Islands, Japan (Pyle *et al.* 2014).

The diminutive *Pelecanoides* diving-petrels do not migrate long distances, but have nevertheless occurred extralimitally with some regularity, with Magellanic Diving-petrel *Pelecanoides magellani* noted as a vagrant to the Falklands (Woods 2017) and once to southern Brazil (Vooren & Fernandes 1989) while there is also an unusual record of a South Georgia Diving-petrel *Pelecanoides georgicus* in the South Atlantic, 2,300km west-north-west of its nearest breeding colony on the Prince Edward Islands (Rollinson *et al.* 2017).

Pachyptila prions wander a little more, but understanding vagrancy in this group is also confounded by the difficulty of at-sea identification. For example, we have little or no idea where MacGillivray's Prions *Pachyptila macgillivrayi* disperse during the non-breeding period. Most confirmed vagrants are thus tideline corpses, including Broad-billed Prions *Pachyptila vittata* in Peru (Hidalgo-Aranzamendi *et al.* 2010), Brazil (Carlos 2005) and the Falklands (Woods 2017); and Antarctic Prion *Pachyptila desolata* on Christmas Island (James & McAllan 2014). Multi-species wrecks of prions are recorded with some regularity in the southern cone of South America, north to southern Brazil, and may involve thousands of birds (Post 2007). Although there are as yet no northern hemisphere records, in one major Brazilian wreck an Antarctic Prion was found dead on the coast of Marajo Island in the eastern Brazilian Amazon, just half a degree from the Equator (Martuscelli *et al.* 1997). Antarctic and Slender-billed Prions *P. belcheri* have also come tantalisingly close to the Northern hemisphere off the coast of Kenya (Fanshawe *et al.* 1992). Similar identification difficulties also confound our knowledge of the distribution of vagrant giant petrels, with several not assigned to species, but Northern Hemisphere records include a Southern Giant Petrel *Macronectes giganteus* seen from a ferry on the Adriatic off Italy (Haas 2012) and a Northern Giant Petrel *Macronectes halli* photographed off Washington state, United States, in December 2019.

▼ Antarctic Prion *Pachyptila desolata*, Tramandaí, Rio Grande do Sul, Brazil, 25 July 2015. Most coastal Brazilian records of prions are of beach-wrecked birds after winter storms (*Paulo Ricardo Fenalti*).

CICONIIDAE
Storks

Of the 19 species of storks, relatively few are obligate migrants although, like many waterbirds, many are nomadic. Most also have very long and broad wings adapted for soaring flight; these greatly reduce their potential for long overwater flights, as is generally the case for soaring raptors. White Stork *Ciconia ciconia* somewhat bucks this trend, however. It is one of the most highly migratory of all broad-winged birds, with some migrating 20,000km annually between the Palearctic and the Afrotropics, travelling around the Mediterranean via the Straits of Gibraltar or the Bosphorus. White Storks are not only regular vagrants north of their current range in Europe but for a large soaring bird, they have an incredible track record of vagrancy to oceanic islands. Not only have they reached all the European Atlantic islands of Iceland, Madeira, the Canaries, the Cape Verdes and the Azores (Snow & Perrins 1998), but they have reached the distant mid-Atlantic outposts of Ascension Island (Bourne & Simmons 1998) and St Helena (Prater 2012) in the Southern Hemisphere, and even reached the Prince Edward Islands (Burger *et al.* 1980). Two records from the Caribbean – from Antigua (Gricks 1994) and Martinique (Leblond 2007) – indicate that future records from the North American or South American mainland seem likely. Oriental Stork *Ciconia boyciana* does not have such an epic migration as its western cousin, but vagrants have occurred in the Philippines (Collar *et al.* 2001) and as far west as Lake Baikal (Mlíkovský 2009). In the Neotropics, Maguari Stork *Ciconia maguari* is somewhat nomadic and has appeared out of range as far north as Costa Rica (Obando-Calderón *et al.* 2013) and west to Peru (CRAP 2017), and is a conceivable vagrant to North America, whilst Asian Woolly-necked *Ciconia episcopus* has wandered widely to northern Vietnam, Bangladesh and China, and notably Iran (Elliott *et al.* 2020).

Several tropical stork species are highly nomadic and extremely abundant. Coupled with their very high detectability, they might be expected to occur as vagrants to temperate latitudes, much like nomadic tropical rail species. However, stork species are also popular with aviculturists, making it difficult to ascertain whether an out-of-range individual is a genuine vagrant or an escaped pet. Records of both Marabou Stork *Leptoptilos crumenifer* and Yellow-billed Stork *Mycteria ibis* from the Western Palearctic tend to cause rarities committees headaches in this way. Marabou Storks have been accepted as wild birds from Morocco – five records, including groups of up to four individuals (Thévenot *et al.* 2003) – twice in Israel (Haas 2012) and one recently photographed in Egypt. By 2010, there were 27 records of Marabou Storks from Spain and seven from Portugal; some were known escapes, but pooling the records there are consistent

► Black Stork *Ciconia nigra*, Moriyamachoshimoimuta, Isahaya-Shi, Nagasaki, Japan, 21 Dec 2019, an annual vagrant to this region (*Yann Muzika*).

◀ Jabiru *Jabiru mycteria* and Great White Egret *Ardea alba*, Arari Marsh, Maranhão, Brazil, 21 January 2013, the only recent record from the state (*Alexander Lees*).

seasonal patterns that attest to genuine vagrancy, with a similar pattern also holding for French records not currently accepted as wild (de Juana & Garcia 2015).

The situation with Yellow-billed Stork *Mycteria ibis* is broadly similar. It is regular in southern Egypt and accepted as a vagrant to Israel (Shirihai 1996), Tunisia (Ouni 2007), Morocco (Thévenot *et al.* 2003), Turkey (Kirwan *et al.* 2008) and even Bulgaria (Ragyov *et al.* 2003). There are also records from Spain, with three placed in a holding category and the rest assumed to be escapes (de Juana & Garcia 2015). Other *Mycteria* storks also wander widely: Milky Stork *Mycteria cinerea* has occurred as a vagrant to Thailand and Vietnam (Elliott *et al.* 2020a) and Wood Stork *Mycteria americana* has occurred in the New World north to British Columbia, Quebec and Manitoba (Campbell *et al.* 1990), with many extralimital records associated with hurricanes (DeBenedictis 1986). Jabiru *Jabiru mycteria* has occurred as a vagrant to three states in the southern United States with 10 records from Texas and singles from Oklahoma, Mississippi and Louisiana (Howell *et al.* 2014). These may emanate from the small population in Mexico or be wanderers from much farther afield. Vagrancy events among remaining stork species are less spectacular, although particularly noteworthy is a single recent Western Palearctic record of African Openbill *Anastomus lamelligerus* from Luxor in Egypt in May 2009 (Steffe 2010).

FREGATIDAE
Frigatebirds

Frigatebirds are mercurial flyers that routinely ride thermals to altitudes of 2,500m or more (Weimerskirch *et al.* 2003). They can even sleep on the wing by shutting down one hemisphere of their brain at a time, or even both hemispheres simultaneously (Rattenborg *et al.* 2016). This strong association with oceanic thermals means these specialist tropical pirates are largely restricted to trade-wind zones (Weimerskirch *et al.* 2003). Moreover, their high- flying habits and inability to land on water may make them uniquely vulnerable as seabirds to extreme weather events such as hurricanes, since they cannot ride out storms on the surface. Inland records of Magnificent Frigatebirds *Fregata magnificens* are annual in Florida and closely tied to the peak of the tropical cyclone season (McNair 2000), and others have been widely recorded inland in North America. Hurricane Wilma displaced many Magnificent Frigatebirds in November 2005; for example, the first two confirmed individuals for New Jersey and a further two unidentified frigatebirds which were probably also Magnificent (Barnes *et al.* 2006). The same storm system displaced one unfortunate individual across the Atlantic, where it was picked up in a field in Shropshire, England, in November (Bradbury *et al.*

▲ Magnificent Frigatebirds *Fregata magnificens*, University Lakes, East Baton Rouge, Louisiana, United States, 13 July 2019, wrecked by Hurricane/Tropical Storm Barry (*Cameron Rutt*).

► Lesser Frigatebird *Fregata ariel* with Lesser Black-backed Gulls *Larus fuscus barabensis*, Haitham, Dhofar, Oman, 1 March 2016 (*Jens Eriksen*).

2008). This species has straggled to Europe on several other occasions, with records in France, Denmark, Spain and the Isle of Man (Lewington *et al.* 1992), and they are regular as vagrants south as far as Punta Rasa in Argentina (Jaramillo 2000). A recent first record for Senegal seems as likely to involve transatlantic vagrancy as it does to involve one of the individuals of the tiny Cape Verde population (Piot & Lecoq 2018).

Field identification of frigatebirds is very challenging, a problem highlighted when the specimen of the first British record of Magnificent Frigatebird from Tiree, Argyll, Scotland, in July 1953 was re-examined and reidentified as an Ascension Frigatebird *Fregata aquila* (Walbridge *et al.* 2003). Two subsequent Western Palearctic records of this species have followed; another on the nearby Hebridean island of Islay, Argyll, in July 2013 (Chalmers, 2013) and a recent photo-documented record from Cape Verde in June 2017. The tendency for this species to wander was recently highlighted by a satellite-tagged Ascension Frigatebird *Fregata aquila* which travelled around 45,000km over a 3.5-month period and passed into Brazilian territorial waters, furnishing the first record for the Americas (Williams *et al.* 2017).

In the Western hemisphere, Great Frigatebird *Fregata minor* has occurred as a vagrant twice to California and once to Oklahoma (Howell *et al.* 2014), as well as to Aruba in the Caribbean (Mlodinow 2006) and once to Fernando de Noronha in north-east Brazil (Silva e Silva & Carlos 2019). The latter probably originated from the tiny population that nests on Trindade Island. Great Frigatebirds occur as vagrants in the Pacific with some regularity north to Japan (Brazil 2018) and south to New Zealand (Miskelly *et al.* 2017). Lesser Frigatebird *Fregata ariel* has also wandered widely in the United States, with singles from Maine, Wyoming, Missouri and California (Howell *et al.* 2014). It would seem unlikely that records from the eastern seaboard of the United States would involve the recently proposed split 'Atlantic' Lesser Frigatebird *Fregata [ariel] trinitatis* from Trindade Island, Brazil (Olson 2017), and are more likely to have involved birds displaced from the Pacific. Lesser Frigatebirds have ventured north up the Red Sea twice as far as Israel and Jordan and up the Persian Gulf to Kuwait (Haas 2012). Further east, the first for Bangladesh was recorded in May 2016 from the Sundarbans (Jahan *et al.* 2016), and in the Pacific this species is regular north to Japan and even Ussuriland in Russia (Brazil 2009). Finally, Christmas Island Frigatebird *Fregata andrewsi* has travelled widely in the Indian and Pacific Ocean regions with three records from Australia (McMaster *et al.* 2015), records from south-east China and Japan (Brazil 2009), multiple records from India (Praveen *et al.* 2013) and an old photo-documented record from Kenya (Fisher & Hunter 2016). Perhaps this is another candidate for future vagrancy to the Red Sea and Persian Gulf?

SULIDAE
Gannets and Boobies

The three species of gannets are extremely wide-ranging in the non-breeding season and are routine vagrants. Not only have vagrant Australasian Gannets *Morus serrator* hybridised with Cape Gannets *Morus capensis* at South African colonies (Dyer *et al.* 2001), but vagrant Cape Gannets *Morus capensis* have also set up territory and hybridised with Australasian Gannets in Australia (Robertson & Stephenson 2005). Both species have also attempted to breed on Saint Paul Island, in the middle of the southern Indian Ocean, a place almost equidistant between the main gannetries of either species (Lequette *et al.* 1995). Incredibly, both 'southern gannets' have also occurred as vagrants to Brazil, a country that can also boast the sole Southern Hemisphere record of Northern Gannet *Morus bassanus,* from Ceará in February 2016 (Bege & Pauli 1989, Vooren 2004, Teixeira *et al.* 2016), and thus complete the gannet set. Cape Gannets have also been recorded from Argentine Patagonia (Rebstock *et al.* 2010), in the Pacific off Peru (García-Godos 2002) and in the northern Indian Ocean off Oman (Eriksen 2004). The species has a chequered history as a vagrant to the Western Palearctic. No sooner had all old records been deemed not proven by Crochet &

◄ A pair of Cape Gannets *Morus capensis* (with the long gular stripe) in a colony of Australasian Gannets *Morus serrator*, Point Danger, Glenelg, Victoria, Australia, 21 December 2017 (*Adrian Boyle*).

Haas (2008) than one was conclusively photographed at sea between Corvo and Flores on the Azores in April 2016. Northern Gannet has also made it all the way to the Pacific – an adult that was first found on Southeast Farallon Island, California, in April 2012 (Garrett 2014), which was still in residence in early 2020. Given the smattering of records of other gannets tens of thousands of kilometres from home, there seems to be little evidence of limits to gannet vagrancy.

The smaller boobies are somewhat less globe-trotting than their larger gannet relatives, and most species are found in tropical and subtropical seas. Better protection of breeding colonies and a release from historical persecution has facilitated rapid expansion of many colonies, and the ranges of several species are now expanding. Concomitantly, vagrancy to northern temperate latitudes has increased markedly in recent years, a phenomenon that might also be helped by warming seas associated with global climate change. Prior to 2010 Red-footed Booby *Sula sula* was only known in the Western Palearctic from four records around Cape Verde (Haas 2012), but the species began breeding on this archipelago in 2016 (Ławicki & van den Berg 2016d). Associated with this expansion have been the first records from north-west Europe, in Britain, Portugal, Spain and France (Cohen 2018). Red-footed Boobies have been widely recorded as vagrants elsewhere, including New Zealand (Miskelly *et al.* 2017), India (Lopes & Kasambe 2016), Peru (García-Godos & Sánchez 2009) and both the Atlantic and Pacific coasts of North America (Howell *et al.* 2014).

A similar phenomenon of increasing vagrancy is also apparent for Brown Booby *Sula leucogaster*, with a major displacement into north-west Europe evident in autumn 2019 when the first records for both Britain and France were recorded. This species has the most ample distribution of any booby and is seemingly most tolerant of cooler temperate waters. It may well become a regular sight in Europe with continued range expansions in the Atlantic islands, and in North America it is routinely wrecked inland by hurricanes. The pattern of extralimital occurrence of Masked Booby *Sula dactylatra* is broadly similar, with Western Palearctic records from six countries, albeit without such a strong recent increase as the preceding two species. Remaining species have more restricted global distributions. Blue-footed Booby *Sula nebouxii* has occurred north to British Columba in Canada (Towers *et al.* 2015) and there are occasional major displacements of this species into the desert south-west of the United States from the Gulf of California in Mexico (Howell *et al.* 2014). Nazca Booby *Sula granti* has undergone a rapid change in status in the North Pacific with many records from California in recent years after the first was confirmed as recently as 2016 (Yang *et al.* 2016). Further records in the Pacific have seen birds wander north to Alaska (Gibson *et al.* 2018), and one was reported off Japan in March 2019. A similar phenomenon of population expansion might explain the seemingly incongruous record of an Abbott's Booby *Papasula abbotti*, a species rarely found far from Christmas Island, from Rota, in the Mariana Islands in April 2007; this was the first contemporary record for the Pacific Ocean (Pratt *et al.* 2009). Abbott's Booby suffered a major historical range contraction due to persecution, and as colonies re-grow then vagrant 'propagules' may eventually re-establish populations at other geographically disjunct sites.

► Red-footed Booby *Sula sula*, Port de l'Estartit, Girona, Spain, 8 December 2010, the second record for Europe (*Stuart Piner*).

ANHINGIDAE
Darters

The four species of darters are mostly sedentary, but occasional significant vagrancy events have been recorded, including eight records of Anhinga *Anhinga anhinga* from California (Hamilton *et al.* 2007), two records of African Darter *Anhinga rufa* from Morocco (Thévenot *et al.* 2003) and four records of Australasian Darter *Anhinga novaehollandiae* from New Zealand (Sagar 2013).

PHALACROCORACIDAE
Cormorants and Shags

Most species in the Phalacrocoracidae are sedentary and long-range vagrancy is relatively rare in the group. A notable exception is Double-crested Cormorant *Phalacrocorax auritus*, which is the only species for which significant transoceanic vagrancy has been reported. This North American species was first recorded in the Western Palearctic as recently as 1989 in the United Kingdom (Williams 1996), and there have been subsequent European records from France (De Broyer 2001), Ireland (Rogers *et al.* 1996) and Tenerife in the Canary Islands (Copete *et al.* 2015), along with records from most of the islands in the Azorean archipelago (Garcia-del-Rey 2011). Beyond Europe, this species – which is highly migratory in the north of its range – has occurred south to Colombia (Donegan & Huertas 2015) and even skipped across the Bering Sea to Chukotka in Russia (Brazil 2008). Records of vagrancy in many cormorants are associated with range and population expansion. For example, Pygmy Cormorants *Microcarbo pygmaeus* are becoming increasingly frequent visitors to northern and western Europe (Ławicki *et al.* 2012). Similarly, the population expansion of Little Black Cormorant *Phalacrocorax sulcirostris* in Australasia is associated with first recent records from the Solomon Islands (Butcher *et al.* 2018) and Fiji (Cherry 2005). The

▼ Double-crested Cormorants *Phalacrocorax auritus*, Vila Franca do Campo, São Miguel, Azores, 28 October 2012. The Azores boasts most of the Western Palearctic records of this species (*Vincent Legrand*).

same is true of Neotropic Cormorant *Phalacrocorax brasilianus* in North America, now recorded as a vagrant regularly as far north as the Great Lakes, quickly rendering distribution maps even in recently published field guides rather obsolete.

Two other controversial cormorants merit a mention. First, a recent re-examination of the sole Spanish record of Pygmy Cormorant collected in Catalunya, Spain, around 1855 resulted in its reidentification as a Long-tailed Cormorant *Microcarbo africanus* (Copete *et al.* 2011). This record has only been admitted to Category D of the Spanish national list because of the possibility that the specimen might have come from elsewhere (Gutiérrez *et al.* 2010). However, the record would fit with a pattern of vagrancy by other Afrotropical wetland species, and future Palearctic records of Long-tailed Cormorant might be expected. Secondly, there is a single record of Antarctic Shag *Phalacrocorax bransfieldensis* from Bahia in Brazil, which is arguably one of the least expected vagrancy events on record, as this species is restricted to the Antarctic Peninsula (Lima *et al.* 2002). The bird in question was an individual found dead on a beach, which carried a ring from the South Shetland Islands in Antarctica. Unfortunately, the carcass was not retained or its identity verified by an ornithologist, so there remains the possibility that another species, such as one of the southern skuas, might have been involved, with the record being the result of a mix-up with the ring number (F. Olmos pers. comm.).

PELECANIDAE
Pelicans

Most pelicans are not highly migratory, but they are highly dispersive and may make rather erratic movements of thousands of kilometres in search of the best seasonal wetlands. Pelicans are another challenging group for rarities committees as many species are kept in captivity, and historically many extralimital records were routinely condemned as escapes. Understanding the status of pelicans in north-west Europe has proven particularly difficult given the presence of known free-flying individuals of captive origin, whilst at the same time wild populations are increasing and natural vagrancy is to be expected. Jiguet *et al.* (2008) explored the occurrence pattern in space and time of three species of pelican – Pink-backed Pelican *Pelecanus rufescens*, Dalmatian Pelican *Pelecanus crispus* and Great White Pelican *Pelecanus onocrotalus* – and despite the

▼ Dalmatian Pelican *Pelecanus crispus*, Land's End, Cornwall, England, 9 May 2016. This, the first accepted British record, was tracked based on plumage features as it passed through Poland, Germany and France, before arriving in England (*Stuart Piner*).

statistical 'noise' presented by known escapes, they found that observed patterns of vagrancy into Europe could be explained by variations in wild population size, breeding success and/or climatic factors. Dalmatian Pelican was recently added to the British List on the basis of an individual first recorded in Poland and then tracked through France and Germany. Given their Afrotropical distribution, Pink-backed Pelicans tend to be viewed with more scepticism, but records are now accepted from Israel and Jordan (Shirihai *et al.* 2000) and Turkey (Kirwan *et al.* 2014) and individuals may end up entrained with 'carrier' flocks of Great White Pelicans returning from East African wintering grounds. Scepticism towards pelicans from some birders is likely to dwindle if vagrancy events start to involve larger numbers as populations grow and wetlands are restored. For example, an invasion of over 60 Great White Pelicans occurred on Sardinia in spring 2008 (Grussu & Atzeni 2012). Jiguet *et al.* (2008) found that birds wandered more widely in years of breeding failure at their Greek colonies.

Beyond Europe, Spot-billed Pelican *Pelecanus philippensis* is not strongly migratory but has been recorded as a vagrant east to Japan (Hisharo *et al.* 2010) and west to the Maldives (Anderson 2007). Australian Pelican *Pelecanus conspicillatus* is not migratory but is nomadic and at least occasionally irruptive – an irruption of this species in 1987 led to records across Indonesia (Elliott *et al.* 2020). Australian Pelicans have also reached as far afield as Nauru (Buden 2008), New Zealand (Miskelly *et al.* 2015) and Mindanao, in the Philippines, in September 2016; the last caused enough of a stir to reach the international press. In the Western Hemisphere it is Brown Pelican *Pelecanus occidentalis* that has wandered most, occurring as a vagrant across most of North America and also in South America, where there are now multiple records, for example from the Amazon Basin (Gomes *et al.* 2016) and one even from the highlands of Ecuador (Jahn *et al.* 2010). The deltaic region at the mouth of the Amazon may be an 'ecological trap' for vagrant Brown Pelicans that hug the coast on migration; the river is so wide in this region that 'following the coast' will swiftly take individuals into the interior of Amazonia. The superficially similar Peruvian Pelican *Pelecanus thagus* is not known to wander so much; a published record from Brazil has subsequently been retracted as fraudulent (Patrial *et al.* 2011) but there are images of an immature from eastern Argentina (eBird) which suggest that vagrancy to the Atlantic is possible. Finally, American White Pelican *Pelecanus erythrorhynchos* is frequently recorded as a vagrant across North America and has been recorded south as far as Costa Rica (Sandoval *et al.* 2010). Records of this species and Brown Pelican from Bermuda indicate that transatlantic vagrancy is not out of the question.

▲ American White Pelican *Pelecanus erythrorhynchos*, Montezuma NWR, Seneca, New York, United States, 25 August 2019, a scarce wanderer to eastern North America (*Jay McGowan*).

► Brown Pelican *Pelecanus occidentalis* Clinton Lake, De Witt County, Illinois, United States, 22 November 2004. A vagrant anywhere away from the coast in North America, this individual was 1,000 km out of range (*Chris Wood*).

BALAENICIPITIDAE
Shoebill

Shoebills *Balaeniceps rex* are regarded as strictly sedentary but there is a record of a vagrant that wandered between three sites in southern Kenya in 1994–95 (Stevenson & Fanshawe 2002).

SCOPIDAE
Hamerkop

There are no extralimital records of Hamerkop *Scopus umbretta*, even from the Seychelles, despite breeding in neighbouring Madagascar.

ARDEIDAE
Herons, Egrets and Bitterns

Incidence of vagrancy in this group, which includes many migratory taxa, is high, and most species have been recorded extralimitally at one time or another; transoceanic vagrancy events are routine in many species. For example, the Azores have hosted 18 species of herons, of which eight are exclusively Nearctic vagrants (Barcelos *et al.* 2015), whilst of 10 species occurring in Hawaii, eight are vagrants, including two exclusively Old World species (Pyle & Pyle 2017). Such extralimital movements are commonest in species that breed at higher latitudes, and geographic quirks may predispose some species to higher rates of vagrancy. For example, as with many species of waders and some landbirds, many North American species regularly overfly part of the western North Atlantic on migration, rendering them susceptible to weather-driven transatlantic vagrancy. It is therefore somewhat surprising that Nearctic herons do not occur on the European side of the Atlantic more frequently. Jackson (2004) suggested three hypotheses to explain

◀ Green Heron *Butorides virescens*, Lost Gardens of Heligan, Heligan, Cornwall, 7 November 2010, the seventh record for Britain (*Stuart Piner*).

this relative rarity: a) their large body size enables them to better cope with their initial displacement during the autumn storms that bring other Nearctic species to Europe; b) these species are displaced by storms but suffer from poor survivorship on the crossing, or c) southerly species avoid entrainment in cyclonic conditions by their habit of migrating close inshore (in contrast to more northern species, which might have a greater transoceanic component to their migration). All three seem highly plausible hypotheses, but critically (and previously unmentioned), population size must also be a key factor. Populations of all these species are much smaller than the comparably more frequent passerines and most (but not all) shorebirds, being limited in effect by the amount of suitable wetland habitat. Moreover, building on hypothesis c), the distributions of most of the species in question lie much further south than the more regular transatlantic vagrants to Europe. Indeed, the occurrence of relatively southern species such as Little Blue Heron *Egretta caerulea* as far north as Ireland is quite surprising. Nevertheless, the simultaneous occurrence in late October 1889 of two American Bitterns *Botaurus lentiginosus* and a Green Heron *Butorides virescens* in Ireland (Hudson 1972), hints that 'falls' of American herons are possible, and suggests that such multiple arrivals are most likely the result of wind drift (and possibly ship assistance) rather than an individual's deviant migratory orientation. Alongside records in Europe, and particularly the Macaronesian Islands of the Azores and Canaries, both Snowy Egret *Egretta thula* and Little Blue Heron have occurred as vagrants to South Africa. Herons are not apparently reticent to rest on ships, and multiple ship-assisted transatlantic

◀ Great Bittern *Botaurus stellaris*, Fljótsdalur, Austurland, Iceland, 3 February 2019, the eighth record for Iceland (*Yann Kolbeinsson*).

crossings have been documented for Great Blue Heron *Ardea herodias* (Gantlett 1998), whilst Grey Herons *Ardea cinerea* have completed the crossing in the opposite direction aboard ships (Pranty *et al.* 2007).

It seems likely that European vagrants of Nearctic species originate from the northern periphery of their respective ranges, where the migration route is longest. The recovery of a Little Blue Heron, ringed as a nestling in June 1964 in New Jersey, United States (close to the northern distributional limit of the species), on the island of Flores in the Azores in November 1964 (Dennis 1981) adds weight to this supposition. The most frequently encountered Nearctic heron species in Britain and Europe is American Bittern, which might appear surprising considering its skulking habits – which ought to preclude regular detection. However, bitterns seem to have a predilection to vagrancy, and American Bittern occurs both further north than other heron species and has a longer migration span, breeding as far north as Hudson Bay and wintering south to Panama. Beyond the aforementioned Grey Heron there are some records of transatlantic heron vagrancy from the Europe to North America with two further records of Grey Heron in Newfoundland, along with two to four records of Western Reef Heron *Egretta gularis* and now fairly regular Little Egrets *Egretta garzetta* in the American north-east (Howell *et al.* 2014). Heron vagrancy to the Western Hemisphere is more prevalent at lower latitudes, with favourable trade winds conveying birds from the Palearctic and Afrotropics to the Caribbean and north-east South America. There are records of Little Bittern *Ixobrychus minutus*, Squacco Heron *Ardeola ralloides*, Purple Heron *Ardea purpurea* and Western Reef Heron from the West Indies, whilst Little Egret has established breeding colonies on Barbados and Antigua (Kushlan & Prosper 2009, Kirwan *et al.* 2019). A recovery of a Spanish-ringed Little Egret on Martinique indicates a likely source of the colonists, and this species has been recorded hybridising with Snowy Egrets *Egretta thula* in French Guiana (Renaudier 2010). A Grey Heron ringed in Charente-Maritime, France, was killed near Belém, Pará, in the Brazillian Amazon in 1973 (Novaes 1978). Further south, the Brazilian island of Fernando de Noronha is also a heron vagrant trap and has hosted Little Egret, Purple, Grey and Western Reef Herons, with Squacco Heron now resident there and probably breeding on the island (Whittaker *et al.* 2019).

Transpacific vagrancy of herons is frequent in the north, facilitated by the Beringian island stepping stones and intensive coverage of some of the Alaskan islands. These have produced some incredible records of vagrant herons. Most notable is surely a single record of Chinese Egret *Egretta eulophotes* from Aggatu, Alaska in June 1974 (Howell *et al.* 2014). This species has also wandered to Bali, Sulawesi and the Andaman Islands. Other stellar vagrants from the Bering Sea region include two Alaskan records of Intermediate Egret *Ardea intermedia*, three Chinese Pond Herons *Ardeola bacchus*, a Great Bittern *Botaurus stellaris* and a Yellow Bittern *Ixobrychus sinensis* (Howell *et al.* 2014). Chinese Pond Herons have been widely recorded elsewhere as vagrants, including in India, the Philippines, Guam, the Cocos Islands, Christmas Island and near Broome in Western Australia (Martínez-Vilalta *et al.* 2020). Records of vagrants in Europe have proven more controversial, with records from Norway,

► Tricoloured Heron *Egretta tricolor*, Playa de las Américas, Tenerife, Canary Islands, Spain, 13 February 2008, the second or third record for the Western Palearctic; conceivably the same individual was present on the Azores in October 2007 (*Chris Batty*).

◄ Snowy Egret *Egretta thula*, River Club Golf Course, Western Cape, South Africa, 9 June 2015, the second record for South Africa (*Niall Perrins*).

◄ Pacific Reef-Herons *Egretta sacra*, St Martin's Island, Cox's Bazar, Bangladesh, 2 January 2010 (*Sayam Chowdhury*).

Hungary and Finland treated as escapes, but two records from England have recently been added to the British List.

Some records of small bitterns in the genus *Ixobrychus* have also proven controversial; there is one accepted record of Schrenck's Bittern *Ixobrychus eurhythmus* from Europe, collected in Piemonte, Italy, in November 1912; there is another rejected 19th-century record from Germany which does not have sufficient detail (Haas 2012). This species has wandered as far as Australian territory, with two records from Christmas Island (James & McAllan 2014). As well as the aforementioned Alaskan record, Yellow Bittern is a vagrant to the Maldives, Chagos Islands and multiple Australian islands and the mainland. A specimen record from Dorset in England has been treated with suspicion and was not admitted to the British List (Melling *et al.* 2008). This species has now been found breeding at the periphery of the Western Palearctic in Egypt (Päckert *et al.* 2014); it is unclear how long the species may have

▲ Dwarf Bittern *Ixobrychus sturmii*, Barranco de Río Cabras, Fuerteventura, Canary Islands, 26 February 2018. This individual had been in residence at this site for three years (*Paul French*).

► Black Heron *Egretta ardesiaca*, Barragem de Poilão, Ilhas de Sotavento, Cape Verde, 24 March 2016. Of ten records of this species from the Western Palearctic, seven have come from Cape Verde (*Josh Jones*).

occupied the region, but long-distance vagrancy has also resulted in its colonisation of the Seychelles and several Pacific islands. Cinnamon Bittern *Ixobrychus cinnamomeus* is perhaps another vagrant possibility to the Western Palearctic or Nearctic and has been widely recorded as a vagrant in the east to Christmas Island (Carter 2003), Western Australia (Hassell & Ward 2012) and Oman (Eriksen & Porter 2017).

Vagrancy is not confined to temperate zone species, although it is less prevalent in the tropics. A notable exception is Dwarf Bittern *Ixobrychus sturmii* an Afrotropical species, which is becoming a regular visitor to the Atlantic islands of the southern Western Palearctic (Haas 2012), in addition to an old record from France and recent records from mainland Spain and Malta (Pitches 2011). Further afield, this species has reached Oman (Eriksen & Porter 2017) and most impressively St Helena in the South Atlantic, requiring a sea crossing of over 2,400km if unaided (Hillman & Clingham 2012). Future records from northern

Europe or even the Western Hemisphere seem likely. Black Heron *Egretta ardesiaca* also wanders north to Cape Verde (Hazevoet 2010), Israel (Shirihai 1999), Greece (van den Berg & Haas 2012) and most recently to Porto Cesareo, Puglia, in Italy (Ławicki & van den Berg 2017d). To the east, the Malayan Night-Heron *Gorsachius melanolophus* has reached as far north as Shikoku Island, Japan, Palau and Christmas Island, Australia (Boles *et al.* 2016). In the Neotropics, the *Tigrisoma* tiger-herons are generally considered to be resident, but Fasciated Tiger-Heron *Tigrisoma fasciatum* has occurred north to Nicaragua (Chavarría-Duriaux 2015) and there is a single record of Bare-throated Tiger-Heron *Tigrisoma mexicanum* in Texas in December 2009–January 2010 (Howell *et al.* 2014).

The ability of vagrant herons to reach the ends of the earth is further underlined by their routine occurrence in the vastness of the Southern Ocean; four species have reached South Georgia – Cattle Egret *Bubulcus ibis*, Cocoi Heron *Ardea cocoi*, Great White Egret *Ardea alba* and Snowy Egret – while Cattle Egret and Black-crowned Night-Heron *Nycticorax nycticorax* have reached Amsterdam and St Paul Islands (Roux & Martinez 1987) and Little Egret, Intermediate Egret and Cattle Egret have reached the Prince Edward Islands (Oosthuizen *et al.* 2009). Of all the herons, Cattle Egret is the most prolific vagrant, having also made it to locations as remote as Easter Island (Klemmer and Zizka 1993), Gough Island (Enticott 1984), Crozet Island (Gauthier-Clerc *et al.* 2002) and the Antarctic Peninsula (Petersen *et al.* 2015). This species is one of the great winners of the Anthropocene, having successfully swapped a close association with a largely extinct megafauna for humanity's domesticated livestock; vagrancy has led to successful colonisation of North, Central and South America in the west and Australia and New Zealand in the east. Cattle Egrets are also spreading north in Europe, perhaps now helped by climate change, and are now breeding as far north as England; populations of many waterbirds are on the rise there with the end of persecution and massive habitat creation.

▲ Cattle Egrets *Bubulcus ibis*, at sea 59.68S 44.05W, 110km north of the South Orkney Islands, Antarctica, a sobering sight as they gradually became weaker aboard the boat. The photographer commented, "when flushed, they would sometimes fly round the ship, lose height and just land/hit the water. GPs [giant petrels *Macronectes* sp.] did the rest" (*Hugh Venables*).

THRESKIORNITHIDAE
Ibises and Spoonbills

In common with many waterbird species, spoonbills and ibises that breed at temperate latitudes, such as Black-faced Spoonbill *Platalea minor* and White Ibis *Eudocimus albus*, are often migratory whilst tropical species like Red-naped Ibis *Pseudibis papillosa* and Spot-breasted Ibis *Bostrychia rara* are typically sedentary, or nomadic. Many species breeding at lower temperate latitudes also undertake post-breeding movements to higher latitudes, including Glossy Ibis *Plegadis falcinellus* and Eurasian Spoonbill *Platalea leucorodia*. Consequently, it is temperate zone species that are most prone to vagrancy, and some trans-oceanic vagrancy events have led to successful cross-continental colonisation. The most celebrated such example is Glossy Ibis, which first colonised the Americas from Europe in the 1880s and transatlantic dispersal is ongoing with birds ringed in Spain recovered on Trinidad & Tobago, Barbados, Bermuda and the Virgin Islands (Santoro *et al.* 2019). The rapidly increasing Spanish population has led to increasing numbers recorded in northern Europe, which may well also be a prelude to further colonisation (Gantlett & Millington 2009). Vagrants have also pushed west as far as the Great Plains and California in the United States. The closely related White-faced Ibis *Plegadis chihi* also wanders extensively; although there are no records outside the Americas yet, vagrants have reached islands as geographically disparate as Hawaii (Pyle & Pyle 2017), Clarion (Wanless *et al.* 2009) and Cuba (Posada *et al.* 2018). Escapes cloud the status of some species, notably Scarlet Ibis *Eudocimus ruber*, for which records from the south-east United States have been deemed unacceptable (Howell *et al.* 2014), although apparently wild vagrants have been widely reported in the West Indies (Kirwan *et al.* 2019). Notable records of vagrants among other ibises include Black-headed Ibis *Threskiornis melanocephalus* west to Lake Baikal (Mlíkovský 2009); Northern Bald Ibis *Geronticus eremita* on Cape Verde (Hazevoet 1995) and Spain (where a reintroduction scheme is now underway; Dies *et al.* 2005); and single records of Puna Ibis *Plegadis ridgwayi* in Ecuador (Williams 2016) and Buff-necked Ibis in Peru (Williams *et al.* 2011).

Among the spoonbills, Eurasian Spoonbill is most prone to vagrancy with multiple records from the West Indies, including a flock of three on Barbados 2008–11 (Kirwan *et al.* 2019) and further south from Trinidad & Tobago (Kenefick 2012) and Fernando de Noronha in Brazil (Schulz-Neto 2004), as well as records east all the way to the Philippines (Jensen *et al.* 2015). As with Glossy Ibis, colonisation of the Western Hemisphere does not seem beyond the realms of possibility, especially as two species of spoonbill – Royal Spoonbill *Platalea regia* and Yellow-billed Spoonbill *Platalea flavipes* – co-occur in Australia. Royal Spoonbill *Platalea regia* colonised New Zealand from Australia

► White Ibis *Eudocimus albus* with Glossy Ibises *Plegadis falcinellus*, Cape Island, Cape May, New Jersey, United States, 16 April 2018. This species bred for the first time in New Jersey in 2020 (*Tom Johnson*).

◄ Roseate Spoonbill *Platalea ajaja* with Cinnamon Teal *Spatula cyanoptera*, Hudsonian Godwit *Limosa haemastica* and Greater Yellowlegs *Tringa melanoleuca*, Río Lluta, Arica, Chile, 7 November 2019. Roseate Spoonbill is an increasing vagrant to Chile (*Alvaro Jaramillo*).

in the 1940s (Schweigman *et al.* 2014) in another major transoceanic dispersal event, whilst Yellow-billed Spoonbill is also a vagrant to New Zealand. Back in the Americas, Roseate Spoonbill *Platalea ajaja* has been recorded as a vagrant as far north as Quebec in Canada and south as far as the Falklands (Woods 2017). A few reports in Europe have been treated as escapes, as have records of African Spoonbill *Platalea alba* in Europe – where apparently feral birds even bred in France in 2001 (Reeber 2002). Records of the latter from Oman and Yemen are understandably treated as wild, and the pattern of records from Spain and France exhibits strong seasonality that is suggestive of natural vagrancy; this is supported by their association with Eurasian Spoonbills, which winter alongside African Spoonbills in Senegal (de Juana & Garcia 2015). Finally, one of the most remarkable incidences of spoonbill vagrancy concerns a recent record of Black-faced Spoonbill *Platalea minor* from Palau in the Pacific between December 2013 and March 2014 (McKinlay 2015). Perhaps this is a possible candidate vagrant to Australia or the Bering Sea region?

CATHARTIDAE
New World Vultures

The only truly migratory members of this group are several subspecies of the Turkey Vulture *Cathartes aura,* which migrate as far as southern Canada to northern South America (Dodge *et al.* 2014). Vagrants have occurred north to Alaska on several occasions (Gibson *et al.* 1992) and one even reached South Georgia (Clarke *et al.* 2012). The almost equally ubiquitous American Black Vulture *Coragyps atratus* is largely resident, but odd individuals in North America have embarked on some epic wanderings which have taken them, for example, to California (Hamilton *et al.* 2007), British Columbia and the Yukon in Canada (Sinclair *et al.* 2003), the West Indies (Kirwan *et al.* 2019) and south as far as Patagonia (Darrieu *et al.* 2008). Records of vagrancy among the remaining species in this family are rare, although there are now photo-documented records of the essentially Amazonian-restricted Greater Yellow-headed Vulture *Cathartes melambrotus* from north-west Argentina, supporting earlier claims from the country (Acosta & Fava 2016). Among the larger species, King Vulture *Sarcoramphus papa* has been observed in active migration in Nicaragua (McCrary & Young 2008), suggesting a potential for vagrancy; this in turn supports a natural origin for a record of an extralimital adult from Baja California Sur, in north-west Mexico in October 1999 (Duncan & Lacroix 2001). Meanwhile, Andean Condors *Vultur gryphus* have wandered out to the chaco in Bolivia (Kratter *et al.* 1993) and the Perijá Mountains of Venezuela (Calchi & Viloria 1991).

▲ American Black Vulture *Coragyps atratus* with Turkey Vultures *Cathartes aura*, Sprague Pond Lakeside Park, Readville, Suffolk, Massachusetts, United States, 27 April 2016. Vagrants are at the vanguard of this species' northward expansion in eastern North America (*Marshall Iliff*).

SAGITTARIIDAE
Secretarybird

No regular movements have been described in Secretarybirds *Sagittarius serpentarius*, although post-breeding dispersal may take individuals over 150km from their breeding sites (Whitecross *et al.* 2019). No extralimital vagrancy has been reported.

PANDIONIDAE
Osprey

High-latitude populations of Ospreys *Pandion haliaetus* are highly migratory, whereas those breeding in the tropics are typically resident although *P. h. ridgwayi* of the West Indies has occurred as a vagrant to Florida. Ospreys typically migrate by flapping rather than soaring and this independence of thermals, unlike many broad-winged raptors, means that the species can routinely undertake transoceanic migrations – as measured with satellite tags – of over 1,500km (Horton *et al.* 2014). This is far from the maximum distance that the species regularly travels, however, as they regularly winter on Bermuda, probably involving an overwater flight of over 2,000km (Bierregaard *et al.* 2020). Amazingly, birds of the North American subspecies *P. h. carolinensis* are annual winter visitors to Hawaii (Pyle & Pyle 2017); this subspecies has also occurred as a vagrant, for example to the Azores (Ławicki & van den Berg 2016d) and Greenland (Boertmann 1994), and it is probably overlooked in Europe.

ACCIPITRIDAE
Kites, Hawks and Eagles

Among the Accipitridae there is huge variation in movement behaviour, from highly migratory species breeding at north temperate latitudes to highly sedentary species living in tropical rainforests. Some species are irruptive, whilst others are nomadic. In some species, juveniles perform migrations while adults are sedentary. Morphology explains much of the variation in migration routes and vagrancy patterns, whilst also influencing the detectability of vagrants. Broad-winged species usually migrate using thermals and avoid energetically costly flapping flight, which may mean huge detours to avoid high mountain ranges or long overwater crossings. Their migration routes thus frequently become geographically very restricted in some regions if they are bottle-necked by mountain ranges or narrow sea straits. These concentrations can enhance the detectability of vagrant individuals at famous 'hawkwatch' spots such as Falsterbo in Sweden or the Veracruz River in Mexico. As a result of these species' reluctance to undertake overwater flights, some islands and archipelagos – including the British Isles – with otherwise rich vagrant faunas have rather modest lists of vagrant raptors. Avoidance of water crossing is most prevalent among the *Aquila* eagles and vultures, for which flapping flight is highly energy-inefficient, and in order to move long distances they must rely on lift from thermals (Pennycuick 1971; Spar & Bruderer 1996). Flapping flight can only be maintained by these species for a finite period (Kerlinger 1985), and if they lose height during water crossings, they can be at significant risk of attack from gulls and other species owing to their restricted manoeuvrability. This makes sea crossings very dangerous for large soaring raptors. As a result, a whole suite of species – including Lesser Spotted Eagle *Clanga pomarina*, Steppe Eagle *Aquila nipalensis*, Eastern Imperial Eagle *Aquila heliaca* and Bonelli's Eagle *Aquila fasciata* – that occur with some regularity as vagrants to the Netherlands, Denmark and Scandinavia are unrecorded in the British Isles. Island-hopping is seemingly easier, however, and one of the most unexpected vagrant raptor records was a Long-legged Buzzard *Buteo rufinus* on St Paul Island, Alaska, in 2018–19, thousands of kilometres from its nearest regular range in Central Asia (Pyle *et al.* 2019).

Large raptors' reluctance to cross waterbodies is well documented, but it does not represent a golden rule governing vagrancy likelihood. The notable run of records of Greater Spotted Eagle *Clanga clanga* from the British Isles in the late 1800s demonstrated the species' capacity to cross the English Channel; its subsequent disappearance as a vagrant was correlated with widespread population reductions and a general fall in the numbers wintering in western Europe (Cramp *et al.* 1980). Despite their large size and soaring habits, Egyptian Vultures *Neophron percnopterus* have also been recorded successfully making relatively long sea crossings, e.g. 120km across the Sicilian Channel (Angostini 2005), and this species has even reached

◄ Swainson's Hawk *Buteo swainsoni*, Lake Ontario Parkway, Monroe, New York, United States, 2 October 2011, a vagrant to eastern North America (*Jay McGowan*).

the Azores (Barcelos *et al.* 2015). Elsewhere, Egyptian Vultures have wandered to Botswana (Dowsett *et al.* 2011), China (Guo & Ma 2012) and Bangladesh (Alam *et al.* 2016). Together with a recent record of Bonelli's Eagle in May 2016 on Madeira (Demey 2016b), this indicates that some broad-winged species do not lack the physiological capacity for overwater flights even if they generally avoid making them. Short-toed Eagle *Circaetus gallicus* is a broad-winged raptor with a noted capacity for overwater flight and is regularly seen crossing 300km of open water on southward migration from central Italy (Agostini *et al.* 2004). However, the majority of individuals from this population (including almost all adults) apparently perform a much more circuitous migration route via the Straits of Gibraltar in order to avoid the water crossing, suggesting that experience may play a role in determining reluctance to cross water (Agostini *et al.* 2002). Nevertheless, this species has now made it to Britain on three occasions, including one individual that spent two months in the summer of 2014 touring almost all the heathland sites available to it in a search for snakes (now increasingly rare in the United Kingdom) across five different southern and eastern English counties (Hudson 2015). Old World vultures tend to avoid overwater crossings and funnel through Palearctic migration bottlenecks (Bildstein 2009). For example, a variety of migrating raptors regularly cross the 120km-wide Sicilian Channel in the central Mediterranean, but Griffon Vultures *Gyps fulvus* almost never attempt a crossing (Thiollay 1977). They undertake a shorter, 12–13km crossing of the Gulf of Suez in Egypt more regularly (Goodwin 1949) but during two years of autumn counts at the 30km Bab-el-Mandeb at the southern end of the Red Sea, only three Griffon Vulture crossings were observed (Welch & Welch 1988). This species is being recorded with increasing frequency in northern Europe, often in flocks in summer; this phenomenon is associated with population increases in southern Europe. Future records from Britain remain a possibility as the 34km English Channel crossing (at its narrowest) should lie within the limits of Griffon Vulture's capacity, particularly in higher air temperatures (and hence increased lift). When not unduly constrained by waterbodies, vulture vagrancy is surprisingly frequent; two Afrotropical species are now recorded with increasing frequency in the Western Palearctic. Rüppell's Vulture *Gyps rueppelli* was first recorded in Europe as recently as 1992 and is now so frequent in Spain it is no longer an official rarity (Rodríguez & Elorriaga 2016), and birds have been observed pairing up with Griffon Vultures. Rüppell's Vultures follow Griffon Vultures north from sub-Saharan wintering grounds and both have been recorded on passage together in Morocco. There are now records of White-backed Vulture *Gyps africanus* from Morocco, Portugal and Spain (Rodríguez & Elorriaga 2016). Rüppell's Vultures are also being recorded with increasing frequency in South Africa and visiting colonies of Cape Vulture *Gyps coprotheres* (Botha & Neethling 2013). Cinereous Vulture *Aegypius monachus* is not migratory to the same extent as Griffon Vulture but nevertheless has wandered in Europe as far north as Norway (Ławicki & van den Berg 2019), as well as east to Bangladesh (Thompson *et al.* 2014); it has even undertaken substantial overwater flights to the Philippines (van der Ploeg & Minter 2004), Taiwan and Japan (Brazil 2009). Hints at the origins of extralimital Cinereous Vultures were provided by the first record for Belgium, at Macon in June 2018,

▲ Griffon Vultures *Gyps fulvus*, Vlaardingen ZH, Netherlands, 30 May 2020. This species is recorded with increasing frequency as a wanderer to western Europe (*Gerjon Gelling*).

◄ Bearded Vulture *Gypaetus barbatus*, Lingshan, Mentougou, China, 12 December 2016, the second record for the Beijing region. The nearest known breeding areas for this species are on the Tibetan Plateau, over 1,200km to the west (*Terry Townshend*).

which accompanied 39 Griffon Vultures, one of which had been ringed as a first-calendar-year bird near Barcelona, Spain, in December 2016 (Baeten *et al.* 2016). An old record from Wales (Vinicombe 1994) has yet to be given the green light for admission to the British List, but powerful arguments have been made for its treatment as a wild individual (Gutierrez 1995).

The *Haliaeetus* sea-eagles are built not that unlike the *Gyps* vultures yet, as their name suggests, they are more adept at long overwater crossings. Most impressively, both White-tailed Eagles and Steller's Sea-Eagles *Haliaeetus pelagicus* have occurred as vagrants to the Hawaiian Islands (Pyle & Pyle 2017), where individuals of both species have been observed preying on adult Laysan and Black-footed Albatrosses at two different colonies. An extinct eagle species, closely related to White-tailed, was formerly present on this archipelago (Hailer *et al.* 2015) and probably disappeared with the human-driven extinction of the large flightless birds formerly found there. Steller's Sea-Eagle and White-tailed Eagle have also both occurred as vagrants to Alaska (Howell *et al.* 2014). In a similar vein, Bald Eagle *Haliaeetus leucocephalus* has occurred as a vagrant to the Old World, across the Pacific to European Russia (Brazil 2009) and recently south to Japan; and also across the Atlantic to Ireland and Northern Ireland (Haas 2012) and the West Indies (Kirwan *et al.* 2019). Bald Eagle is thus one of a select few raptors that has occurred as a transatlantic vagrant to Europe; the group also includes Swallow-tailed Kite *Elanoides forficatus* on the Azores (Haas 2012) and several Northern Harriers *Circus hudsonicus* and Rough-legged Buzzards *Buteo lagopus* of the North American form *sanctijohannis*. The last two species have also reached Hawaii (Pyle & Pyle 2017). Arguably the champion vagrant raptor is Black Kite *Milvus migrans*, which has made it to many disparate oceanic islands including Hawaii (Pyle & Pyle 2017), the Marianas (Clapp *et al.* 1999), New Zealand (Miskelly *et al.* 2013), the West Indies (Bahamas, Virgin islands and Lesser Antillies: Kirwan *et al.* 2019), Trinidad (Kenefick *et al.* 2019) and once in South America (just) on the Brazilian mid-Atlantic archipelago of São Pedro & São Paulo in April–May 2014 (Nunes *et al.* 2014). The first for continental North America was on St Paul Island, Alaska, in January 2017. Raptors such as kites and harriers have relatively long and narrow wings (i.e. higher aspect ratios) compared with vultures and eagles, giving them greater efficiency of flapping flight; they can soar on minimal updraft, making them more capable of long water crossings (Kerlinger 1985). Even so, there are records of several species aboard ships (Durand 1972); the ability of raptors to find food whilst resting on vessels, predating seabirds or other migrants that seek shelter on board, indicates the particularly high potential for this group to make long ship-assisted journeys (Kerlinger *et al.* 1983).

The status of some raptors has potentially been occluded by identification difficulties. Crested Honey Buzzard *Pernis ptilorhynchus* is an interesting example. This species was first detected in the Western Palearctic based on a retrospectively identified photo from Turkey (Laine 1996). The species has subsequently proved to be a regular but scarce passage migrant through Israel (Alon *et al.* 2004), and there have followed records from Italy (Scuderi & Corso 2011), Cyprus (Harrison

► Northern Harrier *Circus hudsonius*, Midway, northwestern Hawaiian Islands, 19 October 2011, one of three birds present on the archipelago in that autumn (*Cameron Rutt*).

2014), Greece (Zannetos *et al.* 2018) and an apparent hybrid Crested x European Honey Buzzard *Pernis apivorus* from Sweden (Swedish Rarities Committee). The annual appearance of migrants moving through countries bordering the Mediterranean suggests some individuals may winter in Africa. Increased awareness precipitated re-evaluation of old reports and the retrospective identification of the first African record of the species from Djibouti in 1987 (Welch & Welch 2017) and subsequently records farther south in Kenya (Kennedy & Marsh 2016) and Tanzania (Fisher & Hunter 2018). This species has also occurred as a vagrant to Christmas Island and even the Northern Territory of the Australian mainland (James & McAllan 2014). Identification difficulties and observer awareness might preclude the discovery of some other candidate vagrant species such as Sharp-shinned Hawk *Accipiter striatus*, whose large migratory populations and frequent offshore presence in the western North Atlantic suggest a very high vagrancy potential to Europe (Kerlinger *et al.* 1983), but it is as yet unrecorded outside of the Americas. Similarly, the highly migratory Chinese Sparrowhawk *Accipiter soloensis* has been recorded in the western Pacific from the Marianas, Yap and Palau (Ferguson-Lees & Christie 2001), and one reached Kure, in the northwestern Hawaiian Islands in September 1991 (Pyle & Pyle 2017); this species might reach the Bering Sea islands.

Several tropical raptors occurring in savanna regions can be quite highly nomadic, and juveniles may undertake extralimital movements. Perhaps the best example is Bateleur *Terathopius ecaudatus*, an Afrotropical species that has occurred north as far as Spain, Turkey and Cyprus (Copete *et al.* 2015).

► Bateleur *Terathopius ecaudatus*, Gal'on Fields, HaDarom, Israel 19 November 2020. Immatures of this Afrotropical raptor are being detected with increasing frequency in the southern fringes of the Western Palearctic (*Yoav Perlman*).

◀ Great Black Hawk *Buteogallus urubitinga*, Deering Oaks Park, Cumberland, Maine, United States, 4 December 2018, the celebrated first North American record (*Tom Johnson*).

◀ Grey-bellied Hawk *Accipiter poliogaster*, Fazenda Mangue Seco, Vigia, Pará, Brazil, 20 August 2014, one of few contemporary records from eastern Pará state. It is unclear whether this species is resident in the region, or even an austral migrant or vagrant. Such uncertainties are common to many tropical raptors (*Alexander Lees*).

Wahlberg's Eagle *Hieraaetus wahlbergi* has also now occurred three times in the southern reaches of the Western Palearctic, with records from Tunisia, Egypt and Mauritania (Waheed 2016). Wandering behaviour is well documented in young raptors, which may undertake epic post-juvenile dispersive movements; this includes tropical birds moving to temperate latitudes and vice versa. Spanish Imperial Eagle *Aquila adalberti* is a useful example; adults of this species are very sedentary, but juveniles wander to France with some regularity (Duchateau 2007) and have occurred once in the Netherlands in May 2007, as well as travelling south to Senegal and even Cameroon (Weenink *et al.* 2011).

One of the most extraordinary episodes of raptor vagrancy involved the first United States record of Great Black Hawk *Buteogallus urubitinga,* which was initially photographed on South Padre Island, Texas, in April 2018 (Pyle *et al.* 2018). This was an anticipated vagrant, breeding as close as south Sonora and south Tamaulipas in Mexico, and its occurrence ties in with records of other largely resident Neotropical raptors like Crane Hawk *Geranospiza caerulescens* and Roadside Hawk *Rupornis magnirostris* in southern Texas (Howell *et al.* 2014). However, the Texas Great Black Hawk shocked birders by reappearing in August 2018 in Maine, where it stayed and attempted to overwinter. Unfortunately, it was eventually taken into care with extensive frostbite on its gangly unfeathered legs – poor physiological adaptations for surviving a harsh north temperate winter – which eventually proved fatal. Two records of Double-toothed Kite *Harpagus bidentatus* in the United States, from Texas and Florida (Howell *et al.* 2014), would also appear unexpected, and the austral migrant Rufous-thighed Kite *Harpagus diodon* would seem a more likely vagrant, with records north to Ecuador (Lees & Martin 2015). Shortfalls in our knowledge of migratory movements, or even broad distribution patterns, in other tropical raptors make it tricky to interpret 'extralimital' records of rare species but, for example, records of Grey-bellied Hawk *Accipiter poliogaster* from Costa Rica are assumed to be vagrants (Araya-H *et al.* 2015). Other notable tropical wanderers include recent records of Red-necked Buzzard *Buteo auguralis* from South Africa (Buij *et al.* 2016) and the first White-eyed Buzzard *Butastur teesa* for the Southern Hemisphere in south Sulawesi, Indonesia (Shagir & Iqbal 2015).

An increase in vagrancy can often be a prelude to colonisation or a change in migration routes. This phenomenon is particularly well illustrated in Black-winged Kite *Elanus caeruleus,* which has undergone a major range expansion in southern Europe and the Middle East, with a concomitant increase in vagrancy. Up to 2016 there had been 88 records from Belgium, Germany, the Netherlands and Switzerland (Ławicki & Perlman 2017). This increase was probably driven by changes in land-use facilitating its colonisation; the species has taken well to the cultivated *dehesas* of Spain, which are structurally similar to African savannas, helped by the fact that this species can routinely have two broods a year, which is very rare among raptors (Balbontin *et al.* 2008). Pallid Harrier *Circus macrourus*, once a near-mythical vagrant in the United Kingdom, is now annual there. This species now regularly breeds in Scandinavia, with these birds moving south to winter in Iberia and North Africa, where records have increased exponentially (Ollé *et al.* 2015, Bergier & Thévenot 2018). Breeding took place for the first time in the Netherlands in summer 2017, and a female raised there went on to breed in Spain in 2019.

Given the popularity of falconry, many raptors regularly escape their owners and may cause headaches for rarities committees, although this is more of an issue with *Falco* falcons. The provenance of many individuals is never resolved, leaving them in assessment limbo. Perhaps the most celebrated such bird was the Golden Eagle *Aquila chrysaetos* that appeared on Kauai in Hawaii in 1967 (Pyle & Pyle 2017). Despite enquiries, there was no evidence of importation, yet Golden Eagles have not been reported previously from oceanic islands, and an overwater crossing of thousands of kilometres would seem unlikely for any *Aquila* eagle. This individual remained on the island for a further 17 years and gained a reputation for attacking tourist helicopters, eventually killing itself in the process in May 1984.

▲ Pallid Harrier *Circus macrourus*, Texel NH, Netherlands, 11 October 2014. The sight of Pallid Harriers quartering arable fields is becoming increasingly frequent in western Europe (*Gerjon Gelling*).

TYTONIDAE
Barn Owls

None of the tytonid owls are migratory, yet many of them show quite remarkable levels of long-distance dispersal. The Barn Owl *Tyto alba* superspecies is one of the world's most cosmopolitan bird taxa and a very successful colonist of remote oceanic islands across the globe (Lees & Gilroy 2014), where much population differentiation has taken place. This species is one of the best-known of all bird species, and studies of post-fledging dispersal have uncovered some impressive movements by ringed individuals – up to 1,600km in Europe, 1,760km in North America and 840km in Australia (Bruce 1999). Unsurprisingly, given this history of oceanic island colonisation, these movements include transoceanic dispersal. For example, one recovered as a vagrant on Bermuda, an island the species colonised in 1931, had been ringed 1,250km away in New Jersey, United States. These oceanic dispersal events may often involve human assistance; Barn Owls are frequently recorded on ships and resting on offshore oil and gas infrastructure. More unusually, there are several records from New Zealand of birds arriving as stowaways in the landing gear of large aircraft (Bruce 1999). This species was first confirmed breeding in New Zealand in 2008 (Hyde *et al.* 2010); the origin of these individuals was unknown but there were only eight previous national records. Barn Owls also colonised the Falklands in the 20th century (Woods 2017). This ongoing range expansion, driven by a tendency towards vagrancy, is marking them as one of the most successful of all bird species, helped by our manifest reshaping of global vegetation to this species' liking.

Other tytonid owls are not so well-travelled, although Australasian Grass-Owl *Tyto longimembris* is highly nomadic, undertaking irruptive movements associated with rodent plagues. Records from Indonesia (Linsley *et al.* 2011) and Fiji (Pernetta & Watling 1978), for example, might conceivably have involved wandering Australian birds. Similarly, African Grass-Owl *Tyto capensis* has occurred at least twice as a vagrant to Ethiopia (Redman *et al.* 2009), which is at least 900km out of range (although it is possible that the species has an undiscovered breeding population there). Finally, there is a single old record of Oriental Bay-Owl *Phodilus badius* from Samar in the eastern Philippines in 1924; if genuine, this might have represented a long-distance vagrant (Warburton 2009).

STRIGIDAE
Owls

Migratory behaviour is known in less than 10 per cent of the family Strigidae, but across the group many species or populations exhibit movement behaviours that vary from nomadism to partial migration. Understanding vagrancy in owls is severely complicated by the difficulty of detecting nocturnal birds extralimitally, especially as vagrants may seldom vocalise. Vagrancy in the group is thus likely to be under-recorded and further complicated in many regions, particularly the tropics, by the difficulty of their field identification. Many of the truly migratory owl species, such as Oriental Scops Owl *Otus sunia* and Eurasian Scops Owl *Otus scops* are insectivorous and can only occupy high latitudes seasonally. Northern populations of Oriental Scops Owl have a long migration span stretching from Russia to South-east Asia; it should not come as a surprise therefore that there are two records from the Aleutians (Howell *et al.* 2014) and a recent record of a bird that boarded a boat operating between the Australian mainland and Barrow Island in Western Australia in May 2013 (Birds Australia Rarities Committee). This species would seem a potential candidate for westward vagrancy towards the Western Palearctic. Eurasian Scops Owl has been widely recorded as an overshooting vagrant to northern Europe, with occasional individuals even holding territory in southern England, and this species is also a vagrant to the Seychelles in the Indian Ocean (Skerrett *et al.* 2006). Many North American populations of Burrowing Owl *Athene cunicularia* are

▲ Tengmalm's Owl *Aegolius funereus*, Tresta, Shetland, Scotland, 9 April 2019, the first 'twitchable' record in Britain for decades (*Rebecca Nason*).

long-distance migrants, and vagrants have reached as far north and east as Ontario and New York; there is also a single record of this species from the Falklands – a remarkable transoceanic movement from the Patagonian mainland (Woods 2017). Another long-range migrant insectivore is Northern Boobook *Ninox japonica*, which moves from south-east Siberia all the way to Indonesia. Vagrants have overshot their wintering areas to appear in Australia (Schodde & van Tets 1981, Johnstone & Darnell 2015) and there are two August records from Alaska, on St Paul Island in 2005 (Yerger & Mohlmann 2008) and on the Aleutians in 2008 (Bond & Jones 2010). There are 32 species of *Ninox* owls, including many island endemics; clearly, a high rate of long-distance dispersal in this genus has helped it become one of the most speciose of owl genera.

Beyond these truly migratory species, vagrancy tends to be most frequently reported among species that specialise on rodents with regular (often three- or four-yearly) population fluctuations. Winter irruptions of species such as Tengmalm's Owl *Aegolius funereus* and Northern Hawk Owl *Surnia ulula* occur during such periods of low abundance of small mammals within the northern parts of their range (Mikkola 1983); resultant movements may take these species south to the temperate zone. There are two instances of transatlantic vagrancy in Hawk Owls of the Nearctic form *caparoch*, both of which boarded ships at sea within European waters – one off the coast of Cornwall in March 1830 (Harrop 2010) and the other off Gran Canaria, in the Canary Islands, in October 1924 (Romay & Roselaar 2013). Movements of Northern Saw-whet Owl *Aegolius acadicus* are complicated, but many individuals vacate the breeding range in winter and irregularly move as far as Oklahoma, Tennessee, Georgia and Arizona (Rasmussen *et al.* 2008). Vagrants have reached Florida, Bermuda, St Lawrence and St Paul Islands in Alaska (Rasmussen *et al.* 2008), and even the Bahamas (Kirwan *et al.* 2019). Historical vagrancy by *Aegolius* to Bermuda led to the evolution of an endemic owl species there, Bermuda Saw-whet Owl *Aegolius gradyi* (Olson 2012). This now-extinct species apparently persisted there until the beginning of the 17th century, but by 1623 Captain John Smith noted that "there were a kinde of small Owles in great abundance, but they are now all slaine or fled" (Lefroy 1981).

Short-eared Owl *Asio flammeus* is an incredible, highly itinerant migrant – one individual fitted with a satellite transmitter was found to have moved 7,057km from breeding grounds in Alaska to Zacatecas in Mexico (Johnson *et al.* 2017). Short-eared Owls are one of the most widely distributed of all owl species and have colonised both the Hawaiian Islands and Pohnpei in Micronesia, with dispersing birds recorded with some regularity in the Marshall Islands (Spenneman 2004). Vagrants have reached most of the islands in the Atlantic, in addition to Spitsbergen, Bear Island and Jan Mayen, and have reached Guinea-Bissau (Demey 2018). Irruptions in another great nomad, Snowy Owl *Bubo scandiacus*, occur every three to five years in North America, driven by food supply (Newton 2002), which may lead to birds occurring considerably further south than normal. Irruptions into the British Isles seem to occur over a longer timescale, and appear to fluctuate according to the relative severity of the winter: records peaked during the relatively cold period in the 1960s and 1970s when an invasion led to extralimital breeding in Shetland. Transatlantic ship assistance is known to have occurred on at least one occasion, with a large group of Snowy Owls boarding a ship off the coast of Canada in October 2001 and disembarking in the southern North Sea in Holland, Belgium and Suffolk in England (Rogers 2002). Records from Bermuda,

▲ Northern Saw-whet Owl *Aegolius acadicus*, Southeast Farallon Island, San Francisco, California, United States, 27 October 2013, a vagrant offshore in California (*Cameron Rutt*).

◄ Eurasian Pygmy Owl *Glaucidium passerinum*, Lettele OV, Netherlands, 29 December 2013, the eighth record for the Netherlands (*Gerjon Gelling*).

the Azores and Hawaii indicate that the species has a prodigious capacity for overwater flights. The single record from Hawaii concerned a first-year male at Honolulu Airport, O'ahu, in November 2011, which was controversially shot as part of bird-strike control efforts (Pyle & Pyle 2017).

Other large *Bubo* owls do not tend to move long distances or over water often, and many are popularly kept in captivity, which makes discerning origins difficult. Eurasian Eagle Owl *Bubo bubo* has an on-again/off-again relationship with the British List, with the latest review concluding that no putative vagrants have a sufficiently strong case for admission (Melling *et al.* 2008b). Stable isotope analysis of a Eurasian Eagle Owl found in Norfolk in November 2006 indicated that the bird had distinct isotopic signatures across two generations of feathers, which would be consistent with an origin from northern Europe, but a northern British origin could not be ruled out (Kelly *et al.* 2010). As such, Britain still awaits an acceptable record backed up by a ringing recovery, although sightings of apparently wild birds from Corsica and Sardinia (Grussu 2012) suggest that a flight across the English Channel is not impossible for this species. Novel scientific techniques were also applied to a specimen from New Mexico apparently of the non-migratory 'Dusky' Great Horned Owl *Bubo virginianus saturatus*, which typically occurs in the Pacific Northwest region of North America (Dickerman *et al.* 2013). Phylogenetic analysis of the specimen was inconclusive as there seemed to be little evidence of population genetic variation across the species' range, but the authors were at least able to prove that the salvaged specimen was not transported after death – as they sequenced the genes of a dead Desert Cottontail *Sylvilagus audubonii* found in its stomach, which was confirmed as matching other local cottontail populations. Other large North American owls are also prone to occasional vagrancy, with recent records of Barred Owl *Strix varia* from Newfoundland and Labrador being examples (Schmelzer & Phillips 2004). Movements are likely less marked in tropical and subtropical species but declaring that with confidence is difficult. There are two records each of Stygian Owl *Asio stygius* and Mottled Owl *Ciccaba virgata* from Hidalgo County in Texas, United States (Howell *et al.* 2014), presumably representing atypical post-juvenile dispersal events from Mexican populations. Finally, the identity of an owl seen in September 1991 on the Prince Edward Islands (Oosthuizen *et al.* 2009) is anyone's guess and may have involved a tytonid species, but indicates that members of the Strigiformes have even penetrated far into the vastness of the Southern Ocean.

COLIIDAE
Mousebirds

Mousebirds are generally highly sedentary, although some species do occasionally undertake some nomadic wandering movements, notably Blue-naped Mousebird *Urocolius macrourus*, which is noted as a vagrant to The Gambia (Borrow & Demey 2019). This species also just scrapes into the southern limits of the Western Palearctic in Mauritania (Crochet & Haas 2013), where it is probably resident, and it was recently recorded for the first time in Algeria.

LEPTOSOMIDAE
Cuckoo-roller

Cuckoo-roller *Leptosomus discolor* is thought to be largely sedentary and there are no extralimital records, although its dispersal to the Comoros Islands (and subsequent speciation) indicates that such vagrancy events have happened over evolutionary time.

TROGONIDAE
Trogons

Most trogons show little or no evidence of migratory behaviour, although several species do undertake altitudinal migrations. An exception is northern-most populations of Elegant Trogon *Trogon elegans*, which are summer migrants to southern Arizona on the United States–Mexico border. A scattering of vagrants has occurred in northern Arizona, New Mexico and central and south Texas (Kunzmann *et al.* 1998). Eared Quetzal *Euptilotis neoxenus* is also an occasional vagrant to the same region where Elegant Trogon breeds, with about 25 records, including one nesting attempt. This species is also thought to be migratory in northern Mexico (Howell *et al.* 2014).

UPUPIDAE
Hoopoes

Northern populations of Eurasian Hoopoe *Upupa epops*, which ranges across the Palearctic, are highly migratory; birds are regularly recorded as overshoots north of their regular range, where occasional breeding takes place, e.g. in southern England (Brown & Grice 2010). Vagrants have reached distant oceanic islands including Iceland (Snow & Perrins 1998), the Azores (Barcelos *et al.* 2015), Diego Garcia (Carr *et al.* 2011) and the Andamans (Rasmussen & Anderton 2005). In the east, birds have straggled once each to Alaska (Dau & Paniyak 1977), Wallacea (McNeill & Lambaihang 2013) and even Western Australia (Onton *et al.* 2012). Future records from the Western Hemisphere, from the Caribbean or northern South America, seem likely with greater observer coverage. Further proof of this species capacity for long over-water flights is provided by prehistoric vagrancy to remote Saint Helena Island in the mid-Atlantic and the evolution of the now-extinct Saint Helena Hoopoe *Upupa antaios* (Olson 1975), a large and probably flightless island form.

◄ Eurasian Hoopoe *Upupa epops*, Fáskrúðsfjörður, Austurland, Iceland, 10 October 2016. There have been fewer than 20 records in Iceland (*Yann Kolbeinsson*).

PHOENICULIDAE
Woodhoopoes and Scimitarbills

We are unable to trace any reports of extralimital movements in woodhoopoes.

BUCORVIDAE
Ground-Hornbills

Both species of ground-hornbills are highly sedentary, and we are unaware of reports of vagrancy.

BUCEROTIDAE
Hornbills

No truly migratory movements are known among the hornbills, and most species are highly sedentary. Nomadism is documented in several species from arid regions, with African Grey Hornbill *Lophoceros nasutus* being the clearest case. For example, 800 were recorded migrating north in an hour along the Niger River near Niamey in Niger in July 2019 (Demey 2020), and this species is considered a vagrant to Liberia (Kemp & Boesman 2020). Another arid zone species, Bradfield's Hornbill *Lophoceros bradfieldi,* is also sometimes reported as far west as the Atlantic coast of Africa (Demey 2020). Hornbills are poor flyers, which may explain the absence of any vagrancy records from oceanic islands; indeed, 1,000 Red-billed Hornbills *Tockus erythrorhynchus* apparently engaged in an unusual mass dispersal event ended up drowning whilst trying to cross Lake Kariba in Zimbabwe (Anon 1998b). Silvery-cheeked Hornbills *Bycanistes brevis* are also somewhat nomadic and have occurred as vagrants to north-east Zambia and north-east South Africa. A recent first record of the sedentary and forest-associated Black-casqued Hornbill *Ceratogymna atrata* from The Gambia (Cross 2019) is less easy to explain, and may be suggestive of a captive origin, but the species is known to wander extensively. For example, Holbrook & Smith (2000) found that tagged individuals moved up to 290km. A Northern Red-billed Hornbill *Tockus erythrorhynchus* photographed in Oman in July 2019 was ruled an escape by the national rarities committee (J. Eriksen *in litt.*).

TODIDAE
Todies

All todies are basically sedentary, although Narrow-billed Tody *Todus angustirostris* may make some altitudinal movements; no vagrancy has been reported (Kirwan *et al.* 2019).

MOMOTIDAE
Motmots

All motmots are thought to be sedentary, with the exception of Turquoise-browed Motmot *Eumomota superciliosa,* which may undertake some local seasonal movements (Murphy 2008). No vagrancy has been reported in this group.

ALCEDINIDAE
Kingfishers

Most kingfishers are non-migratory, although many do undertake some significant long-range dispersal events, which have over time has led to the evolution of many island species and underpinned the spread of this family across the globe (Andersen *et al.* 2015). True migration is most prevalent in species occurring in the most seasonal landscapes, especially at northern temperate latitudes where waterbodies freeze over.

◄ Belted Kingfisher *Megaceryle alcyon*, Kylemore Lough, Galway, Ireland 6 October 2012, the fourth Irish record (*Josh Jones*).

◄ Ringed Kingfisher *Megaceryle torquata*, Lake Martin, Cypress Island Preserve, St Martin, Louisiana, United States, 15 February 2015, the easternmost record in North America of this largely Neotropical species (*Cameron Rutt*).

Belted Kingfishers *Megaceryle alcyon* are partial migrants in the north of their range and regularly move south as far as Panama. Vagrants have reached the Galapagos on several occasions (Lévêque *et al.* 1966) and recently also mainland Ecuador (Cisneros-Heredia 2016). This species has a prodigious capability for overwater flights and has occurred around 30 times on Hawaii (Pyle & Pyle 2016); to the east they have reached Greenland (Feilberg 1985) and several European countries, including the Netherlands, Iceland, Britain and Ireland (Lewington *et al.* 1991).

Northern populations of Common Kingfisher *Alcedo atthis* are migratory, with some Russian individuals moving south over 3,000km; one ringed in Korea was recovered two months later in the northern Philippines (Woodall 2020). Vagrants have reached Palau and Guam in the Pacific (VanderWerf *et al.* 2018), and there is even a report of transatlantic vagrancy to the West Indies (Rodríguez *et al.* 2005), albeit in odd circumstances, and this record was not accepted by Kirwan *et al.* (2019). Five records from Christmas Island are the only ones from Australian territory (James & McAllan 2014). That there is only a single record from Iceland, compared with seven Belted Kingfishers, suggests that this species has a more limited capacity for overwater vagrancy. In the United States, northward-dispersing Ringed Kingfishers *Megaceryle torquata* have wandered as far north as Oklahoma (Dole 1999) and as far east as western Florida (Pranty *et al.* 1996). Amazon Kingfisher *Chloroceryle amazon* has wandered to the United States border on three occasions (Howell *et al.* 2014). American Pygmy Kingfisher *Chloroceryle aenea* is also considered sedentary, but a record from Curaçao, in the Netherlands Antilles, in February 2020 (eBird) would be a first for the West Indies.

Pied Kingfisher *Ceryle rudis* is increasingly recorded as a vagrant in Europe, with records from Cyprus, Greece, Poland and Italy (Lewington *et al.* 1991, van den Berg 2015), whilst White-throated Kingfisher *Halcyon smyrnensis* has also made it to Greece, Cyprus and, recently, Bulgaria (Spasov 2006). To the east, this species has occurred as a vagrant to Taiwan and Japan (Brazil 2009). In Africa most species are fairly sedentary but even so, some significant vagrancy events have been recorded among intratropical migrants such as Grey-headed Kingfisher *Halcyon leucocephala,* which has occurred as a vagrant to the Seychelles (Skerrett *et al.* 2006). African Pygmy-Kingfisher *Ispidina picta* undertakes northward movements of up to 2,500km and might be expected to also occur extralimitally. Other small tropical kingfishers in Asia have occurred as vagrants: Black-backed Dwarf-Kingfisher *Ceyx erithaca,* a species that undertakes some complicated intratropical movements, has reached Taiwan (Brazil 2009); and Little Paradise-Kingfisher *Tanysiptera hydrocharis* has strayed to Australian territory in the Torres Strait (Clarke 2010). Collared Kingfisher *Todiramphus chloris* has a long history of extralimital movements, with records from Japan and Taiwan (Brazil 2009) and Christmas Island (James & McAllan 2014). Sacred Kingfisher *Todiramphus sanctus* also wanders; a recent record from February–March 2019 at Phetchaburi is the first record for Thailand and South-east Asia (eBird), and others have been recorded south all the way to the Chatham Islands (Miskelly *et al.* 2006).

MEROPIDAE
Bee-eaters

Temperate zone bee-eaters and many large tropical savanna bee-eaters are highly migratory, whilst the smaller species and forest-associated bee-eaters tend to be sedentary. European Bee-eaters *Merops apiaster* have a huge migration span, breeding north to southern Russia and wintering as far south as South Africa, which may predispose them to long-range vagrancy. This species is a regular vagrant and occasional breeder in northern Europe and has occurred as a vagrant to Iceland and also in the Indian Ocean, in the Seychelles (Skerrett *et al.* 2011). More amazingly, an adult was photographed in February 2014 on Saint Lucia, in the West Indies – this is the only record of a bee-eater in the Western Hemisphere (Kirwan *et al.* 2019). The more southerly distributed Blue-cheeked Bee-eater *Merops persicus* has also been widely recorded as a vagrant to north and north-west Europe as far north as Norway (Ebels & Van der Laan 1994) and there is a recent record from Taiwan. Also in the east, a Blue-tailed Bee-eater *Merops philippinus* on Hachijo-jima Island in July 2013 was the first record for Japan (Yamamoto *et al.* 2015). Huge movements of bee-eaters have been observed in southern Thailand near Chumphon in spring, mostly of Blue-tailed Bee-eaters but also smaller numbers of Blue-throated Bee-eaters *Merops viridis* (Decandido *et al.* 2010). The

► European Bee-eater *Merops apiaster*, Siglufjörður, Norðurland eystra, Iceland, 16 May 2011, the second record for Iceland (*Yann Kolbeinsson*).

▲ White-throated Bee-eater *Merops albicollis*, Eilat KM20 Flamingo Pools, HaDarom, Israel, 24 August 2019, the first for Israel and third for the Western Palearctic (*Yoav Perlman*).

latter has been recorded once in India, a bird visiting a Blue-tailed Bee-eater colony in Kerala in May 2013 (Manekara 2012), which was initially treated with some scepticism – though a vagrant of this species travelling with Blue-tailed Bee-eaters could potentially overshoot to India. Rainbow Bee-eaters *Merops ornatus* are also widely recorded as vagrants, with birds overshooting their New Guinea wintering areas and occurring north to Japan, South Korea and Taiwan (Ikenaga *et al.* 2009), out into the Pacific to Palau and the Caroline Islands and to remote Lord Howe Island (McAllan *et al.* 2004).

Several African bee-eaters are also highly migratory, most notably White-throated Bee-eater *Merops albicollis*, which has now occurred three times as a trans-Saharan vagrant – twice to Western Sahara in 2013 and 2017 (Jacobs *et al.* 2015) and once to Israel in August 2019, as well as south to South Africa where there have been around 20 records (although the provenance of the Palearctic records has been questioned). Vagrancy among the remaining species is more modest: White-fronted Bee-eater *Merops bullockoides* is a vagrant to the Namibian and South African coast and a Red-throated Bee-eater *Merops bulocki* at Grumeti Game Reserve in August 2016 was the first record for Tanzania (Fisher & Hunter 2017), whilst Madagascar Bee-eater *Merops superciliosus* has straggled to the Seychelles (Skerret *et al.* 2011).

CORACIIDAE
Rollers

Rollers are highly migratory landbirds; most tropical species undertake significant intratropical migrations and the two species that have colonised temperate latitudes – European Roller *Coracias garrulus* and Dollarbird *Eurystomus orientalis* – both undertake epic intercontinental migrations. European Rollers are relatively routine vagrants to northern Europe and even peninsular India (Narayanan *et al.* 2008). Beyond these relatively modest overshoots and perhaps errant post-juvenile dispersal events, this species has also reached Ascension Island, a journey requiring a minimum 1,500km overwater flight (Bourne & Simmons 1998), and most incredibly there is a single record from the Cocos Islands in the Indian Ocean in January 2011 (Palliser & Carter 2013); it is a mystery which route this bird might have taken. Dollarbirds are very powerful flyers and move from the Russian Far East as far south as Australia; one individual collided with a plane at an altitude of 2,600m over Queensland, Australia (Blackman & Locke 1978).

▲ Broad-billed Roller *Eurystomus glaucurus*, Kibbutz Karmiya, HaDarom, Israel, the first for Israel and third for the Western Palearctic (*Yoav Perlman*).

Vagrant Dollarbirds have reached New Zealand and Lord Howe Island (McAllan *et al.* 2004), Micronesia (Pratt *et al.*1977), western India (Maulick & Adhurya 2017) and most recently Pakistan (eBird). Dollarbird would seem a possible vagrant to the Aleutians or even to the eastern limits of the Western Palearctic.

Movements and vagrancy events among the remaining roller species are more modest; Indian Roller *Coracias benghalensis* has wandered out to the Maldives and once to north-east Afghanistan, for example (Rasmussen & Anderton 2005), and also once to Socotra (Redman *et al.* 2009). Afrotropical rollers undertake some very significant movements, Lilac-breasted Roller *Coracias caudatus* has made it as far as Eritrea (Ash & Atkins 2009), Oman and Yemen (Porter & Aspinall 2010). Abyssinian Roller *Coracias abyssinicus* has migratory populations in the Sahel; vagrants have been widely recorded in North Africa and the Middle East, including the Canary Islands and Mauritania (Haas & Ławicki 2018), Morocco (Bergier *et al.* 2011), Egypt (Pfützke & Halley 1995) and Saudi Arabia (Meadows 2003). Broad-billed Roller *Eurystomus glaucurus,* meanwhile, has occurred twice north-west to the Cape Verde Islands (Haas 2012), once north-east to Israel in September 2019, east all the way to Mauritius (Safford & Hawkins 2013) and, incredibly, south to the Kerguelen Islands, where one was killed by a Brown Skua *Stercorarius antarcticus* in November 1961 (Ausilio & Zotier 1989). The last record involved an epic overwater flight of at least 3,600km and ought to put this species firmly on the radar of vagrant-hunters in Europe or the Caribbean.

BRACHYPTERACIIDAE
Ground-Rollers

The ground-rollers are thought to be largely sedentary, although there are anecdotal reports of local movements, and an old record of Pitta-like Ground-Roller *Atelornis pittoides* from a garden in the capital of Madagascar – Antananarivo – might be considered nominally a vagrant if it was of natural provenance (Safford & Hawkins 2013).

BUCCONIDAE AND GALBULIDAE
Puffbirds and Jacamars

We are unaware of any well documented records of extralimital movements in either of these groups, most of which have a very limited dispersal capacity (e.g. Lees & Peres 2009) and the ranges of many species are delimited by major rivers, which are barriers for dispersal for many Neotropical species (Wallace 1852, Haffer 1969). Migration has been reported in the southernmost populations of White-eared Puffbird *Nystalus chacuru* (Sick 1997), but this is as yet unconfirmed and is not obvious from a cursory examination of eBird data.

LYBIIDAE, MEGALAIMIDAE, CAPITONIDAE AND SEMNORNITHIDAE
African Barbets, Asian Barbets, New World Barbets and Toucan-Barbets

None of the barbets apparently undertakes any regular migrations, barring some reported altitudinal movements, and we are unaware of any reports of significant extralimital occurrences (Short & Horne 2001). White-headed Barbet *Lybius leucocephalus*, which usually occurs in upland areas 600–2,300m above sea level, has been reported as a vagrant to sea level in Kenya (Short & Horne 2001).

RAMPHASTIDAE
Toucans

No toucans, aracaris and allies are known to undertake any regular seasonal migrations apart from some altitudinal movements, such as those made by Red-breasted Toucan *Ramphastos dicolorus* in the Atlantic Forest of Brazil (Aleixo & Galetti 1997). Sick (1997) reported irruptions of hundreds or even thousands of three species of toucans – White-throated Toucan *Ramphastos tucanus*, Channel-billed Toucan *Ramphastos vitellinus* and Toco Toucan *Ramphastos toco* – in the Amazon delta region in the Brazilian states of Pará and Amapa in the first half of the 20th century, particularly around the city of Belém, Pará; these were perhaps driven by fluctuations in the local abundance of fruit.

INDICATORIDAE
Honeyguides

There is no evidence for any migratory behaviour in honeyguides, and we are unaware of any extralimital occurrences.

PICIDAE
Woodpeckers

Most woodpeckers are highly sedentary and their reluctance to cross environmental barriers such as mountain ranges and waterbodies has led to a lack of woodpeckers on many island systems, including Madagascar, New Guinea and Australia. The most migratory picid is the Eurasian Wryneck *Jynx torquilla*, which in autumn moves from across the Palearctic south into sub-Saharan Africa and southern Asia. Exceptional vagrants have reached Iceland, as well as Alaska on three occasions (Howell *et al.* 2014, eBird) and San Clemente Island in Los Angeles County, California in September 2017 (Rottenborn *et al.* 2018). Most extralimital vagrancy events among the remaining woodpeckers occur in north-temperate zone species. The most migratory Western Hemisphere woodpecker is Yellow-bellied Sapsucker *Sphyrapicus varius*, which breeds in Canada and parts of the northern and eastern United States and migrates south as far as Central America in winter. Vagrants have reached Colombia in South America (Luna *et al.* 2011) and have crossed the Atlantic to the Azores, Britain, Iceland and Ireland (Haas 2012). The remaining sapsucker species undertake more modest seasonal movements, with correspondingly less extreme vagrancy, although there are still surprises, with Williamson's Sapsucker *Sphyrapicus thyroideus* recorded east to Illinois and even New York, and south to Michoacán in Mexico (Buckley & Mitra 2003, Gyug *et al.* 2012). The *subrufinus* subspecies of Rufous-bellied Woodpecker *Dendrocopos hyperythrus* is highly migratory, moving from north-east China and Russian Ussuriland to winter in southern China and northern Vietnam. Vagrants have reached Japan, South Korea (Brazil 2009) and even Mongolia (Reading *et al.* 2011), marking out this species as a potential vagrant to Alaska. Northern populations of Northern Flicker *Colaptes auratus* are also highly migratory; this species has occurred three times on the Azores, in 2010, 2013 and 2016, and on three other occasions in Europe, but all are known to have been (or suspected to have been) ship-assisted (Haas 2012).

Northern populations of several woodpeckers may be irruptive, especially in the boreal belt. Great Spotted Woodpecker *Dendrocopos major* irruptions, triggered by poor crops of spruce or pine seeds, may involve tens of thousands of individuals in Scandinavia and cause them to travel thousands of kilometres. Such irruptions have been linked with the 10 individuals of this species recorded in the United States, in Alaska on the Aleutians, Probilofs and the mainland (Howell *et al.* 2014). Great Spotted Woodpecker was driven to extinction in Ireland because of historic deforestation, but recolonised in the 20th century with afforestation; genetic data indicates that this recolonisation involved the British subspecies *pinetorum* rather than the more dispersive nominate Scandinavian subspecies

► Yellow-bellied Sapsucker *Sphyrapicus varius*, Grímsnes, Suðurland, Iceland 2 September 2018 the 3rd record for Iceland and 6th for the Western Palearctic was first discovered when its sap wells were spotted on birch trees (*Yann Kolbeinsson*).

▲ Williamson's Sapsucker *Sphyrapicus thyroideus*, Southeast Farallon Island, San Francisco, California, United States, 4 October 2013 (*Cameron Rutt*).

▲ Lesser Spotted Woodpecker *Dryobates minor*, Scalloway, Shetland, Scotland, 15 October 2012, the first for Scotland (*Rebecca Nason*).

(Coombes & Wilson 2015). Individuals of the latter subspecies do occasionally cross the North Sea to occur in the United Kingdom, and yet this relatively narrow sea crossing is a major barrier to several other woodpecker species occurring on the adjacent European mainland. There are no British records of Middle Spotted Woodpecker *Dendrocoptes medius* or Black Woodpecker *Dryocopus martius,* for example, despite the occasional appearance of dispersing individuals at continental migration bottlenecks such as Falsterbo in Sweden. A recent record of Lesser Spotted Woodpecker *Dryobates minor* from Shetland – the first for Scotland that undoubtedly originated from Scandinavia – does suggest that there remains a chance for future woodpecker additions to the British List. Even White-backed Woodpecker *Dendrocopos leucotos* may associate with Great Spotted Woodpecker irruptions; the second to fifth records for Switzerland occurred during such invasions in 1997–98 (Sharrock & Davies 2002). Eurasian Three-toed Woodpecker *Picoides tridactylus*, American Three-toed Woodpecker *Picoides dorsalis* and Black-backed Woodpecker *Picoides arcticus* also wander widely as part of irruptions in search of suitable burned forests, resulting in all three being occasionally recorded significantly out of range.

The *Picus* woodpeckers are less prone to vagrancy. Eurasian Green Woodpecker *Picus viridis* has only ever occurred in Ireland on three occasions, all in the 19th century (Hobbs 2015), while Grey-headed Woodpecker *Picus canus* is a vagrant to the Netherlands (van den Berg & Bosman 1999). Several North American *Melanerpes* species are more migratory and do occasionally undertake rather drastic extralimital movements, such as Red-headed Woodpecker *Melanerpes erythrocephalus* appearing in California (Hamilton *et al.* 2007) and a wide scatter of Acorn Woodpeckers *Melanerpes formicivorus* in the Midwest. Perhaps the most exceptional is Lewis's Woodpecker *Melanerpes lewis,* of western North America, which has occurred right across the northeastern United States and even east to Labrador in Canada and south to Chihuahua in Mexico (Torres-Vivanco *et al.* 2015).

Movements in the tropical and subtropical woodpeckers are less well understood. Intensive observations at the southern border of the Western Palearctic have recorded the occasional presence of African Grey Woodpeckers *Chloropicus goertae* in northern Mauritania, and there are also marginally extralimital

► Lewis's Woodpecker *Melanerpes lewis*, Livonia, Ontario County, New York, United States, 23 December 2010, the fifth record for New York (*Jay McGowan*).

records of that species from southern Nigeria and coastal Ghana (Winkler & Christie 2020). Red-bellied Woodpecker *Melanerpes carolinus* has occurred once as a vagrant to the Bahamas (Kirwan *et al.* 2019) and there is a single recent record of Sunda Pygmy Woodpecker *Picoides moluccensis* from the Turtle Islands of the Philippines, but otherwise there is precious little evidence of any extralimital tropical records.

CARIAMIDAE
Seriemas

No migrations or extralimital vagrancy events are known in the seriemas, although Red-legged Seriema *Cariama cristata* is apparently expanding its distribution as a result of forest loss in the Amazon–Cerrado ecotonal region in Brazil (Gonzaga 1996).

FALCONIDAE
Caracaras and Falcons

Several species of falcon undertake dramatic intercontinental migrations, including substantial sea crossings. Perhaps the most impressive is that undertaken by Amur Falcons *Falco amurensis,* which move between India and East Africa via a two- to three-day overwater flight across the Indian Ocean (Dixon *et al.* 2011, Meryburg *et al.* 2017). Amur Falcons have been observed exploiting ocean thermals that develop over warm waters in trade-wind zones to achieve this (Bildstein 2006). Migrants may be drifted off this route south all the way to the Chagos archipelago (Carr 2011) and are regular on the Seychelles (Skerrett 2008). Amur Falcons have been recorded as vagrants east to Japan (Brazil 2009), the Philippines (Jensen *et al.* 2015) and the Mariana Islands (Stinson *et al.* 1997) and west to several European countries including the Britain, Hungary, Italy, Sweden and Romania (Haas 2012). Even more incredibly, this species has reached the mid-Atlantic on St Helena (Rowlands *et al.* 1998) and the Azores (Barcelos *et al.* 2015). The closely related Red-footed Falcon *Falco vespertinus* has also made it to various Atlantic island outposts including Iceland, the Canary Islands, São Tomé and the Azores; one individual made it to Martha's Vineyard in Massachusetts, United States, in August 2004 (Howell *et al.* 2014). Red-footed Falcon is also noteworthy for its periodic largescale displacements into western Europe – 125 were recorded in Britain in spring 1992, for example (Nightingale & Allsopp 1994). Eurasian Hobby *Falco subbuteo* has both a broader breeding Palearctic distribution and spread of vagrant records than the previous two insectivorous falcons; vagrancy records include Perth, Western Australia (Greatwich 2019), the Philippines (Jensen *et al.* 2015), and both eastern and western North America (Howell *et al.* 2014). Lesser Kestrel *Falco naumanni* also has a huge Holarctic breeding range and has occurred to island groups as disparate as Japan (Brazil 2009), the Comoros (White 2011) and Seychelles (Safford & Hawkins 2013), but most notable was one found dead in July 1996 onboard a fishing vessel in the Barents Sea, south-east of Svalbard (Mobakken 2005). Given this species' intercontinental migration, it is perhaps surprising that Lesser Kestrel has not reached the Americas, unlike Common Kestrel *Falco tinnunculus,* which has occurred as a vagrant to Greenland (Boertmann 1994), both eastern and western North America (Howell *et al.* 2014), the West Indies (Kirwan *et al.* 2019), French Guiana (Renaudier 2010) and on the Brazilian São Pedro and São Paulo archipelago in the mid-Atlantic (Bencke *et al.* 2005).

Two extremely unusual mid-latitude migrant falcons merit a special mention – Eleonora's Falcon *Falco eleonorae* and Sooty Falcon *Falco concolor,* both of which specialise in predating migrant birds and both of which winter on Madagascar (Gschweng *et al.* 2008, Javed *et al.* 2012). Eleonora's Falcon is a highly sought-after vagrant to northern Europe; Sweden has 22 records, which is exceptional given that some countries, e.g. the Netherlands, have only just notched up their first (Rijmenams & Ebels 2011). One satellite telemetry study found that adult Eleonora's Falcons make dispersive forays up to 400km inland into Spain, France and Italy before they start breeding, visiting a variety of habitats hundreds of kilometres from the respective breeding colonies (Mellone *et al.* 2013). All three tagged birds undertook these movements, which might explain some early season records from northern Europe; a single record from the Azores (Barcelos *et al.* 2015) also highlights the potential for transatlantic vagrancy in this species. Ship assistance also seems a strong possibility, and there is a single record of Eleonora's Falcon from Amsterdam Island in the remote Indian Ocean, 3,200km south-east of Madagascar in January 1982 (Rou & Martinez 1987). Intriguingly, perhaps the same individual was seen again two weeks later aboard a boat travelling from Amsterdam Island to Reunion Island. Sooty Falcon *Falco concolor* has a similar life history but has a more south-easterly distribution and is consequently rarer away from the Red Sea and the eastern Mediterranean region. However, there are records from Turkey (Kirwan *et al.* 2008), Malta (Snow & Perrins 1998), Tunisia (Isenmann *et al.* 2005), the Algerian Sahara

◄ Amur Falcon *Falco amurensis*, Polgigga, Cornwall, England, 7 July 2017, the second record for Britain (*Stuart Piner*).

► Red-footed Falcon *Falco vespertinus*, Gibraltar Point NNR, Lincolnshire, England, 29 September 2015. Autumn occurrences of juveniles are considerably rarer in the UK than spring occurrences of adults and first-summer birds (*Alexander Lees*).

► Eurasian Hobby *Falco subbuteo*, Cockburn, Western Australia, Australia, 4 March 2016, the first Australian mainland record (*Rohan Clarke*).

(Isenmann & Moali 2000) and Morocco (Bergier *et al.* 2009), as well as Italy and one record from southern France (Kleis 2008). Northern and eastern European records seem a distinct possibility and this species has also wandered out in the Indian Ocean as far as Mauritius (Safford & Basque 2007).

Vagrancy events by two Holarctic species, Merlin *Falco columbarius* and Peregrine Falcon *Falco peregrinus*, are harder to track, as field identification of subspecies is more challenging. The nominate subspecies of Merlin has occurred as a vagrant to Britain and Ireland and Hawaii (Pyle & Pyle 2017) and has also wandered as far south as Bangladesh (Thompson *et al.* 2014), Peru (Schulenberg *et al.* 2010) and central Brazil (Dornas & Pinheiro 2014). Peregrines take transoceanic movements to an extreme for a landbird; they are a regular winter visitor to Hawaii (Pyle & Pyle 2017), and frequently reach other island groups as remote as South Georgia (Clarke *et al.* 2012). Peregrines have been observed using highly energy-efficient shearing flight, similar to that used by tubenoses (Kerlinger 1985), which probably facilitates such long-range movements over water. The arrival of Peregrines on remote islands can have major consequences for insular species that may lack adaptations to avoid such prodigious predators (see page 68). Peregrines may be relatively regular transatlantic vagrants, but given the difficulties surrounding field identification of subspecies, we only know about these movements through ringing recoveries. These have included at least three individuals ringed in North America and subsequently recovered in Europe, including most recently one ringed in Wisconsin and recovered in the Ticino Mountains in Switzerland (Doolittle *et al.* 2013).

Understanding vagrancy in species that are popular with falconers is fraught with difficulty, compounded by a high incidence of hybridisation among captive-bred birds, which may make field identification impossible. Saker Falcon *Falco cherrug* is a good case in point with records from north-west Europe such as the British Isles often assumed to relate to escapes. However, recent studies with satellite-tracked juvenile Sakers have indicated that long-range post-fledging (pre-migration) dispersal can occur in any compass direction, and has taken individuals from Hungary all the way to Germany (Prommer *et al.* 2012). Juvenile females have been found to routinely winter as far south as the Sahel, with vagrants south to Tanzania (Stevenson & Fanshawe 2002) and the Seychelles (Skerrett *et al.* 2006). Records of vagrant Lanner Falcons *Falco biarmicus* are equally problematic, but this species also probably reaches northern Europe naturally on occasion, and there are 55 records from France (up to 2010) as well as increasing records from southern Spain (de Juana & Garcia 2015). Lanner has even been reported from the Chagos Islands in the Indian Ocean (Carr 2015). Similarly, tagged wild Gyrfalcons *Falco rusticolus* have made epic movements in the Arctic (Burnham & Newton 2011) and the predictable appearance of birds in late winter along the 'Celtic fringe' of western Europe is indicative of natural vagrancy. Greenland Gyrfalcons are known to spend long periods hunting offshore over pack ice, and as such merit inclusion as an 'honorary seabird'. In the United States, Gyrfalcons have been recorded south as far as California, Texas and North Carolina (Booms *et al.* 2008), and in Europe south to Spain (de Juana & Garcia 2015). American Kestrel *Falco sparverius* is also recorded as an escape in Europe with some regularity and is common in captivity, but there are accepted records of this highly migratory species from the Azores (three), Britain (two), Denmark (one) and, incredibly, one from Malta (Haas 2012).

At lower latitudes, movements by falcons tend to be less dramatic. Interesting instances of apparent vagrancy include: a record of Greater Kestrel *Falco rupicoloides* from Niger, the first for West Africa (Rabeil & Wacher 2011); Red-headed Falcon on Sri Lanka (Rasmussen & Anderton 2005); Aplomado Falcons *Falco femoralis* on the West Indies (Mathys 2011) and Falklands (Woods 2017); Nankeen Kestrels *Falco cenchroides* on Tasmania (Ferguson-Lees & Christie 2001); and Brown Falcons *Falco berigora* on Lord Howe Island (McAllan *et al.* 2004) and New Zealand (Miskelly *et al.* 2011). Beyond these, most species of falcons occurring in the tropics, such as caracaras, forest-falcons and falconets, tend to be rather sedentary. Genera such as the *Micrastur* forest-falcons are probably very sedentary, but there is a single record of a Collared Forest-Falcon *Micrastur semitorquatus* from Hidalgo County, Texas, in January–February 1994 (Howell *et al.* 2014) which was around 200km out of range. Arguably far more unusual was a Spot-winged Falconet *Spiziapteryx circumcincta* in Santa Cruz province, southern Argentina, around 1,000km south of its regular range (Bornschein 1996). Two species of caracaras are known for occasionally occurring as vagrants: there are records of Crested Caracara *Caracara cheriway* from across North America – as geographically disparate as British Columbia and Nova Scotia (Nelson & Pyle 2013); whilst Chimango Caracara *Milvago chimango* has occurred as a vagrant to the Falklands (Woods 2017) and southern Bolivia (Herzog *et al.* 2016).

◄ Crested Caracara *Caracara cheriway*, Scotts Corners Golf Course, Orange, New York, United States, 11 April 2015, the third record for New York state (*Jay McGowan*).

STRIGOPIDAE
New Zealand Parrots

There is no evidence for any extralimital movements in the three species of New Zealand Parrots.

CACATUIDAE
Cockatoos

No species of cockatoos show evidence of regular population-wide seasonal migrations, although many are nomadic and, over evolutionary timescales, the group has proven to be good at island-hopping in Australasia (White *et al.* 2011).

PSITTACULIDAE
Old World Parrots

Migration is rare in parrots. Only Swift Parrot *Lathamus discolor* and Orange-bellied Parrot *Neophema chrysogaster* are fully migratory, both moving from Tasmania to mainland Australia, and they can occur extralimitally in southern Australia. The former has been recorded once as a vagrant to Lord Howe Island (McAllan *et al.* 2004), a movement requiring a 600km sea crossing. Most instances of parrot vagrancy are tied to occasional long-distance dispersal events in otherwise sedentary or nomadic populations, which have also facilitated colonisation and speciation in many tropical islands and island chains.

PSITTACIDAE
New World and African Parrots

In common with most of the Old World parrots, regular migrations are rare and instances of vagrancy very limited in this group. Records of some vagrants might reflect individuals at the expanding fringe of growing populations – such as a recent first Peruvian record of Turquoise-fronted Parrot *Amazona aestiva* from the Madre de Dios (Avendaño 2005). In Argentina, records of Blue-crowned Parakeets *Thectocercus acuticaudatus* and Burrowing Parakeets *Cyanoliseus patagonus* from Punta Rasa in Argentina, a migration hotspot, have been attributed as vagrants (Jaramillo 2000) although neither is massively out of range. There are also historic records of invasions of large flocks of parakeets to the Falklands (Woods 2017); these probably involved Austral Parakeet *Enicognathus ferrugineus* but descriptions do not rule out Slender-billed Parakeet *Enicognathus leptorhynchus*. It seems likely that vagrancy might have been prevalent in the now-extinct Carolina Parakeet *Conuropsis carolinensis* of eastern North America, which was apparently quite erratic in its appearances in some areas, but there is no evidence of consistent seasonality in its occurrence patterns (Snyder & Russell 2002).

ACANTHISITTIDAE
New Zealand Wrens

This most ancient group of passerine birds is highly sedentary (Gill 2019) and there are no known instances of extralimital occurrences in the two extant species.

CALYPTOMENIDAE
African and Green Broadbills

Migratory behaviour in this recently erected family of broadbills is also very limited. There is, however, one case of apparent vagrancy involving Whitehead's Broadbill *Calyptomena whiteheadi*, with a bird seen outside of its normal high-altitude forest range at just *c.*75m above sea level at Maau, on the River Bengalun in east Kalimantan, a movement perhaps associated with severe drought conditions in 1997–98 (Bruce 2019).

EURYLAIMIDAE
Asian and Grauer's Broadbills

There is relatively little evidence of broadbill migration beyond altitudinal movements in some species, such as those associated with seasonal movements in the Himalayan foothills involving Long-tailed Broadbill *Psarisomus dalhousiae* and Silver-breasted Broadbill *Serilophus lunatus* (Bruce 2019).

SAPAYOIDAE
Sapayoa

Sapayoa *Sapayoa aenigma* occupies its own family and has been shown recently to be the sole living New World member of an ancient radiation of Old World suboscine passerines (Fjeldså *et al.* 2003), implying a major historical dispersal event. Today, like most Neotropical forest passerines it is highly sedentary and there is no evidence of vagrancy (Snow *et al.* 2020).

PHILEPITTIDAE
Asities

No asities are suspected of making any seasonal movements of any significance that might predispose them to vagrancy, but a record of two male Common Sunbird-Asities *Neodrepanis coruscans* at the Ambohitantely Reserve, Madagascar, in February 1995 – a 150km range extension – might represent vagrancy, or even the discovery of a relict population (Andrianarimisa 1995).

PITTIDAE
Pittas

This sought-after group of rainforest understorey birds includes four highly migratory species: Fairy Pitta *Pitta nympha*, Blue-winged Pitta *Pitta moluccensis*, Indian Pitta *Pitta brachyura* and African Pitta *Pitta angolensis*. The *longipennis* subspecies of African Pitta may move up to 2,000km, migrating at night, and like other pittas they are frequently attracted to lights where small 'falls' may occur (and also frequently die colliding with windows in cities). Overshooting vagrants have occurred north to southern Ethiopia and south as far as south-east South Africa (Erritzoe 2020). Fairy Pittas breed in eastern China, Taiwan, South Korea and southern Japan and migrate to winter in southern China, Laos, Vietnam and Borneo. This species has twice overshot its wintering areas to reach Australia – once to Christmas Island in November 2012 and once to Derby, Western Australia, in December 2007 (James & McAllan 2014). Vagrants have also occurred east to the Philippines (Jensen *et al.* 2015), Lao PDR (Coudrat & Nanthavong 2016), and there was a recent spring overshoot to Bolshoy Pelis Island in southern Russia (Gluschenko Yu & Korobov 2015). Fairy Pitta is conceivably a vagrant to the Aleutian Islands and thus arguably one of the most exciting bird species that *might* occur in North America, although the species' numerical rarity (the global population estimate is just 1,500–7,000 individuals) makes this less likely than for some other species sharing a similar ecology.

Blue-winged Pittas do not have such a long migration span but winter over a broader area of Southeast Asia and have also reached mainland Western Australia four times and Christmas Island twice (James & McAllan 2014). Indian Pittas migrate from northern India to winter in southern India and Sri Lanka; there is a single extralimital record of a juvenile female collected in November 1968 in south-west Iran (van Els & Brady 2014). This record raises the prospect of pitta vagrancy to the southeastern limits of the Western Palearctic. The *cucullata* subspecies of Hooded Pitta *Pitta sordida* is also largely migratory and has occurred north to Yaeyama Island, Japan, in June 1994 (Brazil 2018) and once to Australia where one was found dead on Barrow Island, Western Australia, in January 2010 (Johnsone *et al.* 2013). The remaining pitta species are largely sedentary, but given a history of colonisation of oceanic islands, even occasional vagrancy events by nominally resident species is perhaps to be expected.

▼ Fairy Pitta *Pitta nympha*, Derby, West Kimberley, Western Australia, Australia, 14 December 2007), the first record for Australia (*Adrian Boyle*).

THAMNOPHILIDAE
Antbirds

Antbirds are a group of notoriously sedentary, mostly forest-dwelling insectivorous passerines. Many species are behaviourally reticent to cross narrow habitat discontinuities such as forest roads (Lees & Peres 2009) and experimental evidence indicates that most are physiologically incapable of sustained flight for more than a few tens of metres (Moore *et al.* 2008). As a result, this group is largely absent from Caribbean islands with the exception of Trinidad & Tobago. Trinidad was connected to the mainland by a land bridge as recently as 11,000 years ago, but Tobago has apparently been isolated for at least two million years (Snow 1985). This suggests that the four antbirds that persist there – White-fringed Antwren *Formicivora grisea*, Plain Antvireo *Dysithamnus mentalis*, Great Antshrike *Taraba major* and Barred Antshrike *Thamnophilus doliatus* – may have a history of island-hopping via overwater flights. With the exception of the antvireo, these species are essentially non-forest birds that may be better adapted to long-distance dispersal than most others in this family, given that they occupy ephemeral non-forest habitats (Mahood *et al.* 2012). Elsewhere, three species – Rufous-capped Antshrike *Thamnophilus ruficapillus*, Variable Antshrike *Thamnophilus caerulescens* and Great Antshrike – have been reported as making seasonal movements at their southern range edges (Zimmer & Isler 2003), although there seems to be no evidence of population-level displacement in eBird data, with birds occupying the southern range edge throughout the austral winter.

Given the general lack of movement in this group, antbird vagrancy is largely unrecorded – but there is one apparent exception. There are multiple documented records of Barred Antshrike away from its core Brazilian distribution. For example, there is a record from Santa Catarina (Evair 2019), which is 170km away from the nearest population in Argentina, and a few scattered records in southern Paraná state (e.g. Hoppen 2018) and eastern Minas Gerais (e.g. Miranda 2012). These areas are well enough inventoried not to be hiding undiscovered populations, so records would seem likely to relate to vagrant individuals, perhaps even as a precursor to range extension, particularly at the southern range edge where habitat change may be providing new opportunities for the species. Even in this case, though, the risk of escapes cannot be dismissed entirely as this species has been found for sale in bird markets in north-east Brazil (Pereira & Brito 2005). However, suboscine

◄ Barred Antshrike *Thamnophilus doliatus*, Ouro Fino, Alta Floresta, Mato Grosso, Brazil, 6 September 2006. First found in this region in 2004, this species has rapidly colonised farmland that was formerly unsuitable rainforest vegetation, a preference for human-modified habitats unlike most of its congeners. This population expansion may lead to modest vagrancy events (*Alexander Lees*).

passerines are generally incredibly rare in captivity owing to a lack of expertise in keeping them alive (Conde *et al.* 2011), which is itself hampering efforts to conserve the last individuals of many rare species (Pereira *et al.* 2014). Against this backdrop of extra-limital records in Brazil, a report of this species from Texas (the only record of any antbird in the United States) perhaps needs another appraisal. The Texas Bird Records Committee accepted Barred Antshrike onto the state list on the basis of a sound-recording of an individual singing at night in 2006 in Harlingen (Lockwood, 2008), but this record was not accepted by the American Birding Association Checklist Committee (Pranty *et al.* 2007). The reasoning was that: 1) nocturnal singing by antbirds is extremely unusual; 2) there was no visual confirmation of the species' identity; and 3) there was at the time no history of vagrancy in the species. In the context of the Brazilian records, however, a movement of 250km from the nearest populations in northern Mexico does not seem like an impossible feat of vagrancy, and future records might still be expected.

MELANOPAREIIDAE
Crescentchests

Crescentchests are highly sedentary with no documented vagrancy events.

CONOPOPHAGIDAE
Gnateaters

Gnateaters are another entirely non-migratory group with no history of vagrancy (Whitney 2003). These species are often so hard to find when not singing that any vaguely out-of-range bird would also be very difficult to detect.

GRALLARIIDAE
Antpittas

Along with many groups of forest-dwelling suboscine passerines in South America, no migrations are known in antpittas (Greeney 2018) and members of the group are notable for being poor fliers with a very low dispersal capacity. A few species may make seasonal altitudinal movements, which could lead to downslope vagrancy, but we are unaware of any such reports.

RHINOCRYPTIDAE
Tapaculos

Tapaculos are among the most dispersal-limited bird species known, typically flying only a few metres at a time, with most movement involving hopping and walking. It should not come as a surprise therefore that no known instances of migratory behaviour or vagrancy have been recorded (Krabbe & Schulenberg

2003a). Their low dispersal capacity has massively contributed to the high species richness in tapaculos, with some genera like *Scytalopus* having unique species on each Neotropical mountain range. Although there is no evidence for any long-distance movement by tapaculos, there is one long-standing ornithological mystery that raises the possibility. Charles Darwin collected the sole known specimen of a Magellanic Tapaculo *Scytalopus magellanicus* from the Falklands and claimed to have seen several other individuals in March 1833 (Gould & Darwin 1839, Woods 2017). The provenance of this specimen has been questioned (Bourne 1992b) but it does appear that the extinct Falklands Wolf *Dusicyon australis* colonised via a narrow, shallow marine strait during a prehistoric glacial phase, so natural colonisation for the tapaculo without an overwater flight does not seem out of the question (Austin *et al.* 2013).

FORMICARIIDAE
Antthrushes

Anthrushes are a highly sedentary group of tropical forest passerines (Krabbe & Schulenberg 2003b) for which there are no extralimital movements on record.

FURNARIIDAE
Ovenbirds and Woodcreepers

Although most members of the Furnariidae are sedentary, there are several different genera that have pushed south into the southern cone of temperate South America and there evolved migratory behaviour that predisposes them to vagrancy. This family has also colonised remote oceanic islands, most spectacularly in the form of Masafuera Rayadito *Aphrastura masafuerae,* which is restricted to the island of Alejandro Selkirk, in the Juan Fernández group, and the endemic *baeckstroemii* form of Grey-flanked Cinclodes *Cinclodes oustaleti* on Robinson Crusoe Island. These colonisation events would have required their vagrant ancestors to make overwater flights of up to 730km from mainland Chile. Highlighting the dispersal capacity of cinclodes, the

◄ Buff-winged Cinclodes *Cinclodes fuscus*, Parque Nacional da Lagoa do Peixe, Tavares, Rio Grande do Sul, Brazil, 6 September 2013. This is a rare austral migrant to the state with a handful of vagrants having reached areas further north (*Paulo Ricardo Fenalti*).

Falklands have an endemic subspecies of Blackish Cinclodes, *Cinclodes antarcticus antarcticus,* and there are three records of vagrant Buff-winged Cinclodes *Cinclodes fuscus* (also called Bar-winged Cinclodes) from the islands (Woods 2017). Buff-winged Cinclodes is a regular austral winter visitor to the extreme south of Brazil and has occurred as a vagrant north as far as São Paulo (Silva 2019). Even more impressive is the single record of Grey-flanked Cinclodes from South Georgia (Clarke *et al.* 2012). Several species from other genera are also austral migrants, such as Sharp-billed Canastero *Asthenes pyrrholeuca,* which has overshot to Brazil on several occasions (Rebelato *et al.* 2011). Pale-breasted Spinetail *Synallaxis albescens* is unrecorded as a vagrant but is a suspected austral migrant and has been detected multiple times offshore in the Atlantic off eastern Argentina (eBird). This suggests a greater capacity for flight than would be expected for a species with such short wings and long tail, and future regional vagrancy is to be expected.

PIPRIDAE
Manakins

Most species of manakins are resident. Altitudinal migrations have been reported in species such as White-ruffed Manakin *Corapipo altera,* but there are no extralimital records (Kirwan & Green 2011).

COTINGIDAE
Cotingas

Most species of cotingas are fairly sedentary, but as with many frugivorous tropical species, they may be prone to wander relatively long distances in search of food. Two species, Rufous-tailed Plantcutter *Phytotoma rara* and White-tipped Plantcutter *Phytotoma rutile,* are partially migratory, with the southernmost part of their distributions abandoned in winter. Rufous-tailed Plantcutter has occurred as a vagrant to the Falklands once (Woods 2017), whilst White-tipped Plantcutter *Phytotoma rutila* is a scarce migrant to southern Brazil and has occurred as a vagrant north to Espírito Santo (Venturini *et al.* 2007). Movements by

► Bare-throated Bellbird *Procnias nudicollis*, Alcatrazes Island, São Paulo, Brazil, 5 September 2013. This species is an occasional vagrant to coastal islands in the region (*Fabio Olmos*).

other cotingas are more erratic; several of the *Procnias* bellbirds make altitudinal movements, which might explain vagrancy of Bare-throated Bellbird *Procnias nudicollis* to the Alcatrazes archipelago off São Paulo (Olmos *et al.* 2014) and White Bellbird *Procnias albus* to lowland Amazonia and Trinidad, away from known breeding locations (Kirwan & Green 2011). The recent discovery of a population of White Bellbirds in lowland rainforest in eastern Amazonia, coupled with a lack of genetic differentiation, hints at long-range dispersal between sub-populations (Dantas *et al.* 2017). Several Amazonian cotingas are so poorly known that apparent records of vagrants may yet prove to be range extensions, but a sight record of Crimson Fruitcrow *Haematoderus militaris* from Alta Floresta in Mato Grosso, Brazil (Lees *et al.* 2013a), and an old specimen of Amazonian Umbrellabird *Cephalopterus ornatus* from Ilha Mexiana, Pará, Brazil (Hagmann 1907), are almost certain to represent genuine vagrants.

TITYRIDAE
Tityras and allies

Most of the species in this group are associated with humid tropical and montane rainforests and are considered sedentary, although the group does include some intratropical migrants, such as Crested Becard *Pachyramphus validus*, which is a scarce migrant in the austral winter north to southern Amazonia (Lees *et al.* 2013a, Harvey *et al.* 2014). This species has occurred as a vagrant to the Alcatrazes archipelago off São Paulo (Olmos *et al.* 2014). In the north, populations of Rose-throated Becard *Pachyramphus aglaiae* in northern Mexico appear to be partially migratory and reach the southern United States border in south-east Arizona and south Texas, while this species has also occurred as a vagrant to Cozumel Island, Mexico (Howell 2004). Grey-collared Becard *Pachyramphus major*, an elevational migrant in northern Mexico, has occurred once as a vagrant to Arizona in 2009 (Howell *et al.* 2014) and has also occurred once on Cozumel Island (Howell 2004). Shrike-like Cotinga *Laniisoma elegans* is known to undertake altitudinal movements (Kirwan & Green 2011), and there is a record of an immature male trapped at Quirinópolis, Goiás, Brazil, in August 2005 which was about 600km out of range (Ferreira *et al.* 2010). Such an extreme movement for a rainforest bird is rare and suggests that there may be more complex patterns of long-range movements in this species. Most tityras are resident, although Black-crowned Tityra *Tityra inquisitor* is partially migratory at its southern range edge. Masked Tityra *Tityra semifasciata* has occurred once as a vagrant to Texas in February–March 1990, an event associated with unseasonally cold weather, and another was recorded on a boat 70km off Costa Rica in April 2008 (Howell *et al.* 2014).

OXYRUNCIDAE
Sharpbill, Royal Flycatcher and allies

Most members of this group are sedentary understorey tropical forest species with the exception of Sharpbill *Oxyruncus cristatus,* an odd forest-canopy species and something of an enigma; there is some evidence that Sharpbill makes altitudinal movements into the Amazonia lowlands, but equally there are questions about the provenance of old specimens (da Silva 1993). A recent photo-documented record from the Tapajós National Forest, Brazil (Lopes 2017), a first record in an area with an intense history of ornithological research (Lees *et al.* 2013b,), seems indicative of vagrancy rather than an undiscovered population.

TYRANNIDAE
Tyrant Flycatchers

Unlike the other species-rich branches of the suboscine passerine radiation, which overwhelmingly consist of resident species with a very limited dispersal capacity, almost one-third of all tyrant flycatchers engage in some form of annual migration (Fitzpatrick 2004). These include two major groups. The first of these are species moving from the northern latitudes south towards the tropics in the non-breeding season – boreal migrants – essentially including all North American species breeding north of the United States–Mexico border. The second group are those species that move north from breeding ranges in southern temperate latitudes – the austral migrants – which include all species breeding south of central Argentina. Even within tropical latitudes, a few species make seasonal movements, involving quite complex patterns of migration which are only now being elucidated, such as the longitudinal movements made by some Ash-throated Casiornises *Casiornis fuscus* from the dry *caatinga* of north-east Brazil towards Amazonia during the dry season (Lees 2016). Understanding vagrancy is complicated by the presence of lots of intraspecific variation in movements, with some species having multiple migratory and sedentary subspecies; this also suggests the need for taxonomic revisions in some cases.

◄ Scissor-tailed Flycatcher *Tyrannus forficatus*, Charleston, South Carolina, United States, 27 December 2013. This species has been recorded widely as a vagrant in North America and was recently documented for the first time in South America, in Colombia (Felix & Lima 2020) (*Chris Wood*).

◄ Eastern Kingbird *Tyrannus tyrannus*, Inishbofin, Galway, Ireland, 24 September 2013, the second record for Galway and the Western Palearctic (*Anthony McGeehan*).

The champion migrant in the boreal group is the Eastern Kingbird *Tyrannus tyrannus*, which has the longest migration of any tyrant-flycatcher – breeding from the Northwest Territories of northern Canada to winter regularly as far south as central Bolivia. In the Western Hemisphere, vagrants have occurred north to Nunavut, Baffin Island, the Pribilofs and the Aleutian Islands, and south all the way to the Falklands where there are four records (Woods 2017). Given that this species had occurred four times on Greenland (Boertmann 1994), twice on the Galapagos (Wiedenfeld 2006), and once each on the remote Atlantic islands of Tristan da Cunha (Bond & Glass 2016) and South Georgia (Clarke *et al.* 2012), the long wait for a Western Palearctic record was perhaps surprising. After the first was finally discovered in Ireland, a first-year bird on Inishmore, County Galway, in October 2012 (Delaney 2012), it was quickly followed by an adult the following year, a little further north in the same county; the first for Britain followed three years later on the Outer Hebrides. It was perhaps even more surprising that the first kingbird for the Azorean archipelago was not this species but a Western Kingbird *Tyrannus verticalis* on Flores in October 2018; this was also the first for the Western Palearctic. This western North American species was an unexpected vagrant to Europe, although it occurs as a regular vagrant to the eastern United States and Bermuda – one of six species of *Tyrannus* flycatcher to have occurred there. There is a scatter of records of several *Tyrannus* species with restricted breeding distributions in the southern United States from across North America, including Grey Kingbirds *Tyrannus dominicensis*, with a predominantly eastern record bias, and a few widely scattered records of both Couch's Kingbird *Tyrannus couchii* and Cassin's Kingbird *Tyrannus vociferans*. Cassin's Kingbird has also reached Cuba (Kirwan *et al.* 2019), which is perhaps a potential point of origin for the recent run of records of Loggerhead Kingbird *Tyrannus caudifasciatus* in Florida (Howell *et al.* 2014). The increase in records of Couch's Kingbirds is surprising given that the species is largely resident, exhibiting only small seasonal movements at the northern edge of its range. However, the species is also expanding its range northwards (Brush 2020), which may be a driver of vagrancy.

Among the austral migrant group, Fork-tailed Flycatchers *Tyrannus savanna* are one of the most celebrated vagrants to North America – almost all have been of the abundant migratory austral-migrant nominate subspecies. They have been recorded right across the United States and southern Canada north to Hudson Bay (Howell *et al.* 2014) as well as south as far as the Falklands (Woods 2017), with one record aboard a ship only 80km from the Antarctic peninsula in November 1994 (Montalti *et al.* 1999). Incredibly, there is also a single European record from Doñana, Spain, in October 2002 (Haas 2012). This is the sole record of transatlantic vagrancy by a South American austral migrant, occurring in the wake of a strong weather system that took an unusually southerly track across the mid-Atlantic.

Other tyrant flycatchers are, on the whole, poorly represented as transatlantic vagrants, with the remaining species including a single Eastern Phoebe *Sayornis phoebe* in England, Acadian Flycatchers *Empidonax virescens* in England and Iceland, a Least Flycatcher *Empidonax minimus* on Iceland, Alder Flycatchers *Empidonax alnorum* on Iceland, in England and in Norway (Haas 2012) and, most recently, a Yellow-bellied Flycatcher *Empidonax flaviventris* in Scotland in September 2020. Two records of Eastern Wood-Pewee *Contopus virens* from Corvo on the Azores

◄ Cassin's Kingbird *Tyrannus vociferans*, Floyd Bennett Field, Kings, New York, United States, 30 November 2014, the second state record (*Jay McGowan*).

► Grey Kingbird *Tyrannus dominicensis*, Genesee, New York, United States, 30 Oct 2016. Records of this species show a strong coastal bias in eastern North America (*Jay McGowan*).

► Acadian Flycatcher *Empidonax virescens*, Dungeness, Kent, England, 22 September 2015, the first British and second Western Palearctic record (*James Lowen*).

(Monticelli *et al.* 2018) remain the only other tyrant flycatcher species reported from the archipelago. An old record of Olive-sided Flycatcher *Contopus cooperi* from Greenland (Boertmann 1994) hints at a future possibility for trans-Atlantic vagrancy, as this boreal migrant species winters south as far as Amazonia.

In the Nearctic, multiple tyrant flycatcher species are endemic breeders of eastern North America and western North America respectively. Vagrants of eastern species such as Great Crested Flycatcher *Myiarchus crinitus* are found with some regularity in the west in places like California, whilst western species such as Ash-throated Flycatcher *Myiarchus cinerascens* occur regularly as vagrants in the east, as far as Bermuda and Newfoundland. Field separation of many species is challenging, and often depends on vocalisations, so the few records of Eastern Wood-Pewees in the west or Western Wood-Pewees *Contopus sordidulus* in the east likely reflects a detection bias rather than genuine mega-rarity. The same may be true for many *Empidonax* flycatchers, although the use of genetic tools to confirm the identity of some vagrants has led to the confirmation of tricky eastern vagrants such as Pacific-slope Flycatcher *Empidonax difficilis* (Goldberg & Mason 2017). Many vagrants have reached the Bering Sea region, including Eastern Kingbird, Pacific-slope Flycatcher, Least Flycatcher, Yellow-bellied Flycatcher and Western Wood-Pewee, but none have been reported from the Russian side of the Bering Sea as yet. Several of these boreal breeders have also overshot their wintering grounds to occur in southern South America, with records of both Olive-sided Flycatcher and Eastern Wood-Pewee as vagrants to Argentina, for example.

◀ Hammond's Flycatcher *Empidonax hammondii*, Central Park, New York, United States, 28 November 2017, the third state record (*Jay McGowan*).

North America is also frequently visited by vagrant Neotropical tyrant flycatchers, and these fall into two broad categories: a) short-distance vagrancy often of largely resident species from Mexico or the Caribbean; and b) long-distance vagrancy by austral migrants such as the aforementioned Fork-tailed Flycatcher. Species in the former category include Tufted Flycatcher *Mitrephanes phaeocercus,* an elevational migrant with at least seven records, all from south-east Arizona; Nutting's Flycatcher *Myiarchus nuttingi,* which is thought to be largely resident, yet has occurred as a vagrant to California, Arizona and Texas (this species' breeding range almost reaches the United States border in Sonora, Mexico); and two records of Social Flycatcher *Myiozetetes similis* from Texas (Howell *et al.* 2014). Meanwhile, from the Caribbean, there are records of La Sagra's Flycatcher *Myiarchus sagrae* in Florida and Alabama, and Cuban Pewees *Contopus caribaeus* in Florida (Howell *et al.* 2014). The single, incredible, record of Greenish Elaenia *Myiopagis viridicata* from Texas in May 1984 (Howell *et al.* 2014) might have come from populations in northern Mexico or have been an overshooting austral migrant from populations further south. The same questions hang over the origin of North American Piratic Flycatchers *Legatus leucophaius* in Texas, New Mexico and Florida, as this species has fully migrant populations in Central and South America, although it would seem most likely that vagrants are overshoots from more northern populations in Mexico (M. Iliff *in litt.*).

From further away in southern Central America and South America come a suite of species that are either partial or full austral migrants. These include Variegated Flycatchers *Empidonomus varius,* recorded north to Toronto, Canada, and a trio of mega-rarities: White-crested Elaenia *Elaenia albiceps* recorded twice, in Texas in February 2008 (Howell *et al.* 2014) and Nebraska in October 2020 (eBird); Small-billed Elaenia *Elaenia parvirostris,* with one record assumed to be this species from Chicago in April 2012; and Crowned Slaty Flycatcher *Empidonomus aurantio-atrocristatus,* with one record from Louisiana in June 2008 (Howell *et al.* 2014). The Louisiana record of Crowned Slaty Flycatcher was preceded by the first record beyond South America in Panama (Robb *et al.* 2009), and this species has also been recorded way to the south of its regular range, alighting on a boat off the Falklands (Woods 2017). Some of these long-distance austral migrants might even be candidates for transatlantic vagrancy to the Palearctic, although the predominantly south-easterly Atlantic trade winds may prove a barrier to such movements. Many other species are candidates for vagrancy to North America, notably both the northern and southern subspecies of Streaked Flycatcher *Myiodynastes maculatus* and Slaty Elaenia *Elaenia strepera,* which migrates north along the Andes (Marantz & Remsen 1991) and has occurred as a vagrant to Trinidad (White & Hayes 2002). Another long-range migrant is Rufous-tailed Attila *Attila phoenicurus,* which breeds in the Atlantic Forest of south-east Brazil and migrates north-west to largely unknown winter quarters somewhere in northern South America (T. Chesser *in litt.*). There are records north to Venezuela (Hilty 2003) and apparent vagrants have also reached Argentina in the south (García *et al.* 2016).

In the southern cone of South America there is a suite of highly migratory species that move towards warmer temperate latitudes or even the tropics in

► Pine Flycatcher *Empidonax affinis*, Aliso Spring, Pima, Arizona, United States, 21 June 2016, the first record for North America (*Tom Johnson*).

► Tufted Flycatcher *Mitrephanes phaeocercus*, Ramsey Canyon, Cochise, Arizona, United States, 4 May 2016. First recorded in North America in Arizona in 2005, this species bred extralimitally in Ramsey Canyon in 2015 (*Tom Johnson*).

► La Sagra's Flycatcher *Myiarchus sagrae*, Everglades NP, Miami-Dade, Florida, United States, 27 February 2020. A rare vagrant from the Caribbean region to the south-eastern United States (*Tom Johnson*).

▲ Rusty-backed Monjita *Xolmis rubetra*, Rio Maipo Wetland, Santo Domingo, Valparaíso, Chile, 19 October 2013, the first record for Chile (*Alvaro Jaramillo*).

the austral winter. These include many Patagonian breeding species that may be partial migrants such as Austral Negrito *Lessonia rufa*, Chocolate-vented Tyrant *Neoxolmis rufiventris* and Rusty-backed Monjita *Xolmis rubetra*, all of which typically retreat north as far as the central pampas in the austral winter. Both Chocolate-vented Tyrant (William 1974) and Rusty-backed Monjita (Bellagamba-Oliveira *et al.* 2013) have been recorded as vagrants to southern Brazil. Austral Negrito is also being discovered with increasing frequency along the coast of southern Brazil north to São Paulo (Simpson & Simpson 2011b), but there was also an incredible record of a vagrant photographed at Jacareacanga in the Brazilian Amazon in March 2018 (Luca 2018). Austral Negritos have also been recorded as vagrants to Bolivia (Herzog *et al.* 2016) and are fairly regular on the Falklands (Woods 2017); they have twice also reached the islands of the Antarctic peninsula (Gryz *et al.* 2015). Dark-faced Ground-Tyrant *Muscisaxicola maclovianus* has a similar strategy but routinely winters north to northern Peru; vagrants have reached Rio Grande do Sul in Brazil (Schwertner *et al.* 2011), Ecuador (Freile and Restall 2018) and even South Georgia in September 1994 (Clarke *et al.* 2012). White-browed Ground-Tyrant *Muscisaxicola albilora* undertakes a similar migration and has occurred as a vagrant to the Falklands (Woods 2017) and the Colombian Isla Gorgona (Halle 1990). Some southerly breeders are wholly migratory, such as Lesser Shrike-Tyrant *Agriornis murinus*, recently recorded for the first time in Chile (Barros 2017) and Cinnamon-bellied Ground-Tyrant *Muscisaxicola capistratus*, a first for Brazil in 2015 (Santos *et al.* 2017). The nominate race of Grey-bellied Shrike-Tyrant *Agriornis micropterus* is also wholly migratory and this species was recorded for the first time recently in Brazil (Bellagamba *et al.* 2014).

Further north, other species inhabiting the more temperate grassland regions have been recorded making some surprisingly epic vagrancy journeys. Notable in this group is the partially migratory Spectacled Tyrant *Hymenops perspicillatus*, which in Brazil winters regularly only in the south of the country but has occurred widely as a vagrant. These range from the north-east coast at Galinhos in Rio Grande do Norte (Sagot-Martin *et al.* 2012), in the southern Amazon at Paranaita in Mato Grosso state (Mafra 2013) and on the Abrolhos archipelago off Bahia (Carletti 2014). Spectacled Tyrant has also occurred as a vagrant to southern Peru (Schulenberg *et al.* 2010) and, given the spread of records in Brazil, future records from northern South America seem likely. Several species of *Knipolegus* black-tyrants are migratory: Hudson's Black-Tyrant *Knipolegus hudsoni* is increasingly recorded as a vagrant to Peru (Schulenberg *et al.* 2010) and Brazil (Somenzari *et al.* 2018); there have been several recent records of Cinereous Tyrant *Knipolegus striaticeps* from Brazil,

► White-crested Elaenia *Elaenia albiceps*, Cape Pembroke, Falklands, 23 January 2019. This migratory species is an occasional vagrant to the Falklands and also north as far as Nebraska in the United States (*Alan Henry*).

► Fire-eyed Diucon *Xolmis pyrope*, Stanley, Falklands, 17 April 2015. Two were present together in Stanley in April–June 2015 (*Alan Henry*).

► Subtropical Doradito *Pseudocolopteryx acutipennis*, Cubatão, São Paulo, Brazil, 17 June 2015. Doraditos are tricky to identify and hard to observe within their wetland haunts. This species is a scarce and probably overlooked austral migrant or vagrant to southern Brazil (*Robson Silva e Silva*).

▲ Austral Negrito *Lessonia rufa*, Cubatão, São Paulo, Brazil, 4 September 2006, the first state record of this austral vagrant and one of the northernmost ever (*Robson Silva e Silva*).

which are barely extralimital (Las-Casas & de Azevedo-Júnior 2010); and more unexpected have been amazing records of the apparently resident Rufous-tailed Tyrant *Knipolegus poecilurus* of the Andes in Paraguay (Smith & Easlet 2019).

Long-distance movements among resident tyrant flycatchers are rare, although detecting them on a vast continent with low observer coverage is obviously a challenge. Offshore islands like the Falklands do offer a window into such movements, and unusual records from those islands include Tufted Tit-Tyrant *Anairetes parulus* and Cattle Tyrant *Machetornis rixosa*, both of which may be partially migratory. As such, island vagrancy can highlight cryptic patterns of migration in some regions. The Alcatrazes archipelago 35km off the coast of São Paulo is another useful barometer of migratory activity; Olmos *et al.* (2014) list 17 species of tyrant flycatcher for these islands, of which only three are considered resident; visitors have included nominally resident species, intratropical migrants and even one boreal migrant – Olive-sided Flycatcher. Arguably one of the most unexpected instances of tyranid vagrancy concerns a record of a Piura Chat-Tyrant *Ochthoeca piurae* from the Santa Eulalia Valley in Lima, Peru, *c.*300km out of range. This species is sedentary and has a narrow ecological niche and must have moved through lots of unsuitable habitat to reach this location (Witt *et al.* 2015).

MENURIDAE, ATRICHORNITHIDAE, PTILONORHYNCHIDAE, CLIMACTERIDAE AND MALURIDAE

Lyrebirds, Scrub-birds, Bowerbirds, Australasian Treecreepers and Australasian Wrens

All members of these five Australasian families are essentially resident, with no records of long-distance dispersal events. However, many bowerbirds show altitudinal or seasonal movement at limited scales (Gregory 2020c). For example, Satin Bowerbirds *Ptilonorhynchus violaceus* are occasionally reported as local vagrants in urban Melbourne, 50–100km outside of their usual range, whilst Spotted Bowerbirds *Chlamydera maculata* are occasional vagrants to the state of Victoria (where they formerly bred), suggesting birds can wander 100km or more south of remaining populations (R. Clarke *in litt.*).

MELIPHAGIDAE
Honeyeaters

Honeyeaters display a variety of movement strategies including strictly sedentary behaviour, altitudinal migration and nomadism in response to available resources, and true long-distance migration (Menkhorst *et al.* 2017). In the vastness of Australasia these movements are often poorly documented. Sedentary species such as Bell Miner *Manorina melanophrys* occur as vagrants after large-scale disturbances such as bushfires, whilst the expanding population of Spiny-cheeked Honeyeater *Acanthagenys rufogularis* has seen vagrant individuals reach novel areas within New South Wales with increasing frequency (McAllan & Lindsay 2016). Brown Honeyeaters *Lichmera indistincta* are widespread across much of continental Australia but occur only as vagrants to Victoria, with just a handful of records in that state (R. Clarke *in litt.*). Beyond these relatively modest movements, there are also records of overwater vagrancy in some species. For example, both Eastern Spinebill *Acanthorhynchus tenuirostris* and Scarlet Honeyeater *Myzomela sanguinolenta* have reached King Island, Tasmania (Bennett *et al.* 2015). The first Scarlet Honeyeater for South Australia appeared 700km out of range for a single day at the Australian Arid Lands Botanic Gardens in November 2018 (Black *et al.* 2020). Yellow Chat *Epthianura crocea*, Pied Honeyeater *Certhionyx variegatus* and Rufous-throated Honeyeater *Conopophila rufogularis* have all wandered across 350km of open ocean to reach Ashmore Reef off northern Australia (R. Clarke *in litt.*). Even more impressive is a record of Noisy Friarbird *Philemon corniculatus* on Lord Howe Island (McAllan *et al.* 2004) and two old records of Red Wattlebird *Anthochaera carunculata* in New Zealand (Oliver 1955).

► Yellow Chat *Epthianura crocea*, Ashmore Reef, Australia, 22 October 2008. This species is highly nomadic in the Australian interior but nevertheless unexpected as a vagrant to this remote oceanic island (*Rohan Clarke*).

► Rufous-throated Honeyeater *Conopophila rufogularis*, Ashmore Reef, Australia, 10 November 2019 (*Rohan Clarke*).

DASYORNITHIDAE
Bristlebirds

The bristlebirds have very poor dispersal capacity and there are no records from outside of their regular, if now substantially fragmented, range in Australia.

PARDALOTIDAE
Pardalotes

Pardalotes have complex dispersal patterns, which are poorly understood, with some movement away from southern temperate latitudes and downslope movements in the austral winter. This includes substantial overwater crossings, as the Tasmanian form of Striated Pardalote *Pardalotus striatus* migrates across Bass Strait to the Australian mainland during winter (Menkhorst *et al.* 2017). We are unaware of any significant instances of vagrancy, although the single record of Forty-spotted Pardalote *Pardalotus quadragintus* from King Island in the Bass Strait in 1887 might relate to a vagrant, or possibly to one of the last remnants of a now-extinct island population (Campbell 1888).

ACANTHIZIDAE
Thornbills and allies

There is evidence for partial migration in five species of Australian warblers (Chan 2001). One, White-throated Gerygone *Gerygone olivacea,* has been recorded extralimitally – as a vagrant to King Island, Tasmania, from mainland Australia (Bennett *et al.* 2015). Also on mainland Australia, Striated Fieldwren *Calamanthus fuliginosus* has been recorded near Sydney, some 160km out of range (Gregory 2020a), suggesting occasional long-distance dispersal. Similarly, in New Zealand the endemic Grey Gerygone *Gerygone igata* has been recorded as a vagrant to the Snares Islands, necessitating a 100km sea crossing (Sagar *et al.* 2001).

POMATOSTOMIDAE
Australasian Babblers

This group of five pseudo-babblers is largely sedentary, although Grey-crowned Babblers *Pomatostomus temporalis* seem to be prone to wander and occasionally appear marginally out of range (Rogers 2008).

ORTHONYCHIDAE
Logrunners

The three rainforest and montane forest understorey species in this group are all highly sedentary and there are no reports of extralimital movements.

CINCLOSOMATIDAE
Jewel-babblers and Quail-thrushes

This group of terrestrial passerines is sedentary and no extralimital records have been reported.

CAMPEPHAGIDAE
Cuckooshrikes

Migratory behaviour is the exception rather than the norm among cuckooshrikes and their allies, but several species are long-distance full migrants and others undertake seasonal altitudinal migrations or nomadic movements. The longest migrations are undertaken by minivets, with the nominate race of Ashy Minivet *Pericrocotus divaricatus* moving from breeding grounds in south-east Siberia all the way to peninsular Malaysia, the Philippines and the Greater Sundas. Ashy Minivet has occurred as a vagrant west to the Indian subcontinent (Sridharan *et al.* 2016), north-east to Mongolia in June 2016 (Ariunbaatar *et al.* 2017) and most impressively to the Northern Mariana Islands, where a flock of four wintered in 2015–16 on northern Saipan (Beer *et al.* 2016); the last necessitated an epic overwater flight of 2,000km. Ashy

▲ White-winged Triller *Lalage tricolor*, Narawntapu National Park, Latrobe, Tasmania, Australia, 28 October 2019 (*Peter Vaughan*).

Minivet seems a likely future vagrant to Australian vagrant traps such as Christmas Island, and is a conceivable vagrant to the Western Palearctic. Of the other two highly migratory minivets, Rosy Minivet *Pericrocotus roseus* is a vagrant to the south of peninsular Thailand and Hong Kong whilst Brown-rumped Minivet *Pericrocotus cantonensis* is a vagrant to Taiwan (Brazil 2009) and is probably overlooked as a more regular vagrant to India (Sridharan *et al.* 2016). Even sedentary species have been recorded as vagrants, including Long-tailed Minivet *Pericrocotus ethologus* on Taiwan (Brazil 2009).

In Australia, White-winged Triller *Lalage tricolor* is a partial migrant, and vagrants have reached New Zealand (Miskelly *et al.* 2008) and Lord Howe Island (McAllan *et al.* 2004). Several of the cuckooshrikes undertake partial and altitudinal migrations, especially in Australasia. Some Black-faced Cuckooshrikes *Coracina novaehollandiae* move out of Australia and head north across Torres Strait to New Guinea and the Timor Sea to East Timor and Indonesia (Clarke 2004, R. Clarke *in litt.*). This species is among the most regular passerine vagrants to New Zealand with around 20 records (Miskelly *et al.* 2013), and it has also occurred as a vagrant a dozen times to Lord Howe Island (McAllan *et al.* 2004). White-bellied Cuckooshrike *Coracina papuensis* has also occurred as a vagrant to Lord Howe Island and to the Lesser Sunda Islands (McAllan *et al.* 2004). Further north, Black-winged Cuckooshrike *Lalage melaschistos* is highly migratory and has occurred as a vagrant to Japan, South Korea (Brazil 2009) and the Philippines (Jensen *et al.* 2015). Movements have even been recorded in some largely resident species; for example, there is a record of two Large Cuckooshrikes *Coracina macei* on the Maldives in October 2002 (Anderson 2007). Such movements by migratory or, less likely, resident species have permitted the colonisation and sometimes speciation of many cuckooshrikes on remote islands in the Indian and Pacific Oceans.

MOHOUIDAE
Whiteheads

These three species from New Zealand are sedentary, and extralimital movements are both unknown and unlikely.

NEOSITTIDAE
Sittellas

No sittellas are known to undertake significant movements and we are unaware of any significant vagrancy.

PSOPHODIDAE
Whipbirds and Wedgebills

All the whipbirds are considered sedentary, but the Chirruping Wedgebill *Psophodes cristatus* of the Australian interior is possibly nomadic, and Eastern Whipbirds *Psophodes olivaceus* are occasionally recorded in urban environments tens of kilometres from the nearest known populations (R. Clarke *in litt.*), although we are unaware of any extralimital reports.

EULACESTOMATIDAE
Ploughbill

Wattled Ploughbill *Eulacestoma nigropectus* is assumed to be sedentary, and extralimital movements are considered unlikely.

OREOICIDAE
Australo-Papuan Bellbirds

There are no extralimital records within this group, none of which undertake any regular migratory activity.

FALCUNCULIDAE
Shriketit

There are no records of extralimital movements in the Crested Shriketit *Falcunculus frontatus* complex.

PARAMYTHIIDAE
Painted Berrypeckers

Extralimital records of either species of Painted Berrypecker are unknown, although both species are apparently nomadic to an extent.

VIREONIDAE
Vireos, Shrike-Babblers and Erpornis

The family Vireonidae now includes not only the vireos, greenlets, shrike-vireos and peppershrikes of the New World, but also now the Old World *Pteruthius* shrike-babblers and White-bellied Erpornis *Erpornis zantholeuca*. Among these, only the vireos are migratory and regularly recorded extralimitally. Red-eyed Vireo *Vireo olivaceus* has the largest migration span of all species, breeding into northern Canada and wintering south to northern Bolivia, and vagrants have reached as far south as Chile (Jaramillo *et al.* 2003). This species regularly undertakes long over-water flights, and its tactic of putting on a high fuel load, plus its long migration and large population size conspire to make it the most frequent transatlantic passerine vagrant to the Old World (McLaren *et al.* 2006) with over 150 records from Britain alone.

▲ Red-eyed Vireo *Vireo olivaceus*, Inishbofin, Galway, Ireland, 29 September 2016, the most frequent Nearctic passerine vagrant to Europe (*Anthony McGeehan*).

◄ Yellow-green Vireo *Vireo flavoviridis*, Southeast Farallon Island, San Francisco, California, United States, 29 September 2013. This was one of 15 recorded in California in autumn 2013, a record year for the species (*Cameron Rutt*).

Red-eyed Vireos are annual in small numbers along the eastern Atlantic seaboard from Iceland (where it is also the commonest Nearctic passerine) south all the way to Morocco (Thévenot *et al.* 2003). Small arrivals of multiple individuals are anticipated annually from locations such as the Isles of Scilly in southwest England (Robinson 2003). Beyond the Atlantic coast, Red-eyed Vireos have reached as far east as the Italian island of Linosa in October 2019 as well as to Poland in October 2000 and Malta in October 1983. Other vireos are very rare as transatlantic vagrants, with a handful of records of Philadelphia Vireos *Vireo philadelphicus* from Britain, Ireland and the Azores; Yellow-throated Vireos *Vireo flavifrons* from Germany (Heligoland), England and the Azores; and five records of White-eyed Vireo *Vireo griseus* from the Azores (Haas 2012). The last-named seems a surprising vagrant, as this species is a relatively short-distance migrant by comparison to Philadelphia Vireo, but it has also occurred as a vagrant on the Revillagigedo Islands off western Mexico (Howell & Webb 1992), in Venezuela (Rodríguez *et al.* 2017) and over 50 times in California, 12 of which came in an unprecedented weather-driven influx of vagrant vireos and New World warblers into the state in spring–summer 1991 (Patten & Marantz 1996). The absence of European records of Warbling Vireo *Vireo gilvus* seems striking in this context, being another migratory species that winters only as far south as Central America, although vagrants have reached Bermuda (Amos 1991).

Two other species cryptically similar to Red-eyed Vireo have also been recorded extensively as vagrants. Black-whiskered Vireo *Vireo altiloquus* migrates from breeding grounds in Florida and the West Indies to winter in northern South America, and vagrants have been recorded up the eastern seaboard of the United States north to Massachusetts and east to Bermuda, and also west to Belize (Chace *et al.* 2002). Intriguingly, there is a record of a mummified corpse found aboard a tanker that arrived in Sullom Voe,

▲ Cuban Vireo *Vireo gundlachii*, Fort Zachary Taylor Historic SP, Monroe, Florida, United States, 24 April 2016, the first record for North America (*Tom Johnson*).

Shetland, in November 1993 which had travelled from the United States via Mexico and Venezuela (Pennington *et al.* 2004). It seems likely that this species could be overlooked in North America, and this seems almost certainly to be the case for Yellow-green Vireo *Vireo flavoviridis*, which is also highly migratory. The majority of United States records come from California (Hamilton *et al.* 2007) where any Red-eyed-'type' vireo is a vagrant and worthy of scrutiny, and there are further records of Yellow-green Vireos thinly scattered across the rest of North America, from Vancouver in British Columbia to Massachusetts, Florida and Bermuda, as well as an old specimen record from Quebec. North American observers should also be aware of the possibility of vagrancy by the cryptically similar austral migrant Chivi Vireo *Vireo chivi* from South America.

The restricted-range Black-capped Vireo *Vireo atricapilla* is fully migratory, but the distances covered between its patchy breeding areas in the south-central United States and its Mexican wintering areas are more modest. Rather incredibly, this species has nevertheless occurred as a vagrant to three different Canadian provinces – British Columbia, Ontario and Quebec, each with one record. Grey Vireo *Vireo vicinior* is also migratory but extralimital reports are extremely rare, with one 1964 record from Wisconsin (Robbins 1991); Unitt (1984) lists only two state reports away from the breeding range in San Diego County, California. The United States has also been visited by three other vagrant vireos from the Caribbean region: more than a dozen records of Thick-billed Vireo *Vireo crassirostris*, all in Florida (Howell *et al.* 2014); a single Yucatan Vireo *Vireo magister* in Galveston County, Texas, in April 1984 (Howell *et al.* 2014); and a single Cuban Vireo *Vireo gundlachii* at Ford Zachary, Florida, in April 2016 (Pyle *et al.* 2017).

PACHYCEPHALIDAE
Whistlers and allies

Most whistlers and allies are fairly sedentary, although both the Golden Whistler *Pachycephala pectoralis* and the Rufous Whistler *Pachycephala rufiventris* are partial migrants, with the latter a vagrant to Tasmania, and many other species may make altitudinal movements. There seems to be little other evidence for significant vagrancy with, for example, minor extralimital movements reported for Gilbert's Whistler *Pachycephala inornata* in South Australia (Joseph & Kernot 1982).

◀ Rufous Whistler *Pachycephala rufiventris*, Geeveston, Huon Valley, Tasmania, Australia, 22 October 18. This species is a vagrant to Tasmania but also perhaps a candidate future colonist (*Peter Vaughan*).

ORIOLIDAE
Old World Orioles

Two members of the Oriolidae are long-distance full migrants: Eurasian Golden Oriole *Oriolus oriolus* of the Western and Central Palearctic, which winters in the Afrotropics; and Silver Oriole *Oriolus mellianus* of the Eastern Palearctic, which migrates between Chinese breeding grounds and wintering areas in Thailand and Cambodia. Another eight orioles are partial migrants, and the remainder sedentary. With its broad Palearctic distribution, Eurasian Golden Oriole has been widely recorded as a vagrant and is a routine overshoot to northern Europe. Records are spread very wide, from the Sakhalin peninsula in the Russian Far East and the Atlantic islands of Iceland, the Azores and Cape Verde (Mason & Allsop 2009) and São Tomé (Demey 2016a) to the Seychelles in the Indian Ocean (Skerrett *et al.* 2006) and the remote Prince Edward Islands (Oosthuizen *et al.* 2009). The Endangered Silver Oriole was recently recorded for the first time in Vietnam (Klingel & Mahood 2015), but it is probably a regular passage migrant there and might be expected to occur more widely as a vagrant if it was not so rare in global numerical terms. Other notable records of vagrant orioles include two extralimital reports of Olive-backed Oriole *Oriolus*

◀ Olive-backed Oriole *Oriolus sagittatus*, Douglas Apsley National Park, Tasmania, Australia, 26 October 2018; regarded as a vagrant to Tasmania, this species bred there for the first time in 2018 (*Peter Vaughan*).

sagittatus from Australasia – on Luang island, in the Lesser Sundas, and on New Britain island, north-east of New Guinea (Eaton *et al.* 2016) – and as a vagrant to Tasmania, where the species was first detected breeding in 2018 (P. Vaughan *in litt.*). Meanwhile, Black-naped Oriole *Oriolus chinensis* is a regular vagrant to Japan and also to Sakhalin island in Russia (Brazil 2009).

MACHAERIRHYNCHIDAE
Boatbills

Both species of boatbill are highly sedentary and there are no records of vagrant individuals.

ARTAMIDAE
Woodswallows, Bellmagpies and allies

Butcherbirds and currawongs are largely sedentary, although occasional long-distance dispersal events must occur, as there are endemic island forms such as the Pied Currawong *Strepera graculina crissalis* of Lord Howe Island. Woodswallows are more dispersive overall and have a significant track record for vagrancy. All the woodswallows occurring in Australia are either nomadic or truly migratory, whereas those occurring elsewhere are thought to be largely resident. Both Masked Woodswallow *Artamus personatus* and White-browed Woodswallow *Artamus superciliosus* have occurred as vagrants to Tasmania, Lord Howe Island and New Zealand, whilst Dusky Woodswallow *Artamus cyanopterus* was also first reported from New Zealand relatively recently and has also occurred on King Island, Tasmania (McAllan *et al.* 2004, Bennett *et al.* 2015, Kakishima & Morimoto 2015). Masked Woodswallows established a small breeding population on Norfolk Island (*c.*1,400km east of continental Australia) in the 2000s following the arrival of a vagrant flock, and a population of 10–20 was still present in 2020. In 2018 this small flock of resident Masked Woodswallows was joined by a vagrant White-browed Woodswallow (R. Clarke *in litt.*). Presumably this kind of itinerant wandering by the largely resident White-breasted Woodswallow *Artamus leucorynchus* led to the colonisation and speciation of Fiji Woodswallow *Artamus mentalis*, and White-breasted Woodswallow has also occurred as a vagrant to Korea and Japan (Brazil 2009). Similarly, the largely resident Ashy Woodswallow *Artamus fuscus* has occurred as a vagrant west to the Maldives at least once (Anderson 2007) and south-east to Malaysia (Bakewell 2009).

► White-browed Woodswallow *Artamus superciliosus*, Waterhouse Conservation Area, Tasmania, Australia, 1 February 2018 (*Peter Vaughan*).

RHAGOLOGIDAE
Mottled Berryhunter

Mottled Berryhunter *Rhagologus leucostigma* is a sedentary montane forest species of New Guinea and extralimital movements are highly unlikely.

PLATYSTEIRIDAE
Wattle-eyes and Batises

Some species in the family Platysteiridae perform altitudinal or local migrations, but most are resident, especially those restricted to equatorial forests. We are unaware of any significant extralimital records.

VANGIDAE
Vangas, Helmetshrikes and allies

Most of the members of this group are residents, although it includes suspected partial migrants such as African Shrike-flycatcher *Megabyas flammulatus* and Black-and-white Shrike-flycatcher *Bias musicus,* which might be expected to occur as vagrants (e.g. records of the latter in South Africa). Some species, such as White Helmetshrike *Prionops plumatus,* are irruption-prone and may be found extralimitally on the South African highveld plateau (Allan 2020). Arguably one of the most surprising instances of tropical vagrancy on record is that of an adult male Rufous Vanga *Schetba rufa* reported from Grand Terre, Aldabra, in the Indian Ocean in January 2015 (Skerrett *et al.* 2017) – an amazing transoceanic movement for an apparently sedentary Malagasy endemic.

PITYRIASEIDAE
Bristlehead

It is possible that Bornean Bristlehead *Pityriasis gymnocephala*, the sole member of its family, undertakes altitudinal migrations, but there are no reports of vagrancy.

AEGITHINIDAE
Ioras

There are no records of vagrancy in ioras, although the wide distribution of Common Iora *Aegithina tiphia* throughout the Indian subcontinent and South-east Asia, including many islands and island groups, indicates a propensity for sea crossings.

MALACONOTIDAE
Bushshrikes and allies

No bushshrikes undertake regular migrations, and we are unaware of any extralimital reports.

RHIPIDURIDAE
Fantails

Most fantails and silktails are sedentary, but several Australian species do undertake regular migrations, particularly Grey Fantail *Rhipidura albiscapa,* which has also colonised remote Norfolk Island and New Caledonia, indicative of some major historic vagrancy events. This must also have been the case for other species such as island populations of Rufous Fantail *Rhipidura rufifrons* and New Zealand Fantail *Rhipidura fuliginosa*. Willie-wagtail *Rhipidura leucophrys* is a partial migrant on mainland Australia, and vagrants have reached King Island, Tasmania, and Lord Howe Island (McAllan *et al.* 2004, Bennett *et al.* 2015); perhaps most amazingly, a vagrant has also reached the Chatham Islands east of the main archipelago, which represents the only New Zealand record of that species (Miskelly *et al.* 2006). Arafura Fantails *Rhipidura dryas* of the subspecies *semicollaris*, which were split by Eaton *et al.* (2016) as 'Supertramp Fantail' are, as the name suggests, excellent small island colonisers and have been recorded as vagrants several times since 2010 to Ashmore Reef in Australia (Menkhorst *et al.* 2017). Intriguingly, a Philippine Pied-Fantail *Rhipidura nigritorquis* was photographed at Dingtouer Coastal Forest, Tainan City, Taiwan, in May 2020 (eBird); this seems an unlikely vagrant but might conceivably have been ship-assisted (S. Mahood *in litt.*).

▲ Arafura Fantail *Rhipidura dryas semicollaris* ('Supertramp Fantail'), Ashmore Reef, Australia, 16 October 2010, the first record for Australia of this taxon, which is considered by some authorities to be a separate species (*Rohan Clarke*).

DICRURIDAE
Drongos

Whereas most drongos are largely resident birds of tropical and subtropical wooded habitats, several Asian species have migratory populations breeding at higher latitudes and are frequently recorded as long-distance vagrants. Ashy Drongos *Dicrurus leucophaeus* of subspecies *leucogenis* breeding in eastern China winter as far south as the Malay peninsula, whilst the *longicaudatus* subspecies, which breeds in the Himalayas, winters throughout the Indian peninsula. Ashy Drongos presumed to be of the *longicaudatus* form have now been recorded with some regularity as vagrants to the Middle East, with around 20 records from Iran, United Arab Emirates, Oman, Kuwait and one in central Israel in December 2014 (Khil *et al.* 2019). This was followed by an incredible record, ostensibly of 'unknown origin', but absolutely a plausible natural vagrant, photographed in Andøya in Norway in early June 2019. These records, all since 2006, represent a striking change in status for such an easy-to-detect bird, suggesting that changes in observer activity alone may not explain this trend – perhaps signs of an increasing migratory population? Elsewhere Ashy Drongo has been recently reported for the first time in Kazakhstan and Mongolia, and there are two records of subspecies *leucogenis* from eastern Russia (Gluschenko & Korobov 2012). Black

◀ Ashy Drongo *Dicrurus leucophaeus*, Jahra Farms, Kuwait, 1 January 2011, the second Western Palearctic record (*Vincent Legrand*).

◀ Black Drongo *Dicrurus macrocercus*, Sall Ala, Musandam, Oman, 24 April 2013. Records of this partially-migratory species may be increasing in the Middle East (*Jens Eriksen*).

Drongo *Dicrurus macrocercus* is also a partial migrant, with far-eastern populations the most migratory; in the west, vagrants have reached the United Arab Emirates and Oman, and most recently the first for the Western Palearctic in Kuwait in November 2015 (Ławicki & van den Berg 2016a). In the east, vagrants have reached the Russian Far East as well as Mongolia and the Philippines (Gluschenko & Korobov 2012, Jensen *et al.* 2015).

Crow-billed Drongo *Dicrurus annectens* is also highly migratory and has occurred as a vagrant to northern Borneo and once to the Cocos (Keeling) Islands in December 2015, the first Australian record (Johnstone & Darnell 2015). Northern populations of Hair-crested Drongo *Dicrurus hottentottus* are migratory and vagrants have reached as far north as southern Ussuriland in south-east Russia as well as Korea, Japan and Taiwan (Brazil 2009). In Australia, Spangled Drongo *Dicrurus bracteatus* is a partial migrant in the south, and vagrants have reached Kangaroo Island in South Australia, Victoria and Tasmania (Baxter 1989).

PARADISAEIDAE
Birds-of-paradise

There is anecdotal evidence for altitudinal movements in some birds-of-paradise, but no evidence for regular movements and no reports of significant extralimital movements (Gregory 2020c).

IFRITIDAE
Ifrita

The sole member of the Ifritidae, Blue-capped Ifrita *Ifrita kowaldi* is a sedentary inhabitant of montane forest in New Guinea.

MONARCHIDAE
Monarch Flycatchers

Most of the monarchs are sedentary tropical species, although their high incidence of occupancy of oceanic islands – including some of the remotest, like French Polynesia – indicates a predisposition towards long-distance dispersal even in non-migratory populations. Eight species in this group are partial or full migrants. Japanese Paradise-Flycatcher *Terpsiphone atrocaudata* is highly migratory, moving from southern Korea and Japan to winter in South-east Asia and has occurred as a vagrant north to Ussuriland in south-east Russia (Brazil 2009) and south-east to Sabah, Borneo (Eaton *et al.* 2016). Chinese Paradise-Flycatcher *Terpsiphone incei* breeds further west and has occurred as a vagrant to the Andaman Islands (Grundsten *et al.* 2018) and Japan (Watanabe & Kagoshima 2016) whilst there is a single record of a returning vagrant Indian Paradise-Flycatcher *Terpsiphone paradisi* from Singapore.

Several Australasian monarch species are partial migrants and have been reported with some frequency as vagrants. For example, Satin Flycatcher *Myiagra cyanoleuca* and Black-faced Monarch *Monarcha melanopsis* have occurred as vagrants to New Zealand (Miskelly *et al.* 2019). Black-faced Monarch has reached subantarctic Macquarie Island (R. Clarke *in litt.*) and Leaden Flycatcher *Myiagra rubecula* has reached Lord Howe Island (McAllan *et al.* 2004) and is a vagrant to South Australia (Black *et al.* 2020). Magpie-lark *Grallina cyanoleuca* appears to undertake some northward movement in the non-breeding season from Australia across the Timor Sea, appearing occasionally on Ashmore

◄ Magpie-lark *Grallina cyanoleuca*, Eaglehawk Neck, Tasman, Tasmania, Australia, 16 January 2018. This species has wandered extensively in Australasia (*Peter Vaughan*).

Reef and also on islands in the far northern Torres Strait (Clarke 2004, R. Clarke *in litt.*). This species has also reached Luang Island in the Lesser Sundas and Tayandu Island in the south-eastern Moluccas (Eaton *et al.* 2016), King Island and the Tasmania mainland (Bennett *et al.* 2015), and the first occurred on New Zealand in April 2008 (Miskelly *et al.* 2019). Another nominally resident species is Island Monarch *Monarcha cinerascens* of the Moluccas and Lesser Sundas, a classic 'supertramp' species (*sensu* Diamond 1974), which is common on small islands without competitors and rare or absent on large ones. Vagrants have reached Australian territory at Ashmore Reef with some regularity (Carter *et al.* 2011), but more impressive was a pair photographed on Cassini Island, off the Kimberley coast of Western Australia in October 2010 (Ekins 2011) and four records from Browse Island, also off the Kimberley coast (R. Clarke *in litt.*).

CORCORACIDAE
Australian Mudnesters

No significant extralimital movements have been reported in either member of this family, White-winged Chough *Corcorax melanorhamphos* or Apostlebird *Struthidea cinerea*.

MELAMPITTIDAE
Melampittas

The two melampittas are sedentary inhabitants of montane forest in New Guinea and there are no extralimital reports.

PLATYLOPHIDAE
Crested Shrikejay

Crested Shrikejay *Platylophus galericulatus* is assumed to be resident and there are no extralimital reports.

LANIIDAE
Shrikes

Tropical and subtropical shrikes tend to be resident, although there is some evidence of wandering in some species, such as records of Taita Fiscal *Lanius dorsalis* from south-west Kenya and Dar es Salaam in Tanzania (LeFranc & Worfolk 1997). In contrast, most temperate zone species are migratory. This difference is accounted for by seasonal resource availability; being specialists on large invertebrates, most shrikes are compelled to migrate in winter when these food resources are no longer available. Some larger species like Chinese Grey Shrike *Lanius sphenocercus* and Great Grey Shrike *Lanius excubitor* may be able to overwinter at cold northern latitudes as they can prey on small vertebrates that are available year-round. Some of the temperate zone species undertake epic migrations; for example, populations of Lesser Grey Shrike *Lanius minor* breeding in north-west China migrate 12,000km to Namibia. This species is a vagrant to north-west Europe and the Malagasy region, including the Seychelles (Skerrett *et al.* 2006). Surprisingly it was recorded for the first time in Portugal only in September 2019, despite breeding formerly as close as southern France. Red-backed Shrike *Lanius collurio* has a similar migration strategy but has been more widely recorded as a vagrant, with records from remote oceanic islands including the Prince Edward Islands (Oosthuizen *et al.* 2009) and Ascension Island (Bourne & Simmons 1998). Such records are perhaps surprising as *Lanius* shrikes do not apparently build up large fat reserves prior to migration, probably because they can feed on route; this might be expected to constrain overwater vagrancy. Elsewhere, Red-backed Shrikes have been reported as vagrants from localities as diverse as Ghana (Hulme *et al.* 2012), Afghanistan (Mallalieu & Kaestner 2015), Japan (Brazil 2009) and, in October 2017, St Lawrence Island, Alaska; the last record was the first for the Western Hemisphere (Pyle *et al.* 2018). Prior to this record, a controversial shrike was found in Mendocino County, California, in March 2015; its identity was eventually resolved as a Red-backed Shrike x Turkestan Shrike *Lanius phoenicuroides* hybrid (Pyle *et al.* 2015). Brown Shrike *Lanius cristatus* also has a very long migration span. This Asian species is seemingly being recorded with increasing frequency both in the Western Palearctic, where it was first recorded as recently as 1985 on Shetland, Scotland (Hume 1993), and in North America, where there are records from Alaska, British Columbia, California and, amazingly, Nova Scotia (Howell *et al.* 2014). Elsewhere, there are records from places as diverse as Turkmenistan, Kuwait and Christmas Island, Australia (Lansley *et al.* 2016).

Other temperate zone shrikes cannot match these three species for migration span but still have an impressive track record for vagrancy. For example, Woodchat Shrike *Lanius senator*, which breeds around the Mediterranean basin east to Iran and winters in Central Africa, is a fairly regular vagrant to northern Europe and has also occurred

► Brown Shrike *Lanius cristatus*, Southeast Farallon Island, San Francisco, California, United States, 24 September 2009, the second record for Southeast Farallon Island and third state record (*Matt Brady*).

◀ 'Steppe' Great Grey Shrike *Lanius excubitor pallidirostris*, Grainthorpe Haven, Lincolnshire, England, 11 November 2008. This individual was incredibly confiding during its stay, routinely using birders and their scopes as hunting perches (*Graham Catley*).

on the Azores (Mitchell 2012), in Maharashtra in India (Nandgaonkar 2013) and Kazakhstan (van den Berg & Ławicki 2015). Masked Shrike *Lanius nubicus* has an even more restricted global breeding range centred on the eastern Mediterranean but has been recorded west to Britain, Germany, France, Spain, Italy, Netherlands, Finland, Norway, Sweden and Romania – and amazingly east to India, where one was photographed in Gujarat in December 2016 (Bharti 2017). Isabelline Shrike *Lanius isabellinus* occurs as a vagrant east to Japan (Brazil 2009) and also regularly west to Europe, with one making it out into the Atlantic as far as Tenerife in November 2017 (Rodríguez 2018). Isabelline Shrike seemingly occurs more widely as a vagrant than Turkestan Shrike *Lanius phoenicuroides,* which does however also reach Europe annually, although identification of this species-pair remains a challenge. Tiger Shrike *Lanius tigrinus* is fully migratory and has strayed to Japan, the Philippines, Christmas Island

▼ Northern Shrike *Lanius borealis*, Lighthouse Valley, Corvo, Azores, Portugal, 29 October 2014, the first record for the Western Palearctic of the nominate North American subspecies (*Richard Bonser*).

and the Cocos (Keeling) Islands, Australia (James & McAllan 2014); it would seem to be a candidate for westward vagrancy towards India at least, if not the Middle East or as an overshoot to Alaska.

Other species that are not wholly migratory are still recorded extralimitally with some regularity, most notably Long-tailed Shrike *Lanius schach,* which has both resident and migratory populations and vagrants of the race *erythronotus* have reached Qatar, Kuwait, Israel, Jordan, Turkey, Sweden, Denmark, Switzerland, Belgium, the Netherlands and Britain (Haas 2012). Similarly, northern populations of Loggerhead Shrike *Lanius ludovicianus* are migratory, and this species has reached as far south as Guatemala (Ericsson 1981) and east to the Bahamas (Kirwan *et al.* 2019) and Bermuda (Amos 1991). Bay-backed Shrikes *Lanius vittatus* of the race *nargianus* are migratory and this species has occurred in several Middle Eastern countries including Saudi Arabia, Oman and most recently Kuwait in September 2020; it has even bred in the United Arab Emirates (Campbell *et al.* 2011). The northern 'grey' shrikes mostly move south at least a little way from the breeding grounds in winter, and Chinese Grey Shrike *Lanius sphenocercus* is a vagrant to Myanmar (Naing *et al.* 2020), Japan and Taiwan (Brazil 2009). Northern Shrike *Lanius borealis* is exceptional in being the only New World shrike to have occurred in the Old World – a record of the nominate subspecies *borealis* on the Azores in October 2014 (van den Berg 2014) – whilst the eastern form *sibiricus* has also occurred as a vagrant to Germany (Brauneis & Alex 2015) and Sweden (Ławicki & van den Berg 2017e).

CORVIDAE
Crows, Jays and Magpies

Most species of crows and their allies are sedentary, whilst populations of some are partial migrants and others are irruptive or altitudinal migrants, but these broadly non-migratory tendencies belie a high rate of long-distance vagrancy even among resident species. For example, Steller's Jay *Cyanocitta stelleri* and Blue Jay *Cyanocitta cristata* replace each other in western and eastern North America respectively, but both have been found as extralimital vagrants thousands of kilometres into each other's ranges, whilst Black-billed Magpie *Pica hudsonia* of western North America has occurred widely as a vagrant in the east. Among the most migratory of corvids are the two Old World jackdaws – with northern populations of both Eurasian Jackdaw *Corvus monedula* and Daurian Jackdaw *Corvus dauuricus* highly migratory and frequently reported as vagrants. Eurasian Jackdaw has reached many of the Atlantic islands including Iceland, the Canaries, Madeira and the Azores, which might make it likely that at least some of the 15 records from

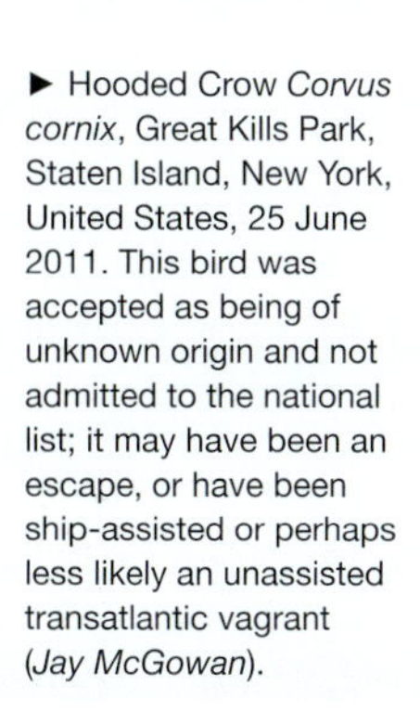

► Hooded Crow *Corvus cornix*, Great Kills Park, Staten Island, New York, United States, 25 June 2011. This bird was accepted as being of unknown origin and not admitted to the national list; it may have been an escape, or have been ship-assisted or perhaps less likely an unassisted transatlantic vagrant (*Jay McGowan*).

the eastern United States (Howell *et al.* 2014) relate to unaided crossings. However, a flock of 52 managed to stow away on a ship to Quebec, Canada, suggesting that ship-assistance may be the usual origin of such records. This species has also occurred eastwards to Japan and Tibet (Brazil 2009). Daurian Jackdaw has occurred as a vagrant to six European countries, albeit with some overlap in the individuals involved and some controversy over their origins (Haas 2012), but there are also records from further east in central Siberia, Uzbekistan and Kazakhstan which support a tendency for genuine vagrancy, as well as records from Taiwan and Hong Kong (Brazil 2009). Northern populations of Hooded Crow *Corvus cornix* are also partially migratory and birds have reached as far south as Malta, Tunisia and Libya, whilst in the north they have reached most of the Arctic archipelagos, including Jan Mayen and Svalbard, as well as two records from Greenland (Boertmann 1994). The species has been reported several times from North America and recent records from New York and New Jersey may represent transatlantic vagrants, presumably ship-assisted (Howell 2014). Records of Rook *Corvus frugilegus* and Eursasian Jackdaw from Greenland also underscore their vagrancy potential, although one of the Rook records was aboard a trawler working off the island (Boertmann 1994) and perhaps again points at the role of shipping in many of these occurrences.

Several corvid species are irruptive migrants, most notably the *macrorhynchos* subspecies of Spotted Nutcracker *Nucifraga caryocatactes*. Episodic failure of the seed crop of Siberian Stone Pine *Pinus sibirica* sometimes drives mass extralimital vagrancy in this species. The most recent major irruption resulted in 315 records in Britain in 1968, for example (Hollyer 1970). Elsewhere, vagrants have reached as far as Spain and Portugal (de Juana & Garcia 2015), Turkey (Kirwan *et al.*, 2010) and Azerbaijan (Heiss 2013). Clark's Nutcracker *Nucifraga columbiana* is also subject to occasional population irruptions, taking birds east as far as Pennsylvania, Illinois and Arkansas. So, too, Pinyon Jay *Gymnorhinus cyanocephalus,* with records from Washington state, Kansas, Iowa and Texas in the United States and Saskatchewan in Canada.

Some crow species make short altitudinal movements, such as those made by Alpine Chough *Pyrrhocorax graculus.* It is still perhaps quite surprising that this highly montane species has occurred widely as a vagrant in Europe, with records from the Balearic Islands and Cyprus involving substantial sea crossings; the first for Sweden was recorded at Halmstad, Halland, in June 2015 (van den Berg & Ławicki 2015). Red-billed Chough *Pyrrhocorax pyrrhocorax* is apparently less prone to such movements but has occurred as a vagrant to Korea (Brazil 2009) and an increasing British population has led to a flurry of recent short-distance dispersal movements within Britain.

Extralimital movements by the more resident species tend to be rare. For example, Siberian Jay *Perisoreus infaustus* has been reported as a vagrant only south to Poland and Slovakia. Canada Jay *Perisoreus canadensis* has been recorded no more than a few hundred kilometres south of its predominantly boreal range, and these events have seemingly become rarer. Brown-necked Raven *Corvus ruficollis* is usually considered to be sedentary, perhaps with some nomadic tendencies, but it is being recorded increasingly as a vagrant to Europe, with first records from: Cyprus in 2016 (Harrison 2016); the Canary Islands,

◄ Red-billed Chough *Pyrrhocorax pyrrhocorax*, Windgather Rocks, Cheshire/Derbyshire, England, 4 August 2019, the first record for Derbyshire; exploratory wandering is associated with population expansion in Britain (*Alexander Lees*).

▲ Pied Crows *Corvus albus*, Café Restaurant Chtoukan, Western Sahara, 22 January 2010, the first record for Western Sahara; this trio later bred (*Alexander Lees*).

on Lanzarote, in 2017 (Ławicki & van den Berg 2017c); Italy, on Pantelleria Island, in 2017 (Fulco & Liuzzi 2019); and mainland Spain in 2018 (Ławicki & van den Berg 2019a). Elsewhere, the species has reached The Gambia, central Syria and northern Pakistan. Perhaps the recent spate of records in southern Europe may reflect northward expansion aided by climate change. Australian corvids tend to be very sedentary, but nevertheless there is a rather remarkable record of Australian Raven *Corvus coronoides* on Lord Howe Island (McAllan *et al.* 2004), perhaps suggestive of ship-assistance.

The names of two species – House Crow *Corvus splendens* and Pied Crow *Corvus albus* – have become synonymous with ship-assistance. Both are commensal species, seldom found outside of human-modified landscapes, and both have a history of stowing away on boats, leading to successful colonisation. Although House Crow has been the most adept at this (see page 59), there are records of the essentially sub-Saharan African endemic Pied Crows from as far afield as Brazil (Lima & Kamada 2009) and India (Saikia & Gaswami 2017), which seemingly must have entailed ship-assistance; the species has also ventured across the Sahara to southern Algeria and Libya (Haas 2012) and even bred in Western Sahara in 2010 (Batty 2010). Records from Europe have been treated as escapes or of origin unknown, rather than as vagrants – ship assisted or otherwise – although recent records from Italy and Spain at least (Haas 2012, Fulco & Liuzzi 2019) may merit re-evaluation in the fullness of time.

CNEMOPHILIDAE
Satinbirds

Altitudinal movements are suspected in several species of satinbird, but there is no evidence of any significant extralimital movements.

MELANOCHARITIDAE
Berrypeckers and Longbills

Members of the Melanocharitidae are typically sedentary, although there is a record of a wandering Mid-mountain Berrypecker *Melanocharis longicauda* in the New Guinea lowlands (Gregory 2020b).

CALLAEIDAE
New Zealand Wattlebirds

This insular group is not known to make any significant movements and there are no extralimital records.

NOTIOMYSTIDAE
Stitchbird

Stitchbirds *Notiomystis cincta* are highly sedentary and extralimital movements are highly unlikely.

PETROICIDAE
Australasian Robins

Most of the Australasian Robins are sedentary, although a few species such as Red-capped Robin *Petroica goodenovii* and Flame Robin *Petroica phoenicea* are partially migratory, and Rose Robin *Petroica rosea* is apparently fully migratory and has been reported extralimitally in South Australia, where Pink Robin *Petroica rodinogaster* is also a regional vagrant (Black *et al.* 2020). Scarlet Robin *Petroica boodang* is apparently also partially migratory and is listed as a vagrant to King Island, Tasmania (Bennett *et al.* 2015). Historic long-range dispersal in this species also led to the colonisation of remote Norfolk Island and the subsequent speciation of Norfolk Robin *Petroica multicolor* (Kearns *et al.* 2015). Flame Robin migrates across the Bass Strait to spend the winter on the Australian mainland, whilst those populations on the Australian mainland migrate to lower altitudes during winter (Menkhorst 2017). Overshoots include records of Flame Robins in semi-arid habitats in north-western Victoria (eBird).

PICATHARTIDAE
Rockfowl

These two spectacular rainforest species are both assumed to be highly sedentary

CHAETOPIDAE
Rockjumpers

Both species of rockjumpers are effectively resident, although some downslope movement has been reported in winter.

EUPETIDAE
Rail-babbler

The Malaysian Rail-babbler *Eupetes macrocerus* is a non-migratory specialist of the understorey of rainforest, and extralimital movements are not anticipated.

HYLIOTIDAE
Hyliotas

All four hylotias are sedentary and there are no extralimital reports.

STENOSTIRIDAE
Fairy Flycatchers

Several fairy flycatchers undertake some form of movement, varying from altitudinal migration in Yellow-bellied Fairy-Fantail *Chelidorhynx hypoxanthus* to some very complex partial migrations in Grey-headed Canary-Flycatcher *Culicicapa ceylonensis*. The latter has been recorded as a vagrant to South Korea and southern China (Park & Oh 2018). Elsewhere, Fairy Flycatcher *Stenostira scita* is also a partial migrant in southern Africa and has been recorded as a vagrant north to Zimbabwe (Taylor 2020).

PARIDAE
Tits, Chickadees and Titmice

Most of the tits and chickadees are resident, although some montane species undertake altitudinal migrations, and populations of many northern species are irruptive. One of the most sought-after vagrants in the group is Azure Tit *Cyanistes cyanus,* which has occurred as an irruptive vagrant to many eastern and northern European countries, including Sweden, Denmark, Germany, Austria and Hungary as well as a 19th-century record from France. Records of vagrant hybrids with Eurasian Blue Tit *Cyanistes caeruleus,* known as 'Pleske's Tit', have outnumbered records of apparently pure birds 2:1 in recent years in Europe (Lawicki 2012), and several western European records have also proven to be escapes from captivity. Crested Tit *Lophophanes cristatus*, which breeds in Scotland, remains an incredibly rare vagrant to England, with the last record being a long-staying individual of the continental subspecies *mitratus* at Hauxley, Northumberland, in August–September 1984 (Brown & Grice 2005). Tits and chickadees seem to be relatively poor at undertaking sea crossings. Irruptive movements have taken Great Tits *Parus major* as far as Iceland and the Canary Islands, but no species of tit has reached the Azores yet. Similarly, no species of chickadee has been reported from Bermuda. Several species of chickadee are irruptive, particularly Black-capped Chickadee *Poecile atricapillus* and Boreal Chickadee *Poecile hudsonicus;* the former has occurred south to Missouri, Oklahoma, Kentucky and Virginia, while the latter has reached Pennsylvania,

▲ Azure Tit *Cyanistes cyanus*, Hiirola, Uurainen, Finland, 5 February 2007, the 24th record for Finland of this enigmatic species, which remains incredibly rare any further west in Europe (*Graham Catley*).

New Jersey, Wyoming, Nebraska, Iowa, Ohio and Virginia. Unlike Black-capped Chickadee, the more southerly distributed Carolina Chickadee *Poecile carolinensis* is not irruptive but there are still some records of vagrants, including two from southern Canada, in Ontario (Holden & Bell 2016).

REMIZIDAE
Penduline Tits

Several of the penduline tits occurring in Eurasia are strongly migratory and have been reported as extralimital vagrants, whilst Verdin *Auriparus flaviceps*, Tit-hylia *Pholidornis rushiae* and several Afrotropical penduline tits are apparently sedentary and have no significant records of vagrancy. Eurasian Penduline Tit *Remiz pendulinus* is migratory in the north of its range, but vagrancy remains rather modest. The species was first recorded as a vagrant to England as recently as October 1966 when one was recorded at Spurn, Yorkshire (Raines & Bell 1967); it is now annual, but there are still no records from Scotland or Ireland. The subsequent upsurge in records in England was associated with a gradual range expansion in the species across Europe (Valera *et al.* 1990). The rather enigmatic Black-headed Penduline Tit *Remiz macronyx* has been recorded as a narrowly extralimital vagrant from European Russia (Bot *et al.* 2011). A report of a White-crowned Penduline Tit *Remiz coronatus* trapped in August 1987 at Lake Neusiedl in eastern Austria has subsequently been ruled to have been an aberrant Eurasian Penduline Tit (Bot *et al.* 2011). Further to the east, Chinese Penduline Tit *Remiz consobrinus* has been recorded as a vagrant to the central Ryukyu Islands of Japan and to Taiwan (Brazil 2009).

ALAUDIDAE
Larks

Among the larks we find some of the most highly migratory and vagrancy-prone species, typically breeding at temperate latitudes, and also some of the most sedentary in the tropics; 90 per cent of Afrotropical species are, for example, considered largely resident. Eurasian Skylark *Alauda arvensis* has one of the largest geographic ranges of any bird species, stretching fully across the Palearctic. Northern populations are highly migratory, wintering in a broad belt at south temperate and subtropical latitudes, with vagrants recorded south to Mauritania, Malaysia (Ławicki & van den Berg 2018c) and northern Borneo (Eaton *et al.* 2016). This is the only lark species to have occurred as a vagrant to the Americas, where it is a regular visitor to the Aleutians, and it has even bred on the Pribilof Islands in the Bering Sea region; it has also occurred as a vagrant to British Columbia, California and amazingly to Lanaudière, Quebec, in eastern Canada in May 2018. The California bird appeared in Marin County in December 1978 and returned to winter at the same site in six subsequent winters (Howell *et al.* 2014). Two birds were present on Kure Atoll in the mid-Pacific Ocean in September–October 1963 (Woodward 1972). Other records from Bermuda in 1850 (Howell *et al.* 2014), the Azores and Bear Island attest to the prodigious capacity for overwater flights in this species. Amazingly, there are three records of Eurasian Skylark from Lord Howe Island, Australia – in September, June and July – assumed to be vagrants from the introduced New Zealand population, which necessitates a 1,300km overwater flight (McAllan *et al.* 2004). Given these epic movements, vagrancy of Skylark subspecies within the Palearctic also seems likely but has proven difficult to confirm. The deep genetic divergence between eastern and western populations could see a future split which may kinder more interest in their field identification, however (Lees & Ball 2011). In contrast to Eurasian Skylark, Raso Lark *Alauda razae,* restricted to Raso Island in the Cape Verde archipelago, has one of the smallest ranges of any bird but has been recorded as a vagrant on the islet of São Nicolau, 15km to the east of Raso, in March 2009 (Hazevoet 2012); its restriction to this single islet may yet prove to be the result of human-driven extinction processes elsewhere in the archipelago, where reintroductions are now underway (Donald *et al.* 2005). Crested Lark *Galerida cristata* has a far larger distribution but is also extremely sedentary. The species formerly bred commonly on the English Channel coast of France and the North Sea coast of the Low Countries but remained an extremely rare vagrant to England even before its range contracted southwards.

Greater Short-toed Lark *Calandrella brachydactyla* has one of the longest migration spans of any lark species, moving between Palearctic breeding and

► Sykes's (or Mongolian) Short-toed Lark *Calandrella dukhunensis* with Bramblings *Fringilla montifringilla* and Oriental Greenfinch *Chloris sinica*, Hegurajima Island, Ishikawa, Japan, 24 Apr 2019. This is an annual vagrant to Japan, mostly to island vagrant traps (*Yann Muzika*).

Afrotropical wintering areas. Vagrants reach northern and western Europe with some regularity and have occurred on Iceland, Madeira and the Canary Islands, as well as south to Tanzania (Fisher & Hunter 2017) and the Seychelles (Skerrett *et al.* 2017). The closely related Hume's (Short-toed) Lark *Calandrella acutirostris* has a more restricted breeding range and a shorter migration span but has occurred once as a vagrant to Israel in February 1986 (Haas 2012). Bimaculated Lark *Melanocorypha bimaculata* winters in East Africa and has a relatively small breeding range, centred around Asia Minor and the Near East, but boasts an impressive spread of vagrant records, albeit very few in number. Bimaculated Lark is an extremely rare vagrant west to western Europe with 25 records from 12 countries, from Sicily to northern Finland (Thom & Täschler 2013, Audevard 2018), east to Japan (Brazil 2009) and south to the Seychelles (Phillips & Phillips 2005) whilst an old record of one collected at Swakopmund, Namibia, in 1930 was treated by Brooke (1988) as probably an escape from captivity, although this conclusion deserves revisting. Two relatively short-distance migrant larks from central Asia – Black Lark *Melanocorypha yeltoniensis* and White-winged Lark *Alauda leucoptera* – are among the most sought-after of vagrants to Europe, with a thin scatter of records from several countries, mostly in late winter and spring. Individuals may be enticed west by prolonged easterly winds and unusual snowfalls within their regular range, and in early May 1993 a Black Lark record from Sweden coincided with a White-winged Lark in Poland, suggesting that their occurrences may be linked to concurrent weather events (Degnan & Croft 2005).

Southern Europe, North Africa and the Middle East host a major swathe of lark diversity and their dispersal tendencies and hence vagrancy potential are at least partially dictated by their life histories. The more nomadic species tend to be gregarious seed-eaters with thicker bills, whereas the sedentary species in the same region tend to be solitary and more insectivorous species with thinner bills (de Juana *et al.* 2004). Temminck's Lark *Eremophila bilopha* falls within the latter group and has been recorded extralimitally very rarely, north to Malta and south to The Gambia (Borrow & Denney 2014), while Greater Hoopoe-Lark *Alaemon alaudipes* has wandered to the Canary Islands, Italy, Greece and Turkey and south to Senegal. Larger-billed larks wander more widely. For example, Calandra Lark *Melanocorypha calandra* has been recorded extralimitally as a vagrant from Norway in the north, west to Madeira and south to eastern Arabia. Bar-tailed Lark *Ammomanes cinctura* has also wandered to Spain, France, Italy, Malta and Cyprus, and records from western or northern Europe would not seem beyond the realms of possibility. The poorly known Dunn's Lark *Eremalauda dunni* is also prone to irruptions and there is a record presumed to be of the African form *E. d. dunni* from Cyprus (Donald & Christodoulides 2018); this suggests that future vagrancy to mainland Europe is not impossible. Among Afrotropical larks, the only species known to undertake regular seasonal migrations is Dusky Lark *Pinarocorys nigricans,* which has been recorded as an overshooting vagrant north to Uganda (Fisher & Hunter 2014) and may be overlooked as a vagrant elsewhere. Some populations of Red-capped Lark *Calandrella cinerea* also apparently engage in regular

◄ Black Lark *Melanocorypha yeltoniensis*, Winterton, Norfolk, England, 20 April 2008, the third British record (*James Gilroy*).

► Shore Lark *Eremophila alpestris* of the eastern North American *alpestris/praticola/hoyti* group ('Horned Lark'), Staines Reservoirs, Greater London, England, 25 January 2018, a remarkable record of a Nearctic vagrant in the British capital (*James Lowen*).

migrations, and vagrants have been reported from northern Gabon and the Democratic Republic of Congo. There is a single exceptional event of extraregional vagrancy by an Afrotropical species – an adult male Chestnut-headed Sparrow-Lark *Eremopterix signatus* was recorded at Eilat, Israel, in May 1983 (Haas 2012). This species occurs regularly no closer than central Ethiopia and underscores the fact that occasional long-distance vagrancy events by nomadic Afrotropical species may involve passerines as well as waterbirds. A single record of Black-crowned Sparrow-Lark *Eremopterix nigriceps* at the Chorokhi Delta in October 2015 was unsurprisingly the first record for Georgia and represents a similarly unexpected movement; this hints at the possibility of vagrancy to Europe in this species.

Finally, the sole regularly breeding lark in the Americas (apart from the introduced Eurasian Skylarks on Vanvouver Island), Shore Lark *Eremophila alpestris*, is also the only documented transatlantic vagrant of the family, with multiple individuals of the North American 'Horned Lark' *Eremophila alpestris alpestris/praticola/hoyti* having now been recorded in Britain (BOURC 2020).

PANURIDAE
Bearded Reedling

Bearded Reedlings *Panurus biarmicus* have a vast distribution across the mid latitudes of Eurasia and, although many populations are fairly sedentary, they do undertake some eruptive post-breeding and wintering movements. An increase in wetland restoration in Europe is leading to rapid population increases and consequently more vagrant dispersing individuals seeking new habitats. Vagrants are recorded with increasing frequency in Ireland, where they now breed, and elsewhere are vagrants to Morocco, Algeria and Egypt, as well as to Kuwait, Pakistan, Korea and Japan (Snow & Perrins 1998, Brazil 2009).

NICATORIDAE
Nicators

All the nicators are sedentary forest-associated species and no extralimital records have been reported.

MACROSPHENIDAE
Crombecs and African Warblers

Some members of this group have been reported undertaking some short-distance movements, but there are no reports of significant vagrancy.

CISTICOLIDAE
Cisticolas and allies

Most of the cisticolas, eremomelas, apalises and allies are resident tropical birds, although some West African species do make local movements in accordance with the rains. There are reports of relatively local vagrancy events in many species, such as a White-winged Apalis *Apalis chariessa* 80km out of range in Malawi (Ryan 2020). Migratory behaviour is strongest in populations of Zitting Cisticola *Cisticola juncidis* at its northern range edge, where it may be partially migratory. This species is a vagrant to several northern European countries including Germany, Austria and Hungary (Németh & Vadász 2008), with increasing frequency from the 1970s as the species spread northwards to breed as far north as the Netherlands (van den Berg & Bosman 1999), although the range subsequently contracted south again. The first British record in 1976 was associated with this expansion and colonisation was anticipated (Ferguson-Lees & Sharrock 1977), but there have only been seven subsequent records, and none since 2010. Only one reached Ireland, in 1985. This may reflect a reticence to cross open water, but there are records from both Sal on Cape Verde and Lanzarote on the Canary Islands (Clarke 2006), suggesting that the English Channel ought not to be a major barrier to dispersal.

▲ Zitting Cisticola *Cisticola juncidis*, Pegwell Bay, Kent, England, 2 November 2009, the seventh British record (*Stuart Piner*).

ACROCEPHALIDAE

Reed Warblers and allies

The Acrocephalidae contains six genera, the *Acrocephalus* and *Arundinax* 'reed warblers', the *Hippolais* and *Iduna* 'tree warblers', the *Chloropeta* 'yellow warblers' and the *Nesillas* 'brush warblers'. *Acrocephalus* is the most diverse group and includes 18 species that are continentally distributed and mostly migratory, whereas the remaining species are island endemics. Nineteen of these island species are found in Micronesia and Australia eastward to the Pitcairn Group, with species like Pitcairn Reed Warbler *Acrocephalus vaughani* occupying among the remotest oceanic islands on the planet. Colonisation of such disparate locations indicates high dispersal ability, and correspondingly we find high vagrancy rates among most of the migratory *Acrocephalus* warblers. The Pacific island taxa are descended from the group of large reed warblers including Oriental Reed Warbler *Acrocephalus orientalis* and Great Reed Warbler *Acrocephalus arundinaceus.* In the Eastern Palearctic, Oriental Reed Warbler *Acrocephalus orientalis* is a long-distance migrant but has been reported relatively rarely as a vagrant. For example, there are just two Western Palearctic records, both from Israel (Haas 2012), and it has occurred more than a dozen times off northern Australia on Ashmore Reef and Christmas Island (James & McAllan 2014). Great Reed Warbler has a more western distribution and is a vagrant to northern Europe, to several Atlantic Islands including the Canary Islands (Garcia del Rey 2011) and most spectacularly to the Prince Edward Islands, 1,700km south-east of South Africa in June 2014 (Kotzé *et al.* 2015). Another member of this large *Acrocephalus* group is Basra Reed Warbler *Acrocephalus griseldis,* which is also a long-distance migrant and has reached Cyprus, Israel and most recently Turkey (Neate-Clegg *et al.* 2019); it is also a reminder that vagrant 'great' reed warblers need to be thoroughly checked in western Europe for both this species and Oriental Reed Warbler.

Several of the *Acrocephalus* warblers are highly enigmatic. For example, the breeding grounds of Streaked Reed Warbler *Acrocephalus sorghophilus* are still unknown, presumed to be somewhere in north-east China, but this species has been recorded as a vagrant to Nansei Shoto in Japan, east of its route through China (Brazil 2009). Likewise, the wintering

► Blyth's Reed Warbler *Acrocephalus dumetorum*, Barton-upon-Humber Gravel Pits, Lincolnshire, England, 15 June 2020, part of a major incursion into western Europe and likely a prelude to colonisation (*Graham Catley*).

grounds of Aquatic Warbler *Acrocephalus paludicola* have only relatively recently been discovered (Salewski *et al.* 2009), and several regions where the species was previously thought to be a vagrant have proven to be regular staging grounds (Salewski *et al.* 2019). Large-billed Reed Warbler *Acrocephalus orinus* was only rediscovered in life in 2006 (Round *et al.* 2007) and has proven to be a long-distance migrant that could be anticipated as a vagrant to the Western Palearctic, albeit rarely given its apparently small population size. Identification difficulties probably hamper our understanding of true patterns of vagrancy in the group, compounded by their skulking behaviour and penchant for dense aquatic vegetation. A combination of both a westward range expansion and greater identification knowledge has seen a dramatic change in status of Blyth's Reed Warbler *Acrocephalus dumetorum* in western Europe; it is now known to be a frequent autumn vagrant west all the way to Iceland (Arnarson & Brynjólfsson 2000) and south to Spain. Five records from Malta may suggest that some birds also winter in Africa rather than on the Indian subcontinent. This pattern of vagrancy is also found in Paddyfield Warbler *Acrocephalus agricola,* which has occurred east to Japan and Korea (Brazil 2009) and west with some regularity to western Europe; there has even been a record from Corvo on the Azores (Alfrey *et al.* 2010), an island group that has only had three records of the perhaps more expected Eurasian Reed Warbler *Acrocephalus scirpaceus.*

Blyth's Reed Warbler and Sedge Warbler *Acrocephalus schoenobaenus* are the only *Acrocephalus* warblers to have occurred in North America, with records of both species from St Lawrence Island in Alaska, and both are also vagrants to Korea and Japan (Brazil 2009). A record of Marsh Warbler *Acrocephalus palustris* in Ussuriland in the Russian Far East is equally unexpected (Ławicki & van den Berg 2019a). Many other species of *Acrocephalus* warblers breed much closer to Alaska yet are unrecorded there, including Blunt-winged Warbler *Acrocephalus concinens* and Manchurian Reed Warbler *Acrocephalus tangorum,* which would seem to be potential vagrants along with Black-browed Reed Warbler *Acrocephalus bistrigiceps,* which has made it as far as the Philippines (Round & Fisher 2009). Moustached Warblers *Acrocephalus melanopogon* are more sedentary than most temperate zone reed warblers, especially in Europe, but have still occurred as vagrants to the Netherlands and Belgium, for example, and this species is anticipated as a likely future vagrant to the British Isles.

Among the remaining members of the group, the monotypic Thick-billed Warbler *Arundinax aedon,* which was formerly placed in *Acrocephalus,* is by comparison relatively easy to identify and has occurred as a vagrant 10 times to the Western Palearctic, in Britain (six, all on Shetland), twice in Norway and once each from Finland and Egypt (Haas 2012), in addition to Saudi Arabia (Ławicki & van den Berg 2018a) and St Lawrence Island, Alaska,

◄ Thick-billed Warbler *Arundinax aedon*, Fair Isle, Shetland, Scotland, 16 May 2003, the fourth record for Britain (*Rebecca Nason*).

► Booted Warbler *Iduna caligata*, North Keraniganj, Dhaka, Bangladesh, 22 November 2018. There are fewer than 10 winter records of this species from Bangladesh (*Sayam Chowdhury*).

in September 2017 (Rosenberg *et al.* 2018). The two Central Asian 'tree warblers', Booted Warbler *Iduna caligata* and Sykes's Warbler *Iduna rama,* are both vagrants west to Europe annually and to the east, with Booted Warbler even reaching Japan (Brazil 2009) and Myanmar (Naing *et al.* 2020), and Sykes's Warbler to Bangladesh (Round *et al.* 2014); in contrast, vagrancy in both Eastern Olivaceous Warbler *Iduna pallida* and Western Olivaceous Warbler *Iduna opaca* is considerably rarer. The former has occurred with greater regularity to north-west Europe, and even to Iceland, and is a longer-distance migrant, but the geographically more proximate Western Olivaceous Warbler has only occurred in the same region once, in Sweden. Similarly, vagrancy is surprisingly rare among the *Hippolais* warblers, which are also long-distance migrants. Two Balkan species are incredibly rare in western Europe: there are single western European records of Upcher's Warbler *Hippolais languida* from the Netherlands in October–November 2019 and Olive-tree Warbler *Hippolais olivetorum* on Shetland, Scotland, in August 2006 (Harrop *et al.* 2008). Upcher's Warbler, which normally winters only as far south as Tanzania, has also recently reached South Africa, with one present near Port Elizabeth, Eastern Cape, in late August and early September 2017 (Demey 2018).

LOCUSTELLIDAE
Grassbirds and allies

The Locustellidae is a diverse group of typically skulking birds of dense habitats including, for example, *Megalurus* grassbirds, *Cincloramphus* songlarks, *Poodytes* fernbirds and *Megalurulus* thicketbirds. Most species are sedentary and tropical in distribution, although some like the songlarks may be nomadic; the high number of island endemics and their broad geographic distribution suggests regular long-distance dispersal even in nominally sedentary genera. Bucking the trend for sedentarism are all the temperate zone *Locustella* warblers, and some *Bradypterus* warblers which are highly migratory; vagrancy seems to be fairly common among the *Locustella* warblers at least. River Warbler *Locustella fluviatilis,* for example, moves from northern Eurasia to winter in South-east Africa and has appeared all the way to North America – once to St Lawrence Island, Alaska, in October 2017 (Lehman 2018). However, our understanding of how common vagrancy is in the group is hampered by the skulking nature of *Locustella* warblers. Lanceolated Warbler *Locustella lanceolata,* for example, has a vast global range in the boreal belt, stretching from

▲ Lanceolated Warbler *Locustella lanceolata*, Fair Isle, Shetland, Scotland, 22 September 2014. Fair Isle is by far the most reliable site for this species in western Europe (*James Lowen*).

◀ Middendorff's Grasshopper Warbler *Locustella ochotensis*, Ashmore Reef, Australia, 19 November 2013. Ashmore Reef has something of a monopoly of records of this vagrant to Australia (*Rohan Clarke*).

southern Finland to the Pacific and it may be abundant, especially in the east. Given this distribution, its relative abundance and long migration span to South-east Asian wintering areas it would be expected to be a regular autumn vagrant to western Europe. As it stands, one single Scottish island – Fair Isle – has had almost 100 records, around two-thirds of all British occurrences; this is due in part to intensive observer coverage and, more importantly, the lack of dense vegetation on the island for this typically fiendishly skulking species to hide in. Lanceolated Warblers must presumably occur at other migrant traps in the British Isles, but there are few records away from the Northern Isles. The detectability issue is nicely illustrated by the sole record from the Isles of Scilly, which comes from the uninhabited and sparsely vegetated islet of Annet in September 2002 (Robinson 2003). Lanceolated Warblers are regular vagrants to the western Aleutians in Alaska and have even bred there (Howell *et al.* 2014). The only United States record away from Alaska was ringed on Southeast Farallon Island, 42km west of San Francisco in September 1995; that autumn was particularly good for Siberian vagrants on the United States west coast, including an Arctic Warbler and two Dusky Warblers (Hickey *et al.* 1996). A single record from Palau in Micronesia (Otobed *et al.* 2018) hints at the species' ability to make long overwater flights.

Pallas's Grasshopper Warbler *Locustella certhiola* has a more eastern distribution but in Europe is also largely a specialty of Shetland and Fair Isle, where most of the 60 or so British records have come from; by contrast there are 26 records from Norway and just one from Sweden. Pallas's Grasshopper Warblers have been reported more widely as vagrants, however, from Poland, Israel and an amazing spring record from Morocco, on the high plains between the Middle Atlas and High Atlas mountain ranges in April 1976 (Bergier *et al.* 2009). In the east, Pallas's Grasshopper Warbler has occurred on both Christmas Island and Ashmore Reef off northern Australia (James & McAllan 2014); the first to be reported in North America was on St Lawrence Island, Alaska, in September 2019. Middendorff's Grasshopper Warbler *Locustella ochotensis* has been recorded as a spring and autumn migrant to the eastern Bering Sea and western Aleutian Islands in Alaska (Howell *et al.* 2014), and Gray's Grasshopper Warbler *Locustella fasciolata* would seem overdue as a vagrant there – this species has been recorded twice historically in Europe, in France in September 1913 and Denmark in September 1955, and future records here also seem overdue (Haas 2012). Baikal Bush Warbler *Locustella davidi* and Sakhalin Grasshopper Warbler *Locustella amnicola* are both long-distance migrants and thus potential vagrants to Alaska or Australia, and perhaps even western Europe. Baikal Bush Warbler has recently been detected for the first time in India (Robson *et al.* 2015) and like many of the members of this group is probably under-recorded; this view is supported by the fact that the wintering grounds of Sakhalin Grasshopper Warbler are still unknown.

DONACOBIIDAE
Black-capped Donacobius

Black-capped Donacobius *Donacobius atricapilla* is sedentary and we are unaware of any significant extralimital movements.

BERNIERIDAE
Madagascan Warblers

All the members of this family are sedentary or presumed to be sedentary and there are no extralimital reports.

PNOEPYGIDAE
Cupwings

Several cupwings undertake altitudinal movements but there are no records of significant extralimital vagrancy.

HIRUNDINIDAE
Swallows and Martins

Temperate zone breeding swallows and martins undertake some of the longest migrations of any passerines; coupled with their adaptations for aerial foraging on active migration, many members of the group seem predisposed to long-range vagrancy. Indeed, there seem to be no global barriers to vagrancy in arguably the champion avian wanderer – Barn Swallow *Hirundo rustica*, which has been recorded more widely as a vagrant than any other bird species (Lees & Gilroy 2014). Barn Swallows have reached the remotest oceanic islands worldwide, including Hawaii (Pyle & Pyle 2017), Tristan da Cunha (Bond & Glass 2016) and the Crozets (Gauthier-Clerc *et al.* 2002) as well as the Arctic from Svalbard, Bear Island and Greenland (Boertmann 1994), and from the Antarctica Peninsula (Korczak-Abshire *et al.* 2011b) and South Georgia (Clarke *et al.* 2012). Barn Swallow is one of the most cosmopolitan of all bird species, but other hirundines restricted to either the Eastern or Western Hemispheres have also reached similarly remote locations. For example, from the Old World, Common House Martin *Delichon urbicum* has reached Bermuda and St Pierre et Miquelon (off Newfoundland) in the western Atlantic (Howell *et al.* 2014) as well as four records in the West Indies (Kirwan *et al.* 2019) and St Helena in the South Atlantic (Hillman *et al.* 2016) where two or three individuals were recorded in November 2013. From the New World, Cliff Swallow *Petrochelidon pyrrhonota* is probably the champion hirundine vagrant; it is the most frequently recorded Nearctic vagrant in the Western Palearctic with 36 records, including birds noted as far east as Sweden and multiple arrivals such as six together at Cabo da Praia, Terceira, on the Azores in September 2019. Cliff Swallow is also the most frequent Nearctic swallow in the Eastern Palearctic with six records involving nine birds from north-east Russia (Arkhipov & Ławicki 2016). In the Pacific, Cliff Swallow has reached Hawaii once (Pyle & Pyle 2017), as well Clipperton Atoll, 1,080km from the Mexican coast; it is also one of the most frequent vagrant hirundines on the Falklands (Woods 2017).

Transatlantic vagrancy among the remaining species is surprisingly rare, involving a handful of records of both Purple Martin *Progne subis* and Tree Swallow *Tachycineta bicolor* in Britain and the Azores (Haas 2012) and Greenland (Boertmann 1994); both species have also reached north-east Russia (Arkhipov & Ławicki 2016). Given the long migration of Purple Martin, its rarity in the Western Palearctic seems unusual, but may reflect its early departure before the main weather window of fast-moving Atlantic depressions. Other temperate zone breeders with shorter migration spans are correspondingly recorded less frequently and less extralimitally as vagrants. For example, Northern Rough-winged Swallow *Stelgidopteryx serripennis* winters only as far south as Central America, although it seems inevitable that it will appear on the Azores in the fullness of time. Similarly, Violet-green Swallow *Tachycineta thalassina*

◄ Red-rumped Swallow *Cecropis daurica*, Arlington Reservoir, Sussex, England 9 May 2010, a regular spring overshoot to southern England (*Rik Addison*).

has a similar migration span and is correspondingly very rare as a vagrant even in eastern North America, but it has reached Puerto Rico in the West Indies (Kirwan *et al.* 2019). The status of Cave Swallow *Petrochelidon fulva* as a vagrant in eastern North America has changed dramatically in recent decades and it is now a predictable late autumn vagrant, a change associated with its range expansion in the south (McNair & Post 2001).

In the Palearctic, Red-rumped Swallow *Cecropis daurica* has a south temperate breeding distribution and a relatively short migration span. There are as yet no records of transatlantic vagrancy, but vagrants have reached the Azores and Iceland, and this species is a regular spring and autumn vagrant to northern temperate latitudes as well as south to South Africa. There are also records of 'Asian Red-rumped Swallow' – either *C. d. daurica* or *C. d. japonica* – as far west as Scotland, where one was recorded on both Orkney and the Western Isles in June 2011 (Thorne & Thorne 2014), and this group also reaches northern Australia with some regularity (James & McAllan 2014). Asian House Martin *Delichon dasypus* is a regular vagrant to Australian territory on Christmas Island (James & McAllan 2014) and has been recorded twice recently from the Western Palearctic in Lithuania and Israel. Eurasian Crag Martin *Ptyonoprogne rupestris* is less migratory still and is a very rare vagrant to northern Europe. South temperate hirundines are also often highly migratory and prone to vagrancy, notably Brown-chested Martin *Progne tapera*, with at least eight records from the United States (Howell *et al.* 2014), and it has also been recorded south-west to Chile (Jaramillo *et al.* 2003). In Australia both Fairy Martin *Petrochelidon ariel* and Tree Martin *Petrochelidon nigricans* undertake some migratory movements, and both have been recorded from New Zealand (Miskelly *et al.* 2017) and Lord Howe Island (McAllan *et al.* 2004). Fairy Martin has also occurred as a vagrant north to Indonesia (Eaton *et al.* 2016).

Identification difficulties may confound knowledge of vagrancy in some hirundine species. Aside from the aforementioned Asian House Martin, which has been reported from western Europe, Chilean Swallow *Tachycineta leucopyga* has been reported as a vagrant to Peru (Schulenberg *et al.* 2010) and Bolivia (Herzog *et al.* 2016) but its separation from other *Tachycineta* swallows can be problematic. Similarly, the rarity and identification challenges of some of the blue *Progne* martins hamper our knowledge of their distribution and hence vagrancy potential. However, now that a Caribbean Martin *Progne dominicensis* has been tracked migrating c.3,550km south-east from Dominica to winter in Bahia, Brazil (Perlut *et al.* 2017), we can anticipate the discovery of substantial vagrancy in this species, which should be sought in North America.

Many tropical hirundines, especially those associated with forested landscapes, are sedentary and are rarely or never reported as vagrants. However, tropical species of arid landscapes often undertake intratropical movements and sometimes even occur as vagrants beyond the tropics. For example, three sub-Saharan African species have been recorded as vagrants outside of the region: Banded Martin *Riparia cincta* in Egypt in November 1988; Ethiopian Swallow *Hirundo aethiopica* in Israel in March 1991 (Haas 2012); and Lesser Striped Swallow *Cecropis*

► Brown-chested Martin *Progne tapera*, Cumberland Farms, Plymouth, Massachusetts, United States, 13 October 2009, the second state record (*Tom Johnson*).

◄ Southern Martin *Progne elegans*, Gypsy Cove, Falklands, 5 April 2020, an occasional vagrant to this archipelago (*Alan Henry*).

◄ Mascarene Martin *Phedina borbonica*, Macaneta, Maputo, Mozambique, 18 May 2019. Post-breeding movements of this species from Madagascar to East Africa are poorly understood but may be more regular than the scant records attest (*Gary Allport*).

◄ Ethiopian Swallow *Hirundo aethiopica*, 23 March 1991, HaZafon, Israel, the first record for the Western Palearctic (*Yoav Perlman*).

abyssinica in Oman (Eriksen & Porter 2017). A fourth has made it to the Western Palearctic – Preuss's Cliff Swallow *Petrochelidon preussi* on Cape Verde. South African (Cliff) Swallow *Petrochelidon spilodera* has now been recorded twice as a vagrant to Kenya (Fisher & Hunter 2019). This species is an austral migrant to western central Africa. White-headed Saw-wing *Psalidoprocne albiceps*, which is partially migratory, may also be a candidate for extratropical vagrancy as it has wandered widely – north to Ethiopia and south to South Africa. In the Western Hemisphere, Bahama Swallow *Tachycineta cyaneoviridis* occasionally wanders to Florida, the same state that holds the sole North American record of Mangrove Swallow *Tachycineta albilinea,* a vagrant from Central America (Howell *et al.* 2014). Records of the ostensibly resident White-winged Swallow *Tachycineta albiventer* wandering from South America to the West Indies, on Martinique and Grenada (Kirwan *et al.* 2019), also reinforce the notion that even very resident hirundines can wander substantial distances.

PYCNONOTIDAE
Bulbuls

Most bulbuls are sedentary, and instances of vagrancy in the group are generally rare and disproportionately involve nomadic species such as Common Bulbul *Pycnonotus barbatus,* which may now be on the cusp of colonising Spain from North Africa (Gil-Velasco *et al.* 2017). Vagrancy in Black-fronted Bulbul *Pycnonotus nigricans* may be anticipated as the species apparently undertakes some regular movements in the austral winter as far as northern Botswana. A recent first report of a pair of Yellow-throated Greenbuls *Atimastillas flavicollis* from Namibia may be vagrants (Demey 2017), but as is often the case in the tropics these may yet prove to indicate a range extension. Black Bulbuls *Hypsipetes leucocephalus* of East Asia on the other hand are among the most migratory of species in the family, particularly the white-headed subspecies group, which have been reported as vagrants from southern Thailand (Robson 2002), Bangladesh (Thompson & Johnson 2003), north-east India (Srinivasan *et al.* 2009) and even as far north as South Korea (eBird) and Japan (Brazil 2009). Northern populations of Light-vented Bulbul *Pycnonotus sinensis* are also migratory, and vagrants have been reported from the Philippines and northern Thailand (Robson 2002).

PHYLLOSCOPIDAE
Leaf Warblers

The leaf warblers are a species-rich group of Old World warblers. They include a high proportion of highly migratory temperate zone species (around 55 per cent), as well as many sedentary tropical and subtropical ones. As a group they have also proven to be workhorses for the study of avian vagrancy, as many are frequent vagrants to regions thousands of kilometres from their regular ranges. Several species have huge Palearctic breeding distributions. Arctic Warbler *Phylloscopus borealis,* for example, breeds from northern Norway east all the way to western Alaska, wintering exclusively in South-east Asia. In North America there are records presumed to be this species south down the west coast as far as Baja California in Mexico (Pyle & Howell 1993) – although possibly some may refer to Kamchatka Leaf Warbler *Phylloscopus examinandus,* which has occurred on the Aleutians. Vagrant Arctic Warblers occur with some regularity in Britain in autumn but are very rare elsewhere in western Europe, e.g. only 15 records from France, one from Spain and none from Italy; the first Portuguese record was in September 2009 (Pacheco *et al.* 2011). In this context records from Iceland and, even more impressively, on Corvo on the Azores in October 2015 (Ławicki & van den Berg 2015b) seem all the more surprising; even they are arguably eclipsed by records from Israel in November 2018, Bermuda in February 2014 (Dobson 2014) and Broome Bird Observatory, Western Australia, in January 1998 (Hassell 2016). Willow Warblers *Phylloscopus trochilus*

◄ Arctic Warbler *Phylloscopus borealis*, Inishbofin, Galway, Ireland, 13 September 2015. This species is remarkably rare as a vagrant in southern and central Europe given its status further north (*Anthony McGeehan*).

◄ Willow Warbler *Phylloscopus trochilus*, Ashmore Reef, Australia 16 November 2013, the first record for Australia (*Rohan Clarke*).

have a not-too-dissimilar breeding range, stretching from eastern Siberia to France, and they winter in central and southern Africa; easternmost populations face a migration of 13,000km to arrive at their wintering grounds. This species is a vagrant to Alaska, with more than 10 records (Howell *et al.* 2014), Greenland (Boertmann 1994), Japan and South Korea (Brazil 2009), Mantanani Island off Sabah, Borneo (Eaton *et al.* 2016), and Ashmore Reef off north-west Australia (Johnstone & Darnell 2015). Three records stand out as among the most extreme for any *Phylloscopus*, however: single Willow Warblers on the Prince Edward Islands (Oosthuizen *et al.* 2009), Kerguelen Islands (Ausilio & Zotier 1989) and Tristan da Cunha (Ryan 2008) raise the possibility of vagrancy even to South America.

Patterns of vagrancy in other *Phylloscopus* with more restricted range sizes seem no less impressive. For example, Wood Warblers *Phylloscopus sibilatrix* are essentially restricted as breeders to the Western Palearctic, extending east to south-central Siberia and wintering in equatorial Africa. Incredibly, there are two records from Alaska (Howell *et al.* 2014) which is 4,500km away from their breeding range, in addition to other eastern records from Japan (Brazil 2009), India (Kang *et al.* 2016) and oceanic islands including the Seychelles, Iceland and Madeira. Pallas's Warbler *Phylloscopus proregulus* on the other hand breeds in central and eastern Siberia and winters in South-east Asia but is a remarkably regular autumn visitor to western Europe as well as Morocco, Israel, Turkey, Georgia, Iran, Azerbaijan, Tajikistan, Japan and

Taiwan, with one record from Alaska on St Lawrence Island in September 2016 (Howell *et al.* 2014). Dusky Warbler *Phylloscopus fuscatus* has a broadly similar range and is also a regular late autumn vagrant to the Western Palearctic, with a major influx in autumn–winter 2020, and has occurred as far south and west as Madeira. Dusky Warbler is also more regular in North America, occurring annually in the Bering Sea region and also more than 20 times in California and twice in Baja California, Mexico (Howell *et al.* 2014). This species has also reached the Australian vagrant trap of Christmas Island (James & McAllan 2014). Radde's Warbler *Phylloscopus schwarzi,* with a narrower breeding range, has yet to be recorded in North America or Australia, but is also a predictable autumn vagrant to western Europe, although unusually virtually unrecorded in winter or spring. The range of Two-barred Warbler *Phylloscopus plumbeitarsus* overlaps broadly with Pallas's Warbler, yet it remains an extremely rare (albeit increasingly detected) vagrant to Europe, most notably to Madeira (Romano *et al.* 2010).

The most intriguing Siberian warbler is arguably Yellow-browed Warbler *Phylloscopus inornatus,* which breeds further west than the preceding species and is now an expected scarce passage migrant right across much of western Europe. Single day totals in the hundreds at coastal and island migration bottlenecks in autumn suggest the potential establishment of a regular migration route (page 32). Small numbers have been found wintering from southern Europe north to southern England, with more further south; 29 were found on Lanzarote in the 2013–14 winter,

► Radde's Warbler *Phylloscopus schwarz*, Inishbofin, Galway, Ireland, 15 October 2015. Almost all Radde's Warblers in Europe have occurred in October (*Anthony McGeehan*).

► Pallas's Warbler *Phylloscopus proregulus*, Grutness, Shetland, Scotland, 14 October 2018, one of the most frequently extralimitally recorded leaf warblers (*Rebecca Nason*).

for example (García Vargas & Sagardía 2014). Some Yellow-browed Warblers may even reach the Afrotropics, with small numbers recorded from the Cap Blanc peninsula in Mauritania (Barcelona *et al.* 2017) and single records from The Gambia (Barlow 2007) and Senegal (Cruse 2004). The species is also becoming more regular in North America away from the Bering Sea region, with records from Wisconsin and California in autumn 2019, and in March 2007 one was discovered at the southern tip of Baja California in Mexico (Mlodinow & Radamaker 2007). Hume's Warbler *Phylloscopus humei* has a very different breeding and wintering range, moving south from central Siberia to winter in the Indian subcontinent and parts of South-east Asia, yet it is also an expected late autumn vagrant to western Europe, albeit rarer than Yellow-browed Warbler; the first national records for Iceland, Cyprus, Bulgaria, Switzerland and Portugal have occurred recently. It is also recorded as a vagrant east to Japan and Korea (Brazil 2009).

The limits of our understanding of vagrancy have been tested by some *Phylloscopus* warblers occurring at either end of the Palearctic. Pale-legged Leaf Warbler *Phylloscopus tenellipes* has now been confirmed as occurring in Europe by DNA, after one died after colliding with the lighthouse on St Agnes, Isles of Scilly, in October 2016 (Headon *et al.* 2018); its demise narrowly prevented what would have been an epic mass twitch to these famous isles had the bird been found alive. This species shares a breeding range in the Far East alongside Eastern Crowned Warbler *Phylloscopus coronatus,* which has now occurred more than a dozen times in Europe, from Sweden, Belgium, France, Netherlands, Germany, Finland, Norway and Britain. The first two records of Eastern Crowned Warbler from Britain were both initially misidentified as Yellow-browed Warblers, which suggests that records may have been overlooked in the past. In the west, the restricted-range Iberian Chiffchaff *Phylloscopus ibericus* is becoming an expected spring overshoot to Britain and elsewhere in western Europe; there have also been records from Greece in June 2010 (Ławicki & van den Berg 2017a) and Lithuania in May 2017 (Ławicki & van den Berg 2018b). Less expected is an amazing but as yet unsubmitted record of one photographed and sound-recorded in late March 2016 at Saya, Musandam in Oman. Incredibly, Iberian Chiffchaff bred extralimitally in the Nedd Valley, south Wales, in 2015 (Hunter 2018). Other competitors in this genus for the spot of 'most unlikely vagrant' include Sulphur-bellied Warbler *Phylloscopus griseolus* in May–June 2016 at Christiansø in Denmark; Plain Leaf Warbler *Phylloscopus neglectus* at Landsort, Sweden, in October 1991; and the only Arabian record of Large-billed Leaf Warbler *Phylloscopus magnirostris* from the United Arab Emirates in October 2014 (Smiles 2016). These remarkable occurrences underline the seemingly

◄ Eastern Crowned Warbler *Phylloscopus coronatus*, Bempton, East Yorkshire, England, 5 October 2016, the fourth record for Britain (*Graham Catley*).

almost limitless potential for vagrancy among the leaf warblers. Tickell's Leaf Warbler *Phylloscopus affinis* has occurred as a vagrant east to Japan and Korea (Brazil 2009) and might also be expected as a westward vagrant; Tytler's Leaf Warbler *Phylloscopus tytleri* also seems a likely vagrant at least to the Middle East. Yellow-streaked Warbler *Phylloscopus armandii* has long been touted as a potential vagrant to the Western Palearctic but identification difficulties mean that this species, and potentially others, go overlooked.

SCOTOCERCIDAE

Bush Warblers and allies

The bush warblers include a mix of migratory and resident species. Most have relatively short wings compared to other Old World warblers, but this does not appear to be a major barrier for vagrancy in some species. Asian Stubtail *Urosphena squameiceps* is not only relatively short-winged but is also very short-tailed, yet this species is wholly migratory, moving from breeding grounds as far north as Ussuriland in south-east Russia to winter south to the Malay Peninsula. At least one individual has long overshot these wintering grounds to reach the vagrant trap of Ashmore Reef, Australia, in April 2012 (Clarke *et al.* 2016) and this species has also occurred west to India (Das 2014) and Bangladesh (Chowdhury 2014), and east to the Philippines (Jensen *et al.* 2015). It would seem a good candidate to appear in the Bering Sea region in future, or even the Western Palearctic. Manchurian Bush Warbler *Horornis canturians* has a similar migration and has appeared as a vagrant in Meghalaya in north-east India, to Sabah (Eaton *et al.* 2016) and at Khurkh Bird Ringing Station in Mongolia in May 2018 (Davaasuren *et al.* 2020). Long-distance movements in the *Horornis* bush warblers in prehistory facilitated their colonisation of remote Pacific islands like Palau and Fiji where Palau Bush Warbler *Horornis annae* and Fiji Bush Warbler *Horornis ruficapilla* occur, respectively.

Several species in the group undertake altitudinal movements and, given the skulking nature of many species, vagrants are probably massively under-recorded. Ringing expeditions to wetlands in Bangladesh have recorded Chestnut-crowned Bush Warbler *Cettia major*, Grey-sided Bush Warbler *Cettia brunnifrons* and Aberrant Bush Warbler *Horornis flavolivaceus*, all of which were national firsts (Round *et al.* 2014) but which may be regular there. Perhaps also in this category is a Slaty-bellied Tesia *Tesia olivea* record from Bangladesh (Thompson & Johnson 2003). Far more surprising was a record of the apparently resident Yellow-bellied Warbler *Abroscopus supercili-*

► Asian Stubtail *Urosphena squameiceps*, Ashmore Reef, Australia, 24 April 2012, the first Australian record of this diminutive long-distance migrant (*Rohan Clarke*).

◄ Manchurian Bush Warbler *Horornis canturians*, Hegurajima Island, Ishikawa, Japan, 16 April 2018, a scarce passage migrant to Japan (*Yann Muzika*).

aris from Odisha in eastern India in May 2009; this location is 650km out of range and perhaps suggests some form of migratory movement in this species (Singh & Panda 2014). Cetti's Warbler *Cettia cetti* is the sole representative of the group in the Western Palearctic – western populations are largely resident, whilst eastern ones are migratory. The species has spread north in Europe in the 20th century, with the vanguard of vagrants first recorded in southern England in 1961 where the species is now common (Brown & Grice 2005). Elsewhere in Europe, it is still a vagrant to Sweden, Germany and Poland.

AEGITHALIDAE
Long-tailed Tits

◄ 'Northern' Long-tailed Tit *Aegithalos caudatus caudatus*, Dymchurch, Kent, England, 30 January 2011. A rare vagrant to western Europe, though occurrences sometimes involve small flocks (*James Lowen*).

Most members in this group are sedentary, although a few species like White-browed Tit-Warbler *Leptopoecile sophiae* make altitudinal movements. The main exception to this rule of sedentarism is the northern nominate race of Long-tailed Tit *Aegithalos caudatus,* which undertakes regular north–south movements in some parts of Siberia whilst other populations may undertake irruptive movements, even as far west as Scandinavia (Bojarinova *et al.* 2016). This white-headed subspecies is a vagrant to western Europe, including Britain, Belgium and the Netherlands (Jansen & Nap 2008). The British and Irish subspecies *rosaceus* is also an occasional vagrant to offshore islands such as the Isles of Scilly, Outer Hebrides, Orkney and Shetland. Long-tailed Tit has also occurred once as a vagrant to Morocco (Thévenot *et al.* 2003) and Tunisia (Harrap & Quinn 1996). In the Americas, Bushtits *Psaltriparus minimus* are highly sedentary but have been recorded marginally out of range in central Kansas (AOU 1998).

SYLVIIDAE
Sylviid Warblers, Parrotbills and allies

This large family contains 12 different genera. In addition to the *Sylvia* warblers familiar to Palearctic and Afrotropical birders there is also a diverse mix of other genera, most of which are distributed in the tropics and subtropics. Most species are sedentary, and vagrancy is unlikely, or at least unrecorded. For example, there are essentially no records of vagrants in the single New World representative of the group – Wrentit *Chamaea fasciata* of western North America (Geupel & Ballard 2020). Species that reach further into the temperate zone seem to be more prone to vagrancy with a record, for example, of Vinous-throated Parrotbill *Sinosuthora webbiana* from Japan (Brazil 2009), but the only group within this family that includes many migratory species and regular vagrants is the genus *Sylvia* itself. This group of warblers includes long intercontinental migrants such as Garden Warbler *Sylvia borin* and Barred Warbler *Sylvia nisoria,* which may move over 6,000km between breeding grounds in Europe and their African winter quarters. At the other extreme are sedentary species such as Layard's Warbler *Sylvia layardi* of southern Africa, which despite being fully resident has been recorded at least 300km out of range in KwaZulu-Natal (Aymí & Gargallo 2020). Barred Warblers are scarce passage migrants to western Europe, subject to periodic influxes of almost exclusively

▲ Eurasian Blackcap *Sylvia atricapilla* with White-backed Mousebirds *Colius colius* and a Cape White-eye *Zosterops virens*, Durbanville, Western Cape, South Africa, 20 July 2017. This was the 26th record for southern Africa and the first for the Western Cape (*Niall Perrins*).

first-winter birds in autumn (e.g. 286 were recorded in Britain in 2002). Vagrants have reached as far west as the Azores, Senegal, The Gambia and Nigeria in West Africa (Borrow & Demey 2014), south all the way to South Africa and east as far as India, Pakistan and, amazingly, South Korea, where a second-calendar-year bird was trapped in Jeju province in April 2013 (Chou *et al.* 2013). Eurasian Blackcap *Sylvia atricapilla* is largely confined to the Western Palearctic as a breeding species and has a shorter migration span than Barred Warbler, but has occurred as a vagrant as far afield as Greenland (Boertmann 1994), South Africa and China (Guo & Ma 2013). A record in unusual circumstances from Cuba in October 2012 (Rodríguez *et al.* 2017) was dismissed by Kirwan *et al.* (2019) given the improbability of vagrancy – and the fact that first New World records of Common Kingfisher and White-winged Snowfinch *Montifringilla nivalis* were reported at the same time. The only accepted record of a *Sylvia* warbler from the New World is the Lesser Whitethroat *Sylvia curruca* that reached St Lawrence Island in the Bering Sea in September 2002 (Howell *et al.* 2014), although a record of one on Corvo in the Azores in October 2018 suggests that substantial sea crossings are within the physiological capacity of this species (Ławicki & van den Berg 2018c).

Assemblages of *Sylvia* warblers are most diverse in the Mediterranean Basin. Most of these are migratory but many move over relatively short distances. Most have been recorded as vagrants to northern Europe, with members of the 'Subalpine Warbler' *Sylvia cantillans* group recorded most frequently; all three members of this group have been recorded in Britain with seemingly increasing regularity. Other strongly migratory species with broadly similar distributions – for example, Western Orphean Warbler *Sylvia hortensis*

◄ Eastern Subalpine Warbler *Sylvia cantillans albistriata*, Mid Yell, Shetland, Scotland, 1 October 2013 (*Rik Addison*).

◄ Ménétries's Warbler *Sylvia mystacea*, International Birding and Research Center Eilat, HaDarom, Israel, 23 March 2011, a rare passage migrant through Israel (*Yoav Perlman*).

► African Desert Warbler *Sylvia deserti*, Alphen aan den Rijn, South Holland, Netherlands, 30 November 2014, the first for the Netherlands and northern Europe (*Vincent Legrand*).

and Eastern Orphean Warbler *Sylvia crassirostris* – are, however, incredibly rare as vagrants in northern Europe. They are far less frequent as vagrants than Sardinian Warbler *Sylvia melanocephala,* which is a much shorter-distance partial migrant. The latter has been widely recorded as a vagrant, reaching as far as Norway, Sweden and Rybachy Biological Station in Russia (Bulyuk & Leoke 2010). Several eastern European and/or Middle Eastern *Sylvia* warblers are heavily coveted and extremely rare vagrants to western Europe, including Ménétries's Warbler *Sylvia mystacea*. This species has occurred recently for the first time in Italy and Greece, and one was trapped in northern Portugal in September 1967 (Lewington *et al.* 1991), hinting at the likelihood of vagrancy to northern Europe. Tristram's Warbler *Sylvia deserticola* has also reached both France and Spain and may be expected in northern Europe before too long.

The two species of 'desert warblers', Asian Desert Warbler *Sylvia nana* of the Middle East and Central Asia and African Desert Warbler *Sylvia deserti* of North Africa, would seem like unlikely vagrants to Europe. However, Asian Desert Warbler has occurred with remarkable frequency in northwest Europe, with 19 records from Sweden, 13 from Britain and 12 from Finland, as well as to Spain and France, for example; late spring birds in Britain and Germany have even built nests. Meanwhile, African Desert Warbler, which is less migratory and considerably rarer, has recently reached northern Europe in the form of a bird found in November 2014 in Zuid-Holland (van den Berg 2014). Marmora's Warbler *Sylvia sarda* is an extremely rare vagrant to Europe away from its breeding grounds in Corsica and Sardinia, but spring migrants returning from North African wintering grounds do sometimes overshoot, and there are now seven records from Britain and two from Denmark. The essentially non-migratory Dartford Warbler *Sylvia undata* is apparently far less dispersive – for example, this species has never been recorded in Denmark, despite breeding as close as southern England.

ZOSTEROPIDAE
White-eyes, Yuhinas and allies

Most species of white-eyes are resident, but four taxa are long-distance migrants: the Chestnut-flanked White-eye *Zosterops erythropleurus*; both the nominate and *simplex* subspecies of the Warbling White-eye *Zosterops japonicus*; and the nominate subspecies of the Silvereye *Zosterops lateralis*. White-eyes often move in quite large flocks, a tendency that facilitated colonisation of New Zealand from Australia in the 1850s. Such events must have occurred prior to human settlement but it seems likely that human modification of the landscape facilitated the establishment of the white-eyes (Clout & Lowe 2000). Subsequently, Silvereyes have reached other remote New Zealand satellite islands as vagrants, such as Macquarie (Copson & Brothers 2008). Elsewhere in Australasia, there is a record of Australian Yellow White-eye *Zosterops luteus* from Sawtell in New South Wales, Australia (Chapman 2009), which is a significant southward incursion by this species which also seems to show some nomadic tendencies outside the breeding season.

Chestnut-flanked White-eyes have the longest migration span of any member of the complex, between 3,000–5,000km from the Russian Far East to Cambodia. Given this migration span, records of vagrants are surprisingly few. Two photographed in April 2018 in the Mishmi Hills of India were a first record for the country and South Asia (Lobo *et al.* 2018), but the species routinely winters to adjacent Myanmar so this record was anticipated. Likewise, the species is a vagrant to islands in the Sea of Japan, which lie only a few hundred kilometres east of the species' migration flightpath (Brazil 2009). A record from the Burgas Valley in Mongolia in September 2012 is more extralimital and significant (Baatargal & Suuri 2013), clearly representing a bird on a westward trajectory. Chestnut-flanked White-eye has occurred several times in western Europe, including at migration hotspots in spring and autumn, but all records predate the European Union ban on the importation of wild birds, and the species was known to be common in the trade at the time; there were also records of escaped Silvereyes in Europe during the same period (Letharby 1998). Genuine vagrancy to the Western Palearctic would appear a possibility, however, given overlap in Chestnut-flanked White-eye's migration routes with those of other far-eastern vagrants.

TIMALIIDAE, PELLORNEIDAE AND LEIOTHRICHIDAE

Tree-Babblers, Scimitar-Babblers and allies, Ground Babblers and allies, and Laughingthrushes and allies

These three families were formerly included in the Timaliidae and overwhelmingly consist of highly sedentary species; altitudinal movements have been reported but often not confirmed in a minority of montane species. We are unaware of any significant vagrancy, but range extensions are common and ascertaining what is a vagrant and what is an undiscovered population is complicated in some cases.

REGULIDAE

Goldcrests and Kinglets

Northern populations of crests and kinglets are migratory across the Holarctic and these migratory tendencies have facilitated colonisation of oceanic islands by endemic forms. Huge falls of Goldcrests *Regulus regulus* are a feature of late autumn in western Europe and this species has occurred as a vagrant south to Morocco (Thévenot *et al.* 2003) and Iraq (Ararat 2016). Common Firecrest *Regulus ignicapilla* has a more southerly breeding distribution and is consequently less migratory. Despite breeding quite widely in southern England, it is still a rare vagrant to Scotland and is less frequent there than many Siberian warblers breeding thousands of kilometres away. In the Nearctic, Golden-crowned Kinglet *Regulus satrapa* is the least migratory of the two kinglets but has occurred in the Bering Sea region as a vagrant, as well as on Bermuda and on ships in the Atlantic (Durand 1963). It would seem an unlikely candidate for transatlantic vagrancy given its body size and life history, yet Ruby-crowned Kinglet *Regulus calendula* has now occurred five times as a vagrant to the Western Palearctic, twice to Iceland (Haas 2012) and once each to Ireland (O'Donnell 2013), Scotland and the Azores. In the Eastern Palearctic there is a single record from Wrangel Island, Russia, in September 1986 (Arkhipov & Ławicki 2016) and a record from Tobishma, Japan, in October 2018. In the Neotropics the species has ventured as far south as Belize (eBird).

► Golden-crowned KInglet *Regulus satrapa*, St Paul Island, Pribilof Islands, Alaska, United States, 1 October 2017, a rarity out into the Bering Sea (*Tom Johnson*).

▼ Goldcrest *Regulus regulus*, Lón, Austurland, Iceland, 7 October 2015, a regular autumn visitor which colonised Iceland as a breeding species during the 1990s (*Yann Kolbeinsson*).

TICHODROMIDAE
Wallcreeper

Wallcreeper *Tichodroma muraria* is a short-distance altitudinal migrant across its wide but narrow range in the mountainous 'mid-latitude belt' of the Eurasian landmass. A few vagrant individuals in Europe have strayed far from the mountains and in some cases have shown site fidelity in multiple winters to buildings, quarries and sea cliffs in the lowlands of France, Germany, Holland, Belgium and England, where they are among the most sought-after of all vagrants. Elsewhere they have appeared extralimitally across north-west Africa and the Middle East, and most recently in Vietnam and Thailand (Harrap & Quinn 1996).

◄ Wallcreeper *Tichodroma muraria*, Ban Thung Quarry, Chiang Rai, Thailand, 18 March 2015, the second record for Thailand (*Wich'yanan Limparungpatthanakij*).

SITTIDAE
Nuthatches

Most nuthatches are extremely sedentary, and instances of vagrancy are consequently rather rare in the family. Notable exceptions concern Red-breasted Nuthatch *Sitta canadensis* and northern populations of Eurasian Nuthatch *Sitta europaea,* which are both irruptive. In some years the predominantly boreal- and montane-breeding Red-breasted Nuthatch may reach as far south as the Gulf Coast and northern Florida. Vagrants have reached Socorro Island and Baja California in western Mexico (Delgado-Fernández & Delgadillo-Nuño 2017), Bermuda on three occasions and, more impressively, twice as transatlantic vagrants – once to Iceland and once to Britain (Haas 2012). Presumably these latter occurrences may have involved a degree of ship-assistance. White-breasted Nuthatch *Sitta carolinensis*, a more strictly resident species, has also been recorded aboard an eastbound ocean liner six hours out of New York (Durand 1963). Eurasian Nuthatch irruptions are less frequent than those of Red-breasted Nuthatch but may involve very

◄ Red-breasted Nuthatch *Sitta canadensis* Chelam Bank, Western Aleutians, United States, 11 September 2006. The first record for the Aleutians came aboard a birder's cruise to Attu (*Marshall Iliff*).

large numbers of birds; in October 1980, 30,000–40,000 were recorded moving south-west in groups of 40–50 birds along the coast of the Sea of Japan in Ussuriland, Russia (Harrap *et al.* 2020). Vagrants of the *asiatica* group of subspecies from Siberia may sometimes occur in Scandinavia west as far as Norway; these movements are triggered by the failure of the Siberian stone pine *Pinus sibirica* crop, which Spotted Nutcrackers *Nucifraga caryocatactes* also depend on. Other populations of Eurasian Nuthatch are very sedentary – the species has yet to be recorded from Ireland, for example, despite breeding commonly in neighbouring Wales, although it has reached the Isle of Man as a vagrant (Harrap & Quinn 1996).

The remaining nuthatches are also very sedentary, but vagrancy events are still well documented. Pygmy Nuthatch *Sitta pygmaea* of western North America has been recorded extralimitally widely in the central United States and Canada, from British Columbia, Montana, North Dakota, eastern Nebraska, Kansas and Iowa. Similarly, there is a scattering of northerly records of the southeastern Brown-headed Nuthatch *Sitta pusilla* from Illinois, Indiana, Michigan, New York, Ohio and Wisconsin. Renfrow (2003) collated these records, spanning nearly 200 years, and suggested that extralimital occurrences of the species corresponded to time periods with episodes of severe deforestation within their normal range.

CERTHIIDAE

Treecreepers

Northern populations of both Eurasian Treecreeper *Certhia familiaris* and Brown Creeper *Certhia americana* are migratory, possibly erratically so, as is Bar-tailed Treecreeper *Certhia himalayana*. Eurasian Treecreepers of the nominate race are vagrants to the Netherlands as well as to the Faroes and the British Isles, which have their own resident race *britannica* (Harrap & Quinn 1996). The more southerly distributed Short-toed Treecreeper *Certhia brachydactyla* is very sedentary; despite breeding just across the English Channel, it is a very rare vagrant to southern England, albeit probably overlooked, and most records come from Dungeness Bird Observatory in Kent, an isolated migrant trap (Brown & Grice 2005). In North America, Brown Creepers are more migratory, and vagrants have even reached Bermuda, a substantial overwater flight, as well as Middleton Island in the Gulf of Alaska. In the tropics African Spotted Creeper *Salpornis salvadori* has been frequently reported 'extralimitally', with a handful of records from, e.g. Senegal, The Gambia, Botswana and north-east South Africa (del Hoyo *et al.* 2020b), but whether these represent vagrants or range extensions is unknown.

► 'Northern' Eurasian Treecreeper *Certhia familiaris familiaris*, Halligarth, Unst, Shetland, Scotland, 7 October 2014, a rare vagrant to the Northern Isles of Scotland from Scandinavia (*Paul French*).

POLIOPTILIDAE
Gnatcatchers

Most gnatcatchers are sedentary. Only northern populations of Blue-grey Gnatcatcher *Polioptila caerulea* are migratory and this is the only species to have occurred as a vagrant significantly out of range. Blue-grey Gnatcatcher underwent a major northward range expansion in the 20th century, and an analysis of vagrant records by Ellison (1993) found that increases in spring vagrants to New England were a better indicator of subsequent range expansion than autumn vagrants. This species has yet to be recorded from Bermuda, but vagrants have straggled all the way north to Hudson Bay in Canada and to Hispaniola and the Cayman Islands in the West Indies (Kirwan *et al.* 2019). Incursions of Black-capped Gnatcatcher *Polioptila nigriceps* into Arizona in the United States are only narrowly extralimital.

TROGLODYTIDAE
Wrens

Most wren species are non-migratory and are found in the New World tropics or subtropics; extralimital occurrences among these are rare, but vagrancy is sometimes associated with species undergoing range expansions, explaining the recent run of records of Sinaloa Wren *Thryophilus sinaloa* from Arizona (Howell *et al.* 2014) and Thrush-like Wrens *Campylorhynchus turdinus* across South America (Hayes *et al.* 2018), for example. The most migratory wrens occur in north temperate latitudes, where House Wren *Troglodytes aedon* of the *aedon* subspecies group is almost entirely migratory. Vagrants have reached Bermuda and north to Hudson Bay and Newfoundland, whilst House Wrens of the southern group have occurred

▲ Sinaloa Wren *Thryophilus sinaloa*, Santa Cruz, Arizona, United States, 2 January 2018. First recorded in Arizona in 2008, this species may yet colonise the state (*Tom Johnson*).

▲ House Wren *Troglodytes aedon*, Cape Pembroke, Falklands, 12 April 2014 – an impressive transoceanic vagrancy event, though ship-assistance is perhaps likely (*Alan Henry*).

as vagrants to the Falklands, with several birds also recorded alighting on ships at sea in the region (Woods 2017). Northern populations of Sedge Wren *Cistothorus platensis* are also wholly migratory, with vagrants recorded widely in western North America and north-east to Nova Scotia. Marsh Wrens *Cistothorus palustris* on the other hand have more complicated patterns of migration, but there also more reports of transoceanic movements in this species: twice as vagrants to Bermuda and a hypothetical record from Cuba (Kirwan *et al.* 2019). In this context, the record of a specimen collected from western Greenland in October 1820 seems plausible (Boertmann 1994).

The Rock Wren *Salpinctes obsoletus* of western North America is also partially migratory, and this species has been widely recorded as a vagrant in eastern North America north to Newfoundland and Nova Scotia (Brewer 2010). This species has also bred extralimitally around Churchill, Manitoba, on the southern edge of Hudson Bay; Seutin & Chartier (1989) suggested that these birds, and indeed other vagrant Rock Wrens, might have reached the region trapped accidentally in railway boxcars, a theory supported by anecdotal observations of these wrens becoming trapped in enclosed spaces. Warning (2016) carried out a quantitative appraisal of the 'boxcar theory' for Rock Wren records and those of other western vagrants and found no evidence of the clustering of records around railway infrastructure. Canyon Wrens *Catherpes mexicanus* share a geographically similar distribution but are generally sedentary and consequently there are far fewer reports of vagrants, although a few have straggled east to the Great Plains region in, for example, Nebraska, Kansas and North Dakota (Brewer 2010).

ELACHURIDAE
Elachura

The sole member of this family – Spotted Elachura *Elachura formosa* – is non-migratory and there are no extralimital reports.

CINCLIDAE
Dippers

The only long-distance migrations undertaken by dippers are those made by the nominate subspecies of White-throated Dipper *Cinclus cinclus*, known to British birders as 'Black-bellied Dipper', which migrate south to central and western Europe in winter. Small numbers reach the United Kingdom almost annually, and vagrants have even reached the Faroe Islands twice in the 19th century (Bloch & Sørensen 1984). American Dippers *Cinclus mexicanus* are largely non-migratory, although some make altitudinal movements and there are a few records of narrowly extralimital movements, including an old record from Attu in the Aleutian Islands during the 1800s (Gabrielson & Lincoln 1959); this raises the intriguing possibility of vagrancy to north-east Russia.

◄ 'Black-bellied' Dipper *Cinclus cinclus cinclus*, Harpham, East Yorkshire, England, 21 February 2015. The nominate subspecies of White-throated Dipper is an annual vagrant to eastern England (*Graham Catley*).

BUPHAGIDAE
Oxpeckers

Both species of oxpeckers are resident but there are periodic records of long-range vagrancy, including Red-billed Oxpecker *Buphagus erythrorynchus* records from Western Cape in South Africa (Craig 2009).

STURNIDAE
Starlings

Few starlings are truly migratory and entirely vacate their breeding grounds in winter – some are partial migrants, some are nomadic but most are sedentary. The most extensive movements in the group are undertaken by Rose-coloured Starling *Pastor roseus,* which is a nomadic breeder across a wide swathe of the central Palearctic from Hungary to western Mongolia, wintering mostly in the Indian subcontinent. The species' nomadism is well illustrated by the recovery of an individual ringed in Hungary and later recovered 4,800km away in western Pakistan just 10 months later (Ali & Ripley 1978). Rose-coloured Starlings are regular vagrants to western Europe and during these westward shifts may breed extralimitally. For example, 6,000–7,000 pairs are estimated to have nested on the castle in Villafranca di Verona in northern Italy in 1875 (Voous 1960). During these incursions, multiples may arrive in European countries bordering the North Atlantic, and vagrants have reached Iceland, Madeira and the Canary Islands (Lewington *et al.* 1991); the first for the Azores was recorded in June 2020, associated with a huge incursion into western Europe. To the south, there are old records in North Africa from Algeria and Tunisia and a more recent first for Morocco in February 2007 (Bergier *et al.* 2011). These are in turn eclipsed by sub-Saharan African records from the Maasai Mara in Kenya in July 2003 (Lindsell & Fisher 2009), Ethiopia in March 2005 (Schollaert 2006) and South Africa in July 2005 (Sagvik 2009). In the Indian Ocean there are several records of Rose-coloured Starlings from the Seychelles (Skerrett *et al.* 2006), the Maldives (Anderson 2007) and the Chagos Islands (Carr 2015). To the east of its breeding and wintering range, Rose-coloured Starlings have reached Cambodia (CBGA 2019), Thailand and Singapore (Robson 2002) as well as Japan, South Korea (Brazil 2009), the Russian Far East (Antonov *et al.* 2012) and the Philippines (Jensen *et al.* 2015). They have reached Sabah, Borneo, three times (Eaton *et al.* 2016), whilst two recent records

► Rose-coloured Starling *Pastor roseus*, Inishbofin, Galway, Ireland, 11 July 2011 (*Anthony McGeehan*).

◄ Daurian Starling *Agropsar sturninus*, Satchari National Park, Sylhet, Bangladesh, 5 March 2019, the fourth for Bangladesh (*Sayam Chowdhury*).

from central Java and Bali in November 2013 and April 2014 respectively are hypothesised to have involved the same individual, possibly an escape from captivity, as captive Rose-coloured Starlings were photographed in a Yogyakarta bird market in April 2014 (Diniarsih *et al.* 2016). Finally, this species has made it all the way to Australia, with records from the Cocos (Keeling) Islands, Broome in Western Australia and Coffs Harbour in New South Wales (Menkhorst *et al.* 2017). Given this spread of records, which encompasses most of the Old World, Rose-coloured Starling would seem a likely vagrant to North America too.

Daurian (Purple-backed) Starling *Agropsar sturninus* is a fully migratory East Asian species, which has also been widely reported as a vagrant. In western Europe two records from the Netherlands and Norway have been accepted as genuine vagrants (Haas 2012), whereas others from Britain and elsewhere have been suspected escapes from the burgeoning trade in Asian passerines in the 1980s and 1990s. Elsewhere, Daurian Starlings are vagrants to India, Bangladesh and Nepal (Basnet & Chaudhary 2003), away from the Andaman Islands where they are regular (Sharma & Sangha 2012). This species is also a vagrant east to Japan (Brazil 2009) and the Philippines (Jensen *et al.* 2015) and has overshot its regular wintering areas in Indonesia to reach Christmas Island and the Cocos (Keeling) Islands of Australia (James & McAllan 2014). Several other East Asian migratory starlings are also frequently reported as vagrants: Chestnut-cheeked Starling *Agropsar philippensis* has occurred south to Ashmore Reef, Australia, and Bali (Kusumanegara & Iqbal 2019) and also to Thailand, Malaysia, Singapore and Cambodia (Mahood 2016), as well as east to Palau in Micronesia (Owen 1977); White-cheeked Starlings *Spodiopsar cineraceus* have wandered north to Ussuriland in the Russian Far East (Antonov *et al.* 2012) and west as far as India (Hatibaruah *et al.* 2017). There are also two North American records of White-cheeked Starling, from Alaska in June 1998 and British Columbia in April 2016; both were assumed to have been ship-assisted and were not accepted onto the North American list (Pyle *et al.* 2018). Elsewhere in the Palearctic, the widely distributed Common Starling *Sturnus vulgaris* has penetrated the Afrotropics as a vagrant to Mauritania, Cape Verde and The Gambia, the last named perhaps as a ship-assisted vagrant (Borrow & Demey 2014); it has also occurred east as a vagrant to Japan and Korea (Brazil 2009) and to the Philippines (Jensen *et al.* 2015). The relatively range-restricted Spotless Starling *Sturnus unicolor* is regarded as highly sedentary, but in addition to wandering to the Atlantic islands of the Canaries and Madeira (Clarke 2006) has also been recorded more recently in northern Europe – in Denmark in 2009 (Olsen *et al.* 2008) and Switzerland in 2011 (Tosoni & Piot 2014).

Few Afrotropical species have been recorded as vagrants with any regularity, except for Wattled Starling *Creatophora cinerea*, which is a nomadic East African insectivore that may move in association with swarms of locusts or moths. This species has occurred as a vagrant to West Africa, to the Central African Republic, Gabon, Congo, Ghana and even as far west as The Gambia (Borrow & Demey 2014). Equally impressive is vagrancy to the Seychelles, 1,500km from the African mainland (Skerrett *et al.*

► Chestnut-cheeked Starling *Agropsar philippensis*, Satchari National Park, Sylhet, Bangladesh, 5 March 2019, the first record for Bangladesh (*Sayam Chowdhury*).

2006), Europa Island in the southern Mozambique Channel (Corre & Probst 1997) and Madagascar (Langrand & Sinclair 1994). Wattled Starling is a regular visitor to the southern tip of the Arabian Peninsula and a vagrant to the United Arab Emirates (Porter & Aspinall 2010); this species would seem like a potential vagrant to Egypt or elsewhere in the Middle East. Another candidate for vagrancy to the Western Palearctic is Chestnut-bellied Starling *Lamprotornis pulcher,* which ranges widely in Mauritania in the non-breeding season and will probably eventually occur in Western Sahara.

MIMIDAE
Mockingbirds and Thrashers

Most members of the family Mimidae are resident, but species breeding at high latitudes in both North and South America are migratory. Grey Catbird *Dumetella carolinensis* is highly migratory in the north of its range and is a regular vagrant to western North America and south as far as Colombia and Venezuela. This species has occurred as a transatlantic vagrant on 10 occasions with records from Germany, Belgium, Britain, Ireland, the Azores and the Canary Islands. One catbird remained aboard a ship which travelled from New York to Southampton in England before continuing through the Mediterranean in October 1998 (Haas 2012). Given that most mimids have short wings and seem physiologically ill-suited for long overwater flights of several days, it seems likely that the three European records of the largely resident Northern Mockingbird *Mimus polyglottos* may also have been at least partially ship-assisted. Durand (1972), recorded mimids on ships at sea in the Atlantic on eight occasions in autumn. Several European records have also been rejected as escapes from captivity (Cobb *et al.* 1996). Within North America, Northern Mockingbird has wandered as far as St Lawrence Island and St Paul Island in the Bering Sea and Hudson Bay. Brown Thrashers *Toxostoma rufum* are also partially migratory with a broad spread of vagrants in North America and to the Bahamas and Cuba (Kirwan *et al.* 2019). There is a single record outside the Americas at Durlston Head, Dorset, England, from November 1966 to February 1967, where again some form of ship assistance seems likely.

Among the other North American thrashers, Sage Thrasher *Oreoscoptes montanus* is wholly migratory, vacating its western breeding areas for Mexican deserts, and it has been recorded widely as a vagrant to the north, south and east of its breeding range. Bendire's Thrasher *Toxostoma bendirei* performs a more modest migration, and vagrants are fairly regular in

▲ Sage Thrasher *Oreoscoptes montanus*, Southeast Farallon Island, San Francisco, California, United States, 17 September 2008 (*Matt Brady*).

◄ Patagonian Mockingbird *Mimus patagonicus*, Shallow Harbour, Falklands, one of six recorded on the Falklands (*Alan Henry*).

coastal southern California, whilst the remaining *Toxostoma* species are sedentary but still occasionally reported as vagrants. In the southern hemisphere, two species – Patagonian Mockingbird *Mimus patagonicus* and White-banded Mockingbird *Mimus triurus* – are both partial migrants in the south of their range. Patagonian Mockingbird has occurred at least six times on the Falklands (Woods 2017). These islands also have three records of White-banded Mockingbird in addition to two other birds known to have been ship-assisted there together in October 2010: Shiny Cowbird *Molothrus bonariensis* and Crowned Slaty Flycatcher *Empidonomus aurantioatrocristatus* (Woods 2017). White-banded Mockingbird is a fairly regular, if scarce, non-breeding visitor both to Chile (Jaramillo *et al.* 2003) and Brazil, where vagrants have reached as far north as the southern Brazilian Amazon at Aripuanã in Mato Grosso state in September 2013 (Veronese 2013) and once to the Peruvian Amazon at Puerto Maldonado in August 2014 (Cieza & Díaz 2014).

Tropical and subtropical species tend to be non-migratory but still prone to wander, albeit with the usual caveats of provenance, with escapes from captivity and ship-assistance often invoked. Tropical Mockingbird *Mimus gilvus* in particular seems to be reported extralimitally with surprisingly frequency; there were no major questions raised over the provenance of the first Peruvian records in Iquitos in August 2017(Cuelo Pizarro 2018), but two records from the southern United States, in Texas and Florida (Kratter 2018), were not accepted as pertaining to wild birds owing to possible captive origin. A Tropical Mockingbird was recorded on Gibraltar in February–March 2012 and it or another was at Algeciras, Spain, in December 2012 (Copete *et al.* 2015), presumably having been ship-assisted to Europe. Deforestation has allowed this species to significantly increase its range and population size within the Neotropics, so increasing numbers of naturally dispersing individuals may be expected. The resident Brown-backed Mockingbird *Mimus dorsalis* of Bolivia and Argentina has recently been recorded as a vagrant to Peru (Ugarte *et al.* 2014) whilst Bahama Mockingbird *Mimus gundlachii* is an occasional vagrant to Florida, where it has even hybridised with Northern Mockingbird. Finally, there is a scatter of Blue Mockingbird *Melanotis caerulescens* records in the southern United States, from California, Arizona, New Mexico and Texas; this species is not known to undertake any seasonal movements, but these records suggest that perhaps the species moves altitudinally, which might drive these occasional extralimital movements (Howell *et al.* 2014).

TURDIDAE
Thrushes and allies

Thrushes include some of the most migratory of all passerines, as well as some of the most sedentary, and in between these extremes there are also many partial migrants and altitudinal migrants. This continuum of movements is well illustrated in the *Myadestes* solitaires. The most migratory of these is Townsend's Solitaire *Myadestes townsendi* of western North America, which has occurred widely as a vagrant in the eastern half of the continent and even out into the Bering Sea on St Lawrence Island (Bowen 2020). Meanwhile, Brown-backed Solitaire *Myadestes occidentalis* is an extreme vagrant to the United States, with two records from south-east Arizona (Howell *et al.* 2014); this species is an altitudinal migrant. It was however Mesoamerican solitaires such as this species, not Townsend's Solitaire, that gave rise to both the endemic solitaires on Hawaii and in the West Indies (Miller *et al.* 2007). The largely resident Eastern Bluebird *Sialia sialis* and Western Bluebird *Sialia mexicana* are not known for widespread vagrancy, but the more migratory Mountain Bluebird *Sialia currucoides* has been widely recorded from eastern North America (Rusk *et al.* 2013) and has even been recorded as a vagrant to Guadalupe Island off Mexico (Barton *et al.* 2004).

► Townsend's Solitaire *Myadestes townsendi*, 4 February 2018, Demarest Lloyd SP, Bristol, Massachusetts, United States (*Marshall Iliff*).

▲ Mountain Bluebird *Sialia currucoides* (right) with Eastern Bluebirds *Sialia sialis*, Tohickon Valley Park, Bucks, Pennsylvania, United States, 4 January 2016 (*Cameron Rutt*).

The highly migratory *Catharus* thrushes of North America evolved from largely sedentary tropical species (Outlaw *et al.* 2003), many of which undertake altitudinal migrations and have occasionally been reported extralimitally. There is a single record of Black-headed Nightingale-Thrush *Catharus mexicanus* and two records of Orange-billed Nightingale-Thrush *Catharus aurantiirostris* from southern Texas; these species breed only a few hundred kilometres away in Mexico. Far more noteworthy is a record of an Orange-billed Nightingale-Thrush in Lawrence County, South Dakota, in July–August 2010, well over 2,000km out of range; a subsequent record from McKinley County, New Mexico, in July 2015 indicates that such incursions may be more regular, but they are still spectacular for an ostensibly sedentary Neotropical passerine. The migratory *Catharus* thrushes and Wood Thrush *Hylocichla mustelina* form the largest group of thrushes that migrate to the tropics. Most other migratory thrushes move from north to south temperate latitudes rather than to the tropics. All of these species, with the exception of Bicknell's Thrush *Catharus bicknelli*, have been recorded as vagrants to the Old World, with Wood Thrush the rarest, recorded only five times – once each from Britain and Iceland and three times on the Azores. Grey-cheeked Thrush *Catharus minimus* is the most frequent, with 65 records from Britain alone, including a fall of eight on the Isles of Scilly, England, in late October 1986. Grey-cheeked Thrush winters in South America, probably mostly in the western Amazon, whilst Veery *Catharus fuscescens* winters even further south in central Brazil and has occurred as a vagrant south to Chile (Jaramillo *et al.* 2003). It is thus somewhat surprising that there are just 14 records of Veery in the Western Palearctic, 11 from Britain, although this may be partially explained by global population sizes, with Grey-cheeked thrush outnumbering Veery 3.7:1 (Rosenberg *et al.* 2019). Swainson's Thrush *Catharus ustulatus* is the commonest of the group in global terms but has a more western Andean wintering range, which may mean its migration route is less likely to take it on a riskier North Atlantic crossing than the preceding species. Grey-cheeked Thrushes extend their breeding range into the Russian Far East, where there are also three records of Hermit Thrush *Catharus guttatus* and one of Swainson's Thrush (Arkhipov & Ławicki 2016); there is also a single record of Grey-cheeked Thrush from Japan on Hegura-Jima in October 2004 (Brazil 2009).

► Grey-cheeked Thrush *Catharus minimus*, Gugh Bar, St Agnes, Isles of Scilly, 30 October 2002. The Isles of Scilly remain the best location for transatlantic vagrants of this species in Europe, and they are often rather atypically confiding, such as this one foraging on the strandline (*Chris Batty*).

Most of the *Zoothera* and *Geokichla* thrushes are sedentary denizens of lowland and montane forests, but two species – White's Thrush *Zoothera aurea* and Siberian Thrush *Geokichla sibirica* – are highly migratory. White's Thrush is an annual vagrant in small numbers to western Europe, mostly to Britain, where there are just under 100 records and also to other countries including Spain, Italy and Malta; the species has also been recorded in Oman (Eriksen & Porter 2017). White's Thrush has strayed out into the Atlantic to Iceland on five occasions and even reached Greenland in October 1954 (Boertmann 1994). A singing White Thrush at Kalajoki, Finland, in June 2009 was only the fifth national record and may portend a westward range expansion in this species. Siberian Thrush is considerably rarer, with most records emanating from vagrant traps in Britain and Norway; this species has also overshot its wintering grounds in South-east Asia to reach Ashmore Reef in Australia (Rohan *et al.* 2016). The remaining members of this group are rarely reported as vagrants, one exception being records of the Himalayan altitudinal migrant Long-billed Thrush *Zoothera monticola* from Bangladesh (Thompson & Johnson 2003). Varied Thrushes *Ixoreus naevius* are partial migrants, with northernmost populations leaving the boreal forests in winter; this species is also irruptive with periodic incursions into central and eastern North America and even two records from Europe – one in Cornwall, England, in November 1982 and one in Iceland in May 2004 (Haas 2012), as well as to Wrangel Island in Russia (Arkhipov & Ławicki 2016). Aztec Thrushes *Ridgwayia pinicola* also seem to be partially migratory in northern Mexico and the species is an occasional vagrant to Arizona and Texas (Howell *et al.* 2014).

The genus *Turdus* has a near-cosmopolitan distribution globally; this was achieved in prehistory by a tendency towards long-distance vagrancy, a trend that continues to the present day. In examining the phylogeny of *Turdus* thrushes, Batista *et al.* (2020) found evidence for a single transatlantic dispersal event by thrushes from the Old World via the Caribbean which gave rise to the New World *Turdus* species. There are no contemporary records of Eurasian or Afrotropical thrushes from the Caribbean region, but in December 2001 a Redwing *Turdus iliacus* was found dead aboard a seismic research vessel operating 150km of the coast of Espírito Santo in south-east Brazil (Brito *et al.* 2013). This is the sole record of transatlantic vagrancy by thrushes in the Southern Hemisphere, but such ocean crossings are relatively routine in the Northern Hemisphere. Redwings have become regular vagrants to northeastern North America, along with the rarer Fieldfare *Turdus pilaris,* and both species have also occurred on the Pacific seaboard in Alaska and also Washington state in the case of Redwing. There are single accepted North American records of Eurasian Blackbird *Turdus merula* from Newfoundland and Song Thrush *Turdus philomelos* from Quebec (Howell *et al.* 2014); less expected was a Mistle Thrush *Turdus viscivorus* that spent December 2017–March 2018 in New Brunswick, Canada (Pyle *et al.* 2018). These thrushes may arrive in eastern North America via staging grounds in Iceland or Greenland. In addition to the White's Thrush, Greenland has been visited by Grey-cheeked Thrush, Hermit Thrush, Eurasian Blackbird, Fieldfare, Redwing, Song Thrush and American Robin *Turdus migratorius,* with the Turdidae accounting for a disproportionate proportion of the passerine vagrants to the island. Mistle Thrush has yet to occur there but has occurred east as a vagrant to Japan (Brazil 2003), and Fieldfare was recorded for the first time in the Indian subcontinent in the Darjeeling district of West Bengal in March 2011 (Banerjee & Inskipp 2013). There are now multiple records of Ring Ouzel *Turdus torquatus* from the Azores since 2009 (Alfrey *et al.* 2010), raising the possibility of a successful transatlantic crossing in that species. American Robins are partial migrants and are almost annual vagrants to Europe. Most records are from Britain although they have also reached central and eastern Europe, including six from Germany and three from Austria (Lewington *et al.* 1991), perhaps indicating that vagrant arrivals on the Atlantic seaboard become entrained in flocks of other *Turdus* thrushes and pulled deeper into Europe.

Eyebrowed Thrush *Turdus obscurus* is the most migratory of the genus, moving from breeding grounds as far north as central and eastern Siberia to winter as far south as Indonesia and the Philippines. The global spread of vagrant records of this species rivals that of any other landbird. Eyebrowed Thrushes have reached the northwestern Hawaiian Islands on two occasions – Midway in 1998 and Kure in 2013 (Pyle & Pyle 2017) – and in North America are uncommon migrants to the western Aleutians, where 180 were recorded on Attu Island on 17 May 1998; they are much rarer on the Pribilofs and Bering Sea islands (Howell *et al.* 2014). There is also a single record of a male at Galileo Hill Park, a desert oasis migrant trap inland in Kern County, California, in May 2001, a site which has also hosted an Arctic/Kamchatka Leaf Warbler. Eyebrowed Thrushes are vagrants to Micronesia, south-west India, Sri Lanka and the Maldives (Clement & Hathway 2000) and have also reached Ashmore Reef off northern Australia. In the Middle East there are records from Oman, UAE and Israel, whilst the species has been widely recorded in Europe, without the same degree of location biases associated with many Siberian vagrants. For instance, there are 24 records from Britain, three from Spain, eight from the Netherlands, nine from France and Poland and six from Finland and Norway. Finally, there are two African records: a first-winter male photographed at Merzouga in Morocco in December 2008 (Messemaker 2009) and, incredibly, one photographed in the Parc National de la Langue de Barbarie in north-west Senegal in December 2015 (Benjumea & Pérez 2016), representing a minimum 10,000km out-of-range movement.

Several other Asian thrushes are vagrants to the Western Palearctic, and in the case of Dusky Thrush *Turdus eunomus* to western North America (Howell *et al.* 2014) as well as the Azores. None, however, occur with anything like the frequency of several Siberian warblers that share similar ranges. Black-throated Thrush *Turdus atrogularis* is perhaps the closest to breaking that rule, a species that normally moves directly south from Siberian breeding grounds to winter around the Indian subcontinent but sometimes irrupts further west to winter in the Middle East – in large numbers in some winters (Clement & Hathway 2000). Such incursions may produce flocks on the doorstep of Europe – with 30 at Gevaş, Van, in Turkey in November 2019, a winter that produced several European records. In addition to being a fairly regular vagrant to the west, Black-throated Thrushes also occur east to Korea and Japan (Brazil 2009). Surely the most surprising record of a vagrant *Turdus* thrush in Eurasia is the adult male Tickell's Thrush

◄ Mistle Thrush *Turdus viscivorus*, Manno, Kagawa, Japan, 10 March 2020. First recorded in Japan in 1984, there are still fewer than a dozen records (*Yann Muzika*).

► Dusky Thrush *Turdus eunomus*, Beeley, Derbyshire, England, 17 December 2016, an extremely popular vagrant to the middle of England in winter (*Alexander Lees*).

► Clay-coloured Thrush *Turdus grayi*, Bentsen-Rio Grande Valley State Park, Hidalgo, Texas, United States, 18 February 2011 (*Tom Johnson*).

Turdus unicolor collected on Heligoland, Germany, in October 1932 (Haas 2012); this relatively short-distance Himalayan migrant has also occurred as a vagrant to Bangladesh. Given that the individual was an adult with some aberrant plumage features, the record does perhaps need revisiting (C. Batty *in litt.*).

There are few records of significant vagrancy by Afrotropical thrushes, although the temperate zone Song Thrush has occurred as a vagrant to the Afrotropics, with records south to Senegal and Chad in the west (Borrow & Demey 2014) and to Djibouti and Ethiopia in the east (Clement & Hathway 2000). There is more evidence of movement in the Neotropics, however: several species are vagrants to the southern United States with Rufous-backed Thrush *Turdus rufopalliatus* recorded widely west all the way to California, whilst White-throated Thrush *Turdus assimilis* and Clay-coloured Thrush *Turdus grayi* are considerably rarer – all three species are largely sedentary (Howell *et al.* 2014). There are also now three records of Red-legged Thrush *Turdus plumbeus* in Florida, these having wandered from the Caribbean. In southern South America, White-necked Thrush *Turdus albicollis* has wandered south to the vagrant trap of Punta Rasa in Argentina (Jaramillo 2000), whilst the highly migratory Creamy-bellied Thrush *Turdus amaurochalinus* has reached Chile once (Jaramillo *et al.* 2003) and the Falklands four times (Woods 2017). Finally, Austral Thrush *Turdus falcklandii* is one of a handful of passerines to have reached Antarctica; a single bird was recorded on King George Island in the South Shetland Islands in September 2002 (Santos *et al.* 2007).

MUSCICAPIDAE
Chats and Old World Flycatchers

The largely Old World chats and flycatchers in the Muscicapidae are a large radiation of 49 genera of insectivorous birds which were previously placed in several other families. Most temperate zone species are migratory, with some moving as far as the tropics. A few species have established bridgeheads in the Bering Sea region of the Nearctic, with both Bluethroat *Luscinia svecica* and Northern Wheatear *Oenanthe oenanthe* breeding in Alaska. The Nearctic wheatears are arguably the champion passerine migrants - Alaskan breeders move 14,500km across the Palearctic to winter quarters in East Africa, whilst wheatears breeding in the eastern Canadian Arctic undertake a 3,500km transoceanic crossing of the North Atlantic en route to wintering areas in West Africa (Bairlein *et al.* 2012). Given this epic migration and broad Holarctic distribution, it shouldn't come as a surprise that some wheatears get their navigation wrong, and the species is a vagrant to the East Asian seaboard of Korea and Japan (Brazil 2009) as well as to India and as far south as South Africa (Borrett & Jackson 1970). Northern Wheatears have been widely recorded across North America as well as in Mexico, Panama (Angehr & Dean 2010), the Caribbean (Kirwan *et al.* 2019) and even twice in French Guiana (Renaudier & Comité d'Homologation de Guyane 2010, Rufray *et al.* 2019). Other wheatear species are also frequent vagrants, although the one species to have reached the New World was a second-year female Pied Wheatear *Oenanthe pleschanka* found on St Paul Island in the Pribilofs in July–August 2017 (Gibson *et al.* 2018). This species, which breeds across a large swathe of Asia and winters in East Africa, has also occurred as a vagrant to Japan and Korea, as well as India, South Africa and the Seychelles (Skerrett *et al.* 2006); it is also a relatively routine vagrant to western Europe, especially in autumn. Isabelline Wheatear *Oenanthe isabellina* has a broadly similar range and migration route and is also a regular, and apparently increasing, vagrant to western Europe, with a major influx in autumn 2016 associated with prolonged easterly winds, coinciding with the arrival of lots of other eastern passerines. Isabelline Wheatears are also vagrants to Japan (Brazil 2009) and out into the Pacific to Palau (Otobed *et al.* 2018) and south to Australia, where one appeared in the Atherton Tablelands of north Queensland in November 2002. Desert Wheatears *Oenanthe deserti* are predictable late autumn vagrants east to Japan and Korea and west to the Atlantic seaboard of Europe, with a few

▼ Northern Wheatear *Oenanthe oenanthe*, with Lark Bunting *Calamospiza melanocorys*, Villa Jesús María, Baja California, Mexico, 24 October 2008. Up to 12 Red-throated Pipits *Anthus cervinus* and a 'Siberian' Buff-bellied Pipit *Anthus rubescens japonicus* were present at this site on the same day (*Steve Howell*).

▲ Desert Wheatear *Oenanthe deserti*, Winterton, Norfolk, England, 6 December 2014, an almost annual vagrant to this stretch of coast (*James Lowen*).

► Seebohm's Wheatear *Oenanthe oenanthe seebohmi*, Solleveld, Den Haag ZH, Netherlands, 22 May 2017, the first record for the Netherlands and northern Europe of this distinctive North African member of the Northern Wheatear group (*Gerjon Gelling*).

overshooting their wintering grounds and reaching as far south as Kenya (Shema & Njoroge 2018). The relative frequency of these species in northern Europe is in stark contrast to the status of the migratory Black-eared Wheatear *Oenanthe hispanica,* which breeds in relatively close proximity within the Mediterranean, but remains a very rare vagrant north of its breeding range, perhaps because the migration span of this species is considerably shorter and its migration route has less of an east–west component (see page 28).

Most of the remaining wheatear species undertake less dramatic migrations and are less frequently reported as vagrants. White-crowned Wheatear *Oenanthe leucopyga* is interesting in this context, as it has been reported several times extralimitally in the Middle East, east to Iran (Rahimi, *et al.* 2019); more controversially, it has been reported in northern Europe, where several birds have been accepted as wild whilst others have proven to be escapes. Feathers were obtained from a White-crowned Wheatear that moved from Denmark to Germany in the summer of 2010 and stable isotope analysis revealed that its deuterium values matched those of specimens from North Africa, suggesting a natural origin (Förschler *et al.* 2018). This species has also been accepted as a wild vagrant in Britain, the Netherlands and Poland. Three other unexpected wheatear vagrancy events necessitate a mention, including a record of Variable Wheatear *Oenanthe picata* in Israel in 1986 (Haas 2012). More mind-boggling are two western European

◄ Pied Bushchat *Saxicola caprata*, Yeruham Lake, HaDarom, Israel, 18 October 2012, the seventh record for Israel and 12th for the Western Palearctic (*Yoav Perlman*).

records: a Kurdish Wheatear *Oenanthe xanthoprymna* photographed in May 2015 at the top of the Puy de Dôme, in Auvergne, France, was the first for western Europe; and a Cyprus Wheatear *Oenanthe cypriaca* was collected on the island of Heligoland, Germany, in May 1867, whose remains were only correctly identified 140 years later (Förschler *et al.* 2010).

The *Saxicola* chats have a habit of swapping hemispheres with regularity. Siberian Stonechat *Saxicola maurus* has reached the New World on several occasions, with most records coming from St Lawrence Island in the Bering Sea. There is also a record from San Clemente Island in California – which has also hosted a Red-flanked Bluetail *Tarsiger cyanurus* – and from Grand Manan Island in New Brunswick, on the Atlantic coast of Canada, in October 1983 (Howell *et al.* 2014). Whinchats *Saxicola rubetra* are vagrants east to Sri Lanka (Steiof *et al.* 2016) and Japan (Brazil 2009). Pied Bushchats *Saxicola caprata* are migratory in the north-west of their range and are increasingly detected as vagrants to the Western Palearctic, with records from Finland (two), Sweden, Italy and Cyprus, for example, as well as occurring east to Japan (Brazil 2009). Many *Phoenicurus* redstarts perform even longer migrations. Common Redstarts *Phoenicurus phoenicurus,* which breed east as far as eastern Mongolia and winter in Africa, have reached both Japan (Brazil 2009) and Alaska (Gibson 2018). Vagrants have also drifted out to the Seychelles. Daurian Redstart *Phoenicurus auroreus* has a chequered history as a vagrant to the Western Palearctic, with European records all deemed to be escapes, occurring at a time when many Siberian passerines were imported into Europe (Parkin & Shaw 1994). A record from Russia in September 2006 on the Upper Pechora is the only accepted regional record (Haas 2012). October records of White-capped Redstart *Phoenicurus leucocephalus* from Germany in 1992 and White-winged (Güldenstädt's) Redstart *Phoenicurus erythrogastrus* from Finland in 1993 were also judged more likely to be escapes (Anon 1995). Plumbeous Redstart *Phoenicurus fuliginosus* has also crept onto the Western Palearctic list in Russia where one was photographed in October 2017 near the Black Sea coast of Crimea. Long-range movements in these relatively sedentary taxa would seem highly unusual, yet Moussier's Redstart *Phoenicurus moussieri* of North Africa has been recorded in Spain, Portugal, France, Italy, Malta and once as far north as Dinas Head in Wales in April 1988.

A record of Blue Rock Thrush *Monticola solitarius* from British Columbia in June 1997 has not been accepted as a genuine vagrant (Howell *et al.* 2014) and some other extralimital records of this species from northern Europe have seemed likely to be escapes; others, however, have had good enough credentials to convince rarities committees. This species has even reached Australia where there are records from Western Australia and Queensland. Common Rock Thrush *Monticola saxatilis* is more migratory and would seem a likely vagrant to the Bering Sea region, with records east to Japan (Brazil 2009). White-throated Rock Thrush *Monticola gularis* is another long-distance migrant that has occurred as a vagrant to Japan; it might be expected as a future vagrant to Alaska or even Europe.

All of the *Larvivora* robins are migratory and include some of the most sought-after rarities; the migrant traps of St Lawrence and Attu islands have each had a Siberian Blue Robin *Larvivora cyane,* but more unexpected was one at Dawson City in the

► Rufous-backed (Eversmann's) Redstart *Phoenicurus erythronotus*, Al Abraq, Kuwait, 3 January 2020, the 16th record for Kuwait (*Paul Dufour*).

► Common Rock Thrush *Monticola saxatilis*, Blaenavon, Gwent, Wales, 29 October 2017. A surprisingly rare vagrant to northern Europe, given the proximity of its breeding range – though this pattern matches a general trend of reduced vagrancy among species that follow a predominantly north-south migration axis (*Rik Addison*).

Yukon in June 2002 (Howell *et al.* 2014). This species was recently first recorded in Australia, predictably at Ashmore Reef in April 2012 (Clarke *et al.* 2016), with recent first records from Bangladesh too (Mohsanin *et al.* 2012). There are six Western Palearctic records of this species and six of Rufous-tailed Robin *Larvivora sibilans*, all of which have come in October. Rufous-tailed Robin has now also occurred twice in the Bering Sea region. Patterns of vagrancy in the spectacular Siberian Rubythroat *Calliope calliope* mirror these two robins, albeit with far more records from western Europe, including 17 from Britain alone, and these records extend west to Iceland. One was found wintering in the Netherlands in January 2016, mirroring the only winter record from North America – one found at the west end of Lake Ontario in Canada in December 1983, the same season that produced the aforementioned Siberian Stonechat in New Brunswick (Howell *et al.* 2014).

European Robin *Erithacus rubecula* is a coveted vagrant to Japan, Korea and China (Brazil 2009) and has also recently reached eastern North America, where one was recorded in the appropriately named North Wales in Bucks County, Pennsylvania, in February–March 2015 (Pyle *et al.* 2018). Another was photographed in Broward County, Florida, in October 2014 – right next to Port Everglades (eBird). Vagrants have also strayed south of the temperate zone to Cape Verde, Mauritania and The Gambia (Borrow & Demey 2014). It would be remiss not to mention Red-flanked Bluetail *Tarsiger cyanurus*, which has undergone a spectacular change in status in western

Europe, from being a near-mythical vagrant in the 1980s to losing its title as an official rarity in Britain by 2016. This change has accompanied a westward range expansion into Finland and there has also been an upsurge in records in western North America.

Two large genera of flycatchers – *Muscicapa* and *Ficedula* – include many long-distance migrants and are very frequent vagrants. Four *Muscicapa* flycatchers – Grey-streaked Flycatcher *Muscicapa griseisticta*, Dark-sided Flycatcher *Muscicapa sibirica*, Asian Brown Flycatcher *Muscicapa dauurica* and Spotted Flycatcher *Muscicapa striata* – have reached Alaska. Of these, the single record of Spotted Flycatcher is most unusual given the species' western distribution, but it has also occurred in Japan (Brazil 2009). The three eastern species have also reached northern Australia, and Grey-streaked Flycatcher – which has the longest migration span in the group – winters regularly south to New Guinea and has reached the Northern Mariana Islands in the Pacific. These Far Eastern flycatchers remain surprisingly rare in the Western Palearctic, with just seven records of Brown Flycatcher and a single record of Dark-sided Flycatcher at Höfn, Iceland, in October 2012 (Brynjólfsson *et al.* 2020). The prize for the most unlikely flycatcher record is still held by the Dark-sided Flycatcher on Bermuda in September 1980, which was at least 9,000km out of range (Wingate 1983). However, a Brown-streaked Flycatcher *Muscicapa williamsoni* at Broome, Western Australia, in October–November 2020 (eBird) was also highly unanticipated given that it is a relatively short-distance migrant.

The black-and-white *Ficedula* flycatchers of the Palearctic are often recorded as vagrants, notwithstanding some considerable identification challenges. European Pied Flycatcher *Ficedula hypoleuca* has the broadest range and vagrants have reached Japan and China in the east (Brazil 2009). Collared Flycatcher *Ficedula albicollis* is a vagrant to north-west Europe and all the way south to South Africa. Three species – Mugimaki Flycatcher *Ficedula mugimaki*, Narcissus

▲ Rufous-tailed Robin *Larvivora sibilans*, Fair Isle, Shetland, Scotland, 23 October 2004, the first for the Western Palearctic (*Rebecca Nason*).

◄ Siberian Rubythroat *Calliope calliope*, Beukenlaan, Noord-Holland, Netherlands, 4 March 2016, the first for the Netherlands (*Richard Bonser*).

► Verditer Flycatcher *Eumyias thalassinus*, Wakimisakimachi, Nagasaki-Shi, Nagasaki, Japan, 14 January 2020. This partially-migratory species from southern Asia is an extremely rare vagrant to southern Japan (*Yann Muzika*).

► Fujian Niltava *Niltava davidi*, Khao Yai NP, Nakhon Ratchasima, Thailand, 3 December 2018. This species is apparently a vagrant to Thailand, although it may be an under-recorded regular migrant (*Wich'yanan Limparungpatthanakij*).

Flycatcher *Ficedula narcissina* and Taiga Flycatcher *Ficedula albicilla* – have occurred in Alaska, and the last-named has even reached California (Howell *et al.* 2014). Taiga Flycatcher is being detected with increasing frequency in Europe, whilst there are two recent records of Mugimaki Flycatcher from Italy and Norway and an older, controversial rejected record from Britain; despite appearing in apparently good circumstances for a natural vagrant, the British record occurred at a time when escapes of Asian passerines were rife, and the species was being traded regularly in the region. Records of Narcissus Flycatcher from Australia, Micronesia and Alaska speak to the vagrancy potential of this species. Green-backed Flycatcher *Ficedula elisae* is a long-distance migrant and has strayed to Japan, while noteworthy vagrancy events from other genera include records of Hill Blue Flycatcher *Cyornis banyumas* from Korea (Brazil 2009) and Blue-and-white Flycatcher *Cyanoptila cyanomelana* from Australia.

Movements among tropical species are fairly limited but are most marked among species of arid environments. Northern Anteater-Chat *Myrmecocichla aethiops* has wandered north over the Western Palearctic boundary at Tibesti in Chad (Haas 2012), while Sooty Chat *Myrmecocichla nigra* has been recorded west to Senegal (Borrow & Demey 2014). Red-capped Robin-Chat *Cossypha natalensis* is a partial migrant with a few records of vagrants in the Eastern Cape in South Africa (Demey 2011).

BOMBYCILLIDAE
Waxwings

The three species of waxwings have divergent patterns of movement, varying from regular migration to erratic irruptive movements, but all move south of their summer breeding ranges and all are prone to long-distance dispersal. Cedar Waxwings *Bombycilla cedrorum* undertake among the most significant movements in the group and regularly winter as far south as Panama. Vagrants have penetrated into northern South America in Venezuela (Hilty 2003), Colombia (Calderón-F. & Agudelo 2014) and the Galapagos (Wiedenfeld 2006). Transatlantic vagrancy has occurred several times, with records from Britain, Ireland, Iceland and the Azores (Haas 2012). Intriguingly, one individual was discovered in Nottingham, England, in February 1996 associated with hundreds of Bohemian Waxwings *Bombycilla garrulus* during an irruption year for this species, prompting thoughts that some of the Bohemian Waxwings might have originated in the Nearctic. Some support for this possibility is provided by a record of three Bohemian Waxwings found on Bermuda in December 2001 (an island where Cedar Waxwings are fairly regular visitors); there have also been four records from Greenland (Boertmann 1994) and records from the Azores (Barcelos *et al.* 2015). During irruption years, Bohemian Waxwings may reach as far south as the southern United States, North Africa, India and Taiwan. Movements may conceivably occur across the entire Holarctic, evidenced by a Bohemian Waxwing ringed one winter in Poland which was recovered the next winter 5,500km away in eastern Siberia (Cornwallis & Townsend 1968). This record is particularly relevant in understanding the vagrancy potential of Japanese Waxwing *Bombycilla japonica,* which is usually considered to be a relatively short-distance migrant. Records of this species from Scotland and the Netherlands that predated the 2007 EU ban on the importation of wild birds have been treated as escapes. However, a first-winter male Japanese Waxwing with Bohemian Waxwings at Białystok, Poland, in January 2009 (Ławicki 2013) seems to have good credentials to be treated as a genuine vagrant. Also supportive of vagrancy is a record of another first-winter bird trapped with Bohemian Waxwings in Ili-Alatau National Park, Kazakhstan, in January 2013 (Wassink 2013) and a first record from the Philippines at Basco, Batanes, in March 2013 (Jensen *et al.* 2015).

▼ Cedar Waxwing *Bombycilla cedrorum*, Inishbofin, Galway, Ireland, 14 October 2009, the first record for Ireland and fifth for the Western Palearctic (*Anthony McGeehan*).

MOHOIDAE
Hawaiian Honeyeaters

The Kioea *Chaetoptila angustipluma* was an extinct Hawaiian endemic and extralimital movements would have been unlikely.

PTILIOGONATIDAE
Silky-flycatchers

Silky-flycatchers do not undertake any long-distance migratory movements. Some species make modest altitudinal movements and Phainopepla *Phainopepla nitens* moves up to a few hundred kilometres between the Sonoran Desert and neighbouring semi-arid woodlands. Incredibly, vagrants have reached as far as the eastern seaboard of the United States, to Rhode Island and Massachusetts as well as Wisconsin and southern Ontario in Canada (Chu & Walsberg 2020). Grey Silky-flycatcher *Ptiliogonys cinereus* is partially migratory in the north of its range and there are two accepted records from southern Texas, along with five others from southern California of uncertain status (Howell *et al.* 2014).

► Phainopepla *Phainopepla nitens*, 29 November 2009, Peel, Ontario, Canada. This, the second record for Ontario, arrived on 9 November, associated with the same weather system that brought two other southwestern species – a Sulphur-bellied Flycatcher *Myiodynastes luteiventris* and an Ash-throated Flycatcher *Myiarchus cinerascens* – into the province. The Phainopepla was taken into care on 9 February but subsequently died (*Jay McGowan*).

DULIDAE
Palmchat

Palmchat *Dulus dominicus* is resident on Hispaniola and has not been reliably reported away from the island (Kirwan *et al.* 2019).

HYLOCITREIDAE
Hylocitrea

The unique Hylocitrea *Hylocitrea bonensis* is assumed to be resident on Sulawesi, and extralimital records are not anticipated.

HYPOCOLIIDAE
Hypocolius

Hypocolius *Hypocolius ampelinus* undertakes short-distance migratory movements to wintering areas south of its breeding grounds in the Middle East. The species has been recorded quite extensively as a vagrant north and west to Turkey and Israel (Kirwan *et al.* 2010), south to Oman, Yemen, Egypt, Sudan and Eritrea, and east to western India where it is fairly regular in Gujarat (Tiwari 2008) and a vagrant south to Maharashtra (Katkar *et al.* 2021).

▲ Hypocolius *Hypocolius ampelinus*, HaDarom, Israel, 30 November 2011 (*Yoav Perlman*).

PROMEROPIDAE
Sugarbirds

Both sugarbirds are largely resident but may undertake local movements, and there are records of extralimital Cape Sugarbird *Promerops cafer* from central South Africa. These are thought to be a result of drought conditions in the Eastern Cape region (Cheke & Mann 2001).

MODULATRICIDAE
Dapple-throat and allies

The three African montane forest specialists in the Modulatricidae family are all largely resident and vagrancy is undocumented in the group.

DICAEIDAE
Flowerpeckers

Most flowerpeckers are assumed to be sedentary but odd instances of apparent vagrancy have been recorded in various species. For example, Blood-breasted Flowerpecker *Dicaeum sanguinolentum* has been recorded on Sumatra and Yellow-bellied Flowerpecker *Dicaeum melanozanthum* in Bangladesh; the latter species apparently makes altitudinal movements (Cheke & Mann 2001).

NECTARINIIDAE
Sunbirds

Around half of all sunbirds are thought to be highly sedentary, but the remaining 60 or so species undertake some form of complex movement; many movements are altitudinal, and others involve intratropical migrations following the rains and flower availability. Extralimital occurrences associated with these movements are, however, apparently fairly rare. Nile Valley Sunbird *Hedydipna metallica,* for example, apparently undertakes movements in the Red Sea region and was recently recorded for the first time in Chad (Demey 2017) but has never occurred in well-watched Israel, only around 300km east of the Nile Valley. Purple Sunbird *Cinnyris asiaticus* on the other hand has now occurred as a vagrant to Kuwait on several occasions

► Purple Sunbird *Cinnyris asiaticus*, Ras Al Subiyah, Kuwait, 13 January 2008, the first record for the Western Palearctic (*Chris Batty*).

(Haas 2012), possibly associated with a range expansion in Iran. Pygmy Sunbird *Hedydipna platura* is a partial migrant in Mauritania and might be a potential vagrant to the Western Palearctic, in Western Sahara or northern Mauritania. Some significant but relatively short-distance vagrancy events include periodic westward incursions by Mouse-coloured Sunbirds *Cyanomitra verreauxii* inland in South Africa and Tanzania; records of Variable Sunbird *Cinnyris venustus* from South Africa; and a record of Scarlet-tufted Sunbird *Nectarinia johnstoni* from the North Pare Mountains, 60km from the nearest known population on Kilimanjaro (Cheke & Mann 2001). This last occurrence highlights the potential importance of occasional long-distance dispersal events in tropical montane birds. In Asia, Fire-tailed Sunbird *Aethopyga ignicauda* undertakes altitudinal movements and has been recorded as a vagrant to Bangladesh (Thompson *et al.* 2014), Myanmar (Bezuijen *et al.* 2010) and Thailand (Round & Kunsorn 2009).

IRENIDAE
Fairy-bluebirds

Both members of this family are sedentary or presumed to be sedentary and there are no extralimital reports. The wide distribution of Asian Fairy-bluebird *Irena puella* suggests that this species does, however, disperse across narrow waterbodies with some regularity.

CHLOROPSEIDAE
Leafbirds

Leafbirds are usually assumed to be sedentary, although Orange-bellied Leafbird *Chloropsis hardwickii* at least engages in altitudinal movements. There are no reports of significant vagrancy in the group.

PEUCEDRAMIDAE
Olive Warbler

Northernmost populations of Olive Warbler *Peucedramus taeniatus* are migratory but there have been surprisingly few reports of vagrants, although they are reported with some regularity in west Texas (Lowther & Nocedal 2012). This is perhaps a likely eventual addition to the California state list.

UROCYNCHRAMIDAE
Przevalski's Pinktail

Przevalski's Pinktail (formerly Pink-tailed Rosefinch) *Urocynchramus pylzowi* is considered to be sedentary and we are unaware of any extralimital reports.

PLOCEIDAE
Weavers and allies

The members of the Ploceidae are largely non-migratory, although some dry-country species undertake movements in response to rains like many granivorous birds. For example, large-scale movements are well-documented in Red-billed Quelea *Quelea quelea* which follow rain fronts, and flocks of this species have even been recorded 100 km off the coast of Senegal (Quantrill 2017). Ringing recoveries indicate that Chestnut Weavers *Ploceus rubiginosus* have been recorded making movements of up to 500 km (Oschadleus 2011) and this species has also been recorded as a vagrant as far south as South Africa (Whittington 2011). A male Village Weaver *Ploceus cucullatus* at Abu Simbel in May 2006 was only a few hundred kilometres out of range but constituted the first Western Palearctic record (el Din *et al.* 2012).

ESTRILDIDAE
Waxbills and allies

Most of the estrildid finches are resident, especially forest associated species, although some with semi-arid seasonal habitats make at least local movements in accordance with the rains and abundance of seeding grasses, whilst others make altitudinal movements. There are a few reports of significant vagrancy: Locust Finches *Paludipasser locustella* have been reported out of range in Cameroon and Nigeria whilst Pale-headed Munia *Lonchura pallida* has occurred on Ashmore Reef, Australia (Forshaw & Shephard 2012), and the nomadic Plum-headed Finch *Neochmia modesta* has wandered to the north-east of South Australia (Dennis 2012). Members of this family are very common as cagebirds and there are naturalised populations of many species across the world, which may in turn generate vagrant individuals. For example, Tricolored Munia *Lonchura malacca* was recently recorded as a vagrant to Florida from feral populations in the West Indies (Pyle 2020).

VIDUIDAE
Indigobirds and Whydahs

No viduid finches are thought to undertake any significant movements, although some species like Broad-tailed Paradise-Whydah *Vidua obtusa* may be somewhat nomadic. Specimens of at least three individuals collected on the Transvaal Drakensberg escarpment in South Africa in February 1961 may represent vagrants (Payne 1967) as the species is otherwise virtually unknown in South Africa.

PRUNELLIDAE
Accentors

▲ Black-throated Accentor *Prunella atrogularis*, Pori, Satakunta, Finland, 6 December 2010 (*Stuart Piner*).

Among the 12 species of accentors, most are resident or undertake altitudinal movements, but several northern species are wholly or partially migratory. Siberian Accentors *Prunella montanella* perform the longest movements between breeding areas in northern Russia and wintering areas in Korea, eastern China and occasionally Japan (Brazil 2009). Vagrants of this species are almost annual on the Alaskan Bering Sea islands and have also occurred south to British Columbia in Canada and Washington state, Oregon, Idaho and Montana in the United States (Howell *et al.* 2014). Siberian Accentors also reach north-west Europe but are rare. Before 2015, there had been 32 individuals, mostly in Finland and Sweden. In October–November 2016 there was an unprecedented influx, involving at least 231 individuals, including first records for Britain, the Netherlands, Germany, Estonia and Hungary (Stoddart 2018). This influx was concurrent with a major westward displacement of other Siberian passerines to Europe, during an extended period of easterly winds that also coincided with major fires in Siberia, potentially impacting migratory behaviour and in particular the capacity of migrants to navigate using the stars as a compass (Sikora & Ławicki 2019, page 13). The remarkable numbers of Siberian Accentors involved in the event also probably reflected a bumper year's breeding success for the species.

The nominate subspecies of Black-throated Accentor *Prunella atrogularis* is also migratory and has also occurred as a vagrant to several countries in western Europe. The majority of records again come from Sweden and Finland, but the species has also reached France, Croatia, Hungary, Slovakia, Croatia and Italy. Further east, vagrants have reached Turkey, Israel, Kuwait and Oman (Alström 1991, Ergen & Barış 2016) as well as the Qinling Mountains of central China (Wang *et al.* 2011). Northern populations of Dunnock *Prunella modularis* are migratory and vagrants have reached Iceland, Svalbard and one each on the Canary Islands (Tenerife) in January 2002 (Clarke 2006) and Madeira in July 2006 (Romano *et al.* 2010). Alpine Accentor *Prunella collaris* is an altitudinal migrant to lower elevations in winter, and vagrants occur irregularly north to Britain, the Netherlands, Belgium, Sweden, Finland and Estonia.

PASSERIDAE
Old World Sparrows and Snowfinches

Most sparrows and snowfinches are resident, although a few species are partial migrants, some are altitudinal migrants and a few dry-country species are nomadic. One of the most vagrant-prone of all sparrows is Spanish Sparrow *Passer hispaniolensis,* with populations ranging from resident to fully migratory. Spanish Sparrows reach north-west Europe occasionally as vagrants north to Britain and Denmark, with the Netherlands claiming about half of all regional records. This includes a flock of 10 (four males and six females) at Eemshaven, Groningen, the Netherlands in April 2009 (Ebels *et al.* 2015). Several records have been in proximity to ports or major shipping lanes, suggesting that ship-assistance may be an important vector for vagrancy in this species. This is certainly the case for other sparrows; perhaps the most celebrated instance concerns a flock of four Cape Verde Sparrows *Passer iagoensis* that arrived at Hansweert, Zeeland, in the Netherlands in May 2013 (van den Berg & Haas 2013b). They were part of an original flock of 11 which landed on Zodiac boats used by birders watching Raso Larks *Alauda razae* close to the shore of the island of Razo in the Cape Verde islands and ended up spending 12 days on board the MV *Plancius* all the way back to the Netherlands. It's possible that some of these individuals may later have contributed their genes into the regional House Sparrow *Passer domesticus* gene pool. House Sparrows are also prone to ship-assisted vagrancy and have reached Japan (Brazil 20009) and even the Prince Edward Islands (Burger *et al.* 1980) by this means. These were trumped by a juvenile House Sparrow which successfully travelled from Los Angeles to Sydney aboard a plane in July 1981 but was summarily euthanised by the Australian Quarantine & Inspection Authorities on arrival in Sydney.

Sudan Golden Sparrows *Passer luteus* are nomadic rains migrants and have now proven to be regular visitors to the fringes of the Western Palearctic (Fareh *et al.* 2019). Birds of unknown origin have recently reached Fuerteventura on the Canary Islands (Gil-Velasco *et al.* 2018) and Sal on Cape Verde (Ławicki & van den Berg 2016c). Given this nomadism, natural or ship-assisted vagrancy seems quite likely. There is a single rather extraordinary record of the largely resident Arabian Golden Sparrow *Passer euchlorus* from near Eilat, Israel, in August 2016, but the record was not accepted as an equivocal new species for the country given the possibility of escape from captivity.

Some snowfinches make altitudinal movements, including White-winged Snowfinch *Montifringilla nivalis,* which has been widely recorded in Europe. Recently, there have been first records for Poland in April 2016 (Komisja Faunistyczna 2017) and Denmark in April 2018 (Olsen *et al.* 2018), as well as the first African record in February 2016 at Kalaat es Senam, in the Kef governorate of Tunisia (Ławicki & van den Berg 2016b). A record of a bird that stayed for three years inland at Lakenheath, Suffolk, England was considered to be an escape (BOU 2009) and the species had been purged from the British List after a review of records that formed part of the fraudulent 'Hastings Rarities' affair (Nelder 1962). With other records from Heligoland, Germany, the Canary Islands and Malta, this species seems likely to eventually make it back onto the British List.

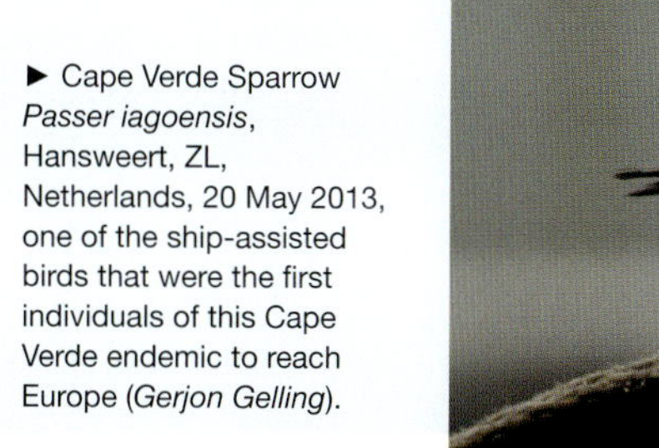

► Cape Verde Sparrow *Passer iagoensis*, Hansweert, ZL, Netherlands, 20 May 2013, one of the ship-assisted birds that were the first individuals of this Cape Verde endemic to reach Europe (*Gerjon Gelling*).

MOTACILLIDAE
Pipits and Wagtails

The pipits and wagtails of the temperate and boreal zones include some of the most highly migratory lineages of all passerines, whilst most tropical species are sedentary. The migratory species, many of which have very large distributions, include species routinely recorded as vagrants thousands of kilometres out of range, and their high dispersal capacity has facilitated colonisation and sometimes speciation in some of the remotest places; this is evidenced by the likes of the endemic South Georgia Pipit *Anthus antarcticus* or the insular subspecies *patriciae* of Grey Wagtail *Motacilla cinerea* on the Azores. Recent phylogenetic studies have confirmed that the endemic São Tomé Short-tail of the Gulf of Guinea – previously thought to be a warbler – is in fact a bizarre island wagtail *Motacilla bocagii* and the sister species to Madagascar Wagtail *Motacilla flaviventris* from the other side of the African continent (Harris *et al.* 2018).

There are occasional reports of relatively modest vagrancy events among the more sedentary tropical species, including in Cape Wagtail *Motacilla capensis* and Fülleborn's Longclaw *Macronyx fuelleborni*, although there is often the question of whether these relate to vagrancy or the discovery of new low-density populations. Apparently vagrant Mountain Wagtails *Motacilla clara* in Botswana, for example, have actually proven to be part of a small local population (Lovett & Brina 2001). A less contentious case for vagrancy in an Afrotropical member of this family concerns records of the partially migratory Golden Pipit *Tmetothylacus tenellus* from Oman and Yemen (Eriksen & Porter 2017).

◄ Citrine Wagtail *Motacilla citreola*, Delta de l'Ebre, Tarragona, Cataluña, Spain, 5 February 2019. Winter records of this species from southern Europe are now becoming more regular (*Brian Sullivan*).

◄ São Tomé Short-tail *Motacilla bocagii*, São Tomé, São Tomé and Príncipe 16 August 2008; this bizarre little bird was formerly classified as the sole member of the genus *Amaurocichla* but phylogenetic analyses revealed that its ancestors were wagtails which succeeded in colonising this tropical island and evolve a very divergent morphology which totally confused taxonomists (*Fabio Olmos*).

► Red-throated Pipit *Anthus cervinus*, Bahía Tortugas, Baja California Sur, Mexico, 17 October 2010, en route to unknown wintering grounds in Central or South America (*Steve Howell*).

Vagrancy in this family is commonest among species of very high latitudes. Few passerines have a more northerly breeding range than Red-throated Pipit *Anthus cervinus* – occurring across a vast swathe of the Arctic from Norway to Alaska, and the wintering range of this species is also large, concentrated in Central Africa, the Middle East and South-east Asia. Red-throated Pipits are regular vagrants along the west coast of the United States in small numbers and have moved as far south as Guatemala in April 2018 (Matías & Eisermann 2018) and, even more impressively, Ecuador in March 2008 (Brinkhuizen *et al.* 2010). Both birds seem likely to have been on spring passage from unknown winter quarters. Elsewhere, vagrant Red-throated Pipits have reached oceanic islands as far afield as the northwestern Hawaiian Islands (Pyle & Pyle 2017), Palau (Pratt *et al.* 2010), the Cocos (Keeling) Islands off northern Australia and the Azores (Barcelos *et al.* 2015). One even reached the Australian mainland as far south as Perth. The closely related Pechora Pipit *Anthus gustavi* has a broadly similar breeding range within Siberia, and its wintering areas remain poorly known, but probably mostly lie in the Philippines, Borneo and Wallacea. Pechora Pipits are annual vagrants to western Europe, although almost all records stem from the Northern Isles of Scotland and from Norway, with around 100 and 50 records respectively. This distribution of records mirrors that of other skulking Siberian passerines such as Lanceolated Warbler, for example, and probably reflects ease of detection in these less-vegetated areas as much as anything. Elsewhere, Pechora Pipits are vagrants to the Aleutians and Bering Sea islands (Howell *et al.* 2014) and from the Australian vagrant traps of Browse Island and Ashmore Reef. Olive-backed Pipit *Anthus hodgsoni* has a more extensive and more southerly breeding range than the previous two species and is likely to be considerably more abundant globally. It is apparently increasing as a vagrant in Europe and is no longer an official rarity in Britain, with more than 45 recorded in 2012 alone. Vagrant records spread right across Europe and the Middle East south to Oman, with counts of multiples now routine – such as four on Linosa Island, Italy, in November 2019. Small numbers are now detected wintering annually in southern Europe, especially on the Canary Islands, while two were also found in Dakhla Bay in Western Sahara in February 2020. Another two together in early November 2016, on the Cap Blanc peninsula, Mauritania (van Bemmelen *et al.* 2017), suggests that Olive-backed Pipits moving through the Western Palearctic may regularly reach Afrotropical wintering grounds. In the Western Hemisphere the species is a regular vagrant to Alaska, where it may have bred on Attu in the Aleutians, and has occurred south to California, Nevada and Mexico (Howell *et al.* 2014). There is a single and completely remarkable record from Hawaii – a flock of 12–15 individuals that reached Kure Atoll in September 1983. Such occurrences give us a window on how birds might colonise such islands through the occasional arrival of flocks of a vagrant species (Pyle & Pyle 2017).

Further south, Blyth's Pipit *Anthus godlewskii* breeds in and around Mongolia, migrating south-west to winter mostly in the Indian subcontinent, whilst Richard's Pipit *Anthus richardi* has more expansive breeding and wintering ranges but still moves largely south-west to Indian and South-east Asian wintering grounds. Blyth's Pipit has been recorded extralimitally both east and west of its normal range and is increasingly detected in Europe and the Middle East as well as wandering east to Japan, Palau and Christmas Island, Australia (Lees & VanderWerf 2011). Blyth's Pipits have also been found wintering in southern Europe, often with Richard's Pipits *Anthus richardi*; the latter has been recognised for decades as a scarce passage migrant and seems to have established viable wintering populations in Europe. High counts of wintering Richard's Pipits in the Western Palearctic include 33 at Cagliari, Sardinia, in February 2004, 14 in the Camargue in January 1998 and 29 at Villafáfila, Spain, in December 2003 (Grussu & Biondi 2004). Intriguingly, neither Blyth's nor Richard's Pipits has been reported as vagrants in North America as yet, nor on Iceland. Interestingly, the similar Tawny Pipit *Anthus campestris* is infrequently reported as a long-range vagrant and is considerably rarer than Richard's Pipits in Britain, despite breeding much closer.

The only transatlantic vagrancy known in pipits concerns records of the nominate race of Buff-bellied Pipit *Anthus rubescens* from North America and Greenland, which is now an expected autumn vagrant to north-west Europe and the Atlantic islands. The Siberian subspecies *japonicus* is rarer in western Europe than its North American counterpart, and has yet to be recorded from Britain, despite records from Norway, Sweden and France. This subspecies is also a rare but regular winter visitor to several countries in the Middle East as close as Turkey and Israel (Kirwan *et al.* 2014); it has also reached Kure Atoll in the north-western Hawaiian Islands once (Pyle & Pyle 2017) as well as down the western North American coastline as far as northern Mexico.

The five species of broadly distributed Palearctic wagtails have been extremely widely reported as vagrants, helped by their relative ease of detection given their bold coloration, strident calls and penchant for open and wetland habitats. Citrine Wagtail *Motacilla citreola* is recorded increasingly frequently in western Europe, associated with a major westward breeding range expansion, and this species may now be establishing new migration routes to African rather than Asian wintering areas. African records extend from Morocco in the north to five records in South Africa, as well as on the Seychelles and in Ethiopia (Underhill 2015). In Australia the species has occurred on Christmas Island, as well as reaching the heart of the continent in New South Wales and South Australia. Surprisingly, there are no records from Alaska, and instead one record from each of British Columbia, California and, most surprisingly, Mississippi – a wintering individual in January–February 1992 (Howell *et al.* 2014).

The more broadly distributed Grey Wagtail *Motacilla cinerea* is a vagrant to South Africa and Australia (including Tasmania) and to Diego Garcia and other Indian Ocean islands in between (Carr 2015). Unlike Citrine Wagtail, this species is a regular vagrant to Alaska and has occurred as far south as Monterey in California (Howell *et al.* 2014). There are also two incredible 'offshore' records of Grey Wagtail from North America: one was photographed on a ship 370km north-west of Prince Patrick Island in the Northwest Territories in September 2009 (at 78 degrees north in the high Arctic); and another – only the second for the United States – was recorded circling a boat during a birders' pelagic 43km off Westport, Washington state, in September 2016.

Records of the diverse White Wagtail *Motacilla alba* complex follow a similar pattern but are even

◄ Sprague's Pipit *Anthus spragueii*, Calipatria State Prison, Imperial, California, United States, 1 January 2012, an annual vagrant to California with several records from this site (*Matt Brady*).

► White Wagtail *Motacilla alba*, Rooisand Nature Reserve, Western Cape, South Africa, 9 January 2018, the first record for South Africa (*Niall Perrins*).

► Forest Wagtail *Dendronanthus indicus*, Alice Springs, Northern Territories, Australia, 8 May 2013, the first Australian record (*Rohan Clarke*).

more widespread. For example, in North America they extend from east to west across the continent south to Mexico and Barbados (Kirwan *et al.* 2019). Vagrants have even reached South America, with records from Trinidad and French Guiana identified as the nominate subspecies from Europe (Ingels *et al.* 2010). This species was also identified for the first time in South Africa in January 2018. The taxonomy of the White Wagtail complex is not fully resolved, but field identification of vagrant forms is relatively easy and is useful in understanding vagrancy patterns. Australia, for example, has recorded three Asian subspecies *baicalensis, lugens, leucopsis,* whilst there are now multiple records of the Asian forms *personata* (Amur Wagtail) and *leucopsis* (Masked Wagtail) from western Europe.

Beyond the *Motacilla* wagtails and *Anthus* pipits, another highly migratory species is Forest Wagtail *Dendronanthus indicus,* which breeds in East Asia and winters in South-east Asia and the Greater Sundas. This species is a regular vagrant to the United Arab Emirates and has reached the Western Palearctic at Kuwait twice (Haas 2012), where more records are perhaps to be expected. It has also strayed to the Maldives (Anderson & Baldock 2001), and there are two Australian records – one from Christmas Island and another discovered at Alice Springs in the Northern Territories in May 2013; the latter remained until September of that year – an incredible record in the heart of the continent at a time when the species should be breeding in the Northern Hemisphere (James & McAllan 2014).

FRINGILLIDAE
Finches, Euphonias and allies

This large family encompasses 49 genera and 227 species. Migratory behaviour is quite common within the group, but mostly concerns partial migrants, altitudinal migrants and irruptive species rather than full migrants. One exception to this rule is Common Rosefinch *Carpodacus erythrinus*, which is fully migratory across a vast Palearctic range extending from the North Sea to the Pacific, with most individuals wintering in the Indian subcontinent and South-east Asia. This species is a regular vagrant to the west of its breeding range in Europe, with more than 80 records from Iceland, and has also reached Alegranza in the Canary Islands (Clarke 2006) and north-east Greenland in 2010 (Olsen *et al.* 2019). To the east it is a regular vagrant to Japan (Brazil 2009) and the Aleutians and Bering Sea islands, with a more noteworthy record from California in September 2007 (Howell *et al.* 2014). This high capacity for vagrancy is further underlined by the fact that colonisation of Hawaii by ancient vagrant rosefinches gave rise to the Hawaiian honeycreepers (page 66).

Bramblings, like many tree-seed eating finches, are highly irruptive, and some species are prone to population-level displacements at huge scales

◄ Common Rosefinch *Carpodacus erythrinus*, Inishbofin, Galway, Ireland, 15 June 2019 (*Anthony McGeehan*).

◄ Grey-crowned Rosy-Finch *Leucosticte tephrocotis*, Talcottville, Lewis, New York, United States, 6 March 2012, the second state record (*Jay McGowan*).

► Hawfinch *Coccothraustes coccothraustes* and Brambling *Fringilla montifringilla*, St Paul Island, Pribilof Islands, Alaska, United States, 1 June 2017 (*Tom Johnson*).

from one year to the next. Most Bramblings move to lower latitudes in the Palearctic in winter, with vagrants reaching as far south as Senegal (Borrow & Demey 2014) and Thailand (Robson 2002), and they have been widely recorded across North America (Howell *et al.* 2014). Movements in Common Redpoll *Acanthis flammea* illustrate the phenomenon of nomadic movements well, with ringing recoveries revealing movements between western Europe and China, and between Michigan in the northern United States and Okhotsk in Russia (Newton 2008). Redpolls are quite capable of long overwater flights, with records of Common Redpoll from the northwestern Hawaiian Islands (Pyle & Pyle 2017) and both Common Redpoll and Arctic Redpoll *Acanthis hornemanni* from the Azores (Barcelos *et al.* 2015). Eurasian Bullfinch *Pyrrhula pyrrhula* and Hawfinch *Coccothraustes coccothraustes* are both regular vagrants to Alaskan islands and the mainland, whilst there are two records of Eurasian Siskin *Spinus spinus* from Attu in the Aleutians (Howell *et al.* 2014). Moving in the other direction, there is a single record of the Nearctic Pine Siskin *Spinus pinus* from Chukotka in Russia (Arkhipov & Ławicki 2016), and one came aboard a trawler operating off southern Greenland in July 2018 (Olsen *et al.* 2018). With records extending south to Bermuda and the Bahamas (Kirwan *et al.* 2019), this highly irruptive species would seem a likely candidate for vagrancy to Europe. A precedent for transatlantic vagrancy in irruptive North American finches comes in the form of two records each of Evening Grosbeak *Coccothraustes vespertinus* from Norway and Britain (Haas 2012). Crossbills *Loxia* spp. are also celebrated for their irruptions, which may take them far out of range: incursions of Two-barred Crossbill *Loxia leucoptera* into western Europe and the United States happen every few years, and the nominate North American subspecies is not averse to considerable sea crossings, with multiple records from Bermuda and, more impressively, at Seltjörn on Iceland in November 2017 – a Western Palearctic first.

Understanding vagrancy in finches is complicated by the fact that many species are popular cagebirds. This was a particular problem in western Europe in the 1980s and 1990s when the opening up of the Chinese bird market led to huge imports of eastern species. A single importer in the Low Countries was reported as possessing about 30,000 Eastern Palearctic passerines (Parkin & Shaw 1994). Around this time there were numerous reports of potential eastern vagrants. For example, there were records of the highly migratory Yellow-billed Grosbeak *Eophona migratoria* from the Faroe Islands, Sweden and Denmark, and Japanese Grosbeak *Eophona personata* from Germany, Sweden and Britain, in addition to Oriental Greenfinch *Chloris sinica* in Germany (Anon 1995). Meanwhile, Yellow-billed Grosbeak is a vagrant to the Philippines (Jensen *et al.* 2015) and Oriental Greenfinch to Alaska, with a record from California considered by Howell *et al.* (2014) to potentially pertain to a wild bird, although not officially accepted as such. In the same period there was also a flurry of records of Long-tailed Rosefinch *Carpodacus sibiricus* and Pallas's Rosefinch *Carpodacus roseus* in Europe, all of which were subsequently rejected. However, Pallas's Rosefinch has since appeared as a vagrant to North America, with one in September 2015 on St Paul Island, Alaska (Pranty *et al.* 2016). Similar assumptions of escaped

◄ Pine Grosbeak *Pinicola enucleator*, Collafirth, Shetland, Scotland, 2 February 2013. The twelfth and most recent British record of this irruptive boreal species (*Rebecca Nason*).

◄ Pallas's Rosefinch *Carpodacus roseus*, St Paul Island, Pribilof Islands, Alaska, United States, 21 September 2015, the first record for North America (*Tom Johnson*).

origin surround many records of European finches in eastern North America, including Common Chaffinch *Fringilla coelebs* and Eurasian Siskin, as well as records of European Goldfinch *Carduelis carduelis* from Japan (Brazil 2009).

Further south in the temperate zone, finches tend to be less prone to long-distance migrations or irruptions, although unusual movements include, for instance, a scatter of records of Lesser Goldfinch *Spinus psaltria* from the eastern United States and extralimital records of Citril Finch *Carduelis citronella* from Britain and Finland. The latter are sometimes touted as potential escapes, but the Finnish individual accompanied a flock of Eurasian Siskins that included a bird ringed in Italy. Additionally, stable isotope analysis of feathers from the first British record indicated a likely origin in the low mountains of the Black Forest, Vosges and Jura in the north of the species' range (Förschler *et al.* 2011).

Several finches of desert regions are nomadic and prone to vagrancy to temperate latitudes, most notably Trumpeter Finch *Bucanetes githagineus,* which is apparently increasing as a vagrant to northern Europe; recent post-breeding influxes have included four in Britain, six in France and four in Sweden in spring 2005 alone (Slack 2009). Both Desert Finch *Rhodospiza obsoleta* and Red-fronted Serin *Serinus pusillus* are vagrants to Cyprus (Richardson & Porter

▲ Trumpeter Finch *Bucanetes githagineus*, Cley and Salthouse Marshes NWT, Norfolk, England, 1 June 2010, the 17th record for Britain (*Alexander Lees*).

► Citril Finch *Carduelis citrinella*, Fair Isle, Shetland, Scotland, 7 June 2008, the first British record (*Stuart Piner*).

2020), but records from western Europe have been ruled as escapes that again coincided with a time when trade was extensive in Europe. We are unaware of any significant movement in Neotropical or Afrotropical finches, although several species are partially migratory; Black-headed Siskin *Spinus notatus*, for example, performs altitudinal movements and there is a single record from Texas, although this was judged to be an escape (Howell *et al.* 2014). Golden-rumped Euphonia *Euphonia cyanocephala* is known to be at least partially migratory (Areta & Bodrati 2010) and there is a scattering of records across central Brazil north as far as the southern Amazon at Parauapebas in Pará (Luz 2019), which seem to relate to vagrants.

CALCARIIDAE
Longspurs and Snow Buntings

This group of buntings breeds at high northern latitudes and most fly south to winter in mid latitudes. Lapland Bunting, or Lapland Longspur, *Calcarius lapponicus* has a very large range across the Holarctic, and vagrants have been widely reported south to Thailand, Bhutan (Chophel & Sherub 2016) and Cuba (Martínez *et al.* 2016), and are frequently recorded on oceanic islands such as the Azores (Barcelos *et al.* 2015) and Madeira (Zino *et al.* 1995). These records may be associated with periodic influxes, which are also characteristic of the other longspurs (Pennington *et al.* 2011). The remaining longspurs are all restricted to North America, following a north–south migration to winter in central regions. All three species – Chestnut-collared Longspur *Calcarius ornatus*, Smith's Longspur *Calcarius pictus* and Thick-billed Longspur (formerly McCown's Longspur) *Rhynchophanes mccownii* – occur periodically as vagrants both to the Atlantic and Pacific coasts. Snow Bunting *Plectrophenax nivalis* has a broad Holarctic distribution, like Lapland Bunting, and also has an impressive capacity for vagrancy; most notably there are several records in the northwestern Hawaiian Islands from Kure, Midway and French Frigate (Pyle & Pyle 2017). Snow Buntings have also

◄ Snow Bunting *Plectrophenax nivalis*, HaZafon, Israel, 26 December 2013, the first for Israel (*Yoav Perlman*).

◄ Smith's Longspur *Calcarius pictus*, Bear Creek Sanctuary, Essex, Massachusetts, United States, 20 January 2016. This species is a rare vagrant to either the east or west coast of North America (*Marshall Iliff*).

occurred south to Morocco (Bergier & Franchimont 2015) and Israel (Anon 2015). McKay's Bunting *Plectrophenax hyperboreus* is an endemic breeder from just two islands in the Bering Sea (St Matthew and Hall islands) and winters mostly in Alaska; vagrants have penetrated south to the north-west United States in Washington state and Oregon, as well as to Chukotka and Kamchatka in Russia (Arkhipov & Ławicki 2016); this species could conceivably be a future vagrant to Japan (Brazil 2009).

RHODINOCICHLIDAE
Thrush-Tanager

The sole member of this family, Rosy Thrush-Tanager *Rhodinocichla rosea* is resident and we are unaware of any extralimital reports.

EMBERIZIDAE
Old World Buntings

All members of the Emberizidae now sit in a single genus *Emberiza*, and this group is ubiquitous across most of Eurasia and Africa, with Northern Hemisphere species being migratory whilst most southern species are resident. There are exceptions to this rule, however. For example, Lark-like Bunting *Emberiza impetuani* of southern Africa is somewhat nomadic and has occurred as a vagrant to Zambia, Angola and DR Congo. Corn Bunting *Emberiza calandra* of Eurasia and North Africa is largely resident, but still has significant wandering tendencies, with vagrants reaching Senegal in the south as well as north to Norway and Finland (Byers *et al.* 1995). Several Palearctic buntings undertake huge migrations; for example, Rustic Bunting *Emberiza rustica* has a vast boreal breeding range stretching from the Baltic Sea to the Pacific, with almost the entire population wintering in eastern China, Korea and Japan. Those breeding in Fennoscandia must undertake an epic migration of over 9,000km, initially moving eastwards

▼ Little Bunting *Emberiza pusilla*, Fair Isle, Shetland, Scotland, 20 September 2014 (*James Lowen*).

towards Siberia before turning south through eastern Mongolia and north-east China. Rustic Buntings are annual vagrants to western Europe and the Middle East and have occurred south to Algeria (Djemadi *et al.* 2018). They are among the most regular Siberian vagrants to North America, with records all the way down the Pacific coast as far as central California, plus a record well inland in Saskatchewan, Canada (Howell *et al.* 2014). The spread of vagrant Little Buntings *Emberiza pusilla* is broadly similar, but this species has a more expansive southerly regular wintering range. Both this species and Rustic Bunting are found wintering annually in western Europe with bunting flocks. Three Little Buntings were found together at the Miño estuary, Spain, in February 2009 (Dies *et al.* 2011) and in Staffordshire/Worcestershire, England, in February 2005, with others recorded south to Morocco (Praus 2016). Little Bunting is rarer in North America than Rustic Bunting but has also been recorded south to California and Baja California Sur in Mexico (Radamaker & Powell 2010). This species is a vagrant to India, east to Sabah and Sarawak (Eaton *et al.* 2016) and out into the Pacific as far as Palau (eBird). A third species, Yellow-breasted Bunting *Emberiza aureola,* once shared a similar distribution, having expanded its range westward into Europe in the 19th century; it had become a regular vagrant to the northwestern seaboard of Europe by the late 20th century, with Shetland claiming 66 per cent of British records (Slack 2009). Yellow-breasted Buntings have also strayed to Alaska (Howell *et al.* 2014) and Brunei, Sabah and Sarawak in Indonesia (Eaton *et al.* 2016). However, the population crashed from the 1990s, as a new market appeared for their usage in traditional medicine, resulting in an estimated harvest of one million buntings in China's Guangdong Province alone in 2001; the species is now Critically Endangered (Wang *et al.* 2019). This collapse led to a range contraction away from Europe, the extinction of the Finnish population and a rapid decrease in the number of vagrant individuals in western Europe. Against this backdrop, the discovery of a Yellow-breasted Bunting in eastern Canada in October 2017, at Forteau Bay in Labrador, constituting the first national record, is perhaps one of the most unlikely vagrancy events on record.

Several highly migratory East Asian buntings are extremely rare vagrants to Europe. These include both more northern species such as Pallas's Reed Bunting *Emberiza pallasi,* Yellow-browed Bunting *Emberiza chrysophrys* and Black-faced Bunting *Emberiza spodocephala,* as well as more southerly species like Chestnut-eared Bunting *Emberiza fucata* and Chestnut Bunting *Emberiza rutila.* The last named is the most frequent to occur, with records from Finland, Britain, Malta, Netherlands, Norway, Sweden, Slovenia and France (Haas 2012). Perhaps somewhat surprisingly, among these only Pallas's Reed Bunting and Yellow-browed Bunting have reached North America, both in Alaska. However, this region has also been visited by Yellow-throated Bunting *Emberiza elegans* (once on Attu in the Aleutians), and there have been four records of Grey Bunting *Emberiza variabilis*; both species are shorter-distance migrants. Another short-distance migrant, Tristram's Bunting *Emberiza tristrami,* is a vagrant east to Japan (Brazil 2009) and west to India (Naniwadekar *et al.* 2013). The only buntings to have reached Australia are vagrant Yellowhammers *Emberiza citrinella* from the introduced population in

◀ Tristram's Bunting *Emberiza tristrami*, Hegurajima Island, Ishikawa, Japan, 9 Oct 2018, a scarce passage migrant to Japan (*Yann Muzika*).

► Pine Bunting *Emberiza leucocephalos*, Wilhelminadorp, Zeeland, Netherlands, 5 March 2016. This species is found wintering in western Europe in some years, sometimes in small flocks (*Richard Bonser*).

▼ Cretzschmar's Bunting *Emberiza caesia*, North Ronaldsay, Orkney, Scotland, 19 September 2008, the fifth British record (*Alexander Lees*).

New Zealand, which have reached Lord Howe Island and Macquarie Island (McAllan *et al.* 2004). Black-faced Bunting has reached as far south as Taliabu in Indonesia in October 1991 (Eaton *et al.* 2016), and a Pallas's Reed Bunting has reached Myanmar (Robson 2002), so future bunting vagrancy to the northern island vagrant traps in Australia seems a possibility.

Pine Bunting *Emberiza leucocephalos* is something of an enigma, having a broad Asian distribution from the Urals to the Pacific but an apparently disjunct wintering range, with the nearest regular wintering grounds to Europe being in Iran. Pine Buntings are now recorded increasingly frequently in north-west Europe as autumn vagrants, with a major incursion in autumn 2016 associated with a prolonged easterly airstream. More noteworthy, however, are flocks of wintering Pine Buntings which have been discovered in southern Europe, most notably in Italy, where more than 50 were present in 1995–96; this suggests an established pseudo-vagrant population analogous to that of Richard's Pipits in the same general region (Lees & Gilroy 2003, Slack 2009). Pine Bunting has also reached North America, where there are two records from Attu and the Pribilofs (Howell *et al.*

2014). Many extralimital Pine Buntings have been found with flocks of Yellowhammers, and the reverse is true of Yellowhammers in flocks of Pine Buntings in China and Japan (Brazil 2009).

Ortolan Bunting *Emberiza hortulana* has a vast breeding area east to central Mongolia and north to northern Finland and migrates to winter in the northern part of sub-Saharan Africa, where vagrants have overshot to Kenya (Stevenson & Fanshawe 2002), Tanzania (Fisher & Hunter 2018) and Namibia (Donald 2013). Others have occurred east as far as Japan (Brazil 2009). Meanwhile the closely related Cretzschmar's Bunting *Emberiza caesia* has a restricted eastern Mediterranean breeding range and winters largely in Sudan, although it still appears with some regularity as a vagrant in western Europe; it has occurred west to Spain, south to the Canary Islands and Chad, north to Finland (five records) and east to Kuwait, Oman and Iran. The third member of this group – Grey-necked Bunting *Emberiza buchanani* – breeds from eastern Turkey to central Mongolia, wintering in India, and has been recorded in the Netherlands, Finland, Norway, Sweden and Germany; all these records have been since 2003, so it must now be one of the most overdue firsts for Britain. It has also been recorded as a vagrant east to Bangladesh (Thompson *et al.* 2014), Korea and Japan (Brazil 2009) and has probably been overlooked as a vagrant given its similarity to other species.

Buntings are another group that often pose a headache to rarities committees because of their presence in the bird trade. This was a significant historic problem in western Europe with records of species such as Meadow Bunting *Emberiza cioides* and Yellow-throated Bunting invariably deemed to be escapes. The most controversial species is Red-headed Bunting *Emberiza bruniceps,* which was frequently recorded at vagrant hotspots prior to the collapse of the bird trade, but records decreased rapidly since then, especially after a 1982 ban on the export of this species in India (Vinicombe 2007b). There have been records since 2016 in Norway, Finland and the Netherlands, and Red-headed Bunting has also occurred as a vagrant east to Korea, China and Japan (Brazil 2009). Given that this species is both highly migratory and shares a migration route and migration span with several other European vagrants, future records are to be expected. Escaped Black-headed Buntings *Emberiza melanocephala* were also a feature of the late 20th century in western Europe, but this species was recorded more frequently than Red-headed Bunting, which is unsurprising given that it breeds considerably closer, and records have continued as the escape risk has fallen substantially. It is also a vagrant in the east to Korea and Japan, as well as south to Mantanani and Pulau Tiga, islands off Sabah, Borneo (Eaton *et al.* 2016), and has even been recorded from Micronesia, on Palau (Otobed *et al.* 2018).

▲ Chestnut Bunting *Emberiza rutila*, Papa Westray, Orkney, Scotland, 25 October 2015, the first accepted record for Britain; previous individuals have been known or suspected escapes (*Stuart Piner*).

PASSERELLIDAE
New World Sparrows

Almost all populations of New World sparrows breeding in northern North America in Canada and Alaska migrate south to avoid the worst of the boreal winter, although sparrows winter much further north than many other Nearctic–Neotropical migrants. Most do not reach much further south than southern Mexico, although a few individuals have even penetrated as far as South America as vagrants. These include recent records of three species from Colombia: a Lincoln's Sparrow *Melospiza lincolnii* in Norte de Santander in March 2017 (Edwards & Scheffers 2018); a Clay-coloured Sparrow *Spizella pallida* at Valle de Cauca in May 2016 (Tigreros *et al.* 2019); and a White-crowned Sparrow *Zonotrichia leucophrys* at the Universidad Nacional de Colombia in June 2019 (eBird). White-throated Sparrow *Zonotrichia albicollis* has also occurred south to Aruba, and Voous (1985) considered this a ship-assisted record, although this species has reached Belize on Half Moon Caye in May 2005, the first Central American record (Komar *et al.* 2006).

Most American sparrows are not long-distance migrants to the same extent as Old World buntings, or indeed New World warblers, but nevertheless they are relatively frequent vagrants outside of the Western Hemisphere. Ten species have reached the Western Palearctic: Lark Sparrow *Chondestes grammacus*, American Tree Sparrow *Spizelloides arborea*, Fox Sparrow *Passerella iliaca*, Dark-eyed Junco *Junco*

▲ LeConte's Sparrow *Ammospiza leconteii*, Bon Portage Island, Shelburne, Nova Scotia, Canada, 13 September 2002, the fourth record for Nova Scotia (*Alexander Lees*).

► White-throated Sparrow *Zonotrichia albicollis*, Höfn, Austurland, Iceland, 13 November 2007, the sixth record for Iceland (*Yann Kolbeinsson*).

hyemalis, White-crowned Sparrow, White-throated Sparrow, Savannah Sparrow *Passerculus sandwichensis*, Song Sparrow *Melospiza melodia*, Lincoln's Sparrow and Eastern Towhee *Pipilo erythrophthalmus*. Of these, White-throated Sparrow and Dark-eyed Junco occur with the greatest frequency, with 48 and 46 British records respectively. Both species have been recorded travelling aboard ships, sometimes even together, and it seems likely that many if not most American sparrows may be ship-assisted to Europe. In May 1986 Gibraltar, which lies on a major shipping lane, hosted both species together (Lewington *et al.* 1991). Several American sparrows have appeared at major ports, with Seaforth in Merseyside, England, having hosted both Song Sparrow and White-crowned Sparrow. Ship-assistance would seem a near-certainty for the two English records of Lark Sparrow, an essentially western North American species, although this species is also a vagrant to the Bahamas and Jamaica (Kirwan *et al.* 2019). Ship-assistance would also seem most likely for the sole Western Palearctic record of Eastern Towhee on Lundy Island, Devon, England, in June 1966 (Lewington *et al.* 1991). Intriguingly, the Azores have hosted only five species, although this does include the only Lincoln's Sparrow for the Western Palearctic (Monticelli *et al.* 2018), a species that has also reached Greenland, along with four other New World sparrows (Boertmann 1994). Given that 18 species have reached Bermuda there seems considerable scope for additions to the Azores list still.

New World sparrows have also reached the Eastern Palearctic; Japan has hosted five species, four of which are also vagrants to the Western Palearctic: White-crowned Sparrow, Fox Sparrow, Savannah Sparrow and Song Sparrow, plus Golden-crowned Sparrow *Zonotrichia atricapilla* of the Pacific coast of North America. The absence of Dark-eyed Junco there is surprising, as there are 16 records, involving 19 birds, from north-east Russia. American Tree Sparrow and Swamp Sparrow *Melospiza georgiana* have also occurred in north-east Russia, and Chipping Sparrow *Spizella passerina* has occurred once on Wrangel Island, Russia (Arkhipov & Ławicki 2016). Savannah Sparrow has been found breeding on Chukotka and this small population, or perhaps others from Alaska, may be responsible for the small number of birds found annually wintering in Japan, a nominal pseudo-vagrant population (Brazil 2009). Savannah Sparrows have also reached South Korea and as far south as Taiwan in November 2018, and there are two records from the northwestern Hawaiian Islands (Pyle & Pyle 2017). Perhaps with the exception of the Savannah Sparrows, ship-assisted passage would seem the most likely origin for the Japanese records too, given the rarity of many sparrows in the Bering Sea region.

Among the exclusively North American species, vagrancy is common among the migratory taxa but seemingly unknown among resident species like Canyon Towhee *Melozone fusca*. Several interior species such as Lark Bunting *Calamospiza melanocorys* are highly migratory and regular vagrants to both the Atlantic and Pacific coasts, as are Clay-coloured

▲ Fox Sparrow *Passerella iliaca*, Haapsalu, Estonia, 8 December 2012, the third record for the Western Palearctic (*Richard Bonser*).

► Lark Bunting *Calamospiza melanocorys*, Myers Point, Tompkins, New York, United States, 9 September 2007. This highly migratory species is a vagrant to both the Atlantic and Pacific coasts of North America (*Chris Wood*).

► Green-tailed Towhee *Pipilo chlorurus*, Southeast Farallon Island, San Francisco, California, United States, 9 September 2013. This is predominantly a scarce autumn drift migrant to coastal California (*Cameron Rutt*).

► Baird's Sparrow *Centronyx bairdii*, Southeast Farallon Island, San Francisco, California, United States, 14 September 2008, the second record for the island that autumn and fifth ever of six state records (*Matt Brady*).

Spizella pallida and Harris's *Zonotrichia querula* Sparrows. Golden-crowned Sparrow has a narrow distribution on the Pacific seaboard but has been recorded as a vagrant throughout North America. Black-throated Sparrow *Amphispiza bilineata* of the western deserts is also a vagrant across the east, north as far as Labrador. Several species with quite restricted distributions have also been recorded widely as vagrants; these include a handful of extralimital records of Baird's Sparrow *Centronyx bairdii* east as far as New York and west to California, where most records have come from Southeast Farallon Island (Hamilton *et al.* 2007). There is a similar spread of records of Cassin's Sparrow *Peucaea cassinii*, but Black-chinned Sparrow *Spizella atrogularis* is unusual in being partially migratory but almost never reported extralimitally.

Meanwhile, South and Central American species are typically sedentary, and little or no vagrancy has been reported in groups like the *Atlapetes* brush-finches, *Arremon* sparrows or *Chlorospingus* bush-tanagers. An exception is Rufous-collared Sparrow *Zonotrichia capensis* – populations of the southernmost subspecies *australis* of Patagonia migrate north in the austral winter, and vagrants are regular on the Falklands (Woods 2017).

CALYPTOPHILIDAE, PHAENICOPHILIDAE AND NESOSPINGIDAE

Chat-Tanagers, Hispaniolan Tanagers and Puerto Rican Tanager

All seven species in these two families are resident endemics of the islands of Hispaniola or Puerto Rico (Kirwan *et al.* 2019).

SPINDALIDAE

Spindalises

◄ Western Spindalis *Spindalis zena zena*, Gulfstream Shores, Monroe, Florida, United States, 5 May 2017 (*Tom Johnson*).

▲ Western Spindalis *Spindalis zena townsendi*, Gulfstream Shores, Monroe, Florida, United States, 5 May 2017 (*Tom Johnson*).

Western Spindalis *Spindalis zena* is a surprising wanderer considering its resident status, having occurred as a vagrant more than 50 times to Florida, most of which were of the Bahamian subspecies *zena*. This species has bred extralimitally once in the state at Long Pine Cay (Howell *et al.* 2014). There is also a single record of the Cuban subspecies *pretrei* from Florida. Other species in this family are even more restricted-range resident endemics and are rare even on satellite islands; for example, there is a single record of Puerto Rican Spindalis *Spindalis portoricensis* from Vieques Island (Kirwan *et al.* 2019).

ZELEDONIIDAE
Wrenthrush

The single representative of this family, Wrenthrush *Zeledonia coronata,* is a restricted-range non-migratory inhabitant of dense bamboo thickets in humid montane forest and no extralimital movements have been reported.

TERETISTRIDAE
Cuban Warblers

Both species in this group are sedentary Cuban endemics (Kirwan *et al.* 2019).

ICTERIIDAE
Yellow-breasted Chat

Most Yellow-breasted Chats *Icteria virens* are highly migratory, moving between North and Central America seasonally. Vagrants occur with some regularity in Atlantic Canada and as far north as Hudson Bay and Labrador, in addition to Bermuda, the Dominican Republic and Jamaica (Eckerle & Thompson 2001, Kirwan *et al.* 2019).

◄ Yellow-breasted Chat *Icteria virens*, Seal Island, Yarmouth, Nova Scotia, Canada, 19 September 2002, a vagrant to Atlantic Canada, lurking in stacked lobster traps (*Alexander Lees*).

ICTERIDAE
Troupials and allies

Most icterids are resident with the exception of some temperate zone species. The most migratory member of the group is the Bobolink *Dolichonyx oryzivorus*, which moves from southern Canada to wintering grounds centred on Paraguay, including a substantial sea crossing over the Caribbean; falls of hundreds on Bermuda are not uncommon. Given this huge migration span, it is unsurprising that Bobolinks are among the more regular transatlantic landbird vagrants, with 34 records from Britain and 32 from the Azores, for example. Others have penetrated Europe as far east as Tuscany, Italy, in September 1989; the sole spring record – on Gibraltar in May 1984 – was perhaps ship-assisted (Slack 2009). Elsewhere, Bobolink is one of the most frequent passerine vagrants to the Galapagos Islands (Wiedenfeld 2006) and has occurred on Clipperton Atoll, more than 1,000km off the Mexican coast (Howell *et al.* 1993), with two records each from Greenland (Boertmann 1994) and Chile (Jaramillo *et al.* 2003). Transatlantic vagrancy is also relatively routine in Baltimore Oriole *Icterus galbula*, which has reached Britain regularly (25 records), as well as Ireland, Norway, the Netherlands, Iceland and the Azores. This species winters south to northern South America and vagrants have straggled to northern Brazil and Ecuador, but most remarkable is a single record of a male at Calama, northern Chile, in June 2002 (Torres-Mura *et al.* 2003).

Transatlantic vagrancy is much rarer among the remaining icterids; Brown-headed Cowbird

► Red-winged Blackbird *Agelaius phoeniceus*, Garso, North Ronaldsay, Orkney, Scotland, 30 April 2017, the first record for the Western Palearctic (*Stuart Piner*).

▼ Bronzed Cowbird *Molothrus aeneus* with a Shiny Cowbird *Molothrus bonariensis* and Red-winged Blackbirds *Agelaius phoeniceus*, West Kendall Agricultural Area, Miami-Dade, Florida, United States, 31 July 2018. Bronzed Cowbirds are in the process of colonising the southern United States (*Tom Johnson*).

Molothrus ater has reached Britain, Norway and Germany on nine occasions, of which six were in the springs of 2009 and 2010 (Haas 2012). There is a single record of Red-winged Blackbird *Agelaius phoeniceus* from Orkney, Scotland, in April 2017 and two accepted European records of Yellow-headed Blackbird *Xanthocephalus xanthocephalus* from Iceland and the Netherlands (Haas 2012), as well as two from Greenland (Boertmann 1994). Records of this species from other countries have been judged to be escapes, or in the case of a record from Fair Isle, Scotland, deemed to be 'of unknown origin' (Haas 2012). Yellow-headed Blackbird is a western North American species and might seem an unlikely vagrant to Europe, but vagrancy in this species has been shown to be increasing in accordance with eastward population expansion (Veit 1997); it is regularly recorded on the east coast of North America and south to Costa Rica, where the first for Isla del Coco – nearly 500km from the mainland – was recorded in October 2010 (Villalobos & Sandoval 2012). Common Grackle *Quiscalus quiscula* is known in Europe from one accepted record, an individual which flew over the migration watchpoint of Kamperhoek, the Netherlands, in April 2013 (Slaterus 2013), as well as rejected records from Denmark and Gibraltar deemed to be escapes. Rusty Blackbird *Euphagus carolinus* is the most frequent vagrant icterid to Greenland, with four records (Boertmann 1994), and it has also occurred three times in north-east Russia in Kamchatka and

Chukotka (Arkhipov & Ławicki 2016). Vagrancy to Europe would have seemed likely prior to the major decline in the population of this species (Greenberg & Matsuoka 2010) but remains a distinct possibility.

The only other icterid species to have been recorded outside of the Americas is Western Meadowlark *Sturnella neglecta*. An adult bird was collected at Providence Bay, Chukotka, Russia, in late June 2002 (Arkhipov & Ławicki 2016). Western Meadowlarks are far more migratory than their counterpart Eastern Meadowlark *Sturnella magna* and in addition to the Russian record they are also regular in eastern North America, east to Newfoundland; they have also reached the Canadian Northwest Territories on the shores of the Beaufort Sea (eBird). Other notable extralimital records of North American species include a wide scatter of records of Hooded Oriole *Icterus cucullatus* in the east as far as Nova Scotia, Quebec and New Brunswick (eBird), and records of Scott's Oriole *Icterus parisorum* north to Ontario in Canada (Denis 1976). More southerly distributed species are far more sedentary, but vagrancy has been reported frequently in several orioles. The partially migratory Streak-backed Oriole *Icterus pustulatus* has been widely reported in the southern border states of the United States, where it has even bred, and there is also one record each from Oregon, Colorado and Wisconsin (Howell *et al.* 2014). Meanwhile, Black-vented Oriole *Icterus wagleri* has reached southern Texas and south-east Arizona (Howell *et al.* 2014). One of the most unusual records of vagrancy in North America concerns an adult male Black-backed Oriole *Icterus abeillei,* which appeared at a bird feeder in Reading, Pennsylvania, in January 2017, staying till April before reappearing in Sutton, Massachusetts, and then Stamford, Connecticut, in May (Pyle *et al.* 2018). This national first is a central Mexican endemic and seemingly unlikely vagrant, but this species is a short-distance migrant and may be under-recorded as a vagrant since first-year birds, which would be more expected as vagrants, are difficult to separate from Baltimore Oriole and Bullock's Oriole *Icterus bullockii*. Both Shiny Cowbird *Molothrus bonariensis* and Bronzed Cowbird *Molothrus aeneus* creep into the southern United States as breeding species and have wandered extensively in North America with records of both as far north as Nova Scotia in Canada.

Vagrancy in South America is more limited and most evident in populations of species that penetrate into southern South America and have evolved to be at least partial migrants. Two examples are Shiny Cowbird and Yellow-winged Blackbird *Agelasticus thilius*, both of which have reached the Falklands as vagrants (Woods 2017). Long-tailed Meadowlark *Leistes loyca* is also thought to undertake local movements but does not entirely abandon southernmost latitudes during the austral winter – which makes the single record from South Georgia in April 1987 all the more extraordinary and presumably of ship-assisted origin (Woods 2017). Further north there are single records of White-browed Meadowlark *Leistes superciliaris*, Greyish Baywing *Agelaioides badius* and Screaming Cowbird *Molothrus rufoaxillaris* from Chile, the latter involving two juveniles in December 2010 (Jaramillo *et al.* 2003, Barros 2015). Some southern populations of Unicoloured Blackbird *Agelasticus cyanopus* are also migratory, and the species recently reached south-east Peru for the first time (CRAP 2012). The few historical records of Carib Grackle *Quiscalus lugubris* from Amapa in northern Brazil (e.g. Novaes 1959) might represent a lost population or perhaps wanderers from further north. However, given the proximity of most

◀ Streak-backed Oriole *Icterus pustulatus*, Prairie Trails subdivision, Larimer, Colorado, United States 13 December 2007, the first record for Colorado (*Chris Wood*).

► Yellow-winged Blackbird *Agelasticus thilius*, Cape Pembroke, Falklands, 11 January 2012, the sole record for the Falklands of this South American species (*Alan Henry*).

records to the state capital, the major port of Macapa, perhaps these birds were also ship-assisted. Grackles have a history of ship-assistance; in addition to the Common Grackles already mentioned near major ports, two Great-tailed Grackles *Quiscalus mexicanus* appeared at Girona in Spain and Gibraltar in October and November 2014, and a Boat-tailed Grackle *Quiscalus major* was present in Antwerp, Belgium, in March 2008.

PARULIDAE
New World Warblers

Broadly speaking, it is possible to separate the New World warblers into two major groups – highly migratory temperate zone species breeding in North America, and generally sedentary tropical species of Middle and South America, including the Caribbean. Given their long-distance and – for some – intercontinental migrations, it is not surprising that rates of vagrancy among temperate zone species are high, but vagrancy is not restricted to these species by any means.

Blackpoll Warblers *Setophaga striata* perform the longest migration of any New World warbler; individuals from Alaska may travel over 8,000km to reach wintering grounds in Amazonia via an epic transoceanic migration leg of around 2,500km, necessitating three days of non-stop flight (DeLuca *et al.* 2015). Given this huge migration span, a high capacity for non-stop flight and a large population size across their vast boreal range, it is unsurprising that Blackpoll Warblers are the most frequent transatlantic vagrants to Europe, with 46 records from Britain, for example (McLaren *et al.* 2006). Blackpoll Warblers are correspondingly also among the most abundant eastern warblers in California, and there are records south all the way to Chile and Argentina.

Eastwards vagrancy out into the Atlantic is relatively routine among the migratory warblers. Thirty-nine species of warblers have reached Bermuda, 26 the Azores, with 18 recorded in Britain and three in the Netherlands. This west–east gradient essentially reflects flight performance and population size, with longer-distance migrants like Blackpoll Warbler and American Redstart *Setophaga ruticilla* – species that winter in the Caribbean or in South America – being more likely to have sufficient fat reserves to survive the Atlantic. Ease of detection probably also plays a role; Blackpoll Warblers share their transatlantic migration route with the very secretive and globally much rarer Connecticut Warbler *Oporornis agilis* (McKinnon *et al.* 2017). Given that the latter species is so hard to detect in the field, both the location and manner of discovery of the first for the Western Palearctic should not come

◄ Cape May Warbler *Setophaga tigrina*, Baltasound, Unst, Shetland, Scotland, 24 October 2013, the second record for Britain and the Western Palearctic (*Stuart Piner*).

◄ Prothonotary Warbler *Protonotaria citrea*, Ribeira do Cantinho, Corvo, Azores, Portugal, 5 October 2019, the first record for the Western Palearctic (*Paul French*).

◄ Black-and-white Warbler *Mniotilta varia*, Sein Island, Finistère, France, 17th October 2019, the first record for France (*Paul Dufour*).

as a surprise – a bird that dropped into a mist-net on the island of Flores on the Azores in October 2019. Connecticut Warbler is also regular on Bermuda, with 75 recorded there on a single day in September 1987, grounded by Hurricane Emily (Pitocchelli *et al.* 2020).

The rarest transatlantic vagrant warblers are species that winter almost exclusively in Central America – such as Golden-winged Warbler *Vermivora chrysoptera,* which has occurred once each in Britain and the Azores, and Hooded Warbler *Setophaga citrina,* with two British records and six from the Azores. The same is true of species that winter mostly in western South America and especially along the Andes, like Cerulean Warbler *Setophaga cerulea,* with one record from Iceland, and Blackburnian Warbler *Setophaga fusca,* with six records between Britain, France, Iceland and the Azores (Haas 2012). Although these species undertake very long migrations, their migration routes are more westerly, and they are less likely to be entrained in fast-moving Atlantic depressions than species that habitually migrate to eastern South America and the Caribbean and face longer overwater flights. Several New World warblers have been discovered wintering in the Palearctic, including Northern Waterthrush *Parkesia noveboracensis*, Black-and-white Warbler *Mniotilta varia*, Blackpoll Warbler, Yellow-rumped Warbler *Setophaga coronata*, Common Yellowthroat *Geothlypis trichas* and the aforementioned Golden-winged Warbler. Spring arrivals of North American warblers in the Palearctic are very rare in comparison with North American sparrows and may involve individuals that have wintered on the 'wrong' side of the ocean undetected; a few have also appeared in port settings, indicating assisted passage, such as the male Blackpoll Warbler at Seaforth Docks in Merseyside, England, in June 2000. Blackpoll Warbler is arguably the passerine most physiologically capable of unassisted passage, yet there is even a record of one completing a full return passage on a boat from the United States to Britain and back in October 1961 (Durand 1972), a reminder of how pervasive ship-assisted movements may be. Greenland has hosted 23 species of New World warbler, including five species – Louisiana Waterthrush *Parkesia motacilla,* Orange-crowned Warbler *Leiothlypis celata*, Nashville Warbler *Leiothlypis ruficapilla*, Mourning Warbler *Geothlypis Philadelphia* and Pine Warbler *Setophaga pinus* – yet to be recorded in the Palearctic but all surely potential future additions (Boertmann 1994).

Eastern North American warblers are also routine vagrants to western North America, with only two extant species – Swainson's Warbler *Limnothlypis swainsonii* and Kirtland's Warbler *Setophaga kirtlandii* – as yet unrecorded in the west (although Swainson's has reached Colorado and New Mexico), in addition to the extinct Bachman's Warbler *Vermivora bachmanii.* Intriguingly, the first two have reached Bermuda, which suggests they may be capable of longer-range vagrancy.

California hosts the lion's share of eastern vagrant warblers, and indeed eastern vagrant landbirds in general, probably a consequence of both coastal concentration effects and greater observer coverage. This is further reflected in biases at coastal observatories – over half of all Californian Connecticut Warbler records come from Southeast Farallon Island, for example, a situation analogous to the detection bias of Lanceolated Warblers on Shetland – and indeed

► Yellow-throated Warbler *Setophaga dominica*, Poço de Agua, Corvo, Azores, Portugal, 17 October 2013, the first for the Western Palearctic (*Josh Jones*).

◄ Connecticut Warbler *Oporornis agilis*, Southeast Farallon Island, San Francisco, California, United States, 14 September 2010. The island the most reliable site in the state for this skulking eastern vagrant (*Matt Brady*).

the only Californian Lanceolated Warbler was also on Southeast Farallon Island (Hamilton *et al.* 2007). Ralph & Wolfe (2018) explored the patterns in vagrancy of eastern warblers to north-west California and southern Oregon and found that vagrancy was strongly correlated with larger North American population size and longer migration spans. In addition to the coastal bias in some species, they also found that young birds were over-represented as vagrants.

Vagrancy to the Palearctic is not exclusively from the west. Four species – Northern Waterthrush, Yellow-rumped Warbler, Wilson's Warbler *Cardellina pusilla* and Orange-crowned Warbler – have reached the Eastern Palearctic in north-east Russia (Brazil 2009, Arkhipov & Ławicki 2016). Further south, two species have been recorded in Japan: Wilson's Warbler on the vagrant trap of Hegurajima and a Common Yellowthroat *Geothlypis trichas* in Ibaraki prefecture (Ławicki & van den Berg 2017b). This contrasts with the high number of American sparrows recorded in the region and may reinforce the supposition that many of these sparrows may be ship-assisted to the region. Warbler diversity is low in Alaska, so it is unsurprising that vagrany to the Eastern Palearctic is more limited, but sparse survey effort is likely also a major contributing factor and many more species have been recorded on the geographically proximate Bering Sea island vagrant traps in American territory.

The eastern warblers are also recorded with some regularity thousands of kilometres south of their regular wintering ranges. The most spectacular example of this are two records of American Redstarts from the Falklands, 6,500km south of their regular winter range (Woods 2017). American Redstarts have also occurred in Chile (Jaramillo *et al.* 2013), one of seven eastern warbler species to have occurred there (Barros *et al.* 2015), also including Black-throated Green Warbler *Setophaga virens*, which has also been recorded in eastern Brazil (Deconto & Vallejos 2017). The Atlantic Forest of south-east Brazil has hosted a number of other vagrant warblers including Blackpoll Warbler, Blackburnian Warbler and most recently the first documented Cerulean Warbler for the biome in April 2018 at Florianópolis, Santa Catarina (de Farias & Dalpaz 2019).

Exclusively western North American warblers tend to have far shorter migration spans than eastern species, especially the birds associated with the desert south-west. Of these species, Townsend's Warbler *Setophaga townsendi* has the largest migration span; it moves from Alaska to Costa Rica and has been recorded south as far as Panama. Townsend's Warbler is a regular vagrant throughout eastern North America, north-east to Newfoundland and to the Bahamas, Cuba and the Cayman Islands (Kirwan *et al.* 2019); it has even been recorded at least nine times in Bermuda – the only western warbler to reach this island. This raises the intriguing possibility of vagrancy as far as the Azores. Two other species – Black-throated Grey Warbler *Setophaga nigrescens* and Hermit Warbler *Setophaga occidentalis* – are also recorded widely in the east, at least in the case of Black-throated Grey Warbler, but both have considerably shorter migrations than Townsend's and are considerably rarer as vagrants. Virginia's Warbler *Leiothlypis virginiae* is a migrant breeder to the south-western deserts and a sporadic vagrant to both seaboards and even to the Bahamas (Kirwan *et al.* 2019).

► American Redstart *Setophaga ruticilla*, Cape Dolphin, Falklands, 4 December 2014, the first record for the Falklands (*Alan Henry*).

► Tennessee Warbler *Leiothlypis peregrina*, Iguana Island on the Madeira River, Amazonas, Brazil, 21 September 2017, the first record for Brazil (*Cameron Rutt*).

It is not only the more northern species that have a habit of long-range vagrancy. The distribution of Painted Redstart *Myioborus pictus* just creeps into Arizona, New Mexico and a slither of Texas, largely the only parts of its distribution where it is migratory. Despite its short migration span and presumably small global population size, Painted Redstart has occurred north to the Canadian provinces of British Columbia, Manitoba, Ontario and Quebec, with a scatter of records throughout the United States, especially in the north-east and Great Lakes region. Meanwhile, Red-faced Warbler *Cardellina rubrifrons,* which breeds in southern Arizona and New Mexico and is if anything more migratory, has not been recorded quite so far afield although it has still reached California, Wyoming, Colorado, Louisiana and Georgia. Also, very worthy of mention here is Golden-cheeked Warbler *Setophaga chrysoparia,* which has a highly restricted breeding range in Texas and winters from southern Mexico south to Nicaragua. Extralimital records are very rare, as might be expected for this globally Endangered species, but there are single records west from Southeast Farallon Island, California, north from St Louis County, Missouri, and east from Pinellas County, Florida, and St Croix in the Virgin Islands (Long *et al.* 2019).

Six species of Neotropical warblers are accepted as vagrants to the southern United States: Mexican Fan-tailed Warbler *Basileuterus lachrymosus*, Rufous-capped Warbler *Basileuterus rufifrons* and Slate-

◄ Townsend's Warbler *Setophaga townsendi*, West Barrier Bar County Park, Cayuga, New York, United States, 2 May 2019. This species has been widely recorded in eastern North America and has even reached Bermuda (*Jay McGowan*).

◄ Slate-throated Redstart *Myioborus miniatus*, Pinery Canyon, Cochise, Arizona, United States, 6 May 2018. Multiple vagrant individuals have occurred at this site in recent years, and hybrid pairings with Painted Redstart *Myioborus pictus* have been observed (*Tom Johnson*).

throated Redstart *Myioborus miniatus* to Arizona, New Mexico and Texas; Crescent-chested Warbler *Oreothlypis superciliosa* to Arizona and Texas; Golden-crowned Warbler *Basileuterus culicivorus* to New Mexico and Texas; and Grey-crowned Yellowthroat *Geothlypis poliocephala* to Texas (Howell *et al.* 2014). Of these, only the yellowthroat (which formerly bred in southern Texas) and Rufous-capped Warbler are non-migratory; the remainder are partial migrants. A record of the non-migratory Red Warbler *Cardellina rubra* from Pima County, Arizona, in April 2018 was rejected as a likely escape because the individual was an adult of the central Mexican nominate subspecies (Pyle *et al.* 2018) rather than the geographically more proximate northwestern subspecies *melanauris*.

Vagrancy in the remaining tropical and southern temperate species is more limited, as is migratory activity in general, although in many cases this may reflect a lack of knowledge of the natural history of many Neotropical species. For example, Tropical Parula *Setophaga pitiayumi* is known to be migratory at the northern edge of its range in the southern United States, with vagrants reaching California, Colorado and Louisiana. However, eBird data also indicates significant migratory activity in southern South America (Fink *et al.* 2020), in which light vagrants to the Falklands in November 2015 (Woods 2017) and to Chile in June 2017, at the campus of the Universidad de Concepción (Barros 2019), make more sense. Masked Yellowthroat *Geothlypis aequinoctialis* is also migratory at its southern range edge (Capllonch & Ortiz 2007); it was presumably a migrant from this population that ended up providing the first Chilean record of the species in May 2017 at Algarrobo (Barros 2019). Even insular species can occasionally move – Adelaide's Warbler *Setophaga adelaidae* was an endemic of Puerto Rico and Vieques Island, but vagrants colonised St John in the Virgin Islands in January 2015, travelling against prevailing easterly trade winds, and now breed there (Veit *et al.* 2016).

MITROSPINGIDAE
Mitrospingid Tanagers

All four of these tropical tanagers are largely sedentary, and extralimital records are unreported and unexpected.

CARDINALIDAE
Cardinals and allies

This large family includes a relatively high number of migratory species, although most representatives are more sedentary species of the tropics and subtropics. The majority of temperate zone species are highly migratory and have a correspondingly high incidence of long-range vagrancy. Transatlantic movements of this exclusively New World family have occurred in four different genera. Rose-breasted Grosbeak *Pheucticus ludovicianus* is among the most frequent of Nearctic vagrants to the Western Palearctic and shares with many frequent transatlantic vagrants a large population size, boreal breeding distribution and wintering grounds that extend to northern South America. The Azores and Britain have hosted most European records, but some vagrants have reached much further east to Croatia, Sweden and amazingly three records from Malta (Lewington *et al.* 1991), perhaps involving ship-assisted individuals if not escaped cagebirds. Rose-breasted Grosbeaks are regular vagrants west to California, and two reached the Galapagos in April 1983 (Wiedenfeld 2006) with the first Brazilian record in May 2018 from Caxias, Maranhão, in the east of the country (Hamada & Rodrigues 2018). Two species of *Piranga* 'tanagers'

► Rose-breasted Grosbeak *Pheucticus ludovicianus*, West Burra, Shetland, Scotland, 3 May 2016. The vast majority of European records of this species arrive in autumn, and this was only the second UK spring record (*Rebecca Nason*).

have reached Europe. There are four records of Summer Tanager *Piranga rubra* – one from Bardsey Island in Wales in September 1957 and three records from the Azores, all from Corvo since 2006 (Haas 2012). Summer Tanagers winter in western South America, with a single incredible southerly record from Chile (Jaramillo *et al.* 2003), and have a more southerly breeding range than Scarlet Tanager *Piranga olivacea,* with which they share a similar winter range. Scarlet Tanager is considerably more common as a vagrant to the Western Palearctic. There were only 12 records up until 2004 from Iceland, France, Ireland and Britain, but this increased to 54 records by 2019 following increased coverage of the Azores, indicating that Scarlet Tanagers must be frequently wind-drifted into the Atlantic but many do not have the energy to reach continental Europe. The same might be true of Dickcissel *Spiza americana*, known from only one pre-2009 Palearctic record – in Norway in 1981 (Haas 2012) – but eight subsequent records from the Azores. Dickcissels are rare on the United States eastern seaboard but also winter in northern South America and may be vulnerable to wind displacement over the Caribbean. Two apparently vagrant individuals were recorded as far south as the southern Amazon in September 2004, on an island in the Madeira River in Amazonas state (Cohn-Haft *et al.* 2007).

Two species of *Passerina* buntings are accepted as vagrants to Europe, but ascertaining wild origin in this genus is complicated by their high prevalence in the bird trade, at least historically. Prior to 2005 the only Indigo Buntings *Passerina cyanea* widely accepted as wild in Europe were singles in County Cork, Ireland, in 1985, Pembrokeshire, Wales, in 1996 and singles in Iceland in 1951 and 1985; all occurred in October. Two spring records from the Netherlands in June–July 1983 and March 1989 were accepted but fall into a broader pattern of European records deemed to be escapes, including four from Germany and five from Finland, most of which form a cluster in space and time. All subsequent records bar one in Wales and one in Denmark – totalling 60 individuals – have come from the Azores. Elsewhere, there are two records from north-east Russia, in June 1975 and April 1982, both on Curonian Spit, Kaliningrad Oblast (Arkhipov & Ławicki 2016), and one from the Galapagos in May 1979 (Wiedenfeld 2006). Some Indigo Buntings may travel up to 4,000km to reach their winter quarters; thus, spring records may be in keeping with records of North American sparrows but also include undoubted escapes from captivity. The only other *Passerina* bunting to be widely regarded as a transatlantic vagrant is Blue Grosbeak *Passerina caerulea*, one of which one was present in October 2018 on Corvo in the Azores (Ławicki & van den Berg 2018c). Previous records from Britain, the Netherlands, Norway and Sweden were deemed unacceptable, including four British records in March, May (two) and August.

Lazuli Bunting *Passerina amoena* is the western counterpart of Indigo Bunting and just as the latter is a vagrant to the west, so Lazuli Bunting is a vagrant in the east – all the way to Newfoundland and Labrador; it is also a vagrant north to Gambell in the Bering Sea region. European records from the Faroe Islands and Finland have been treated as escapes, as have European records of Painted Bunting *Passerina ciris,* which remains a relatively frequent escape.

◄ Western Tanager *Piranga ludoviciana*, Cornell University, Tompkins, New York, United States, 26 February 2016, a vagrant to the eastern United States and a county first (*Alexander Lees*).

► Blue Grosbeak *Passerina caerulea*, Poço de Agua, Corvo, Azores, 17 October 2018, the first acceptable record for the Western Palearctic (*Vincent Legrand*).

▼ Painted Bunting *Passerina ciris*, Prospect Park, Kings, New York, United States, 4 December 2015, a candidate 'reverse migrant' wintering well to the north of the normal range (*Jay McGowan*).

Notwithstanding the escape risk, Painted Buntings are also frequent vagrants east and west of their North American range and are far more common in the east than Lazuli Bunting, having reached Bermuda on multiple occasions. Varied Bunting *Passerina versicolor* only breeds locally along the southern border of the United States, where it is migratory; vagrants have reached California on several occasions, as well as more exceptionally to Long Point, Ontario, Canada, and Sarasota County, Florida (Woolfenden & van Deventer 2006), with another at Allegheny, Pennsylvania, in 2018 (eBird). Two highly migratory western species, Black-headed Grosbeak *Pheucticus melanocephalus* and Western Tanager *Piranga ludoviciana,* have been widely recorded in eastern North America and the Caribbean, and the first Western Tanager for South America was photographed at Reserva Otongachi in Pichincha province, Ecuador, in May 2019; this is a classic time of year for individuals of short-range Nearctic species to have migrated in the 'wrong' direction from Central American wintering areas (Freile *et al.* 2019).

Vagrancy is also widely documented in several largely sedentary species. This includes extralimital records of Northern Cardinal *Cardinalis cardinalis* from, for example, California although many records

◄ Flame-coloured Tanager *Piranga bidentata*, Big Bend NP, Brewster, Texas, United States, 25 July 2016, a vagrant to the United States border in Texas and Arizona (*Jay McGowan*).

probably relate to escapes (Hamilton *et al.* 2007). Another such example is Pyrrhuloxia *Cardinalis sinuatus,* with a few remarkable accepted extra-limital records from Oregon, United States, and Ontario, Canada. Several other essentially resident tropical members of the Cardinalidae have reached the southern United States, including Blue Bunting *Cyanocompsa parellina* in Texas, Crimson-collared Grosbeak *Rhodothraupis celaeno* in southern Texas and Flame-coloured Tanager *Piranga bidentata* in Arizona (where they have even bred) and Texas (Howell *et al.* 2014).

As a group, the *Pheucticus* grosbeaks seem to be particularly prone to vagrancy. In addition to the two species already named, the northern subspecies of Yellow Grosbeak *Pheucticus chrysopeplus dilutus* is migratory and an occasional vagrant to Arizona and New Mexico; records in other US states have been treated as escapes (Howell *et al.* 2014). In South America, southern populations of Black-backed Grosbeak *Pheucticus aureoventris* are austral migrants and are regular in south-western Brazil, with apparent vagrants occurring east to São Paulo state (Carlos 2018) and north to Goias and southern Amazonas (Serpa *et al.* 2014).

THRAUPIDAE
Tanagers and allies

With the recent taxonomic rearrangements informed by the latest phylogenetic research, the Thraupidae now encompasses 104 genera and 377 species of largely resident and mostly tropical bird species. Only one species, Morelet's Seedeater *Sporophila morelleti* is native to the United States, breeding along the Mexican border in southern Texas, but four species are vagrants to Florida: Red-legged Honeycreeper *Cyanerpes cyaneus,* of which there are several records, currently viewed with suspicion by rarities committees but regarded by Howell *et al.* (2014) as likely genuine vagrants; Bananaquit *Coereba flaveola*; Black-faced Grassquit *Melanospiza bicolor;* and Yellow-faced Grassquit *Tiaris olivaceus*, which is also a vagrant to southern Texas. None of these species are particularly migratory, but island-hopping dispersal is prevalent among many such insular species, so these Florida records of Caribbean species match a pattern seen in other families.

Migratory activity in the tanagers is most prevalent in the southern cone of South America, where several species formerly classed as emberizids are at least partial austral migrants. These include Patagonian Sierra-Finch *Phrygilus patagonicus*, Mourning Sierra-Finch *Rhopospina fruticeti* and Common Diuca-Finch *Diuca diuca,* whose southern populations retreat north in winter. All three have reached the Falklands (Woods 2017) and the last two species have occurred as vagrants to Rio Grandse do Sul in southern Brazil (Franz *et al.* 2018). Grassland Yellow-Finch *Sicalis luteola* has also occurred on the Falklands, along

► Mourning Sierra-Finch *Rhopospina fruticeti*, Cape Pembroke, Falklands, 26 June 2019, one of around a dozen records of this widespread South American species from the Falklands (*Alan Henry*).

with Double-collared Seedeater *Sporophila caerulescens,* which has now occurred twice there (Woods 2017) and once in Chile (Barros *et al.* 2014). Woods (2017) speculated that these Falklands records may have been ship-assisted, and intriguingly six Double-collared Seedeaters, a White-banded Mockingbird, a Mourning Sierra-Finch and an Eared Dove all came aboard a cruise liner off Rio Grande do Sul in April 2013 (Tamarozzi 2013). Many other *Sporophila* seedeaters are highly migratory and have been recorded frequently as vagrants, often in association with flocks of other seedeaters. There is a single record of Lined Seedeater *Sporophila lineola* south to Chile: a single male in December 2011 in the Antofagasta region at an altitude of 4,250m, which is quite remarkable for this lowland species (González 2013); there are also records north to Guadeloupe in the West Indies (Kirwan *et al.* 2019) and Costa Rica (Garrigues *et al.* 2016). Yellow-bellied Seedeater *Sporophila nigricollis* is a vagrant to Argentina and Paraguay (Fast *et al.* 2019).

Away from austral latitudes, relatively few tanagers are known to make migratory movements, and vagrancy events have been documented from even fewer. Black-backed Tanager *Stilpnia peruviana* is a partial austral migrant at least, retreating from the extreme south of the Atlantic Forest of Brazil in the austral winter; vagrants have reached as far north as southern Bahia. Movements of Fawn-breasted Tanager *Pipraeidea melanonota* are poorly known (Ortiz & Capllonch 2008) but there are odd records well out of range in Brazil, north to Pernambuco (WikiAves) and to Alta Floresta in the southern Amazon (Lees *et al.* 2013), and south to coastal Argentina including at the vagrant trap at Punta Rasa (Jaramillo 2000). Meanwhile Black-goggled Tanager *Trichothraupis melanops,* which is thought to be sedentary, has been recorded extralimitally several times recently in extreme southern Brazil and Argentina (Gomes *et al.* 2020, Pearman & Areta 2020). Movements of Swallow Tanagers *Tersina viridis* are poorly understood and have been classified in some texts as nomadic – large flocks may be observed in visible migration in some areas, and apparent vagrants have been recorded with some frequency. Zelaya *et al.* (2013) reported records of vagrants from the Chaco Seco and the Córdoba and Formosa regions of Argentina, whilst other Argentinian records of vagrant Swallow Tanagers include one from Los Berros in departamento Sarmiento, San Juan (Lucero 2013), and there is at least one record from Punta Rasa on the coast (Jaramillo 2000). Even more unexpected, a female Swallow Tanager – the first for Chile – was photographed in April 2015 at high and dry Salar de Aguas Calientes, Antofagasta (Barros *et al.* 2016), and to the north there is a single record from the Cayman Islands (Kirwan *et al.* 2019).

One of the most enigmatic migratory South American birds is Black-and-white Tanager *Conothraupis speculigera*, which appears to be restricted as a breeder to a relatively small region of arid south-west Ecuador and north-west Peru, migrating across the Andes to the southwestern Amazon in the non-breeding season. Two records would seem to relate to vagrants. The first is a female-plumaged bird collected in October 1969 in Putumayo, Colombia, the only national record and around 300km out of range (Lobo-y-Henriques *et al.* 2011). The second, a male, was mist-netted in March 2011 at Awala-Yalimapo in north-west French Guiana, a staggering 2,800km out of range (Claessens *et al.* 2012); it represents one of the most remarkable instances of vagrancy recorded in any tropical bird.

AVIAN VAGRANCY IN AN ERA OF GLOBAL CHANGE

Humans have been reorganising the biosphere for millennia by extinguishing species, altering habitats and changing ecological processes, but the pace and magnitude of the change has increased in the last century – the 'Great Acceleration' (Steffen *et al.* 2007). These changes will favour some species but threaten far more. In order to survive in our anthropogenic landscapes, organisms need to be able to adapt to changing conditions or move with changing climate 'envelopes' of suitable environmental conditions. Vagrants are at the vanguard of such change and may be extremely important for range expansion and thus in facilitating resilience to human impacts. Although it is hardly a conservation success story, non-forest species have quickly colonised agricultural areas formerly covered in rainforest in the Brazilian Amazon (Mahood *et al.* 2012). In some instances this has involved impressive long-range dispersal events. One good example is the arrival in 2013 of Guira Cuckoos *Guira guira* in newly deforested areas in south-east Peru (CRAP 2019), where dispersing vagrants crossed a wide belt of unsuitable rainforest vegetation to arrive in the recently cleared pastures in Madre de Dios. We therefore need to ensure that natural vagrancy processes are left as uninhibited as possible.

We already have a reasonable idea of the future of changes in patterns of vagrancy we might anticipate. Moss (1998) made a number of predictions about the impacts of climate change on avian vagrancy to Britain, for example. These are: 1) spring overshooting to increase, including more North African species; 2) vagrancy from the Eastern Palearctic to decrease; 3) spring appearances of both transatlantic and Eastern Palearctic vagrants to increase due to greater survival rates resulting from milder winters; 4) transatlantic vagrancy to decrease and shift further north; and 5) vagrancy from 'unexpected quarters' to increase, especially irruptive behaviour from the north and east. Twenty years on we can evaluate some of these predictions and make some of our own. It seems likely that records of some Iberian species have increased in Britain, although we need to be careful to tease out the effects of enhanced observer awareness for species such as Iberian Chiffchaff and range expansions for species like Black-winged Stilt, which may be driven by habitat creation. Likewise, many southern European 'farmland' species are declining rapidly due to habitat degradation and as such we are unlikely to see increases in numbers of species like Little Bustard, Tawny Pipit and Ortolan Bunting, which have contracted in range and population sizes towards regions where agricultural modernisation has been less aggressive.

▼ Tawny Pipit *Anthus campestris*, Bryher, Isles of Scilly, 27 September 2018. This species is becoming rarer as a vagrant to the British Isles, the result of Europe-wide population declines (*Rik Addison*).

► Ivory Gull *Pagophila eburnea*, Plymouth Town Wharf, Plymouth, Massachusetts, United States, 20 January 2009. Ivory Gulls are threatened by both climate change and bioaccumulation of contaminants – especially organochlorines – and future declines are likely (*Ian Davies*).

We have yet to see decreases in many Eastern Paleartic species, other than those directly impacted by overharvesting, such as Yellow-breasted Buntings. Moreover, Jiguet & Barbet-Massin (2013) found that climate change was likely to trigger both increases and decreases in vagrancy rates in different Siberian species, depending on their current ranges. Spring records of many Siberian species have increased, and this may be facilitated by milder winters allowing species to survive further north. Equally, however, this may reflect greater numbers of vagrants or pseudo-vagrants arriving in autumn. There appears to be limited evidence of a shift in arrivals of Nearctic vagrants. Their arrivals are dependent on the position of the jet stream and associated storm tracks, and vagrants are still arriving on a broad front from the Azores to Iceland. Likewise, changes in 'unexpected vagrancy' are rather confounded by greater observer coverage and greater observer skill.

That said, Moss's predictions still seem intuitive, and more severe weather seems likely to cause more vagrancy or different vagrancy, especially in the case of hurricane-driven seabirds, which may become a more regular phenomenon in the eastern North Atlantic and South Atlantic. However, the 'bounce' in records is likely to be short-lived, as severe weather events can have major impacts on avian populations (Newton 2007) and storm intensity from one year to the next has been shown to be associated with trends in populations in North American passerines (Butler 2000). Long-distance migrants are likely to be most negatively affected by climate change. Those species that are able to reduce their migration span are likely to fare better and there is already evidence for these micro-evolutionary changes in European bird species (Visser *et al.* 2009). A reduction in long-distance migratory behaviour will result in a corresponding decrease in long-distance vagrancy. The environmental teleconnections of these declines will be felt across hemispheres and amplified by land-use change impacts. Vagrants therefore represent a powerful means for people to connect with nature, with their occurrence tied to the fate of habitats in distant lands. This becomes a compelling reason to support conservation initiatives in places that most birders will never themselves physically visit, but where conservation success can still be measured tens of thousands of kilometres away by the continued movement of vagrants.

The future of rare bird discovery and 'twitching'

With impacts from global climate change on ecosystems, biodiversity and people being felt increasingly (Pecl *et al.* 2017), there is a pressing need to reduce the carbon footprint of amateur and professional ornithology (Caletrio 2018). Vagrant hunting and – even more so – vagrant chasing is often a very high-carbon hobby, especially when it involves long-haul flights from one corner of a biogeographic realm to another, chartered short-haul flights aboard small aircraft or epic long-distance car journeys, all of which we have ourselves done in the quest to see rare birds. Reducing as much as possible our dependence on high-carbon travel need not, however, spell an end to the excitement of birding. Many birders have taken to using public transport or cycling to chase rare birds, and a combination of bike, train and taxi will put many continental European rarities within range of British birders and vice versa. Unfortunately, many parts of the world do not have such good public transport links as Europe, but clearly we should be striving to convince politicians that these are a necessity.

▲ Birders attending a twitch for a Japanese Robin *Larvivora akahige* (inset), Lianhuachi Park, Beijing, China, 24 November 2012, the second record of this species for Beijing (*Terry Townshend*).

Avoiding planes, the highest-carbon form of travel, incurs longer travel times and will likely necessitate more overnight stays. The economic value of twitching as a form of ecotourism will thus increase if travel is greener, as people will spend longer at sites geographically distant from their homes. The value of such activity is potentially large. For example, Callaghan *et al.* (2018) found that the Black-backed Oriole in Pennsylvania attracted 1,824 observers from across the United States, stimulating travel activity valued at about US$223,000 over 67 days. A similar study by the same author team found that expenditure by birders twitching Aleutian Terns in Australia ranged from AU$199,000–$363,000 (Callaghan *et al.* 2020) and that the birders were 'cumulatively willing to donate upwards of AU$30,000 to a non-governmental conservation organisation in order to have viewed the terns'. It is now relatively routine to have on-site charity collections at sites hosting vagrant birds in the United Kingdom, and these spontaneous collections do help to diffuse any ill-feeling towards hordes of visiting birders, especially in built-up areas.

Beyond changing the ways in which we travel to see birds, there is also an opportunity to change the way in which we go birding. Local patch birding has always attracted stalwarts but in an age when emissions need to be reduced, stimulating more patch birding represents a major opportunity to reduce the activity's carbon footprint. Someone who stays local will usually stand a lower chance of encountering rare birds than someone who is prepared to travel thousands of kilometres overnight, yet rarity in itself is of course relative. Species can be vagrants at local, subnational and continental scales. Collapsing one's horizons and focussing at a local level, your reference point for rarity will have to be realigned – and with enough effort there is still scope to find nationally rare species almost anywhere. The pull of islands in attracting vagrants is well understood by birders, and this book's examples are very biased towards oceanic islands; we previously described the 'Fair Isle Effect' (Lees & Gilroy 2014) in recognition of the draw that small, isolated oceanic islands have on vagrants. However, oceanic and coastal islands are not the only islands to concentrate migrant and vagrant birds. Urban parks are 'islands' of high-quality habitat in a sea of concrete and glass; so too are desert oases or woodland copses surrounded by intensive agriculture or scrubby hilltops amidst barren moorland. Finding and identifying such local hotspots will result in the discovery of many more vagrants. So too will an increasingly bird-literate public. Equipped with access to free identification tools such as Merlin (https://merlin.allaboutbirds.org) anyone with a smartphone could potentially find and identify vagrants in their gardens.

▶ White-billed Diver *Gavia adamsii*, River Witham between Stixwould and Kirkstead, Lincolnshire, England, 28 January 2017. This same stretch of the River Witham has twice hosted White-billed Divers, otherwise recorded inland in England only twice before – an amazing coincidence (*Alexander Lees*).

The internet and applications like Google Earth are not only tools to work out where rare birds might be found, they are also tools to find rare birds directly. Even if we can't physically travel, webcams provide a gateway to rare bird finding across the globe. Famous ones such as on Southeast Farallon Island routinely host vagrants – sometimes even the webcam watchers find them before the birders on site. Likewise, browsing through vast online archives of bird images on sites like eBird https://www.ebird.org/ and WikiAves http://www.wikiaves.com/ has proven a fruitful source of rare bird discoveries, and between us we have found considerably more national firsts on the internet than in the flesh. Granted, that is not as exciting as finding a bird in the field, but it is still *exciting*. In the future, remote autonomous recording stations using artificial intelligence may routinely record and identify rare birds independently of human mentors, but for the time being human processing of camera traps, autonomous sound recorders and even webcams have produced records of many vagrants. And the uptake of such technology is increasing rapidly.

Advances in the scientific understanding of vagrancy and future research directions

Recent years have seen a steady growth in scientific interest in the study of vagrancy as a biological phenomenon, and this growth has been underpinned by a number of important advances. The crowd-sourcing of large open-access biodiversity datasets containing 'complete checklist' data permits broad-scale analyses of where birds occur in space and time (e.g. Sullivan *et al.* 2014). Combined with detailed environmental data layers and meteorological information, they provide novel opportunities to understand what drives patterns of vagrancy at global scales (e.g. Farnsworth *et al.* 2015). Advances in molecular genetic techniques now mean that the identification of cryptic bird taxa that may be impossible to identify with certainty (based on current knowledge) under field conditions is now routine, based on faeces or shed feathers collected non-invasively (Collinson 2017).

DNA also offers a window on the origins of vagrants. For example, Engel *et al.* (2011) carried out a genetic assay of the mitochondrial cytochrome b gene of three vagrant Cave Swallow *Petrochelidon fulva* specimens from Illinois, New York and New Jersey; this indicated that they pertained to populations from the southwestern United States or Mexico. As range-wide sampling for taxa increases, both DNA and stable isotopes offer unique chances to ascertain broad geographic origins of individual vagrants. To understand the origins and ultimately the fate of individual vagrants, which remains a major point of contention for competing theories, then newer technologies offer some promise of resolution of some of these mysteries. Historically we have relied on bird ringing to give us an insight into the movement of vagrants, but recoveries are always very rare and typically only give us two data points, e.g. a point of capture as a vagrant in western Europe and a point of recovery. The instigation of colour-ringing/marking schemes for some species has shed more light as some individual vagrants of larger species can now be field recognised without recourse to recapture. For example, a vagrant Lesser Scaup

◄ Pacific-slope Flycatcher *Empidonax difficilis*, Central Park, New York, New York, United States, 22 November 2015 (*Jay McGowan*). The first state record was confirmed by voice and DNA by Goldberg & Mason 2017 (inset image of the flycatcher's faecal matter; *Alexander Lees*).

▲ Richard's Pipit *Anthus richardi*, Lespignan, Hérault, Finistère, France, 12 November 2020. This bird was fitted with this geolocator in the previous winter at this site and should now reveal where it spent the summer and hence the geographic origins of the apparent pseudo-vagrant population of Richard's Pipits in south-west Europe (*Paul Dufour*).

Aythya affinis, which was fitted with a nasal saddle at São Jacinto in Portugal in December 2013 was relocated at Llangorse Lake, Powys, in Wales in October 2014 before reappearing in West Yorkshire, England, in May 2015 and then in Perth/Kinross, Scotland, in July 2016. In order to gain fine temporal-scale understanding of movements, tracking technologies are needed. These include radio-tracking, which may give a relatively brief understanding of dispersal vectors; data loggers (which necessitate recapture) for vagrant individuals that have shown multi-year site fidelity and can be reliably recaptured; and finally satellite tags, which give the most accurate data and do not require recapture but also come with minimum size limits and substantial financial costs. As these technologies are refined it seems likely that we will soon answer some of the more intriguing questions about the origins and fate of vagrants, which will in turn open up a raft of new questions to beguile us.

REFERENCES

Abbasi, E., Moazeni, M. & Khaleghizadeh, A. 2019. First record of Pheasant-tailed Jacana *Hydrophasianus chirurgus* in Iran. *Sandgrouse* 41: 214–215.

Abhinav, C. & Dhadwal, D.S. 2014 European Golden Plover *Pluvialis apricaria* at Pang Lake, Himachal Pradesh, India. *Indian Birds* 9: 149–151.

Able, K.P. & Belthoff, J.R. 1998. Rapid 'evolution' of migratory behaviour in the introduced house finch of eastern North America. *Proceedings of the Royal Society of London. Series B: Biological Sciences*, 265: 2063–2071.

Acosta, J.C. & Fava, G.A. 2016. Registro documentado de Jote Grande de cabeza amarilla (*Cathartes melambrotus*) en Argentina. *Nuestras Aves* 60: 51–52.

Agostini N., Baghino L., Pannuccio, M., Premuda, G. & Provenza, A. 2004 The autumn migration strategies of adult and juvenile short-toed eagles *Circaetus gallicus* in the central Mediterranean. *Avocetta* 28: 37–40.

Ahmad, M. 2005. Franklin's Gulls and Laughing Gulls in Britain & Ireland in November 2005. *Birding World* 18: 461–464.

Åkesson, S., Morin, J., Muheim, R. & Ottosson, U., 2005. Dramatic orientation shift of displaced birds in response to the geomagnetic field. *Current Biology* 15: 1591–1597.

Åkesson, S. & Hedenström, A. 2007. How migrants get there: migratory performance and orientation. *BioScience* 57: 123–133.

Åkesson, S., Ilieva, M., Karagicheva, J., Rakhimberdiev, E., Tomotani, B. & Helm, B. 2017. Timing avian long-distance migration: from internal clock mechanisms to global flights. *Philosophical Transactions of the Royal Society of London. Series B: Biological Sciences* 372: 20160252.

Åkesson, S. & Helm, B. 2020. Endogenous programs and flexibility in bird migration. *Frontiers in Ecology and Evolution* 8: 78.

Alam, A.B.M.S., Ahmed, S. & Ahammed, R. 2016 Egyptian Vulture *Neophron percnopterus*, the second record for Bangladesh. *BirdingASIA* 25: 104–105.

Alam, A.B.M.S., Ahmed, S., Azmiri, K. Z. & Tareq, O. 2019. Sykes's Nightjar *Caprimulgus mahrattensis*: first record for Bangladesh. *BirdingASIA* 31: 112–114.

Albano, C. 2015. WA1670766, *Glareola pratincola* (Linnaeus, 1766) http://www.wikiaves.com/1670766

Aldabe, J., Rocchi, A. & Mondón, G., 2010. Primer registro de *Chlidonias leucopterus* (Charadriiformes: Sternidae) para Brasil y Sudamérica. *Revista Brasileira de Ornitologia* 18: 261–262.

Aleixo, A. & Galetti, M. 1997. The conservation of the avifauna in a lowland Atlantic forest in south-east Brazil. *Bird Conservation International* 7: 235–261.

Alerstam, T. 1978. Analysis and a theory of visible bird migration. *Oikos* 30: 273–349.

Alerstam, T. 1981. The course and timing of bird migration. Pp 9–54 in: Aidley D. J., ed. *Animal Migration*. Society for Experimental Biology Seminar Series. Vol. 13. Cambridge University Press. Cambridge, United Kingdom.

Alerstam, T., 1990. Ecological causes and consequences of bird orientation. *Experientia* 46: 405–415.

Alerstam, T., Gudmundsson, G.A., Green, M. & Hedenström, A. 2001. Migration along orthodromic sun compass routes by Arctic birds. *Science* 291: 300–303.

Alferez, J. & Delgado, S. 2019. Primer registro de *Chrysolampis mosquitus* para el departamento de Madre de Dios y el Perú. *Boletín de la Unión de Ornitólogos del Perú* 14: 15–18.

Alfrey, P. & Ahmad, M. 2007. Short-billed Gull on Terceira, Azores, in February–March 2003 and identification of the 'Mew Gull complex'. *Dutch Birding* 29: 201–212.

Alfrey, P., Buckell, S., Legrand, V. & Monticelli, D. 2010. Birding on Corvo, Azores, and Nearctic vagrants in 2005–09. *Dutch Birding* 32: 299–315.

Alfrey P., Monticelli, D. & Legrand, V. 2018. Nearctic vagrants on Corvo, Azores, 2005–2017. *Dutch Birding* 40: 297–317.

Ali, S. & Ripley, S.D. 1978. *Handbook of the Birds of India and Pakistan: together with those of Bangladesh, Nepal, Bhutan and Sri Lanka*. Vol. 1. Second edition. Oxford University Press, Delhi.

Allan, D.G. & Perrins, N.D. 2019. First confirmed African record of Tahiti Petrel *Pseudobulweria rostrata*. *Marine Ornithology* 47: 89–91.

Allan, D. 2020. White Helmetshrike (*Prionops plumatus*), version 1.0. In *Birds of the World* (J. del Hoyo, A. Elliott, J. Sargatal, D. A. Christie, and E. de Juana, Editors). Cornell Lab of Ornithology, Ithaca, NY, USA. https://doi.org/10.2173/bow.whihel1.01

Allen, D., Española, C.P., Broad, G., Oliveros, C. & Gonzalez, J.C.T. 2006. New bird records for the Babuyan Islands, Philippines, including two first records for the Philippines. *Forktail* 22: 57–70.

Allport, G. 2018. First record of Pectoral Sandpiper *Calidris melanotos* for Mozambique. *Bulletin of the African Bird Club* 25: 73–77.

Allport, G. 2020. First record of White-rumped Sandpiper *Calidris fuscicollis* for Mozambique. *Bulletin of the African Bird Club* 27: 240–245.

Al-Sirhan, A. & Al Hajji, R. 2010. The first Asian Koel for Kuwait. *Phoenix* 25: 8.

Alström, P. 1991 A Radde's Accentor *Prunella ocularis* from Oman reidentified as Black-throated Accentor *P. atrogularis*. *Sandgrouse* 2: 106–108.

Alves, F. 2013. WA1194475, *Chauna torquata* (Oken, 1816). http://www.wikiaves.com/1194475

American Ornithologists' Union 1998. Check-list of North American Birds, 7th edition. American Ornithologists' Union, Washington, DC, USA.

Ammitzboell, N.P., Werner, S., Marques, D.A. & Schweizer, M. 2017. Pacific Loon at Silvaplanersee, Switzerland, in December 2015, with notes on genetics, identification and WP records. *Dutch Birding* 39: 228–238.

Amos, E.J.R. 1991. *The Birds of Bermuda*. Corncrake Press, Bermuda.

Andersen, M. J., Shult, H. T., Cibois, A., Thibault, J. C., Filardi, C. E. & Moyle, R. G. 2015. Rapid diversification and secondary sympatry in Australo-Pacific kingfishers (Aves: Alcedinidae: *Todiramphus*). *Royal Society Open Science* 2: 140375.

Anderson, R.C. 2007. New records of birds from the Maldives. *Forktail* 23: 135–144.

Anderson, R.C. & Baldock, M. 2001. New records of birds from the Maldives, with notes on other species. *Forktail* 17: 67–74.

Andrews, I.J., Khoury, F. & Shirihai, H. 1999. Jordan Bird Report 1995–97. *Sandgrouse* 21: 10–35.

Andrianarimisa, A. 1995 A record of the Sunbird-asity *Neodrepanis coruscans* in the Réserve Spéciale d'Ambohitantely. *Working Group on Birds of the Madagascar Region Newsletter* 5: 8–9.

Angehr, G.R. & Dean, R. 2010. *The Birds of Panama: a field guide*. Cornell University Press, Ithaca, NY.

Anon. 1993. Finland's first Madeiran Petrel. *Birding World* 6: 65–66.

Anon. 1995. European News. *British Birds* 88: 26–45.

Anon. 1998a. Bird News February 1998. *Birding World* 11: 48.

Anon. 1998b. Africa round-up. Mass kill of Red-billed Hornbills in Zimbabwe. *Bulletin of the African Bird Club* 5: 8.

Anon. 2015. Bulletin 9:01 Rare Birds in Israel https://www.israbirding.com/irdc/bulletins/bulletin_9/

Antonov A.I., Avdeyuk S.V., Leader P., Carey, G. & Stanton D. 2012. New data on some of protected and rare birds in Northeastern Primorie. *Far-Eastern Journal of Ornithology* 3: 77–79.

Ararat, K. 2016. First record of Lesser Spotted Eagle *Clanga pomarina*, first breeding record of Eurasian Penduline Tit *Remiz pendulinus* and first records of Eastern Bonelli's Warbler *Phylloscopus orientalis*, Olive Tree Warbler *Hippolais olivetorum* and Goldcrest *Regulus regulus*, for Iraq. *Sandgrouse* 38: 106–109.

Araya-H., D., Contreras, C. & Sandoval, L. 2015. Gray-bellied Hawk, *Accipiter poliogaster* (Temminck, 1824) (Accipitriformes: Accipitridae), in Costa Rica. *Check List* 11: 1559.

Areta, J.I. & Bodrati, A. 2010. A longitudinal migratory system within the Atlantic Forest: Seasonal movements and taxonomy of the Golden-rumped Euphonia (*Euphonia cyanocephala*) in Misiones (Argentina) and Paraguay. *Ornitologia Neotropical* 21: 71–86.

Areta, J.I., Bodrati, A., Thom, G., Rupp, A.E., Velazquez, M., Holzmann, I., Carrano, E. & Zimmermann, C.E., 2013. Natural history, distribution, and conservation of two nomadic *Sporophila* seedeaters specializing on bamboo in the Atlantic forest. *The Condor* 115: 237–252.

Arévalo, J.E. & Sánchez, J. 1999. Oilbird *Steatornis caripensis* in Costa Rica. *Cotinga* 11: 96.

Ariunbaatar, B., Buuveibaatar, B. & Otgonsuren, A. 2017 First record of Ashy Minivet *Pericrocotus divaricatus* for Mongolia. *BirdingASIA* 27: 118–119.

Arizaga, J., Crespo, A., Telletxea, I., Ibáñez, R., Díez, F., Tobar, J. F., Minondo, M., Ibarrola, Z., Fuente, J. J. and Pérez, J. A. 2015. Solar/Argos PTTs contradict ring-recovery analyses: Woodcocks wintering in Spain are found to breed further east than previously stated. *Journal of Ornithology* 156: 515–523.

Arkhipov, V.Y., Haas, M. & Crochet, P.A. 2010. Western Palearctic list updates: Yellow-eyed Dove. *Dutch Birding* 32: 191–193.

Arkhipov V.L & Blair M.I. 2007. Skua (*Catharacta, Stercorarius*) occurrence in the OSME region. *Sandgrouse.* 29: 183–204.

Arkhipov, V.Y. & Ławicki, Ł. 2016 Nearctic passerines in Russia. *Dutch Birding* 38: 201–214.

Arnarson, B. & Brynjólfsson, B. 2000. Elrisongvari i Sudursveit. [Blyth's Reed Warbler *Acrocephalus dumetorum*: new to Iceland]. *Bliki* 21: 59–60.

Ash, J.S. & Miskell, J.E. 1998. *Birds of Somalia*. Pica Press, Robertsbridge, UK.

Ash, J.S. & Atkins, J. 2009. *Birds of Ethiopia and Eritrea: an Atlas of Distribution*. Christopher Helm, London.

Audevard, A. 2018. Observation d'une Alouette monticole *Melanocorypha bimaculata* à Hyères en 2015: première mention française de l›espèce. *Ornithos* 25: 312–317.

Ausilio, E. & Zotier, R. 1989. Vagrant birds at iles Kerguelen, southern Indian Ocean. *Marine Ornithology* 17: 9–18.

Austin, J.J., Soubrier, J., Prevosti, F. J., Prates, L., Trejo, V., Mena, F. & Cooper, A. 2013. The origins of the enigmatic Falkland Islands wolf. *Nature Communications* 4: 1552.

Avendaño, J.L.M. 2015. Primer registro del Loro de Frente Turquesa (*Amazona aestiva*) para el Perú. *Boletín de la Unión de Ornitólogos del Perú* 10: 19–21.

Aymí, R. & Gargallo, G. 2020. Layard's Warbler (*Sylvia layardi*), version 1.0. In *Birds of the World* (J. del Hoyo, A. Elliott, J. Sargatal, D. A. Christie, and E. de Juana, Editors). Cornell Lab of Ornithology, Ithaca, NY, USA. https://doi.org/10.2173/bow.laywar2.01

Baatargal, O. & Suuri, B. 2013. Chestnut-flanked White-eye *Zosterops erythropleurus*: first record for Mongolia. *BirdingASIA* 20: 111–112.

Babbington, J. & Roberts, P. 2014. Further records of Small Buttonquail *Turnix sylvaticus* and 'mangrove white-eye' *Zosterops* sp in southwest Saudi Arabia. *Sandgrouse* 36: 50–52.

Baeten S., Vanhove F., Lebrun R., Demeulemeester M. & the members of the BRBC. 2019. *Rare Birds in Belgium in 2018 Report of the Belgian Rare Birds Committee (BRBC)*: 1–16.

Baha el Din, M. & Baha el Din, S. 1996. The first Oriental Pratincole *Glareola maldivarum* in Egypt. *Sandgrouse* 18: 64–65.

Bailey, R.S., Pocklington, R. & Willis, P.R. 1968. Storm-petrels *Oceanodoma* spp. in the Indian Ocean. *Ibis* 110: 27–34.

Bairlein, F., Norris, D.R., Nagel, R., Bulte, M., Voigt, C. C., Fox, J.W., Hussell, D.J. & Schmaljohann, H. 2012. Cross-hemisphere migration of a 25g songbird. *Biology Letters* 8: 505–507.

Baker-Gabb, D.J., Benshemesh, J. & Maher, P.N. 1990. A revision of the distribution, status and management of the Plains-wanderer *Pedionomus torquatus*. *Emu* 90: 161–168.

Bakewell, D. 2009. Malaysian bird report 2008. *Malayan Nature Journal* 61: 243–256.

Balbontin, J., Negro, J.J., Sarasola, J.H., Ferrero, J.J. & Rivera, D. 2008. Land-use changes may explain the recent range expansion of the Black-shouldered Kite *Elanus caeruleus* in southern Europe. *Ibis* 150: 707–716.

Banerjee, M. & Inskipp, T. 2013. First photographic record of Fieldfare *Turdus pilaris* from the Indian Subcontinent. *Indian Birds* 8: 77.

Bannerman, D.A. 1919. VI.–List of the Birds of the Canary Islands, with detailed reference to the Migratory Species and the Accidental Visitors. Part 1. Corvidæ–Sylviidae. *Ibis* 61: 84–131.

Barcelos, L. M., Rodrigues, P. R., Bried, J., Mendonça, E. P., Gabriel, R. & Borges, P. A. V. 2015. Birds from the Azores: An updated list with some comments on species distribution. *Biodiversity Data Journal* 3: e6604.

Barlow, C.R. 2007. First record of Yellow-browed Warbler *Phylloscopus inornatus* for The Gambia. *Bulletin of the African Bird Club* 14: 74–75.

Barlow, M. 1976. Breeding of the Hoary-headed Grebe in Southland. *Notornis* 23: 183–187.

Barnes, S., Burgiel, J., Elia, V., Hanson, J., Larson, L. & Lehman, P. 2006. New Jersey Bird Records Committee–Annual Report 2006. *New Jersey Birds* 32: 66–76.

Barré, N. & Bachy, P. 2003. Complément à la liste commentée des oiseaux de Nouvelle-Calédonie. *Alauda* 71: 31–39.

Barros, R. 2015. El Mirlo de pico corto *Molothrus rufoaxillaris*, una nueva especie para Chile. *La Chiricoca* 19: 36–44.

Barros, R. & Schmitt, F. 2013. Resumen de avistamientos marzo–agosto 2012. *La Chiricoca* 16: 24–37.

Barros, R., Schmitt, F. & la red de observadores de aves. 2014. Resumen de avistamientos marzo – agosto 2013. *La Chiricoca* 18: 20–31.

Barros R., Jaramillo A. & Schmitt, F. 2015. Lista de las aves de Chile 2014. *La Chiricoca* 20: 79–100.

Barros & la red de observadores de aves. 2016. Resumen de Avistamientos, Enero – Diciembre 2015. *La Chiricoca* 21: 21–46.

Barros, R. & la red de observadores de aves. 2017. Resumen de Avistamientos, Enero – Junio 2016. *La Chiricoca* 22: 28–48.

Barros, R. & la red de observadores de aves. 2019. Resumen de Avistamientos, Enero– Diciembre 2017. *La Chiricoca* 24: 25–56.

Barthel, P.H., Bezzel, E., Krüger, T., Päckert, M. & Steinheimer, F.D. 2018: Checklist of the birds of Germany 2018: updates and changes. *Vogelwarte* 56: 205–224.

Barton, D.C., Lindquist, K.E., Henry III, R.W. & Luna Mendoza, L.M. 2004. Landbird and waterbird notes from Isla Guadalupe, Mexico. *Western Birds* 35: 186–196.

Basnet, S. & Chaudhary, B. 2003. Purple-backed Starling *Sturnus sturninus*: a new species for Nepal. *Forktail* 19: 129.

Batchelor, A.L., 1979. Records of Red-tailed and White-tailed Tropicbirds in South African waters. *Marine Ornithology* 7: 21–23.

Batista, R., Olsson, U., Andermann, T., Aleixo, A., Ribas, C.C. & Antonelli, A. 2020. Phylogenomics and biogeography of the world's thrushes (Aves, *Turdus*): new evidence for a more parsimonious evolutionary history. *Proceedings of the Royal Society of London. Series B Biological Sciences* 287: p.20192400.

Batty, C. 2010. Pied Crows in Western Sahara, Morocco. *Dutch Birding* 32: 329–332.

Baxter, C. 1989. A Spangled Drongo on Kangaroo Island. *South Australian Ornithologist* 30: 197–198.

Beck, W. & Wiltschko, W. 1982. The magnetic field as a reference system for genetically encoded migratory direction in pied flycatchers (*Ficedula hypoleuca* Pallas). *Zeitschrift für Tierpsychologie* 60: 41–46.

Beekman, J., Koffijberg, K., Wahl, J., Kowallik, C., Hall, C., Devos, K., Clausen, P., Hornman, M., Laubek, B., Luigujõe, L. & Wieloch, M. 2020. Long-term population trends and shifts in distribution of Bewick's swans *Cygnus columbianus bewickii* wintering in northwest Europe. *Wildfowl* 69: 73–102.

Beer, K. C. J., Rehm, E. M. & Savidge, J. A. 2016. First record of the Ashy Minivet (*Pericrocotus divaricatus*), a passerine bird, for Saipan, Commonwealth of the Northern Mariana Islands. *Micronesica* 2: 1–4.

Bege, L.A.R. & Pauli, B.T. 1988. As aves nas Ilhas Moleques do Sul – Santa Catarina: Aspectos da ecologia, etologia e anilhamento de aves marinhas. Florianópolis. FATMA, 64p.

Bell, M. & Bell, D. 2000. First wrybill (*Anarhynchus frontalis*) record from the Chatham Islands. *Notornis* 47: 6.

Bellagamba, G., Bellagamba-Oliveira, D. & Dias, R.A. 2014. The Grey-bellied Shrike Tyrant (*Agriornis micropterus*), a new tyrant flycatcher for Brazil. *Revista Brasileira de Ornitologia* 22: 303–304.

Bellagamba-Oliveira, D., Bellagamba, G. & Rocchi, A. 2013. First record of the Rusty-backed Monjita, *Xolmis rubetra* (Passeriformes: Tyrannidae) for Brazil. *Revista Brasileira de Ornitologia* 21: 144–146.

Bellebaum, J., Bock, C., Garthe, S., Kube, J., Schilz, M. & Sonntag, N. 2010. Vorkommen des Gelbschnabeltauchers *Gavia adamsii* in der deutschen Ostsee. *Vogelwelt* 131: 179–184.

Belton, W. 1984. Birds of Rio Grande do Sul, Brazil. Part 1. *Bulletin of the American Museum of Natural History* 178: 369–636.

Bencke, G.A., Ott, P., Moreno, I., Tavares, M. & Caon, G. 2005. Old World birds new to the Brazilian territory recorded in the Archipelago of São Pedro and São Paulo, equatorial Atlantic Ocean. *Ararajuba* 13: 26–129.

Bencke, G.A. & de Souza, F.J. 2013. Upland Goose *Chloephaga picta* (Anseriformes, Anatidae): first Brazilian record. *Revista Brasileira de Ornitologia* 2: 292–294.

Benjumea, R. and Pérez, B., 2016. First record of Eyebrowed Thrush *Turdus obscurus* for Senegal and sub-Saharan Africa. *Bulletin of the African Bird Club* 23: 215–216.

Bennett, M., Burgess, N. & Woehler, E. J. 2015. Interim checklist of King Island birds, July 2015. *Tasmanian Bird Report* 37.

Bent, A. C. 1926. Life histories of North American marsh birds. *Bulletin of the United States National Museum* 135.

Bent, A. C. 1932. Life histories of North American gallinaceous birds. *Bulletin of the United States National Museum* 162: 178–202.

Bergier, P., Franchimont, J. & Thévenot, M. 2005. Les oiseaux rares au Maroc. Rapport de la Commission d'Homologation Marocaine Numéro 10 (2004). *Go-South Bulletin* 2: 23–30.

Bergier, P., Zadane, Y. & Qninba, A. 2009. Cape Gull: a new breeding species in the Western Palearctic. *Birding World* 22: 253–256.

Bergier, P., Franchimont, J.& Thévenot, M. 2009 Rare birds in Morocco: report of the Moroccan Rare Birds Committee (2004-2006) *Bulletin of the African Bird Club* 16: 23–36.

Bergier, P., Franchimont, J. & Thevenot, M. 2011. Les oiseaux rares au Maroc. Rapport de la Commission d'Homologation Marocaine Numéro 16 (2010). *Go-South Bulletin* 8: 1–20.

Bergier, P. & Thévenot, M. 2018. Le Busard pâle *Circus macrourus* au Maroc. *Go-South Bulletin* 15: 69–76.

Bernadon, B. & Valsecchi, J. 2014. First record of the Andean Flamingo in the Brazilian Amazon. *Revista Brasileira de Ornitologia* 22: 285–287.

Bertram, B.C.R., Boesman, P.F.D. & Kirwan G.M. 2020. Masked Finfoot (*Heliopais personatus*), version 1.0. In *Birds of the World* (J. del Hoyo, A. Elliott, J. Sargatal, D. A. Christie, and E. de Juana, Editors). Cornell Lab of Ornithology, Ithaca, NY, USA. https://doi.org/10.2173/bow.masfin3.01

Best, H.A. 1976. First sightings of hoary-headed grebe (*Podiceps poliocephalus*) in New Zealand. *Notornis* 23: 182–183.

Bezuijen, M.R., Eaton, J.A., Gidean, Hutchinson, R.O. & Rheindt, F.E. 2010. Recent and historical bird records for Kalaw, eastern Myanmar (Burma), between 1895 and 2009. *Forktail* 26: 49–74.

Bharti, H. 2017 Masked Shrike near Vyara: an intriguing first record for India. *Flamingo* 15: 6.

Bierregaard, R.O., Poole, A.F., Martell, M.S., Pyle, P. & Patten, M.A. 2020. Osprey (*Pandion haliaetus*), version 1.0. In Birds of the World (P. G. Rodewald, Editor). Cornell Lab of Ornithology, Ithaca, NY, USA. https://doi.org/10.2173/bow.osprey.01

Bigonneau, R., Bazire, R., Marteau, C., Barbraud, C., Ferrer-Obiol, J. & Delord, K. 2020. First record of Black-winged Pratincole *Glareola nordmanni* for Amsterdam Island, Indian Ocean. *Bulletin of the African Bird Club* 27: 94–95.

Bildstein, K.L. 2006. *Migrating Raptors of the World: Their Ecology and Conservation.* Cornell University Press, Ithaca, NY.

Bildstein K.L., Bechard, M.J., Farmer, C. & Newcomb, L. 2009. Narrow sea crossings present major obstacles to migrating Griffon Vultures (*Gyps fulvus*). *Ibis* 151: 382–391.

Birch, A. & Lee, C.T. 1995. Identification of Pacific Diver, a potential vagrant to Europe. *Birding World* 8: 458–466.

BirdLife International 2008. Nearly half of all bird species are used directly by people. In BirdLife State of the world's birds website. http://www.birdlife.org/datazone/sowb/casestudy/98

Biro, D., Sumpter, D.J., Meade, J. & Guilford, T. 2006. From compromise to leadership in pigeon homing. *Current Biology* 16: 2123–2128.

Black, A., Rogers, C. & Fennell, J. 2020. The Birds SA Rare Bird Committee (SARC): a report on submissions 2006-2019. *South Australian Ornithologist* 45: 12–22.

Blackman, J. G. & Locke, D. K. 1978. Nocturnal flying height of a Dollarbird. *Emu* 78: 163–163.

Blamire S. 2004. Red-billed Tropicbird: new to Britain. *British Birds.* 97: 231–7.

Blechman, A. 2007. *Pigeons – The fascinating saga of the world's most revered and reviled bird.* University of Queensland Press, St Lucia, Queensland.

Bloch, D. & Sørensen, S. 1984. *Checklist of Faroese Birds.* Fnroya Skulab6kagrunnur, Thorshavn.

Blokpoel, H., Naranjo, L.G. & Tessier, G.D. 1984. Immature little gull in South America. *American Birds* 38: 372–374.

Bloom, P.H., Scott, J.M., Papp, J.M., Thomas, S.E. & Kidd, J.W. 2011. Vagrant western Red-shouldered Hawks: origins, natal dispersal patterns, and survival. *The Condor* 113: 538–546.

Boertmann, D. 1994. An annotated checklist of the birds of Greenland. *Bioscience* 38: 1–63.

Boertmann, D. 2008. The lesser black-backed gull, *Larus fuscus*, in Greenland. *Arctic* 61: 129–133.

Bojarinova, J., Babushkina, O., Shokhrin, V. and Valchuk, O., 2016. Autumn migration of the Long-tailed Tit (*Aegithalos c. caudatus*) at the opposite sides of the Eurasian continent. *Ornis Fennica* 93: 235–245.

Boles, W.E., Smith, M.J., Tsang, L. & Sladek, J. 2016. The first specimen of a Malayan Night-Heron '*Gorsachius melanolophus*' from Australia. *Australian Field Ornithology* 33: 148–150.

Bolton, M., Watt, R., Ellick, G. & Scofield, P. 2009. Evidence of breeding White-faced Storm-petrel *Pelagodroma marina* on St Helena Island, South Atlantic: vagrancy or a relict from human recolonisation. *Seabird* 23: 135–138.

Bonaccorsi, G. 2003. Les Procellariiformes (Diomedeidae, Procellariidae et Hydrobatidae) non nicheurs en Méditerranée occidentale: une synthèse. *Alauda* 71: 1–7.

Bond, A.L. & Jones, I.L. 2010. A Brown Hawk-Owl (*Ninox scutulata*) from Kiska Island, Aleutian Islands, Alaska. *Western Birds* 41: 107–110.

Bond, A.L., McClelland, G.T.W., Glass, T., Herian, K. & Malan, L. 2015. Multiple records of a Red-tailed Tropicbird *Phaethon rubricauda* on Inaccessible Island, Tristan da Cunha. *Bulletin of the British Ornithologists' Club* 135: 190–192.

Bond, A. L. & Glass, T. 2016. First record of Yellow-billed Cuckoo *Coccyzus americanus* on Tristan da Cunha, South Atlantic Ocean. *Bulletin of the British Ornithologists' Club* 136: 214–216.

Bonilla, S., Sandoval, L., Sánchez, J.E. & López-Pozuelo, F. 2014. Two new *Larus* gulls recorded in Costa Rica. *Marine Ornithology* 42: 3-4.

Booms, T.L., Cade, T. J. & Clum, N. J. 2008. Gyrfalcon (*Falco rusticolus*), version 2.0. In *The Birds of North America* (A. F. Poole, Editor). Cornell Lab of Ornithology, Ithaca, NY, USA. https://doi.org/10.2173/bna.114

Bornschein, M.R., Maurício, G.N. & Sobânia, R.L.M. 2004. First records of the Silvery Grebe *Podiceps occipitalis* Garnot, 1826 in Brazil. *Ararajuba* 12: 61–63.

Bornschein, M.R. 1996. Extralimital record of the Spot-winged Falconet *Spiziapteryx circumcinctus. Bulletin of the British Ornithologists' Club* 116: 197–198.

Borrett, R.P. & Jackson, H.D. 1970. The European Wheatear *Oenanthe Oenanthe* (L.) in southern Africa. *Bulletin of the British Ornithologists' Club* 90: 124–129.

Borrow, N. & Demey, R. 2001. *Birds of Western Africa.* Christopher Helm, London.

Borrow, N. & Demey, R. 2011. *Birds of Senegal and The Gambia.* Christopher Helm, London.

Bostock, N. 2000. Australian Avocet in Irian Jaya. *Kukila* 11: 152.

Bot, S., Brinkhuizen, D., Pogány, Á., Székely, T. & van Dijk,

R. 2011. Penduline tits in Eurasia: distribution, identification and systematics. *Dutch Birding* 33:177–187.

Botha, A. & Neethling, M. 2013. Record of Rüppell's Griffon *Gyps rueppellii* in southern KwaZulu-Natal, South Africa. *Vulture News*, 63: 50–53.

Bourass, E., Baccetti, N., Bashimam, W., Berbash, A., Bouzainen, M., De Faveri, A., Galidan, A., Saied, A.M., Yahia, J. & Zenatello, M. 2013. Results of the seventh winter waterbird census in Libya, January–February 2011. *Bulletin of the African Bird Club* 20: 20–26.

Bourliere, F. 1950. Esquisse ecologique. Pp. 1089-1099 in *Traite de zoologie*. Volume XV. Oiseaux. (Grasse, P. Ed.). Masson et Cie, Paris.

Bourne, W.R.P. 1967. Long-distance vagrancy in the petrels. *Ibis* 109: 141–167.

Bourne, W.R.P. & Simmons, K. E. L. 1998. A preliminary list of the birds of Ascension Island, South Atlantic Ocean. *Sea Swallow* 47: 42–56.

Bourne, W.R.P. 1992a. Debatable British and Irish seabirds. *Birding World* 5: 382–390.

Bourne, W.R.P. 1992b. FitzRoy's foxes and Darwin's finches. *Archives of Natural History* 19: 29–37.

Bouwman, R.G. 1987. Crab Plover in Turkey in July 1986. *Dutch Birding* 9: 65–67.

Bowen, R.V. 2020. Townsend's Solitaire (*Myadestes townsendi*), version 1.0. In *Birds of the World* (A. F. Poole and F. B. Gill, Editors). Cornell Lab of Ornithology, Ithaca, NY, USA. https://doi.org/10.2173/bow.towsol.01

Bowlin, M.S., Enstrom, D.A., Murphy, B.J., Plaza, E., Jurich, P. & Cochran, J. 2015. Unexplained altitude changes in a migrating thrush: long-flight altitude data from radio-telemetry. *The Auk* 132: 808–816.

Bowman, N., Roberts, I. & Dawson, M. 2017. Chinese Pond Heron: new to Britain. *British Birds* 110: 335–344.

Boyle Jr, W.J. 2011. *The Birds of New Jersey: status and distribution*. Princeton University Press, Princeton, New Jersey.

Boyle, W.A., Norris, D.R. & Guglielmo, C.G. 2010. Storms drive altitudinal migration in a tropical bird. *Proceedings of the Royal Society of London. Series B: Biological Sciences* 277: 2511–2519.

Bradbury, R., Eaton, M., Bowden, C. & Jordan, M., 2008. Magnificent Frigatebird in Shropshire. *British Birds* 101: 317–321.

Brady, M.L., Hiller, A.E., Rumiz, D.I., Herzog-Hamel, N.L. & Herzog, S.K. 2019. First Bolivian record of Laughing Gull *Leucophaeus atricilla*, and two noteworthy records of *Fulica* coots from Laguna Guapilo, dpto. Santa Cruz. *Cotinga* 41: 98–100.

Braun, M.J. & Wolf, D.E. 1987. Recent records of vagrant South American land birds in Panama. *Bulletin of the British Ornithologists' Club* 107: 115–117.

Brauneis, J. & Alex, U. 2015. Nachweis eines Sibirien-Raubwürgers *Lanius excubitor* sibiricus Bogdanov, 1881 bei Witzenhausen (Nordhessen) im Jahr 1878. *Ornithologische Mitteilungen* 67: 29–32.

Bräunlich, A. & Rank, M. 1998. Notes on the occurrence of the Corncrake (*Crex crex*) in Asia and in the Pacific region. In *Proceedings 3rd Workshop on Corncrakes*, Hilpoltstein.

Brazil, M.A. 2003. Mistle Thrush *Turdus viscivorus*: New for Japan. *Journal of the Yamashina Institute for Ornithology* 34: 320–324.

Brazil, M. 2009. *Birds of East Asia: China, Taiwan, Korea, Japan, and Russia*. Christopher Helm, London.

Brazil, M. 2018. *Birds of Japan*. Bloomsbury Publishing, London.

Bretagnolle, V., Carruthers, M., Cubitt, M., Bioret, F. & Cuillandre, J-P. 1991. Six captures of a dark-rumped, fork-tailed storm-petrel in the northern Atlantic. *Ibis* 133: 351–356.

Brewer, D., 2010. *Wrens, Dippers and Thrashers*. Christopher Helm, London.

Bried, J. 2003. Impact of vagrant predators on the native fauna: a short-eared owl (*Asio flammeus*) preying on Madeiran storm petrels (*Oceanodroma castro*) in the Azores Arquipelago. *Life and Marine Sciences* 20: 57–60.

Brinkhuizen, D. M., Brinkhuizen, L., Keaveney, A. & Jane, S. 2010. Red-throated Pipit *Anthus cervinus*: a new species for South America. *Cotinga* 32: 98–100.

British Ornithologists' Union 2009. British Ornithologists' Union Records Committee 37th Report (October 2008). *Ibis* 151: 224–230.

British Ornithologists' Union 2015. British Ornithologists' Union Records Committee 43rd Report (October 2014). *Ibis* 157: 186–192.

British Ornithologists' Union 2020. British Ornithologists' Union Records Committee 51st Report (January 2020). *Ibis* 62: 600–603.

Brito, G.R., Nacinovic, J.B. & Teixeira, D.M. 2013. First record of Redwing *Turdus iliacus* in South America. *Bulletin of the British Ornithologists' Club* 133: 316–317.

Brooke, R.K. 1968. On the distribution, movements and breeding of the Lesser Reedhen *Porphyria alleni* in southern Africa. *Ostrich* 39: 259–262.

Brooke, R.K. 1988. Is the Bimaculated Lark *Melanocorypha bimaculata* a valid member of the southern African avifauna? *Ostrich* 59: 76–77.

Brooks, D.M., Cancino, L. & Pereira, S.L. 2006. Conserving cracids: the most threatened family of birds in the Americas. *Miscellaneous Publications of the Houston Museum of Natural Science* 6: 1–169.

Brown, A. & Grice, P. 2010. *Birds in England*. T & AD Poyser, London.

Brown, J. M. & Taylor, P. D. 2015. Adult and hatch-year blackpoll warblers exhibit radically different regional-scale movements during post-fledging dispersal. *Biology Letters* 11: 20150593.

Browne, S. 1997. The first White-rumped Sandpiper *Calidris fuscicollis* in Turkey and the Middle East. *Sandgrouse* 19: 16–147.

Bruce, M.D. 2019. Typical Broadbills (Eurylaimidae). In del Hoyo, J., Elliott, A., Sargatal, J., Christie, D.A. & de Juana, E. (Eds.). *Handbook of the Birds of the World Alive*. Lynx Edicions, Barcelona. https://www.hbw.com/node/52287

Bruce, M.D. 1999. Family Tytonidae (barn owls). Pp. 34–75 in del Hoyo, J, Elliott, A & Sargatal, J. (Eds.). *Handbook of the Birds of the World*. Volume. 5. Lynx Edicions, Barcelona.

Bruinzeel, L.W. 1986. Roodsnavelkeerkringvogel bij Egmond aan Zee in Januari 1985. *Dutch Birding* 8: 45–48.

Brush, T. 2020. Couch's Kingbird (*Tyrannus couchii*), version 1.0. In *Birds of the World* (A. F. Poole and F. B. Gill,

Editors). Cornell Lab of Ornithology, Ithaca, NY, USA. https://doi.org/10.2173/bow.coukin.01

Bruun, B. 1971. North American waterfowl in Europe. *British Birds* 64: 385–408.

Brynjólfsson, B., Arnarson B.G., Kolbeinsson, Y., Ebels, E. B. & Ławicki, L. 2020. Dark-sided Flycatcher at Höfn, Iceland, in October 2012. *Dutch Birding* 42: 113–115.

BTO 2021. Summary of all recoveries for Barn Owl (*Tyto alba*). https://app.bto.org/ring/countyrec/resultsall/rec7350all.htm

Buckley, P.A. & Mitra, S.S. 2003. Williamson's Sapsucker, Cordilleran Flycatcher, and other long-distance vagrants at a Long Island, New York stopover site. *North American Birds* 57: 292–304.

Buden, D. W. 2008. The birds of Nauru. *Notornis* 55: 8–19.

Bugoni, L., Sander, M. & Costa, E.S. 2007. Effects of the first southern Atlantic hurricane on Atlantic petrels (*Pterodroma incerta*). *The Wilson Journal of Ornithology* 119: 725–729.

Buij, R., Clark, W.S. & Allan, D.G. 2016. First records of Red-necked Buzzard *Buteo auguralis* for southern Africa, with notes on identification of *Buteo* buzzards in the subregion. *Bulletin of the African Bird Club* 23: 46–63.

Buitrón-Jurado, G. 2014. An aberrant record of Hoatzin *Opisthocomus hoazin* (Statius Muller, 1976) (Aves: Opisthocomidae) in Venezuela. *Check List* 10: 153–155.

Bulyuk, V.N. & Leoke, D. 2010. The Sardinian Warbler, *Sylvia melanocephala* (JF Gmelin, 1789), a new species for Russia's fauna. *Avian Ecology and Behaviour* 17: 23–24.

Burgas, A. & Ollé, À. 2017 Sykes's Nightjar at Muntasar oasis, Oman, in December 2016. *Dutch Birding* 39: 329–332.

Burger, A.E., Williams, A.J. & Sinclair, J.C. 1980. Vagrants and the paucity of land bird species at the Prince Edward Islands. *Journal of Biogeography* 7: 305-310.

Burgos, K. & Olmos, F. 2013. First record of Corncrake *Crex crex* (Rallidae) for South America. *Revista Brasileira de Ornitologia* 21: 205–208.

Burnham, K.K. & Newton, I. 2011. Seasonal movements of Gyrfalcons *Falco rusticolus* include extensive periods at sea. *Ibis* 153: 468–484.

Burnham, K.K., Burnham, J.L., Johnson, J.A., Konkel, B.W., Stephens, J. & Badgett, H., 2020. First record of horned puffin in the North Atlantic and tufted puffin in High Arctic Greenland. *Polar Research* 39: 4458.

Butcher, S.C., Crossland, A.C., Crutchley, P. & Mugan, N.D. 2018. First confirmed records of little black cormorant (*Phalacrocorax sulcirostris*) for the Solomon Islands. *Notornis* 65: 109–112.

Butler, R.W., Delgado, F.S., de la Cueva, H., Pulido, B. & Sandercock, B.K. 1996 Migration routes of the Western Sandpiper. *Wilson Bulletin* 108: 662–672.

Butler, R.W. 2000. Stormy seas for some North American songbirds: Are declines related to severe storms during migration? *The Auk* 117: 518–522.

Byers, C., Olsson, U. & Curson, J. 1995. *Buntings and Sparrows: a guide to the buntings and North American sparrows.* Pica Press, Robertsbridge, UK.

Cabot, J. 1992. Family Tinamidae (tinamous). Pp. 111–138 in del Hoyo, J., Elliott, A. & Sargatal, J. (Eds.). *Handbook of the Birds of the World.* Volume 1. Lynx Edicions, Barcelona.

Cadbury, C.J. 1975. Populations of swans at the Ouse Washes, England. *Wildfowl* 26: 148–159.

Cade B. S. & Hoffman, R. W. 1993. Differential migration of Blue Grouse in Colorado. *The Auk* 110: 70–77.

Calchi, R. & Viloria, A. L. 1991. Occurrence of the Andean Condor in the Perijá Mountains of Venezuela. *Wilson Bulletin* 103: 720–722.

Calderón-F., D. & Agudelo, L. 2014. Primer registro de *Bombycilla cedrorum* en los Andes de Colombia. *Boletín SAO* 23: 18–21.

Caletrío, J. 2018. Are we addicted to high-carbon ornithology? *British Birds* 111: 182–185.

Callaghan, C.T., Slater, M., Major, R.E., Morrison, M., Martin, J.M. & Kingsford, R.T. 2018. Travelling birds generate eco-travellers: The economic potential of vagrant birdwatching. *Human Dimensions of Wildlife* 23: 71–82.

Callaghan, C.T., Benson, I., Major, R.E., Martin, J.M., Longden, T. & Kingsford, R.T. 2020. Birds are valuable: the case of vagrants. *Journal of Ecotourism* 19: 82–92.

Camacho, I. & Accorsi, M. 2016. Confirmação da sora, *Porzana carolina*, em território brasileiro e contribuições para a conservação das áreas úmidas da Área de Proteção Ambiental de Maricá (RJ) para espécies migratórias neárticas. *Atualidades Ornitológicas* 191: 60–66.

Cambodia Bird Guide Association – CBGA 2019. *Birds of Cambodia.* Lynx Edicions, Barcelona.

Campbell, A.J. 1888. Field Naturalists Club of Victoria. Expedition to King Island, November, 1887. Official Report. *Victorian Naturalist* 4: 129–164.

Campbell, O. & Berhe, S. 2020. First record of Franklin's Gull *Leucophaeus pipixcan* for Ethiopia. *Bulletin of the African Bird Club* 27(2): 232–233.

Campbell, O., Hare, W. & Milius, N. 2011. The first confirmed breeding record of Bay-backed Shrike *Lanius vittatus* in the United Arab Emirates, with comments on field characters of juveniles. *Sandgrouse* 33: 7–11.

Campbell, O & O'Mahoney, D. 2013. White-rumped Sandpiper *Calidris fuscicollis* in the United Arab Emirates: the first records for the Arabian peninsula. *Sandgrouse* 35: 36–38.

Campbell, R.W., Dawe, N.K., McTaggart-Cowan, I., Cooper, J.M., Kaiser, G.W. & McNall, M.C.E. 1990. *The Birds of British Columbia* – Volume 1 (Nonpasserines [Introduction, Loons through Waterfowl]). Royal British Columbia Museum, Victoria.

Capllonch, P. & Ortiz, D. 2007. ¿Migra el Arañero cara negra (*Geothlypis aequinoctialis velata*). *Ornitología Neotropical* 18: 195–208.

Carboneras, C. 1992. Family Anhimidae (screamers). Pp. 528–535 in del Hoyo, J., Elliott, A. & Sargatal, J. (Eds.). *Handbook of the Birds of the World.* Volume 1. Lynx Edicions, Barcelona.

Carey, M.P., Sanderson, B.L., Barnas, K.A. & Olden, J.D., 2012. Native invaders – challenges for science, management, policy, and society. *Frontiers in Ecology and the Environment* 10: 373–381.

Carletti, A. 2014. WA1335827, *Hymenops perspicillatus* (Gmelin, 1789)]. http://www.wikiaves.com/1335827

Carlos, C.J. 2005. Notes on the specimen record of the Broad-billed Prion *Pachyptila vittata* from Rio Grande do Sul, south Brazil. *Ararajuba*, 13: 124–125.

Carlos, J. 2018. WA2998389, *Pheucticus aureoventris* (d'Orbigny & Lafresnaye, 1837) http://www.wikiaves.com/2998389

Carmody, M. & Hobbs, J. 2015. Irish rare bird report 2014. *Irish Birds* 10: 235–263.

Carr, P. 2015. Birds of the British Indian Ocean Territory, Chagos archipelago, central Indian Ocean. *Indian Birds* 10: 57–70.

Carter, M. 2003. Cinnamon Bittern *Ixobrychus cinnamomeus* on Christmas Island: a new bird for Australian territory. *Australian Field Ornithology* 20: 55–58.

Carter, M. 2009. Grey Nightjar *Caprimulgus indicus*: first Australian record. *Australian Field Ornithology* 26:116–122.

Carter, M., Clarke, R.H. & Swann, G. 2011. Island Monarch *'Monarcha cinerascens'* on Ashmore Reef: First records for Australian Territory. *Australian Field Ornithology* 28: 150–161.

Castro, F., Castro, J., Ramos Ferreira, A., Crozariol, M.A. & Lees, A.C. 2012. A first documented Brazilian record of Least Seedsnipe *Thinocorus rumicivorus* Eschscholtz, 1829 (Thinocoridae). *Revista Brasileira de Ornitologia* 20: 455–457.

Cestari, C., 2008. Aves, Charadriidae, *Charadrius modestus*: Geographic distribution and a recent record to state of São Paulo, Brazil. *Check List* 4: 464–466.

Chace, J.F., Woodworth, B.L. & Cruz, A. 2002. Black-whiskered Vireo (*Vireo altiloquus*), version 2.0. In *The Birds of North America* (Poole, A. F. & Gill, F. B. Eds). Cornell Lab of Ornithology, Ithaca, NY, USA. https://doi.org/10.2173/bna.607

Chadwick, P.I. 1991. Third specimen record of the Antarctic Petrel *Thalassoica antarctica* for Africa. *Marine Ornithology* 19: 69–70.

Chalco, L.J.J. 2013. Primer registro para el Perú de Gallareta Cornuda (*Fulica cornuta*). *Boletín de la Unión de Ornitólogos del Perú* 8: 14–15.

Chalmers, M.L. 2013. Ascension Frigatebirds *Fregata aquila* in Great Britain. *Seabird* 26: 100–103.

Chan, K. 2001. Partial migration in Australian landbirds: a review. *Emu* 101: 281–292.

Chantler, P. & Driessens, G. 2000. *Swifts: A Guide to the Swifts and Treeswifts of the World*. 2nd edition. Pica Press, Robertsbridge.

Chapman, G.E. 2009. A Yellow Zosterops Sighting at Sawtell. *Sunbird: Journal of the Queensland Ornithological Society* 39: 14–17.

Charles, D. 2008. Ring-billed Gull breeding with Common Gull on Copeland Islands Co. Down. The first confirmed breeding record for Ring-billed Gull in the Western Palearctic. *Northern Ireland Bird Report* xviii: 122.

Chavarría-Duriaux, L. 2015. Garza Tigre de Río (Fasciated Tiger-Heron) *Tigrisoma fasciatum*: nuevo reporte para Nicaragua. *Zeledonia* 19: 116–117.

Cheke, R.A & Mann, C.F. 2001. *Sunbirds: A Guide to the Sunbirds, Flowerpeckers, Spiderhunters and Sugarbirds of the World*. Christopher Helm, London.

Chernetsov N., Berthold P. & Querner U. 2004. Migratory orientation of first-year white storks (*Ciconia ciconia*): Inherited information and social interactions. *Journal of Experimental Biology* 207: 937–943.

Chernetsov, N., Pakhomov, A., Kobylkov, D., Kishkinev, D., Holland, R.A. & Mouritsen, H. 2017. Migratory Eurasian reed warblers can use magnetic declination to solve the longitude problem. *Current Biology* 27: 2647–2651.

Cherry, B. 2005. A little black cormorant (*Phalacrocorax sulcirostris*) from Kadavu Island, Fiji. *Notornis* 52: 249.

Chesser, R.T., Burns, K.J., Cicero, C., Dunn, J.L., Kratter, A.W., Lovette, I.J., Rasmussen, P.C., Remsen, J.V., Rising, J.D., Stotz, D.F. & Winker, K. 2017. Fifty-eighth supplement to the American Ornithological Society's check-list of North American birds. *The Auk*, 134: 751–774.

Childress, B. & Hughes, B. 2007. Evidence of interchange between African Lesser Flamingo populations. Proc. XI Pan-Afr. Orn. Congr. 2004. *Ostrich* 78: 507.

Choi, C.-Y., Kang, C.-W., Ji, N.-J., Kang, H.-M., Oh, M.-R. & Park, J.-Y. 2013. Barred Warbler *Sylvia nisoria*, a new species for Korea and east Asia. *BirdingASIA* 20: 110–111.

Chophel, T. & Sherub. 2016. Lapland Longspur *Calcarius lapponicus* in Bhutan: a first record for the Indian subcontinent. *Indian Birds* 12: 24.

Chowdhury, S.U. 2014. First photographic record of the Asian Stubtail *Urosphena squameiceps* from the Indian Subcontinent. *Indian Birds* 9: 25.

Chu, M. & G. Walsberg. 2020. Phainopepla (*Phainopepla nitens*), version 1.0. In *Birds of the World* (Poole, A. F. & Gill, F. B. Eds). Cornell Lab of Ornithology, Ithaca, NY, USA. https://doi.org/10.2173/bow.phaino.01

Chupil, H., Marques, V., Nagaoka, S. & Murro, R.S. 2019. First record of Grey Gull *Leucophaeus modestus* in Brazil. *Revista Brasileira de Ornitologia* 27: 140–142.

Cieza Ponce, J. & Díaz Villalobos, O. 2014. Primer registro documentado del Calandria de Ala Blanca (*Mimus triurus*) en Perú. *Boletín de la Unión de Ornitólogos del Perú* 9: 49–51.

Cisneros-Heredia, D.F. 2016. First documented record of the Belted Kingfisher, *Megaceryle alcyon* (Linnaeus, 1758), in mainland Ecuador. *Check List* 12: 1–4.

Claessens, O., Cambrézy, C., Cobigo, M., Maillé, S. & Renaudier, A. 2012. So far from the Andes: the Black-and-white Tanager *Conothraupis speculigera*, an unexpected vagrant to French Guiana. *Bulletin of the British Ornithologists' Club* 132: 55–59.

Clapp, R.B., Wiles, G.J., Aldan, D.T. & Pratt, T.K. 1990. New migrant and vagrant bird records for the Mariana Islands, 1978-1988. *Micronesica* 23: 67–89.

Clark, N.J., Adlard, R.D. & Clegg, S.M. 2014. First evidence of avian malaria in Capricorn Silvereyes (*Zosterops lateralis chlorocephalus*) on Heron Island. *The Sunbird* 44: 1–11.

Clarke, A., Croxall, J.P., Poncet, S., Martin, A.R. & Burton, R. 2012. Important bird areas: South Georgia. *British Birds* 105: 118–144.

Clarke, R.H. 2004 The avifauna of northern Torres Strait: notes on a wet season visit. *Australian Field Ornithology* 21: 49–66.

Clarke, R.H. 2007. An orange-bellied fruit-dove *Ptilinopus iozonus* on Boigu Island, Torres Strait: the first record for Australian territory. *Australian Field Ornithology* 24: 44–48.

Clarke, R.H., Davis, K. & Davis, B. 2008. A Grey-headed Lapwing *Vanellus cinereus* at Burren Junction, New South

Wales: the first record for Australia. *Australian Field Ornithology* 25: 194–197.
Clarke, R.H., Gosford, R., Boyle, A., Sisson, L. & Ewen, J.G. 2010. A Specimen Record of the Little Paradise-Kingfisher *'Tanysiptera hydrocharis'* from Torres Strait, Queensland: A New Bird for Australian Territory. *Australian Field Ornithology* 27: 165.
Clarke, R.H., Carter, M., Swann, G. & Herrod, A. 2016. A record of the Siberian Blue Robin *'Larvivora cyane'* at Ashmore Reef off north-western Australia, April 2012. *Australian Field Ornithology* 33: 41–43.
Clarke, T. 2006. *A Field Guide to the Birds of the Atlantic Islands: Canary Islands, Madeira, Azores, Cape Verde.* Christopher Helm, London.
Clay, R.P., Lesterhuis, A.J. & Smith, P. 2017. Status and distribution of the suborder Lari in Paraguay, including new country records. *Revista Brasileira de Ornitologia* 25: 128–136.
Cleere, N. 1998. *Nightjars: A Guide to Nightjars and Related Nightbirds.* Pica Press, Robertsbridge, UK.
Clement, P. & Hathway, R. 2000. *Thrushes.* Helm identification guides. Christopher Helm, London.
Clements, J.F., Schulenberg, T.S., Iliff, M.J., Billerman, S.M., Fredericks, T.A., Sullivan, B.L. & Wood, C.L. 2019. *The eBird/Clements Checklist of Birds of the World:* v2019. https://www.birds.cornell.edu/clementschecklist/download/
Clout, M.N. & Lowe, S.J. 2000. Invasive species and environmental changes in New Zealand. Pp. 369–384 in Mooney, H.A., Hobbs, R.J. eds. *Invasive Species in a Changing World.* Island Press, Washington, D.C.
Cobb, P.R., Rawnsley, P., Grenfell, H.E., Griffiths, E. & Cox, S., 1996. Northern Mockingbirds in Britain. *British Birds* 89: 347–356.
Coffey, J., 2012. Black-browed Albatross (*Thalassarche melanophrys*) Sighting off Northern Labrador, Canada. *Northeastern Naturalist* 19: 130–135.
Cohen, C. & Winter, D. 2003. Great Knot *Calidris tenuirostris*: a new species for sub-Saharan Africa. *Bulletin of the African Bird Club* 10: 120–121.
Cohen, G. 2018. Red-footed Booby in Sussex: New to Britain. *British Birds* 111: 42–46.
Cohn-Haft, M., Pacheco, A.M.F., Bechtoldt, C.L., Torres, M.F.N.M., Fernandes, A.M., Sardelli, C.H. & Macêdo, I.T. 2007. Inventário ornitológico. Pp. 145–178 in Rapp Py-Daniel, L., Deus, C.P. Henriques, A.L., Pimpão, D.M. & Ribeiro, O. M. (orgs.) *Biodiversidade do médio Madeira: bases científicas para propostas de conservação.* Ed. INPA, Manaus.
Cohn-Haft, M. 1999. Family Nyctibiidae (potoos). Pp. 288–301 in del Hoyo, J., Elliott, A. & Sargatal, J. (Eds). *Handbook of the Birds of the World.* Lynx Edicions, Barcelona.
Collar, N.J., Andreev, A.V., Chan, S., Crosby, M.J., Subramanya, S. & Tobias, J.A. eds. 2001. *Threatened Birds of Asia: the BirdLife International Red Data Book.* BirdLife International, Cambridge, UK.
Collar, N. 1996. Bustards (Otididae). Pp. 240–273 in del Hoyo, J., Elliott, A. & Sargatal, J. (Eds). *Handbook of the Birds of the World.* Volume 3. Lynx Edicions, Barcelona.
Collinson, J.M. 2017. CSI: Birding–DNA-based identification. *British Birds* 110: 8–26.
Colwell, M.A. & Jehl, J. R. Jr. 1994. Wilson's Phalarope (*Phalaropus tricolor*), The Birds of North America Online (Poole, A. Ed.). Birds of North America Online: http://bna.birds.cornell.edu.libraryproxy.amnh.org:9000/bna/species/083
Colyn, R.B., Campbell, A. & Smit-Robinson, H.A. 2019. Camera-trapping successfully and non-invasively reveals the presence, activity and habitat choice of the Critically Endangered White-winged Flufftail *Sarothrura ayresi* in a South African high-altitude wetland. *Bird Conservation International* 29: 463–478.
Combreau, O., Riou, S., Judas, J., Lawrence, M. & Launay, F. 2011. Migratory Pathways and Connectivity in Asian Houbara Bustards: Evidence from 15 Years of Satellite Tracking. *PLOS One* 6: e20570.
CRAP – Comité de Registros de Aves Peruanas 2012. Reporte del Comité de Registros de Aves Peruanas del periodo 2010 – 2011. *Boletín Informativo de la Unión de Ornitólogos del Perú* 7: 51–62.
CRAP – Comité de Registros de Aves Peruanas 2014. Reporte del Comité de Registros de Aves Peruanas del periodo 2013. *Boletín de la Unión de Ornitólogos del Perú* 9: 45–54.
CRAP – Comité de Registros de Aves Peruanas 2016. Reporte del Comité de Registros de Aves Peruanas del periodo 2015. *Boletín de la Unión de Ornitólogos del Perú* 11: 71–81.
CRAP – Comité de Registros de Aves Peruanas 2017. Reporte del Comité de Registros de Aves Peruanas del periodo 2016. *Boletín de la Unión de Ornitólogos del Perú* 12: 49–56.
CRAP – Comité de Registros de Aves Peruanas 2019. Reporte del Comité de Registros de Aves Peruanas del periodo 2018. *Boletín de la Unión de Ornitólogos del Perú* 14: 63-74.
Commission European Communities 2007. Commission Regulation (EC) No 318/2007 of 23 March 2007 laying down animal health conditions for imports of certain birds into the Community and the quarantine conditions thereof. http://eur-lex.europa.eu/legal-content
Conde, D.A., Flesness, N., Colchero, F., Jones, O.R. & Scheuerlein, A. 2011. An emerging role of zoos to conserve biodiversity. *Science* 331: 1390–1391.
Conklin, J.R. & Battley, P.F. 2011. Impacts of wind on individual migration schedules of New Zealand bar-tailed godwits. *Behavioral Ecology* 22: 854–861.
Conlin, A. & Williams, E. 2018. *Rare and Scarce Birds of Cheshire & Wirral.* Privately published.
Coombes, R.H. & Wilson, F.R. 2015. Colonisation and breeding status of the Great Spotted Woodpecker *Dendrocopos major* in the Republic of Ireland. *Irish Birds* 10: 183–196.
Cooper, J. & Underhill, L.G. 2002. Blacksmith Plover *Vanellus armatus* on Prince Edward Island, Southern Ocean. *Wader Study Group Bulletin* 97: 47–48.
Copete, J.L., Ferrer, X., Bigas, D., Sanpera, C. & Vieites, D. R. 2011. Reed Cormorant collected in Catalunya, Spain, in c.1855. *Dutch Birding* 33: 40–43.
Copete, J.L., Lorenzo, J.A., Amengual, E., Bigas, D.,

Fernández, P., López-Velasco, D., Rodríguez, G. & García-Tarrasón, M. 2015. Observaciones de aves raras en España, 2012 y 2013. *Ardeola* 62: 453–508.

Copson, G.R. & Brothers, N. 2008. Notes on rare, vagrant and exotic avifauna at Macquarie Island, 1901–2000. *Papers and Proceedings of the Royal Society of Tasmania*. 142: 105–115.

Cornwallis, R.K. & Townsend, A.D. 1968. Waxwings in Britain and Europe during 1965/66. *British Birds* 61: 97–118.

Corre, M.L. & Probst, J.M. 1997. Migrant and vagrant birds of Europa Island (southern Mozambique Channel). *Ostrich* 68: 13–18.

Correia, E.C. 2017. WA2461382, *Anhima cornuta* (Linnaeus, 1766)]. http://www.wikiaves.com/2461382

Correia-Fagundes, C. & Romano, H. 2011. A Black-bellied Storm-petrel off Madeira – a new Western Palearctic bird. *Birding World* 24: 326.

Cottridge, D.M. & Vinicombe. K. 1996. *Rare Birds in Britain and Ireland: A. Photographic Record.* HarperCollins Publishers, London.

Coudrat, C.N.Z. & Nanthavong, C. 2016. First record of Fairy Pitta *Pitta nympha* for Lao PDR. *BirdingASIA* 25: 112.

Couto, G.S., Interaminense, L.J. & Morette, M.E. 2001. Primeiro registro de *Phaethon rubricauda* Boddaert, 1783 para o Brasil. *Nattereria* 2: 24–25.

Couve, E. & Vidal, C. 2003. *Birds of Patagonia, Tierra del Fuego & Antarctic Peninsula: the Falkland Islands & South Georgia*. Fantástico Sur, Punta Arenas, Chile.

Craig, A. 2009. Family Buphagidae (Oxpeckers). Pp. 642–653 in del Hoyo, J., Elliott, A. & Christie, D. (Eds.). *Handbook of the Birds of the World.* Volume 14. Lynx Edicions, Barcelona.

Cramp, S. & Reynolds, J.F. 1972. Studies of less familiar birds, 168: Cream-coloured Courser. *British Birds* 65: 120–124.

Cramp, S. & Simmons, K.E.L. 1980. *Handbook of the Birds of Europe, the Middle East and North Africa*. Volume II. Hawks to Bustards. Oxford University Press, Oxford.

Cramp, S. 1985 *Handbook of the Birds of Europe, the Middle East and North Africa*. Volume IV. Terns to Woodpeckers. Oxford University Press, Oxford.

Cresswell, W. 2014. Migratory connectivity of Palaearctic–African migratory birds and their responses to environmental change: the serial residency hypothesis. *Ibis* 156: 493–510.

Crochet, P.-A. & Haas, M. 2008. Western Palearctic list update: deletion of Cape Gannet. *Dutch Birding* 30: 103–104.

Crochet, P-A. & Haas, M. 2013. Western Palearctic list updates: re-evaluation of five species from continental Mauritania. *Dutch Birding* 35: 28–30.

Cross, C.J., Gregory, L.V., Walsh, R. & Fox, O.J.L. 2017. First record of American Wigeon *Mareca americana* in The Gambia, and associated records from Kartong Bird Observatory. *Bulletin of the African Bird Club* 24: 85–87.

Cross, J.H. 2019. First record of Black-casqued Hornbill *Ceratogymna atrata* for The Gambia. *Bulletin of the African Bird Club* 26.103.

Crozariol, M.A. & Nacinovic, J.C. 2017. A historical Australasian Shoveler *Spatula rhynchotis* specimen from southern South America. *Bulletin of the British Ornithologists' Club* 137: 312–314.

Cruse, R. 2004. Yellow-browed Warbler *Phylloscopus inornatus* in Senegal in December 2003. *Bulletin of the African Bird Club* 11: 147–148.

Cubitt, M.G. 1995. Swinhoe's Storm-petrels at Tynemouth: new to Britain and Ireland. *British Birds* 88: 342–348.

Cuelo Pizarro, W. 2018. Primer registro documentado de Calandria Tropical (*Mimus gilvus*) para el Perú. *Boletín de la Unión de Ornitólogos del Perú* 13: 7–9.

Cusa, N.W. 1949. Birds of the North Atlantic in March. *British Birds* 42: 33–41.

da Silva, J.M.C. 1993. The Sharpbill in the Serra Dos Carajás, Pará, Brazil, with Comments on Altitudinal Migration in the Amazon Region. *Journal of Field Ornithology* 64: 310–315.

Dale, S., Steifetten, Ø., Osiejuk, T.S., Losak, K. & Cygan, J.P. 2006. How do birds search for breeding areas at the landscape level? Interpatch movements of male ortolan buntings. *Ecography* 29: 886–898.

Dalvi, S., Kataria, G., Shah, M., Jabestin, A., James, C. & Vishnupriya, S. 2017. First record of Grey Nightjar *Caprimulgus indicus* jotaka/hazarae from the Nicobar Islands, India. *Indian Birds* 12: 162–164.

Daly, C.K. 2017. First record of Franklin's Gull *Leucophaeus pipixcan* for Seychelles. *Marine Ornithology* 45: 223–224.

Dánjalsson & Ryggi, M. 1951. *Fuglabókin*. Tórshavn.

Dann, P., Cullen J.M., Thoday R. & Jessop, R. 1992. Movements and patterns of mortality at sea of Little Penguins *Eudyptula minor* from Phillip Island, Victoria. *Emu* 91:278–286.

Dantas, S.M, de Sousa Miranda, L., Ravetta, A.L. & Aleixo, A. 2017. A new population of the White Bellbird *Procnias albus* (Hermann, 1783) from lowland southern Brazilian Amazonia, with comments on genetic variation in bellbirds. *Revista Brasileira de Ornitologia*, 25: 71–74.

Darlaston, M. & Langman, M. 2016. Yelkouan Shearwater in Devon: new to Britain. *British Birds* 109: 448–456.

Darrieu, C.A., Camperi, A.R. & Imberti, S. 2008. Avifauna (Non Passeriformes) of Santa Cruz province, Patagonia (Argentina): annotated list of species. *Revista del Museo Argentino de Ciencias Naturales* 10: 111–145.

Darwin, C. 1873. Origin of certain instincts. *Nature* 7: 417–418.

Das, S. 2014. Asian Stubtail *Urosphena squameiceps* in Rabindrasarobar, Kolkata: a first record for India. *Indian Birds* 9: 26–27.

Dau, C.P. & Paniyak, J., 1977. Hoopoe, a first record for North America. *The Auk* 94: 601–601.

Davaasuren, B., Erdenechimeg, T. & Sukhbaata, T. 2020. Annual report of Mongolian bird ringing program.

Davis, R.A. & Watson, D.M. 2018. Vagrants as vanguards of range shifts in a dynamic world. *Biological Conservation* 224: 238–241.

Davy, C.M., Ford, A.T. & Fraser, K.C. 2017. Aeroconservation for the fragmented skies. *Conservation Letters* 10: 773–780.

de Almeida, B.J.M., Rodrigues, R.C., Mizrahi, D. & Lees, A.C. 2013. A Lesser Black-backed Gull *Larus fuscus* in Maranhão: the second Brazilian record. *Revista Brasileira de Ornitologia* 21: 213–216.

de Boer, M.N., Saulino, J.T. & Williams, A.C. 2014. First documented record of Common Swift *Apus apus* for Surinam and South America. *Cotinga* 36: 107–109.

De Broyer, A. 2001. Double-crested Cormorant on Ouessant, France, in October 2000. *Dutch Birding* 23: 136–139.

de Brum, A. C., Benemann, V. R. F., Bezerra, A. L., Corrêa, L. L. C. & Petry, M. V. 2018 Record of ship-assisted displacement of *Chionis albus* (Gmelin, 1789) (Aves: Chionidae) from Antarctica to South America. *Oecologia Australis* 22: 89–92.

de Juana, E., 2008. Where do Pallas's and Yellow-browed Warblers (*Phylloscopus proregulus, Ph. inornatus*) go after visiting Northwest Europe in Autumn? An Iberian perspective. *Ardeola* 55:179–192.

de Juana, E. 2011. The Sociable Lapwing in Europe. *British Birds* 104: 84–90.

de Juana, E. & Garcia, E. 2015. *The Birds of the Iberian Peninsula*. Christopher Helm, London.

de Juana, E. & Kirwan, G. M. 2019. Western Capercaillie (*Tetrao urogallus*). In del Hoyo, J., Elliott, A., Sargatal, J., Christie, D. A. & de Juana, E. (Eds.). *Handbook of the Birds of the World Alive*. Lynx Edicions, Barcelona. https://www.hbw.com/node/53328

de Juana, E., Suárez, F. & Ryan, P. 2004. Larks (Alaudidae). Pp. 496–541 in del Hoyo, J., Elliott, A. & Christie, D. (Eds.). *Handbook of the Birds of the World*, Volume 9. Lynx Edicions, Barcelona.

de Jong, A., Torniainen, J., Bourski, O.V., Heim, W. & Edenius, L. 2019. Tracing the origin of vagrant Siberian songbirds with stable isotopes: the case of Yellow-browed Warbler (*Abrornis inornatus*) in Fennoscandia. *Ornis Fennica* 96: 90–99.

de la Puente, J., Pinilla, J. & Lorenzo, J.A. 1999. Noticiario ornitológico 1999. *Ardeola* 46: 149–162.

De Sante, D. 1973. *Analysis of the fall occurrences and nocturnal orientations of vagrant warblers (Parulidae) in California*. Unpublished Ph. D. dissertation. Stanford University, Stanford, California.

del Hoyo, J., Collar, N. & Kirwan, G. M. 2020a. Australian Painted-Snipe (*Rostratula australis*), version 1.0. In *Birds of the World* (J. del Hoyo, A. Elliott, J. Sargatal, D. A. Christie, and E. de Juana, Editors). Cornell Lab of Ornithology, Ithaca, NY, USA. https://doi.org/10.2173/bow.auspas1.01

del Hoyo, J., Collar, N. & Kirwan, G. M. 2020b. African Spotted Creeper (*Salpornis salvadori*), version 1.0. In *Birds of the World* (J. del Hoyo, A. Elliott, J. Sargatal, D. A. Christie, and E. de Juana, Editors). Cornell Lab of Ornithology, Ithaca, NY, USA. https://doi.org/10.2173/bow.spocre2.01

Dean, W.R.J. 2004. *Nomadic Desert Birds*. Springer Science & Business Media.

DeBenedictis, P.A. 1986. The changing seasons. *American Birds* 40: 75–82.

Decandido, R., Nualsri, C. & Allen, D. 2010. Mass northbound migration of Blue-tailed *Merops philippinus* and Blue-throated *M. viridis* Bee-eaters in southern Thailand, spring 2007–2008. *Forkail* 26: 2–48.

Deconto, L.R. & Vallejos, M.A.V. 2017. Primeiro registro documentado de *Setophaga virens* (Aves: Parulidae) no Brasil. *Atualidades Ornitológicas* 198: 14–15.

Degnan, L. & Croft, K. 2005. Black Lark: new to Britain. *British Birds* 98: 306–313.

Dehnhard, N., Ludynia, K. & Almeida, A. 2012. A Royal Penguin *Eudyptes schlegeli* in the Falkland Islands? *Marine Ornithology* 40: 95–98.

de Farias, F.B. & Dalpaz, L. 2019. First documented record of Cerulean Warbler *Setophaga cerulea* (Parulidae) in Brazil. *Ornithology Research* 27: 132–134.

Deighton, D. 2005. First record of Asiatic Dowitcher *Limnodromus semipalmatus* for Africa. *Bulletin of the African Bird Club* 12: 156–157.

Delaney, H. 2012. The Eastern Kingbird in Co. Galway – a new Western Palearctic bird. *Birding World* 25: 430–432.

Delaney, S. & Scott, D. 2006. *Waterbird population estimates: fourth edition*. Wetlands International, Wageningen, The Netherlands.

Delgado-Fernández, M. & Delgadillo-Nuño, M.A. 2017. Primer registro de bajapalos pecho canela (*Sitta canadensis*) en el bosque de Los Attenuatas, Baja California. *Huitzil* 18: 7–10.

Delmore, K.E. & Irwin, D.E. 2014. Hybrid songbirds employ intermediate routes in a migratory divide. *Ecology Letters* 17: 1211–1218.

DeLuca, W.V., Woodworth, B.K., Rimmer, C.C., Marra, P.P., Taylor, P.D., McFarland, K.P., Mackenzie, S.A. & Norris, D.R. 2015. Transoceanic migration by a 12g songbird. *Biology Letters* 11: 20141045.

Dementiev, G.P. & Gladkov, N.A., 1969. *Birds of the Soviet Union*. Israel Program for Scientific Translations, Jerusalem.

Demey, R. 2011. Recent reports. *Bulletin of the African Bird Club* 18: 228–243.

Demey, R. 2012. Recent reports. *Bulletin of the African Bird Club* 19: 221–236.

Demey, R. 2016a. Recent reports. *Bulletin of the African Bird Club* 23: 104–123.

Demey, R. 2016b. Recent reports. *Bulletin of the African Bird Club* 23: 231–249.

Demey, R. 2017. Recent Reports. *Bulletin of the African Bird Club* 24: 229–252.

Demey, R. 2018. Recent Reports. *Bulletin of the African Bird Club* 25: 230–251.

Demey, R. 2020. Recent Reports. *Bulletin of the African Bird Club* 27: 102–123.

Demey, R. & Fishpool, L. D. C. 1991. Additions and annotations to the avifauna of Côte d'Ivoire. *Malimbus* 12: 61–86.

Demongin, L., Poisbleau, M., Strange, G. & Strange, I.J. 2010. Second and third records of Snares penguins (*Eudyptes robustus*) in the Falkland Islands. *The Wilson Journal of Ornithology* 122: 190–193.

Denis, K. 1976. Scott's Oriole near Thunder Bay, Ontario. *The Canadian Field Naturalist* 90: 500–501.

Dennis, J.V. 1981. A summary of banded North American birds encountered in Europe. *North American Bird Bander* 6: 88–96.

Dennis, J.V. 1990. Banded North American Birds Encountered in Europe: An Update. *North American Bird Bander* 15: 130–133.

Dennis, M.C. 1998. Redhead in Nottinghamshire: new to Britain and Ireland. *British Birds* 91: 149–154.

Dennis, T. 2012. Recent record of Plum-headed Finch on Cooper Creek in the north-east of South Australia. *South Australian Ornithologist* 38: 28–29.

Densley, M. 1999. *In Search of Ross's Gull.* Peregrine Books, Leeds, UK.

Deutschlander, M.E. & Muheim, R. 2009. Fuel reserves affect migratory orientation of thrushes and sparrows both before and after crossing an ecological barrier near their breeding grounds. *Journal of Avian Biology* 40: 85–89.

Devisse, R. 1992. Première observation au Sénégal du Martinet marbré *Tachymarptis aequatorialis. Malimbus* 14: 16.

Diamond, J. M. 1974. Colonization of exploded volcanic islands by birds: the supertramp strategy. *Science* 184: 803–806.

Dias, R.A. & Cardozo, J.B. 2014. First record of the Puna Flamingo *Phoenicoparrus jamesi* (Sclater, 1886) (Aves: Phoenicopteridae) for the Atlantic coast of South America. *Check List* 10: 1150–1151.

Dias, M.P., Alho, M., Granadeiro, J.P. & Catry, P. 2015. Wanderer of the deepest seas: migratory behaviour and distribution of the highly pelagic Bulwer's petrel. *Journal of Ornithology* 156: 955–962.

Dickerman, R.W., McNew, S.M. & Witt, C.C. 2013. Long-distance movement in a Dusky Great Horned Owl and limits to phylogeography for establishing provenance. *Western North American Naturalist* 73: 401-408.

Dies, J.I., Gutiérrez, J.A.L., Gutiérrez, R., García, E., Gorospe, G., Martí-Aledo, J., Gutiérrez, P. & Vidal, C. 2007. Observaciones de aves raras en España, 2005. *Ardeola* 54: 405–446.

Dies, J.I., Gutiérrez, J.A.L., Gutiérrez, R., García, E., Gorospe, G., Martí-Aledo, J., Gutiérrez, P., Vidal, C., Sales, S. & López-Velasco, D. 2010. Observaciones de aves raras en España, 2008. *Ardeola* 57: 481–516.

Dies, J.I., Lorenzo, J.A., Gutiérrez, R., García, E., Gorospe, G., Martí-Aledo, J., Gutiérrez, P., Vidal, C., Sales, S. & López-Velasco, D. 2011. Observación de aves raras en España, 2009. *Ardeola* 58: 441–480.

Diller, L.V., Hamm, K.A., Early, D.A., Lamphear, D.W., Dugger, K.M., Yackulic, C.B., Schwarz, C.J., Carlson, P.C. & McDonald, T.L., 2016. Demographic response of northern spotted owls to barred owl removal. *The Journal of Wildlife Management* 80: 691–707.

Dinets, V. 2018. First record of the Chinese Crested Tern *Thalasseus bernsteini* in Japan. *Yamashina Institute for Ornithology*. 50: 1–3.

Ding, J.-Q. & Ma, M. 2012. White-tailed Lapwing *Vanellus leucurus* – A newly recorded of bird in China. *Zoological Research* 33: 545–546.

Diniarsih, S., Jones, S., Setiyono, J. & Noske, R. 2016. Rosy Starling *Pastor roseus*: a new species for Indonesia. *Kukila* 19: 60–64.

Dionne, M., Maurice, C., Gauthier, J. & Shaffer, F. 2008. Impact of Hurricane Wilma on migrating birds: the case of the Chimney Swift. *The Wilson Journal of Ornithology* 120: 784–793.

Divoky, G.J. 1976. The pelagic feeding habits of Ross's and Ivory Gulls. *The Condor* 78: 85–90.

Dixey, A.E., Ferguson, A., Heywood, R. & Taylor, A.R. 1981. Aleutian Tern: new to the Western Palearctic. *British Birds* 74: 411–416.

Dixon, A., Nyambayar, B. & Gankhuyag, P. 2011. Autumn migration of an Amur Falcon *Falco amurensis* from Mongolia to the Indian Ocean tracked by satellite. *Forktail* 27: 86–89.

Djemadi, I., Draidi, K. & Bouslama, Z. 2018. First record of Rustic Bunting *Emberiza rustica* for Algeria. *Bulletin of the African Bird Club* 25: 211–212.

Dobbs, R.C., Carter, J. & Schulz, J.L. 2019. Limpkin, *Aramus guarauna* (L., 1766) (Gruiformes, Aramidae), extralimital breeding in Louisiana is associated with availability of the invasive Giant Apple Snail, *Pomacea maculata* Perry, 1810 (Caenogastropoda, Ampullariidae). *Check List* 15: 497–507.

Dobos, R.Z. 1997. Ontario Bird Records Committee report for 1996. *Ontario Birds* 146: 51–80.

Dobson, A. 2014. Arctic Warbler – New to Bermuda and the east coast of North America. *Bermuda Audubon Society Newsletter Summer 2014.*

Dodge, S., Bohrer, G., Bildstein, K., Davidson, S.C., Weinzierl, R., Bechard, M.J., Barber, D., Kays, R., Brandes, D., Han, J. & Wikelski, M., 2014. Environmental drivers of variability in the movement ecology of turkey vultures (*Cathartes aura*) in North and South America. *Philosophical Transactions of the Royal Society of London. Series B: Biological Sciences* 369: 20130195.

Dokter, A.M., Liechti, F., Stark, H., Delobbe, L., Tabary, P. & Holleman, I. 2011. Bird migration flight altitudes studied by a network of operational weather radars. *Journal of the Royal Society Interface* 8: 30–43.

Dokter, A.M., Shamoun-Baranes, J., Kemp, M. U., Tijm, S. & Holleman, I. 2013. High altitude bird migration at temperate latitudes: a synoptic perspective on wind assistance. *PLOS One* 8: e52300.

Dole, J. 1999. First Ringed Kingfisher in Oklahoma: northernmost record for the United State. *Bulletin of the Oklahoma Ornithological Society* 32: 9–12.

Donald, P. F. 2013. First record of Ortolan Bunting *Emberiza hortulana* for southern Africa, in Namibia. *Bulletin of the African Bird Club* 21: 228–229.

Donald, P.F., Brooke, M.D.L., Bolton, M.R., Taylor, R., Wells, C.E., Marlow, T. and Hille, S.M. 2005. Status of Raso Lark *Alauda razae* in 2003, with further notes on sex ratio, behaviour and conservation. *Bird Conservation International* 15: 165–172.

Donald, P.F. & Christodoulides, S. 2018. The 2007 record of 'Dunn's Lark' on Cyprus revisited, with notes on the separation of Dunn's Lark *Eremalauda dunni* and Arabian Lark *E. eremodites. Sandgrouse* 40: 17–24.

Donald, P.F., Geraldes, P. & Garcia-del-Rey, E. 2019. First record of Buff-breasted Sandpiper *Calidris subruficollis* for the Cape Verde Islands, with a brief review of African and Macaronesian records. *Bulletin of the African Bird Club* 26: 222–229.

Donegan, T. & Huertas, B. 2015. Noteworthy bird records on San Andrés island, Colombia. *Conservación Colombiana* 22: 8–12.

Donegan, T.M., Ellery, T., Pacheco, J.A., Verhelst, J.C. & Salaman, P. 2018. Revision of the status of bird species occurring or reported in Colombia 2018. *Conservación Colombiana* 25: 4–47.

Doolittle, T.C.J., Berger, D.D. & Van Stappen, J.F. 2013. Easternmost recovery in Europe of a Peregrine Falcon banded in North America. *Journal of Raptor Research* 47: 75–76.

Dornas, T. & Pinheiro, R.T. 2014. First record of Merlin *Falco columbarius* from Tocantins and a review of previous Brazilian records. *Ornithology Research* 22: 49–52.

Dos Remedios, N., Küpper, C., Székely, T., Zefania, S., Burns, F., Bolton, M. & Lee, P.L., 2018. Genetic structure among *Charadrius* plovers on the African mainland and islands of Madagascar and St Helena. *Ibis* 162: 104–118.

Dowsett, R.J., Brewster, C.A. & Hines, C. 2011. Some bird distributional limits in the Upper Zambezi Valley. *Bulletin of the African Bird Club* 18: 17–30.

Draffan, R.D.W., Garnett, S.T. & Malone, G.J. 1983. Birds of the Torres Strait: An annotated list and biogeographic analysis. *Emu* 83: 207–234.

Drovetski, S.V., Zink, R.M., Rohwer, S., Fadeev, I.V., Nesterov, E.V., Karagodin, I., Koblik, E.A. & Red'kin, Y.A. 2004. Complex biogeographic history of a Holarctic passerine. *Proceedings of the Royal Society of London. Series B: Biological Sciences* 271: 545–551.

Duchateau, S. 2007 Le statut de l'Aigle ibérique *Aquila adalberti* en France. *Alauda* 75: 33–42.

Duckworth, R.A. & Badyaev, A.V. 2007. Coupling of dispersal and aggression facilitates the rapid range expansion of a passerine bird. *Proceedings of the National Academy of Sciences* 104: 15017–15022.

Dudley, S.P., Gee, M., Kehoe, C. & Melling, T.M. 2006. The British list: a checklist of the birds of Britain, 7th Edition. *Ibis* 148: 526–563.

Dufour, P., Pons, J.M., Collinson, J.M., Gernigon, J., Dies, J.I., Sourrouille, P. & Crochet, P.A. 2017. Multilocus barcoding confirms the occurrence of Elegant Terns in Western Europe. *Journal of Ornithology* 158: 351–361.

Duncan, R.B. & Lacroix, J.V. 2001. First sight record of the King Vulture in Baja California, Mexico. *Journal of Raptor Research* 35: 74.

Dunn, P.O., May, T.A., McCollough, M.A. & Howe, M.A. 1988. Length of stay and fat content of migrant Semipalmated Sandpipers in eastern Maine. *The Condor* 90: 824–835.

Dunn, P.J. 2015. A White-necked Petrel (*Pterodroma cervicalis*) off Baja California Sur, Mexico: first record for North America. *North American Birds* 68: 305–307.

Durand, A.L. 1963. A remarkable fall of American landbirds on the 'Mauretania', New York to Southampton, October 1962. *British Birds* 56: 157–164.

Durand, A.L. 1972. Landbirds over the North Atlantic: unpublished records 1961–65 and thoughts a decade later. *British Birds* 65: 428–442.

Dutson, G. & Watling, D. 2007. Cattle egrets (*Bubulcus ibis*) and other vagrant birds in Fiji. *Notornis* 54: 54–55.

Dyer, B.M., Upfold, L., Kant, W. & Ward, V.L. 2001. Breeding by and additional records of Australasian Gannet, *Morus serrator*, in South Africa. *Ostrich* 72: 124–125.

Eaton, J.A., van Balen, S., Brickle, N. W. & Rheindt, F.E. 2016. *Birds of the Indonesian Archipelago: Greater Sundas and Wallacea.* Lynx Edicions, Barcelona.

Ebels, E.B. & Van der Laan, J. 1994. Occurrence of Blue-cheeked Bee-eater in Europe. *Dutch Birding* 16: 95–101.

Ebels, E.B., Versteeg, G.-J. & Bakker, G. 2015. Spaanse Mussen in Nederland in 1997–2014. *Dutch Birding* 37: 69–79.

Eckerle, K.P. & Thompson, C.F. 2001. Yellow-breasted Chat (*Icteria virens*), version 2.0. In *The Birds of North America* (Poole, A. F. & Gill, F. B. Eds). Cornell Lab of Ornithology, Ithaca, NY, USA. https://doi.org/10.2173/bna.575

Edwards, D.P. & Scheffers, B.R. 2018. Lincoln's Sparrow *Melospiza lincolnii*: first record for Colombia and second record for South America. *Cotinga* 40: 92–93.

Ekins, M. 2011. Island Monarch *Monarcha cinerascens*: First record on an island close to Mainland Australia. *Australian Field Ornithology* 28: 162.

Elkins, N. 1979. Nearctic landbirds in Britain and Ireland: a meteorological analysis *British Birds* 72: 417–433.

Elkins, N. 1988. Recent transatlantic vagrancy of landbirds and waders. *British Birds* 81: 484–491.

Elkins, N. 1988. *Weather and Bird Behaviour.* 2nd Edition. T & AD Poyser, Calton, UK.

Elliott, A., Garcia, E.F.J. & Boesman, P. F. D. 2020a. Milky Stork (*Mycteria cinerea*), version 1.0. In *Birds of the World* (J. del Hoyo, A. Elliott, J. Sargatal, D.A. Christie, and E. de Juana, Editors). Cornell Lab of Ornithology, Ithaca, NY, USA. https://doi.org/10.2173/bow.milsto1.01

Elliott, A., Christie, D.A., Jutglar, F. & Kirwan, G. M. 2020. Australian Pelican (*Pelecanus conspicillatus*), version 1.0. In *Birds of the World* (J. del Hoyo, A. Elliott, J. Sargatal, D. A. Christie, and E. de Juana, Editors). Cornell Lab of Ornithology, Ithaca, NY, USA. https://doi.org/10.2173/bow.auspel1.01

Ellison, W.G. 1993. Historical Patterns of Vagrancy by Blue-Gray Gnatcatchers in New England (Patrones Históricos de Movimientos Errantes de Individuos de Polioptila caerulea en New England). *Journal of Field Ornithology* 64: 358–366.

Emlen, S.T. 1970. Celestial rotation: its importance in the development of migratory orientation. *Science* 170: 1198–1201.

Engel, J.I., Hennen, M.H., Witt, C.C. & Weckstein, J.D., 2011. Affinities of three vagrant Cave Swallows from eastern North America. *The Wilson Journal of Ornithology* 123: 840–845.

Engilis Jr, A., Pyle, R.L. & David, R.E. 2004. Status and occurrence of migratory birds in the Hawaiian Islands: Part 1–Anseriformes: Anatidae (Waterfowl). *Bishop Museum Occasional Paper*, 81.

England, M.D. 1974. A further review of the problem of 'escapes'. *British Birds* 67: 177–197.

Enticott, J.W. 1984. New and rarely recorded birds at Gough Island, March 1982–December 1983. *Marine Ornithology* 12: 75–81.

Ergen, A.G. & Barış, S.Y. 2016. First record of Black-throated Accentor, *Prunella atrogularis* (Brandt, 1884), in Turkey. *Zoology in the Middle East* 62: 367–369.

Ericsson, S. 1981. Loggerhead Shrike in Guatemala in December 1979. *Dutch Birding* 3: 27–28.

Ericsson, P. 2015. Black-fronted Dotterel *Elseyornis melanops* near Waingapu, Sumba, Lesser Sunda Islands, the first record for Indonesia. *BirdingASIA* 24: 143.

Eriksen, H. 2004. The first Cape Gannet *Sula capensis* in Oman and the Middle East. *Sandgrouse* 26: 146–148.

Eriksen, J. & Victor, R. 2013. *Oman Bird List: The Official List of the Birds of the Sultanate of Oman*. Seventh edition. Sultan Qaboos University, Muscat.

Eriksen, J. & Porter, R. 2017. *Birds of Oman*. Helm, London.

Erikstad, K.E., Fauchald, P., Tveraa, T. & Steen, H. 1998. On the cost of reproduction in long-lived birds: the influence of environmental variability. *Ecology* 79: 1781–1788.

Erritzøe, J., Mann, C.F., Brammer, F. & Fuller, R.A. 2012. *Cuckoos of the World*. Christopher Helm, London.

Erritzoe, J. 2020. African Pitta (*Pitta angolensis*), version 1.0. In *Birds of the World* (J. del Hoyo, A. Elliott, J. Sargatal, D. A. Christie, and E. de Juana, Editors). Cornell Lab of Ornithology, Ithaca, NY, USA. https://doi.org/10.2173/bow.afrpit1.01

Evans, M.I., Hawkins, F., Duckworth, J.W. & Kirwan, G.M. 2020. Brown Mesite (*Mesitornis unicolor*), version 1.0. In *Birds of the World* (J. del Hoyo, A. Elliott, J. Sargatal, D. A. Christie, and E. de Juana, Editors). Cornell Lab of Ornithology, Ithaca, NY, USA. https://doi.org/10.2173/bow.bromes1.01

Ewins, P.J. & Weseloh, D.V. 1999. Little Gull (*Hydrocoloeus minutus*), version 2.0. In *The Birds of North America* (Poole, A.F. & Gill, F.B. Eds. Cornell Lab of Ornithology, Ithaca, NY, USA. https://doi.org/10.2173/bna.428

Fanshawe, J.H., Prince, P. & Irwin, M. 1992. Black-bellied Storm Petrel *Fregetta tropica*, Antarctic Prion *Pachyptila desolata*, and Thin-billed Prion *P. belcheri*: three species new to Kenya and East Africa. *Scopus* 15: 102–108.

Fareh, M., Maire, B., Laïdi, K. & Franchimont, J. 2019. Rapport de la Commission d'Homologation Marocaine Numéro 24 (2018). *Go-South Bulletin* 1: 21–45.

Farnsworth, A. 2005. Flight calls and their value for future ornithological studies and conservation research. *The Auk* 122: 733–746.

Farnsworth, A., La Sorte, F.A. & Iliff, M.J. 2015. Warmer summers and drier winters correlate with more winter vagrant Purple Gallinules (*Porphyrio martinicus*) in the North Atlantic region. *The Wilson Journal of Ornithology* 127: 582–592.

Fast, M., Ríos, S.D., Clay, R.P. & Smith, P. 2019. *Sporophila nigricollis* (Aves, Thraupidae) en Paraguay: documentación fotográfica y comentarios sobre su estado en el país. *Boletin Museo Nacional de Historia Natural del Paraguay* 23: 14–17.

Favero, M. & Silva, M.P. 1999. First record of the Least Seedsnipe *Thinocorus rumicivorus* in the Antarctic. *Ornitología Neotropical* 10: 107–109.

Fayet, A.L. & Becciu, P. 2018. Easternmost record of an Atlantic Puffin *Fratercula arctica* in the Mediterranean Sea on the coast of Israel. *Seabird* 31: 84–87.

Feilberg, J. 1985. Belted Kingfisher in Godhavn, W. Greenland. *Dansk Ornithologisk Forenings Tidsskrift* 79:34.

Félix, F. & Haase, B. 2016. A new record of the Blackish Oystercatcher, *Haematopus ater* (Vieillot and Oudart, 1825), in the Gulf of Guayaquil, Ecuador. *Check List* 12: 1857.

Felix, R. & Lima, B.M.M. 2020. First confirmed record of Scissor-tailed Flycatcher *Tyrannus forficatus* for Colombia and South America. *Conservación Colombiana* 26: 42–43.

Ferguson-Lees, I.J. & Sharrock, J.T.R. 1977. When will the Fan-tailed Warbler colonise Britain? *British Birds* 70: 152–159.

Ferguson-Lees, J. & Christie, D.A. 2001. *Raptors of the World*. Christopher Helm, London.

Ferreira, A. A., Neto, A. F., Paula, J. P., Machado, N., Azevedo, P. L. & Laranjeiras, T. O. 2010. Ocorrência do *Laniisoma elegans* (Passeriformes: Tityridae) no estado de Goiás. *Atualidades Ornitológicas* 156: 9.

Ferroni, M.L., Earnshaw, A. & Regio, N. 2017. Picaflor nuca blanca (*Florisuga mellivora*), una nueva especie para la avifauna argentina. *Nótulas Faunísticas* 216: 1–7.

Figuerola, J. 2007. Climate and dispersal: black-winged stilts disperse further in dry springs. *PLOS One* 2: e539.

Fijn, R.C., Hiemstra, D., Phillips, R.A. & van der Winden, J. 2013. Arctic Terns *Sterna paradisaea* from The Netherlands migrate record distances across three oceans to Wilkes Land, East Antarctica. *Ardea* 101: 3–12.

Fink, D., Auer, T., Johnston, A., Strimas-Mackey, M., Robinson, O., Ligocki, S., Hochachka, W., Wood, C., Davies, I., Iliff, M., Seitz, L. 2020. eBird Status and Trends, Data Version: 2019; Released: 2020. Cornell Lab of Ornithology, Ithaca, New York. https://doi.org/10.2173/ebirdst.2019.

Fisher, D. & Hunter, N. 2014. East African Rarities Committee Report 2010–2013. *Scopus* 33: 87–91.

Fisher, D. & Hunter, N. 2016. East African Rarities Committee Report 2013–2015. *Scopus* 36: 57–64.

Fisher, D. & Hunter, N. 2017. East African Rarities Committee (EARC) Rarities Report. *Scopus* 37: 46–48.

Fisher, D. & Hunter, N. 2018. East African Rarities Committee Report for 2017. *Scopus* 38: 25–29.

Fisher, D. & Hunter, N. 2019. East African Rarities Committee Report for 2018. *Scopus* 39: 44–48.

Fitzpatrick, J. 2004. Tyrant Flycatchers (Tyrannidae). Pp. 170–462 in del Hoyo J., Elliott A. & Christie, D. A. (Eds.). *Handbook of the Birds of the World*. Volume 9: Cotingas to pipits and wagtails. Lynx Edicions, Barcelona.

Fjeldså, J., Zuccon, D., Irestedt, M., Johansson, U.S. & Ericson, P. G. P. 2003. *Sapayoa aenigma*: a New World representative of "Old World suboscines". *Proceedings of the Royal Society of London. Series B: Biological Sciences* 270: 238–241.

Flack, A., Pettit, B., Freeman, R., Guilford, T. & Biro, D. 2012. What are leaders made of? The role of individual experience in determining leader–follower relations in homing pigeons. *Animal Behaviour* 83: 703–709.

Flegg, J. 2004. *Time to Fly: Exploring Bird Migration*. British Trust for Ornithology, Thetford, UK.

Flood, R.L. 2009. 'All-dark' *Oceanodroma* storm-petrels in the Atlantic and neighbouring seas. *British Birds* 102: 365–385.

Flood, R.L. 2012. Madeiran Storm-petrel off Scilly: new to Britain. *British Birds* 105: 2–10.

Flood, R.L., Simon, J., Tribot, J. & Pineau, K. 2017. A Swinhoe's Storm-petrel (*Hydrobates monorhis*) in French Guiana: the first record for South America. *Revista Brasileira de Ornitologia* 25: 227–231.

Flood, R.L. & Wilson, A.C. 2017. A New Zealand Storm Petrel *Fregetta maoriana* off Gau Island, Fiji, in May 2017. *Bulletin of the British Ornithologists' Club* 137: 278–286.

Folch, A., Christie, D. A., Jutglar, F. & Garcia, E.F.J. 2020a. Common Ostrich (*Struthio camelus*), version 1.0. In *Birds of the World* (J. del Hoyo, A. Elliott, J. Sargatal, D. A. Christie, and E. de Juana, Editors). Cornell Lab of Ornithology, Ithaca, NY, USA. https://doi.org/10.2173/bow.ostric2.01

Folch, A., Christie, D.A. & Garcia, E.F.J. 2020b. Emu (*Dromaius novaehollandiae*), version 1.0. In *Birds of the World* (J. del Hoyo, A. Elliott, J. Sargatal, D. A. Christie, and E. de Juana, Editors). Cornell Lab of Ornithology, Ithaca, NY, USA. https://doi.org/10.2173/bow.emu1.01

Folvik, A. & Mjøs, A. T. 1995. Spring migration of White-billed Divers past southwestern Norway. *British Birds* 88: 125–129.

Fonseca, V. S. S., Azevedo, M.S. & Petry, M.V. 2000. Nota sobre a ocorrência da pomba-antártica, *Chionis alba* (Gmelin, 1789), no litoral norte do Rio Grande do Sul, Brasil. *Acta Biologica Leopoldensia* 22: 133–135.

Forewind 2014. Dogger Bank Teesside A & B. Environmental statement. Chapter 11: Marine and coastal ornithology. F-OFC-CH-011_Issue 4.1. Forewind Ltd.

Forone, P. 2004. WA460605, *Podiceps occipitalis* Garnot, 1826. http://www.wikiaves.com/460605

Förschler, M.I., Randler, C., Dierschke, J. & Bairlein, F. 2010. Morphometric diagnosability of Cyprus Wheatears *Oenanthe cypriaca* and an unexpected occurrence on Helgoland Island. *Bird Study* 57: 396–400.

Förschler, M.I., Shaw, D.N. & Bairlein, F. 2011. Deuterium analysis reveals potential origin of the Fair Isle Citril Finch *Carduelis citrinella*. *Bulletin of the British Ornithologists' Club* 131: 189–191.

Förschler, M.I., Voigt, C.C. & Bairlein, F. 2018. Potential origin of White-crowned Wheatear in Denmark and Germany in 2010. *Dutch Birding* 40: 400–405.

Forshaw, J.M. & Shephard, M. 2012. *Grassfinches in Australia*. CSIRO Publishing, Collingwood, Australia.

Fox, T.A.D., Christensen, T.K., Bearhop, S. & Newton, J. 2007. Using stable isotope analysis of multiple feather tracts to identify moulting provenance of vagrant birds: a case study of Baikal Teal *Anas formosa* in Denmark. *Ibis* 149: 622–625.

Fox, T.A., Hobson, K.A., Ekins, G., Grantham, M. & Green, A.J. 2010. Isotope forensic analysis does not support vagrancy for a Marbled Duck shot in Essex. *British Birds* 103: 464–467.

Franz, I., Agne, C.E., Bencke, G.A., Bugoni, L. & Dias, R.A. 2018. Four decades after Belton: a review of records and evidences on the avifauna of Rio Grande do Sul, Brazil. *Iheringia Série Zoologia* 108.

Fraser, P.A. 2007. Report on rare birds in Great Britain in 2006. *British Birds* 100: 614–675.

Fraser, D.F., Gilliam, J.F., Daley, M.J., Le, A.N. & Skalski, G.T. 2001. Explaining leptokurtic movement distributions: intrapopulation variation in boldness and exploration. *The American Naturalist* 158: 124–135.

Freile, J.F. & Restall, R. 2018. *Birds of Ecuador.* Helm, London.

Freile, J.F., Ahlman, R., Brinkhuizen, D.M., Greenfield, P.J., Solano-Ugalde, A., Navarrete, L. & Ridgely, R.S. 2013. Rare birds in Ecuador: first annual report of the Committee of Ecuadorian Records in Ornithology (CERO). *Avances en Ciencias e Ingenierías* 5: B24–B41.

Freile, J.F., Solano-Ugalde, A., Brinkhuizen, D.M., Greenfield, P.J., Lysinger, M., Nilsson, J., Navarrete, L. & Ridgely, R.S. 2017. Rare birds in Ecuador: third report of the Committee for Ecuadorian Records in Ornithology (CERO). *Revista Ecuatoriana de Ornitología* 1: 8–27.

Freile, J.F., A. Solano-Ugalde, D.M. Brinkuizen, P.J. Greenfield, M. Lysinger, J. Nilsson, L. Navarrete & R.S. Ridgely. 2019. Fifth report of the Committee for Ecuadorian Records in Ornithology (CERO) with comments on some published, undocumented records. *Revista Ecuatoriana de Ornitología* 6: 103–133.

Fuglei, E., Blanchet, M.A., Unander, S., Ims, R.A. & Pedersen, Å.Ø. 2017. Hidden in the darkness of the Polar night: a first glimpse into winter migration of the Svalbard rock ptarmigan. *Wildlife Biology* 2017: wlb-00241.

Fulco, E. & Liuzzi, C. 2019. Commissione Ornitologica Italiana (COI) – Report 28. *Avocetta* 43.

Furness, R. W. 1978. Movements and mortality rates of Great Skuas ringed in Scotland. *Bird Study* 25: 229–238.

Furness, R.W. & Monteiro, L.R. 1995. Red-billed Tropicbird *Phaethon aethereus* in the Azores: first breeding record for Europe. *Bulletin of the British Ornithologists' Club* 115: 6–8.

Gabrielson, I.N. & Lincoln, F.C. 1959. *The Birds of Alaska*. Stackpole Company, Harrisburg, PA, USA.

Galbraith, M., Tennyson, A., Shepherd, L., Robinson, P. & Fraser, D. 2013. High altitude New Zealand record for a long-tailed skua (*Stercorarius longicaudus*). *Notornis* 60: 245–248.

Galvez, X., Guerrero, L. & Migoya, R. 2016. Evidencias físicas de la estructura metapoblacional en el Flamenco Caribeño (*Phoenicopterus ruber ruber*) a partir de avistamientos de individuos anilados. *Revista Cubana de Ciencias Biológicas* 4: 92–97.

Gantlett, S. 1998. Identification of Great Blue Heron and Grey Heron. *Birding World* 11: 12–20.

Gantlett, S. & Millington, R. 2009. The Glossy Ibis invasion in autumn 2009. *Birding World* 22: 424–434.

Gao, L., Mi, C. & Guo, Y. 2019. Expansion of sandhill cranes (*Grus canadensis*) in east Asia during the non-breeding period. *PeerJ* 7: e7545

García-Borboroglu, P., Boersma, P.D., Ruoppolo, V., Pinho-da-Silva-Filho, R., Corrado-Adornes, A., Conte-Sena, D., Velozo, R., Myiaji-Kolesnikovas, C., Dutra, G., Maracini, P. & Carvalho-do-Nascimento, C. 2010. Magellanic penguin mortality in 2008 along the SW Atlantic coast. *Marine Pollution Bulletin* 60: 52–1657.

Garcia-del-Rey, E. 2011. *Field Guide to the Birds of Macaronesia*. Lynx Edicions, Barcelona.

García-Godos, I. 2002. First record of the Cape Gannet *Morus capensis* for Peru and the Pacific Ocean. *Marine Ornithology* 30: 51–54.

García-Godos, I. & Sánchez, R. 2009. First photographic records of the Red-footed Booby *Sula sula* in Perú. *Boletín de la Unión de Ornitólogos* 4: 7–8.

García Vargas, F.J. & Sagardía, J. 2014. El influx de Mosquitero Bilistado *Phylloscopus inornatus* en Lanzarote (Islas Canarias) durante el otoño-invierno 2013-2014. Reservoir Birds (web). URL: http://www.reservoirbirds.com/Articles/RBAR_000013.pdf

Gardner, M. & Gilfedder, M. 2011. First records of Silver

Gull *Larus (Chroicocephalus) novaehollandiae* for Bali and Oriental Region. *Kukila* 15: 118–121.

Garrett, K.L. 2014. The 38th annual report of the California Bird Records Committee: 2012 records. *Western Birds* 45: 246–275.

Garrido, O. & Kirkconnell, A. 1993. *Checklist of Cuban Birds.* Havana.

Garrigues, R. & Costa Rican Rarities Committee. 2016. Lista Oficial de las Aves de Costa Rica. Actualización 2016. *Zeledonia.* 20: 3–12.

Gaston, A.J. & Hipfner, J.M. 2000. Thick-billed Murre (*Uria lomvia*), version 2.0. In *The Birds of North America* (Poole, A. F. and Gill, F. B. Eds). Cornell Lab of Ornithology, Ithaca, NY, USA. https://doi.org/10.2173/bna.497

Gauthier-Clerc, M., Jiguet, F. & Lambert, N. 2002. Vagrant birds at Possession Island, Crozet Islands and Kerguelen Island from December 1995 to December 1997. *Marine Ornithology* 30: 38–39.

Geupel, G.R. & G. Ballard. 2020. Wrentit (*Chamaea fasciata*), version 1.0. In *Birds of the World* (Rodewald, P. G. Ed.). Cornell Lab of Ornithology, Ithaca, NY, USA. https://doi.org/10.2173/bow.wrenti.01

Ghizoni-Jr, I.R. & Piacentini, V.D.Q. 2010. The Andean Flamingo *Phoenicoparrus andinus* (Philippi, 1854) in southern Brazil: is it a vagrant? *Revista Brasileira de Ornitologia* 18: 263–266.

Gibson, D.D. & Kessel, B. 1992. Seventy-four new avian taxa documented in Alaska 1976–1991. *The Condor* 94: 454–467.

Gibson, D.D., DeCicco, L.H., Gill, R.E., Jr., Heinl, S.C., Lang, A.J., Tobish, T.G., Jr. & Withrow, J.J. 2018. Fourth report of the Alaska Checklist Committee, 2013–2017. *Western Birds* 49: 174–191.

Gilfedder, C., Gilfedder, M., Brickle, N. & Ichsan, A. K. 2011. First record of Pied Oystercatcher *Haematopus longirostris* for Bali and the Greater Sundas. *Kukila* 15: 124–125.

Gilg, O., Moe, B., Hanssen, S.A., Schmidt, N.M., Sittler, B., Hansen, J., Reneerkens, J., Sabard, B., Chastel, O., Moreau, J. & Phillips, R.A. 2013. Trans-equatorial migration routes, staging sites and wintering areas of a high-arctic avian predator: the Long-tailed Skua (*Stercorarius longicaudus*). *PLOS One* 8: e64614

Gill Jr, R.E., Tibbitts, T.L., Douglas, D.C., Handel, C.M., Mulcahy, D.M., Gottschalck, J.C., Warnock, N., McCaffery, B.J., Battley, P.F. & Piersma, T., 2008. Extreme endurance flights by landbirds crossing the Pacific Ocean: ecological corridor rather than barrier? *Proceedings of the Royal Society of London. Series B: Biological Sciences* 276: 447–457.

Gill, B. 2019. New Zealand Wrens (Acanthisittidae). In del Hoyo, J., Elliott, A., Sargatal, J., Christie, D. A. & de Juana, E. (Eds.). *Handbook of the Birds of the World Alive.* Lynx Edicions, Barcelona. https://www.hbw.com/node/52301

Gilpin, M.E. & Diamond, J.M., 1982. Factors contributing to non-randomness in species co-occurrences on islands. *Oecologia* 52: 75–84.

Gilroy, J.J. & Lees, A.C. 2003. Vagrancy theories: are autumn vagrants really reverse migrants? *British Birds* 96: 427–438.

Gilroy, J.J., Avery, J.D. & Lockwood, J.L. 2017. Seeking international agreement on what it means to be 'native'. *Conservation Letters* 10: 238–247.

Gil-Velasco, M., Rouco, M., Ferrer, J., García-Tarrasón, M., García-Vargas, F. J., Gutiérrez, A., Hevia, R., López, F., López-Velasco, D., Ollé, A., Rodríguez, G., Sagardía, J. & Salazar, J.A. 2018. Observaciones de aves raras en España, 2016. *Ardeola* 65: 97–139.

Gil-Velasco, M., Rouco, M., García-Tarrasón, M., García-Vargas, F. J., Hevia, R., Illa, M., López, M., López-Velasco, D., Ollé, À., Rodríguez, G., Rodríguez, M. & Sagardía, J. 2019. Observaciones de aves raras en España, 2017. *Ardeola* 66: 169–204.

Gluschenko Yu. N. & Korobov D.V. 2012. Ashy Drongo Vieillot, 1817 – a new species in the avifauna of Russia. *Far-Eastern Journal of Ornithology* 3: 61–64.

Gluschenko Yu. N. & Korobov D.V. 2015. New data on study of the fauna of the birds of Far Eastern Marine Reserve. *Biodiversity and Environment of Far East Reserves* 5: 22–45.

Gokulakrishnan, G., Sivaperuman, C. & Dash, M. 2018. Greater Painted-snipe *Rostratula benghalensis*, and Yellow-breasted Bunting *Emberiza aureola* in the Andaman Islands. *Indian Birds* 14: 153–154.

Goldberg, N.R. & Mason, N.A. 2017. Species identification of vagrant *Empidonax* flycatchers in Northeastern North America via non-invasive DNA sequencing. *Northeastern Naturalist* 24: 499–504.

Goldstein, M.I., Duffy, D.C., Oehlers, S., Catterson, N., Frederick, J. & Pyare, S. 2019. Interseasonal movements and non-breeding locations of Aleutian Terns *Onychoprion aleuticus. Marine Ornithology* 47: 67–76.

Gomes, F.B.R., Bernardes, V.C.D., Lacava, R., Lopes Baptista, L.A.M. & Oliveira e Silva, M.B. 2016. Records of Brown Pelican *Pelecanus occidentalis* inland in northern Brasil: distributional expansion or juvenile dispersal? *Marine Ornithology* 44: 247–248.

Gomes, G.C., de Freitas, T.C., da Cunha, H.N., Jacobs, F.P. & Wall, M.S. 2020. New records and range extension of Black-goggled Tanager, *Trichothraupis melanops* (Vieillot, 1818) (Passeriformes, Thraupidae), in extreme southern Brazil. *Check List* 16: 67–73.

Gonzaga, L.P. 1996. Family Cariamidae (Seriemas). Pp. 234–239 in del Hoyo, J., Elliot, A. & Sargatal, J. (Eds.). *Handbook of the Birds of the World.* Volume 3. Lynx Edicions, Barcelona.

González C. G. 2013. Primer registro documentado del Corbatito Overo (*Sporophila lineola*) en Chile. *El Hornero* 28: 75–78.

Goodwin, D. 1949. Notes on the migration of birds of prey over Suez. *Ibis* 91: 59–63.

Gordinho, L. O. & Martins, G. M. 2009. The first Audouin's Gull for the Azores just 1,500 miles off Newfoundland. *Birding* 46: 1–4.

Gorman, G. 1996. *The Birds of Hungary.* Christopher Helm, London.

Gorman, K.B., Erdmann, E.S., Pickering, B.C., Horne, P.J., Blum, J.R., Lucas, H.M., Patterson-Fraser, D.L. & Fraser, W.R. 2010. A new high-latitude record for the macaroni penguin (*Eudyptes chrysolophus*) at Avian Island, Antarctica. *Polar Biology* 33: 1155–1158.

Gould J. & Darwin, C. 1839. *The zoology of the voyage of*

H.M.S. Beagle, under the command of Capt. Fitzroy, R. N., during the years 1832 to 1836. Part 3: birds, issue 4. Smith, Elder, London, pp 57–96.

Granit, B. & Smith, J. P. 2004. The first Red-necked Stint *Calidris ruficollis* in Israel. *Sandgrouse* 26: 53–55.

Grantham, M. 2004. Ringing in December. *Birding World* 17: 482.

Grantham, M. 2014. Orkney Rosefinch to Germany. BTO Demog Blog. http://btoringing.blogspot.com/2014/05/

Greatwich, B. 2019. Eurasian Hobby *Falco subbuteo* at South Lake, Perth, February 2016: The first Australian mainland record. *Australian Field Ornithology* 36: 71–73.

Green, A.J. & Sánchez, M.I., 2006. Passive internal dispersal of insect larvae by migratory birds. *Biology Letters* 2: 55–57.

Greenberg, R. & Matsuoka, S. M. 2010. Rusty Blackbird: mysteries of a species in decline. *The Condor* 112: 770–777.

Greeney, H. 2018. *Antpittas and Gnateaters.* Helm, London.

Greenlaw, J.S. 2013. Twenty-second report of the Florida Ornithological Society Records Committee: 2012. *Florida Field Naturalist* 42: 153–171.

Gregory, P. 2020a. Striated Fieldwren (*Calamanthus fuliginosus*), version 1.0. In *Birds of the World* (J. del Hoyo, A. Elliott, J. Sargatal, D. A. Christie, and E. de Juana, Editors). Cornell Lab of Ornithology, Ithaca, NY, USA. https://doi.org/10.2173/bow.strfie1.01

Gregory, P. 2020b. Mid-mountain Berrypecker (*Melanocharis longicauda*), version 1.0. In *Birds of the World* (J. del Hoyo, A. Elliott, J. Sargatal, D. A. Christie, and E. de Juana, Editors). Cornell Lab of Ornithology, Ithaca, NY, USA. https://doi.org/10.2173/bow.lebber1.01

Gregory, P. 2020c. *Birds of Paradise and Bowerbirds.* Helm, London.

Greig, E.I., Wood, E.M. & Bonter, D.N. 2017. Winter range expansion of a hummingbird is associated with urbanization and supplementary feeding. *Proceedings of the Royal Society of London. Series B: Biological Sciences* 284: 20170256.

Gricks, N.P. 1994. Vagrant White Stork *Ciconia ciconia* (Aves: Ciconiidae) found in Antigua: A first record for the West Indies. *Journal of Caribbean Ornithology* 7: 2.

Grieve, A., Hill, B. J. N., Lassey, P.A. & Wallace, D.I.M. 2005. The first American Golden Plover *Pluvialis dominica* in Oman and Arabia. *Sandgrouse* 27: 75–77.

Grinnell, J., 1922. The role of the "accidental". *The Auk* 39: 373–380.

Grønningsaeter, E. 2005. White-rumped Sandpiper summering in Arctic Norway. *Birding World* 18: 349–350.

Grundsten, M., Sångberg, O. & Thankachan, D. 2018. First documented record of Chinese Paradise Flycatcher *Terpsiphone incei* on the Andaman Islands, India. *BirdingASIA* 29: 44–45.

Grussu, M. 2012. Occurrence of the Eagle Owl in the Mediterranean. *British Birds* 105: 100.

Grussu, M. & Biondi, M. 2004. Record numbers of wintering Richard's Pipits in the Western Palearctic. *British Birds* 97: 194–195.

Grussu, M. & Atzeni, A. 2012. Influx of Great White Pelicans in Sardinia, Italy, in spring 2008. *Dutch Birding* 34: 289–293.

Gryz, P., Korczak, M. & Gerlée, A. 2015. First record of the Austral Negrito (Aves: Passeriformes) from the South Shetlands, Antarctica. *Polish Polar Research* 36: 297–304.

Gschweng, M., Kalko, E.K., Querner, U., Fiedler, W. & Berthold, P. 2008. All across Africa: highly individual migration routes of Eleonora's falcon. *Proceedings of the Royal Society of London. Series B: Biological Sciences*, 275: 2887–2896.

Guilherme, E., Aleixo, A.L.P., Guimarães, J.D.O., Dias, P.R.D.F., Amaral, P.P.D., Zamora, L.M. & Souza, M.S.D. 2005. Primeiro registro de *Phoenicoparrus jamesi* (Aves, Phoenicopteriformes) para o Brasil. *Revista Brasileira de Ornitologia* 13: 212–214.

Guilherme, E. & Dantas, S. 2008. First record of Ruby Topaz *Chrysolampis mosquitus* in Acre, Brazil. *Cotinga* 30: 84–85.

Gunn, W.W. & Crocker, A.M. 1951. Analysis of unusual bird migration in North America during the storm of April 4–7, 1947. *The Auk* 68: 139–163.

Guo, H. & Ma, M. 2012. The Egyptian Vulture (*Neophron percnopterus*): record of a new bird in China. *Chinese Birds* 3: 238–239.

Guo, H. & Ma, M. 2013. Eurasian Blackcap (*Sylvia atricapilla*) – a new bird record in China. *Zoological Research* 34: 507–508.

Gutiérrez, R. 1995. Vagrancy likelihood of the Welsh Monk Vulture. *British Birds* 88: 607-607.

Gutiérrez, R., Lorenzo, J.A., Elorriaga, J., Gorospe, G., López-Velasco, D., Martí-Aledo, J., Rodríguez, G. & Sales, S. 2013. Observaciones de aves raras en España, 2011. *Ardeola* 60: 437–506.

Gwinner, E. & Wiltschko, W. 1980. Circannual changes in migratory orientation of the garden warbler, *Sylvia borin. Behavioral Ecology and Sociobiology* 7: 73–78.

Gyug, L.W., Dobbs, R.C., Martin, T.E. & Conway, C.J. 2012. Williamson's Sapsucker (*Sphyrapicus thyroideus*), version 2.0. In *The Birds of North America* (Poole, A. F. Ed. Cornell Lab of Ornithology, Ithaca, NY, USA. https://doi.org/10.2173/bna.285

Haas, M. 2012. *Extremely Rare Birds in the Western Palearctic.* Lynx Edicions, Barcelona.

Hachenberg, A., Cruz-Flores, M. & Militão, T. 2017. Bulwer's Petrel at Kressbachsee, Germany, in July 2015. *Dutch Birding* 39: 183–191.

Haffer, J. 1969. Speciation in Amazonian forest birds. *Science* 165: 131–137.

Hagmann, G. 1907. Die Vogelwelt der Insel Mexiana, Amazonenstrom. *Zoologische Jahrbücher. Abteilung für Systematik* 26: 11–62.

Hagstrum, J.T. 2000. Infrasound and the avian navigational map. *Journal of Experimental Biology* 203: 1103–1111.

Hailer, F., James, H.F., Olson, S.L. & Fleischer, R.C. 2015. Distinct and extinct: Genetic differentiation of the Hawaiian eagle. *Molecular Phylogenetics and Evolution* 83: 40–43.

Hake M., Kjellén N., Alerstam T. 2003. Age-dependent migration strategy in honey buzzards *Pernis apivorus* tracked by satellite. *Oikos* 103: 385–396.

Hall, R., Erdélyi, R., Hanna, E., Jones, J.M. & Scaife, A.A. 2015. Drivers of North Atlantic polar front jet stream variability. *International Journal of Climatology* 35: 1697–1720.

Halle, B.O.V. 1990. Adiciones a la avifauna de Colombia de especies arribadas a la Isla Gorgona. *Caldasia* 16: 209–214.

Hallgrimsson, G., Van Swelm, N.D., Gunnarsson, H.V., Johnson, T.B. & Rutt, C.L. 2011. First two records of European-banded lesser black-backed gulls *Larus fuscus* in America. *Marine Ornithology* 39: 137–139.

Halpern, B.S., Walbridge, S., Selkoe, K.A., Kappel, C.V., Micheli, F., D'Agrosa, C., Bruno, J.F., Casey, K.S., Ebert, C., Fox, H.E. & Fujita, R. 2008. A global map of human impact on marine ecosystems. *Science* 319: 948–952.

Halpin, L.R. & Willie, M.M.C. 2014. First Record of Dovekie in British Columbia. *Northwestern Naturalist* 95: 56–60.

Hamada, F.H. & Rodrigues, T. 2018. Primeiro registro de *Pheucticus ludovicianus* (Passeriformes: Cardinalidae) no Brasil. *Atualidades Ornitológicas* 205: 75.

Hameed, S., Norwood, H.H., Flanagan, M., Feldstein, S. & Yang, C.H. 2009. The influence of El Niño on the spring fallout of Asian bird species at Attu Island. *Earth Interactions* 13: 1–22.

Hamilton, R.A., Patten, M.A. & Erickson, R.A. 2007. *Rare Birds of California: a work of the California Bird Records Committee.* Western Field Ornithologists, Camarillo, California.

Hannon, S.J., Eason, P.K. & Martin, K. 1998. Willow Ptarmigan (*Lagopus lagopus*), version 2.0. In *The Birds of North America* (Poole, A. F. & Gill, F. B. Eds.). Cornell Lab of Ornithology, Ithaca, NY, USA. https://doi.org/10.2173/bna.369

Harrap, S. & Quinn, D. 1996. *Tits, Nuthatches & Treecreepers.* A&C Black, London.

Harrap, S., Christie, D. A. & Kirwan, G. M. 2020. Eurasian Nuthatch (*Sitta europaea*). In del Hoyo, J., Elliott, A., Sargatal, J., Christie, D. A. & de Juana, E. eds. *Handbook of the Birds of the World Alive.* Lynx Edicions, Barcelona. https://www.hbw.com/node/59914

Harris, M.P. & Murray, S. 1989. *Birds of St. Kilda.* HMSO, London.

Harris, R.B., Alström, P., Ödeen, A. & Leaché, A.D. 2018. Discordance between genomic divergence and phenotypic variation in a rapidly evolving avian genus (*Motacilla*). *Molecular Phylogenetics and Evolution* 120: 183–195.

Harrison, I. 2014. From the rarities committees. *Sandgrouse* 36: 110–116.

Harrison, I. 2015. From the rarities committees. *Sandgrouse* 37: 203–210.

Harrison, I. 2016. From the rarities committees. *Sandgrouse* 38: 214–222.

Harrison, I. & Grieve, A. 2012. Around the region. *Sandgrouse* 34: 189–203.

Harrison, P. 1983. Laysan Albatross *Diomedea immutabilis*: new to the Indian Ocean. *Cormorant* 11: 39–44.

Harrop, A. H. 2008. The rise and fall of Bulwer's Petrel. *British Birds* 101: 676–681.

Harrop, A. H. 2010. Records of Hawk Owls in Britain. *British Birds* 103: 276–283.

Harrop, H. R., Mavor, R. & Ellis, P. M. 2008. Olive-tree Warbler in Shetland. *British Birds* 101: 82–88.

Harvey, M.G., Seeholzer, G.F., Cáceres A.D., Winger, B.M., Tello, J.G., Camacho, F.H., Aponte Justiniano, M.A., Judy, C.D., Ramírez, S.F., Terrill, R.S. & Brown, C.E. 2014. The avian biogeography of an Amazonian headwater: the Upper Ucayali River, Peru. *The Wilson Journal of Ornithology* 126: 179–191.

Harvey, R. 2018. Western Swamphen in Suffolk and Lincolnshire: new to Britain. *British Birds* 111: 512–518.

Hassell, C.J. & Ward, N. 2012. Cinnamon Bittern (*Ixobrychus cinnamomeus*) at Anna Plains Station, Western Australia, the first record for mainland Australia. *Amytornis* 4: 9–12.

Hassell, C.J. 2016. Arctic Warbler: first confirmed record for mainland Australia. *Australian Field Ornithology* 17: 365–369.

Hatibaruah, B., Ovalekar, S. & Ghosh, S. 2017. First record of White-cheeked Starling *Spodiopsar cineraceus* from India. *Indian Birds* 13: 73–74.

Hayes, F.E., Lecourt, P. & del Castillo, H. 2018. Rapid southward and upward range expansion of a tropical songbird, the Thrush-like Wren (*Campylorhynchus turdinus*), in South America: a consequence of habitat or climate change? *Revista Brasileira de Ornitologia* 26: 57–64.

Hayman, P. & Thorns, D. 2015. Long-tailed Skua *Stercorarius longicaudus*: first record for Uganda. *Scopus* 35: 47–49.

Hays, C. 1986. Effects of the 1982–1983 El Nino on Humboldt Penguin colonies in Peru. *Biological Conservation* 36: 169–180.

Hazevoet, C.J. 1995. *The Birds of the Cape Verde Islands. An Annotated Checklist. BOU Check-list 13.* British Ornithologists' Union, Tring, UK.

Hazevoet, C.J. 1997. On a record of the Wattled Crane *Bugeranus carunculatus* from Guinea-Bissau. *Bulletin of the British Ornithologists' Club* 117: 56–59.

Hazevoet, C.J. 2010. Sixth report on birds from the Cape Verde Islands, including records of 25 taxa new to the archipelago. *Zoologia Caboverdiana* 1: 3–44.

Hazevoet, C.J. 2012. Seventh report on birds from the Cape Verde Islands, including records of nine taxa new to the archipelago. *Zoologia Caboverdiana* 3: 1–28.

Headon, J., Collinson, M. & Cade, M. 2018. Pale-legged Leaf Warbler: new to Britain. *British Birds* 111: 438–440.

Hedenström, A., Norevik, G., Warfvinge, K., Andersson, A., Bäckman, J. & Åkesson, S. 2016. Annual 10-month aerial life phase in the common swift *Apus apus. Current Biology* 26: 3066–3070.

Heiss, M. 2013. Spotted Nutcracker *Nucifraga caryocatactes*: a new species for Azerbaijan. *Sandgrouse* 35: 39–42.

Helbig, A.J. 1992. Ontogenetic stability of inherited directions in a nocturnal bird migrant: comparison between the first and second year of life. *Ethology, Ecology and Evolution* 4: 375–388.

Henshaw, I., Fransson, T., Jakobsson, S., Lind, J., Vallin, A. & Kullberg, C. 2008. Food intake and fuel deposition in a migratory bird is affected by multiple as well as single-step changes in the magnetic field. *Journal of Experimental Biology* 211: 649-653.

Hermes, C., Döpper, A., Schaefer, H.M. & Segelbacher, G. 2016. Effects of forest fragmentation on the morphological and genetic structure of a dispersal-limited, endangered bird species. *Nature Conservation* 16: 39–58.

Herzog P.W. & Keppie, D.M. 1980. Migration in a local population of Spruce Grouse. *The Condor* 82: 366–372.

Herzog, S.K., Terrill, R.S., Jahn, A.E., Remsen, J.V., Maillard, O.Z., García-Solíz, V.H., MacLeod, R., Maccormick, A. & Vidoz, J.Q. 2016. *Birds of Bolivia: Field Guide.* Asociacíon Armonía, Santa Cruz de la Sierra, Bolivia.

Hickey, C.M., Capitolo, P. & Walker, B. 1996. First record of a Lanceolated Warbler in California. *Western Birds* 27: 197–201.

Hidalgo-Aranzamendi N., Alfaro-Shigueto J. & Zavalaga C. 2010. New records of Broad-billed Prions (*Pachyptila vittata*) in southern Peru. *Notornis 57*: 39–42.

Higgins, P.J. & Davies, S.J.J.F., eds. 1996. *Handbook of Australian, New Zealand, and Antarctic Birds.* Volume 3. Oxford University Press, Melbourne.

Hill, G.E., Sargent, R.R. & Sargent, M.B. 1998. Recent change in the winter distribution of Rufous Hummingbirds. *The Auk* 115: 240–245.

Hill, N.P & Bishop, K.D. 1999. Possible winter quarters of the Aleutian Tern? *Wilson Bulletin* 111: 559–560.

Hillman, J.C. & Clingham, E. 2012. First record of Dwarf Bittern *Ixobrychus sturmii* for St. Helena, South Atlantic. *Bulletin of the African Bird Club* 19: 213–214.

Hillman, J.C., Hillman, S.M., Ellick, G., George, K., Higgins, D., Lambdon, P. & Beard, A. 2016. Swifts *Apus* sp. and Common House Martins *Delichon urbicum* on St. Helena, South Atlantic, in 2012–13. *Bulletin of the African Bird Club* 23: 95–98.

Hilty, S.L. 2003. *Birds of Venezuela.* Christopher Helm, London.

Himmel, C. 2019. First record of White-rumped Sandpiper *Calidris fuscicollis* for Azerbaijan. *Sandgrouse* 41: 78–79.

Hisahiro, T., Mikio, T. & Takashi, S. 2010. A record of a juvenile Spot-billed Pelican *Pelecanus philippensis* on Amami-Oshima Island, Kagoshima Prefecture. *Japanese Journal of Ornithology* 59: 65–68.

Hjerppe, S., Jones, S. & Trainor, C. 2015. First record of Buff-breasted *Sandpiper Calidris subruficollis* from Indonesia and South-East Asia at Serangan Island, Benoa harbour, Bali, Indonesia. *BirdingASIA* 24: 137–139.

Hjort, C. 2005. Siberian Pectoral Sandpipers seen migrating towards the southwest. *British Birds* 98: 257–262.

Hobbs, J. 2015. A list of Irish birds. Version 7.1 http://www.southdublinbirds.com/notes/notes/List-of-Irish-Birds-v7.1(JH).pdf

Hockey, P., Kirwan, G.M. & Boesman, P. 2020. Eurasian Oystercatcher (*Haematopus ostralegus*), version 1.0. In *Birds of the World* (S.M. Billerman, B.K. Keeney, P.G. Rodewald, and T.S. Schulenberg, Editors). Cornell Lab of Ornithology, Ithaca, NY, USA. https://doi.org/10.2173/bow.euroys1.01

Høgsås, T.E., Vizcarra, J.K., Hidalgo Aranzamendi, N. & Málaga Arenas, E. 2010. Primeros registros documentados de *Phoenicoparrus andinus* en la costa sur de Perú. *Cotinga* 32: 120–121.

Holbrook, K.M. & Smith, T.B. 2000. Seed dispersal and movement patterns in two species of *Ceratogymna* hornbills in a West African tropical lowland forest. *Oecologia* 125: 249–257.

Holden, B.R. & Bell, D.M. 2016. Carolina Chickadee: Second record for Ontario and Canada. *Ontario Birds* 34: 2–13.

Holland, R.A., Thorup, K., Gagliardo, A., Bisson, I.A., Knecht, E., Mizrahi, D. & Wikelski, M., 2009. Testing the role of sensory systems in the migratory heading of a songbird. *Journal of Experimental Biology* 212: 4065–4071.

Hollyer, J.N. 1970. The invasion of Nutcrackers in autumn 1968. *British Birds* 63: 353–373.

Holt, P. 2005. Franklin's Gull *Larus pipixcan* at Tanggu, Tianjin: first record for China. *Forktail* 21: 171–173.

Holt, P.I., England, A.S., Beaton, R.E. & Bloss, J. 2013 Franklin's Gull *Leucophaeus pipixcan* in Goa: a new species for India. *BirdingASIA* 20: 119–120.

Hoogendoorn, W. & Steinhaus, G.H. 1990. Nearctic gulls in the West Palearctic. *Dutch Birding* 12: 109–164.

Hopkin, P. 2011. Black-tailed Gull – New to Bermuda. *Bermuda Audubon Society Newsletter* 22.

Hoppen, R.C. 2016. WA2173409, *Thamnophilus doliatus* (Linnaeus, 1764)]. http://www.wikiaves.com/2173409

Horton, T.W., Bierregaard, R.O., Zawar-Reza, P., Holdaway, R.N. & Sagar, P. 2014. Juvenile osprey navigation during trans-oceanic migration. *PLOS One* 9: e114557

Horton, K.G., Van Doren, B.M., Stepanian, P.M., Hochachka, W.M., Farnsworth, A. & Kelly, J.F. 2016. Nocturnally migrating songbirds drift when they can and compensate when they must. *Scientific Reports* 6: 21249.

Horton, K.G., Van Doren, B.M., La Sorte, F.A., Fink, D., Sheldon, D., Farnsworth, A. & Kelly, J.F. 2018. Navigating north: how body mass and winds shape avian flight behaviours across a North American migratory flyway. *Ecology Letters* 21: 1055–1064.

Hosner, P.A., Tobias, J.A., Braun, E.L. & Kimball, R.T. 2017. How do seemingly non-vagile clades accomplish trans-marine dispersal? Trait and dispersal evolution in the landfowl (Aves: Galliformes). *Proceedings of the Royal Society of London. Series B: Biological Sciences* 284: 20170210.

Howell, S.N.G. 2004. An update on status of birds from Isla Cozumel, Mexico. *Cotinga* 22: 15–19.

Howell, S.N.G., Pyle, P., Spear, L.B. & Pitman, R.L. 1993. North American migrant birds on Clipperton Atoll. *Western Birds* 24: 73–80.

Howell, S.N.G. 2007. Chilean records of Royal Tern *Sterna maxima*, Black Tern *Chlidonias niger*, Grey-hooded Gull *Larus cirrocephalus* and Manx Shearwater *Puffinus puffinus*. *Cotinga* 27: 84–85.

Howell, S.N.G. & Webb, S. 1992. Observations of North American migrant birds in the Revillagigedo Islands. *Euphonia* 1: 27–33.

Howell, S.N.G. & Dunn, J. 2017. *Gulls of the Americas.* Houghton Mifflin, Boston & New York.

Howell, S.N.G., Erickson, R.A., Hamilton, R.A. & Patten, M.A. 2001. *An annotated checklist of the birds of Baja California and Baja California Sur.* Birds of the Baja California Peninsula: status, distribution, and taxonomy, pp.171-203.

Howell, S.N.G., Lewington, I. & Russell, W. 2014. *Rare Birds of North America.* Princeton University Press, Princeton, New Jersey.

Hudson, R. 1972. Green Heron in Cornwall in 1889. *British Birds* 65: 424–427.

Hudson, R. 1974. Allen's Gallinule in Britain and the Palearctic. *British Birds* 67: 405–413.

Hudson, N. & the Rarities Committee 2013. Report on rare birds in Great Britain in 2012. *British Birds* 106: 570–641.

Hudson, N. & the Rarities Committee 2015. Report on rare birds in Great Britain in 2014. *British Birds* 108: 565–633.

Hulme, M., Riley, A. & Sansom, A. 2012. First records of Red-backed Shrike *Lanius collurio* for Ghana. *Bulletin of the African Bird Club* 19: 204–205.

Hume, R.A. 1993. Brown Shrike in Shetland: new to Britain and Ireland. *British Birds* 86: 600–604.

Hunter, E. 2018. Iberian Chiffchaff: a new breeding species for Great Britain. *British Birds* 111: 100–108.

Hyde N.H., Matthews K.E., Thompson, M.A. & Gale, R.O. 2010. First record of barn owls (*Tyto alba*) breeding in the wild in New Zealand. *Notornis 56:* 169–175.

Ikenaga, H., Sweet, P. & Hart, M. 2009. Analysis and reconfirmation of a Bee-eater specimen from Japan. *Ornithological Science* 8: 169–171.

Ingels, J., Claessens, O., Luglia, T., Ingremeau, P. & Kenefick, M. 2010. White Wagtail *Motacilla alba*, a vagrant to Barbados, Trinidad and French Guiana. *Bulletin of the British Ornithologists' Club* 130: 224–226.

Ingels, J., Dechelle, M. & Bøgh, R. 2011. Little Wood Rail *Aramides mangle*, a Brazilian endemic, found in French Guiana. *Bulletin of the British Ornithologists' Club* 131: 200–202.

Interaminense, L.J.L., Almeida, L.B.A. & Hortêncio, C.A. 1996. Nota sobre o primeiro registro de *Chionis alba* para o Arquipélago de Abrolhos, BA. In Vielliard, J.M.E., Silva, M.L. & Silva, W.R. eds. *V Congresso Brasileiro de Ornitologia, Resumos*. Campinas: Sociedade Brasileira de Ornitologia.

Iqbal, M. & Albayquni, A.A. 2016. First record of a Slaty-backed Gull *Larus schistisagus* for Indonesia. *Marine Ornithology* 44: 135–136.

Isenmann, P. & Moali, A. 2000. *Oiseaux d'Algérie*. Ornithologiques de France, Paris.

Isenmann, P., Gaultier, T., El Hili, A., Azafzaf, H., Dlensi, H. & Smart, M. 2005. *Oiseaux de Tunisie*. Société d'Études Ornithologiques de France, Paris.

Jackson, C. 1997. First record of Northern Lapwing *Vanellus vanellus* in East Africa. *Scopus* 19: 113–114.

Jackson, B. 2004. Snowy Egret in Argyll & Bute: new to Britain. *British Birds* 97: 270–275.

Jacobs, A., Herman, B. & Bertrands, J. 2018. White-throated Bee-eaters in Western Sahara, Morocco, in December 2013 and February–May 2017. *Dutch Birding* 40: 29–32.

Jahan, I., Haque, M.O., Kamal, G.M. & Rahman, N.A. 2016. Lesser Frigatebird *Fregata ariel* in the Sundarban mangrove forest: the first record for Bangladesh. *BirdingASIA* 26: 134–135.

Jahn, A E., Davis, S.E. & Zankys, A.M.S. 2002. Patterns of austral bird migration in the Bolivian Chaco. *Journal of Field Ornithology* 73: 258–267.

Jahn, O., Cosgrove, P., Cosgrove, C., Cevallos, T.M. & García, T.S. 2010 First record of Brown Pelican *Pelecanus occidentalis* from the Ecuadorian highlands. *Cotinga* 32: 108.

James, D.J. & McAllan, I.A. 2014. The birds of Christmas Island, Indian Ocean: A review. *Australian Field Ornithology* 31: S1–S175.

Jansen, J.J.F.J. & Nap, W. 2008. Identification of White-headed Long-tailed Bushtit and occurrence in the Netherlands. *Dutch Birding* 30: 293–308.

Jaramillo, A.P. 2000. Punta Rasa, South America's first vagrant trap. *Cotinga* 14: 33–38.

Jaramillo, A. 2003. *Field Guide to the Birds of Chile including the Antarctic Peninsula, the Falkland Islands and South Georgia*. Christopher Helm, London.

Javed, S., Douglas, D.C., Khan, S., Shah, J.N. & Hammadi, A.A. 2012. First description of autumn migration of Sooty Falcon *Falco concolor* from the United Arab Emirates to Madagascar using satellite telemetry. *Bird Conservation International* 22: 106–119.

Jay, R.S. 1993. The Plymouth sheathbill. *Sea Swallow* 42: 72.

Jayson E.A., Babu S. & Govind, S.K. 2013. Recovery of White Tern *Gygis alba* at Athirapilly, Kerala, India. *Indian Birds* 8: 163.

Jensen, A.E., Fisher, T.H. & Hutchinson, R.O. 2015. Notable new bird records from the Philippines. *Forktail* 31: 24–36.

Jeong, M.S., Kim, H.K., Park, J.Y. & Lee, W.S., 2014. The first record of the White Tern (*Gygis alba*) in South Korea. *Korean Journal of Ornithology* 21: 69–73.

Jiguet, F., Robert, A., Micol, T. & Barbraud, C. 2007. Quantifying stochastic and deterministic threats to island seabirds: last endemic prions face extinction from falcon peregrinations. *Animal Conservation* 10: 245–253.

Jiguet, F., Doxa, A. & Robert, A. 2008. The origin of out-of-range pelicans in Europe: wild bird dispersal or zoo escapes? *Ibis* 150: 606–618.

Jiguet, F. & Barbet-Massin, M. 2013. Climate change and rates of vagrancy of Siberian bird species to Europe. *Ibis* 155: 194–198.

Jiguet, F., el Din, B., Bonser, R., Crochet, P.A., Grieve, A., Hoath, R., Haraldsson, T., Riad, A. & Megalli, M. 2012. Second report of the Egyptian Ornithological Rarities Committee – 2011.

Jiménez, S. 2013. First record of Salvin's albatross (*Thalassarche salvini*) in Uruguayan waters. *Notornis* 60: 313–314.

Johnson, A.R. 1989. Movements of greater flamingos (*Phoenicopterus ruber roseus*) in the Western Palearctic. *Revue d'écologie* 4: 75–94.

Johnson, K.P., Clayton, D.H., Dumbacher, J.P. & Fleischer, R.C. 2010. The flight of the Passenger Pigeon: Phylogenetics and biogeographic history of an extinct species. *Molecular Phylogenetics and Evolution* 57: 455–458.

Johnson, T.H. & Stattersfield, A.J., 1990. A global review of island endemic birds. *Ibis* 132: 167–180.

Johnson, J.A., Booms, T.L., DeCicco, L.H. & Douglas, D.C. 2017. Seasonal movements of the Short-eared Owl (*Asio flammeus*) in western North America as revealed by satellite telemetry. *Journal of Raptor Research* 51: 115–128.

Johnstone, R.E. & Darnell, J.C. 2015. Checklist of the Birds of Western Australia. In *Western Australian Museum* http://www.museum.wa.gov.au/collections/databases/fauna.asp

Johnstone, R.E., Burbidge, A.H. & Darnell, J.C. 2013. Birds of the Pilbara region, including seas and offshore islands, Western Australia: distribution, status and historical changes. *Records of the Western Australian Museum* 78: 343–441.

Jones, D. & Birks, S. 1992. Megapodes: recent ideas on origins, adaptations and reproduction. *Trends in Ecology & Evolution* 7: 88–91.

Jones, H.L. & Komar, O. 2008 Central America. *North American Birds* 61: 648–651.

Jönsson, O. 2011. Great Black-backed Gulls breeding at Khniffis Lagoon, Morocco and the status of Cape Gull in the Western Palearctic. *Birding World* 24: 68–76.

Joseph, L. & Kernot, R. 1982. Range extensions of Gilbert's Whistler. *South Australian Ornithologist* 28: 217–218.

Joubert, S.C.J. 1977. Avian marine vagrants to the Kruger National Park. *Koedoe* 20: 185.

Juáres, M.A., Libertelli, M.M., Santos, M.M., Negrete, J., Gray, M., Baviera, M., Moreira, M.E., Donini, G., Carlini, A. & Coria, N.R., 2010. Aves, Charadriiformes, Scolopacidae, *Limosa haemastica* (Linnaeus, 1758): first record from South Shetland Islands and Antarctic Peninsula, Antarctica. *Check List* 6: 559–560.

Jukema, J. & Piersma, T. 2002. Occurrence of Pacific Golden Plovers in The Netherlands; Historical Perspectives from the "Wilsternetters". *Waterbirds* 25: 93–100.

Kakishima, S. & Morimoto, T. 2015. First record of dusky woodswallow (*Artamus cyanopterus*) in New Zealand. *Notornis* 62: 231–232.

Kalwij, J.M., Medan, D., Kellermann, J., Greve, M. & Chown, S.L. 2019. Vagrant birds as a dispersal vector in transoceanic range expansion of vascular plants. *Scientific Reports* 9: 1–9.

Kang, M., Dhillon, R., Kahlon, N. & Singh, N. 2016. Wood Warbler from Ladakh, India. *Indian Birds* 12: 135–136.

Kasius, B., Witkamp, C. & Ingels, J. 2019. Observation of Audouin's Gull *Ichthyaetus audouinii* in Suriname on South American mainland. *Revista Brasileira de Ornitologia* 27: 59–60.

Kasozi, S.N. & Byaruhanga, A. 2009. Records: First record of the Wattled Crane *Grus carunculatus* in Uganda. *Scopus* 29: 21–22.

Katkar, P.S., Katkar, P.P. and Adhikari, O.D., 2021. Sighting of Grey Hypocolius *Hypocolius ampelinus* Bonaparte 1850 in Palghar District, Maharashtra, India. *Journal of the Bombay Natural History Society*, 118.

Kearns, A.M., Joseph, L., White, L.C., Austin, J.J., Baker, C., Driskell, A.C., Malloy, J.F. & Omland, K.E. 2015. Norfolk Island Robins are a distinct endangered species: ancient DNA unlocks surprising relationships and phenotypic discordance within the Australo-Pacific Robins. *Conservation Genetics* 17: 321–335.

Kelly, A., Leighton, K. & Newton, J. 2010. Using stable isotopes to investigate the provenance of an Eagle Owl found in Norfolk. *British Birds* 103: 213–222.

Kelly, A.G., Coombes, R.H., O'Mahony, D. & Porter, B. 2014. First record of Audouin's Gull *Ichthyaetus audouinii* for Ghana. *Bulletin of the African Bird Club* 21: 226–228.

Kemp, A.C. & Boesman, P. 2020. African Gray Hornbill (*Lophoceros nasutus*), version 1.0. In *Birds of the World* (J. del Hoyo, A. Elliott, J. Sargatal, D. A. Christie, and E. de Juana, Editors). Cornell Lab of Ornithology, Ithaca, NY, USA. https://doi.org/10.2173/bow.afghor1.01

Kenefick, M. 2012. Report of the Trinidad and Tobago Rare Birds Committee: rare birds in 2008–10. *Cotinga* 34: 166–171.

Kenefick, M., Restall, R. & Hayes, F. 2019. *Birds of Trinidad & Tobago*. 3rd edition. Helm, London.

Kennedy, A.S. & Marsh, R.M. 2016. First record of Crested Honey Buzzard *Pernis ptilorhynchus* for Kenya and East Africa. *Bulletin of the African Bird Club* 136: 145–146.

Kerlinger, P., Cherry, J.D. & Powers, K.D. 1983. Records of migrant hawks from the North Atlantic Ocean. *The Auk* 100: 488–490.

Kerlinger, P. 1985. Water-crossing behaviour of raptors during migration. *Wilson Bulletin* 97: 109–113.

Kessler, A.E., Batbayar, N., Natsagdorj, T., Batsuur', D. & Smith, A.T. 2013. Satellite telemetry reveals long-distance migration in the Asian Great Bustard *Otis tarda dybowskii*. *Journal of Avian Biology* 44: 311–320.

Khaleghizadeh, A., Scott, D.A., Tohidifar, M., Musavi, S. B., Ghasemi, M., Sehhatisabet, M.E., Ashoori, A., Khani, A., Bakhtiari, P., Amini, H., Roselaar, C., Ayé, R., Ullman, M., Nezami, B. & Eskandari, F. 2011. Rare birds in Iran in 1980–2010. *Podoces* 6: 1–48.

Kharitonov S.P. 1999. The first record of the Atlantic puffin (*Fratercula arctica*) in the Pacific, and routes of alcid vagrancy between the Atlantic and Pacific oceans. *Bulletin of the North-Eastern Research Center of FEB RAS* 3: 105–107.

Khil, L., Ullman, M. & Ławicki, Ł., 2019. Ashy Drongos in Iran in 2014–18 and status in the WP. *Dutch Birding* 41: 23–28.

King, J. & Minguez, E. 1994. Swinhoe's Petrel: the first Mediterranean record. *Birding World* 7: 271–273.

Kinsky, F.C. 1968. An unusual seabird mortality in the southern North Island (New Zealand), April 1968. *Notornis* 15: 143-155.

Kinsky, F.C. & Jones, E.B. 1972. Northern shovelers (*Anas clypeata*) in New Zealand. *Notornis* 19: 105–110.

Kirwan, G.M. 1994. American Golden Plover, *Pluvialis dominica*, in Turkey. *Zoology in the Middle East* 10: 23–26.

Kirwan, G.M., Demirci, B., Welch, H., Boyla, K., Özen, M., Castell, P. & Marlow, T. 2010. *The Birds of Turkey*. Christopher Helm, London.

Kirwan, G.M. & Green, G. 2011. *Cotingas and Manakins*. Christopher Helm, London.

Kirwan, G.M., Özen, M., Ertuhan, M. & Atahan, A. 2014. Turkey Bird Report 2007–2011. *Sandgrouse* 36: 146–175.

Kirwan, G.M., Pacheco, J.F. & Lees, A.C. 2015. First documented record of the Sapphire Quail-Dove *Geotrygon saphirina* Bonaparte, 1855, in Brazil, an overlooked specimen from the Klages expedition to Amazonia. *Revista Brasileira de Ornitologia* 23: 354–356.

Kirwan, G. M., Levesque, A. Oberle, M., Sharpe, C.J. 2019. *Birds of the West Indies*. Lynx Edicions, Barcelona.

Kiss, J.B., Szabó, L. & Duquet, M. 2001. White-tailed Lapwing, a new breeding species for Europe. *Ornithos* 8: 100–107.

Kitching, M. 2002. The Wilson's Petrel off Northumberland – the first British North Sea record. *Birding World* 15: 390–391.

Kleis, J. L. 2008. Note: Sooty falcon, new to France. *Ornithos* 15: 144–145.

Klemmer, K. & Zizka, G. 1993. The terrestrial Fauna of Easter Island. In Fischer, S. R. (Ed.), *Easter Island studies: Contributions to the history of Rapanui in memory of William T. Mulloy*. Oxbow Books, Oxford.

Klingel, F. & Mahood, S.P. 2015. Silver Oriole *Oriolus mellianus*: first record for Vietnam. *BirdingASIA* 23: 131.

Knowlton, W.H. 2016. Sharp-tailed Sandpiper *Calidris acuminata* in Bolivia: first documented record for South America. *Cotinga* 38: 20–22.

Knox, A., Mendel, H. & Odin, N. 1994. Red-billed Tropicbird in Suffolk. *British Birds* 87: 488–491.

Komar, O., de Gelder, A., Boyla, K.A., Angehr, G., Balderamos, P. & Allport, G. 2006. The first Central American record of White-throated Sparrow *Zonotrichia albicollis*. *Cotinga* 26: 43–45.

Komisja Faunistyczna. 2017. Rzadkie ptaki obserwowane w Polsce w roku 2016. *Ornis Polonica* 58: 83–116.

Konyukhov, N.B. & Kitaysky, A.S. 1995. The Asian race of the Marbled Murrelet. Pp. 23–29 in Ralph, C.J., Hunt, G.L., Raphael, M.G. & Piatt, J.F. (eds.), *Ecology and Conservation of the Marbled Murrelet*. Albany.

Korczak-Abshire, M., Angiel, P. J. & Wierzbicki, G. 2011a. Records of white-rumped sandpiper (*Calidris fuscicollis*) on the South Shetland Islands. *Polar Record* 47: 262–267.

Korczak-Abshire, M., Lees, A. C. & Jojczyk, A. 2011b. First documented record of barn swallow (*Hirundo rustica*) in the Antarctic. *Polish Polar Research* 4: 355–360.

Korobov, D.V. & Glushchenko, Y.N. 2014. The first record of the White-necked Petrel *Pterodroma cervicalis* in the territorial waters of Russia. *Russian Journal of Ornithology* 1074: 3715–3716.

Körtner, G., Brigham, R.M. & Geiser, F. 2000. Metabolism: winter torpor in a large bird. *Nature* 407: 318.

Kotzé, L.J.D., Green, D.B. & Dyer, B.M. 2015. First record of a Great Reed Warbler on the Prince Edward Islands. *Ornithological Observations* 6: 16–18.

Krabbe, N.K. & Schulenberg, T.S. 2003a. Family Rhinocryptidae (Tapaculos). Pp. 748–787 in del Hoyo, J., Elliott, A. & Christie, D.A. (Eds.). *Handbook of the Birds of the World*, Volume 8. Lynx Edicions, Barcelona.

Krabbe, N.K. & Schulenberg, T.S. 2003b, Family Formicariidae (Ground-Antbirds). Pp. 682–73 in del Hoyo, J., Elliott, A. & Christie, D.A. (Eds.). *Handbook of the Birds of the World,* Volume 8. Lynx Edicions, Barcelona.

Kramer, G. 1952. Experiments on bird orientation. *Ibis* 94: 265–285.

Kratter, A.W., Sillett, T.S., Chesser, R.T., O'Neill, J.P., Parker, T.A. & Castillo, A. 1993. Avifauna of a Chaco locality in Bolivia. *The Wilson Bulletin* 105: 114–141.

Kratter, A.W. & Brothers, M. 2010. First record of Ancient Murrelet (*Synthliboramphus antiquus*) for Florida. *Florida Field Naturalist* 38: 106–109.

Kratter, A.W. 2018. Twenty-seventh report of the Florida Ornithological Society records committee: 2016–2017. *Florida Field Naturalist* 46: 96–117.

Krüger, T. 2018: The occurrence of the Demoiselle Crane *Grus virgo* in Germany. *Vogelwarte* 56: 225–245.

Kullberg, C., Henshaw, I., Jakobsson, S., Johansson, P. & Fransson, T. 2007. Fuelling decisions in migratory birds: geomagnetic cues override the seasonal effect. *Proceedings of the Royal Society of London. Series B: Biological Sciences* 274: 2145–2151.

Kunzmann, M.R., Hall, L.S. & Johnson, R.R. 1998. Elegant Trogon (*Trogon elegans*), version 2.0. In *The Birds of North America* (Poole, A. F. & Gill, F. B. Eds). Cornell Lab of Ornithology, Ithaca, NY, USA. https://doi.org/10.2173/bna.357

Kushlan, J.A. & Prosper, J.W. 2009. Little Egret (*Egretta garzetta*) nesting on Antigua: a second nesting site in the Western Hemisphere. *Journal of Caribbean Ornithology* 22: 108–111.

Kusumanegara, H. & Iqbal, M. 2019. A southernmost record of chestnut-cheeked starling, *Agropsar phillippensis*, in Bali, Indonesia. *Journal of Indonesian Natural History* 3: 44–46.

La Sorte, F.A., Fink, D., Hochachka, W.M., Farnsworth, A., Rodewald, A.D., Rosenberg, K.V., Sullivan, B.L., Winkler, D.W., Wood, C. & Kelling, S. 2014. The role of atmospheric conditions in the seasonal dynamics of North American migration flyways. *Journal of Biogeography* 41: 1685–1696.

La Sorte, F.A., Fink, D., Hochachka, W.M. & Kelling, S. 2016. Convergence of broad-scale migration strategies in terrestrial birds. *Proceedings of the Royal Society of London. Series B: Biological Sciences* 283: p.20152588.

Laine, L.J. 1996. The 'Borcka Puzzle' – the first Western Palearctic Crested Honey Buzzard. *Birding World* 9: 324–325.

Lallsingh, N. 2018. A vagrant from the Old World: a mysterious gull in Trinidad. *Neotropical Birding* 22: 54–58.

Lambert, K. 2001. Sightings of new and rarely reported seabirds in southern African waters. *Marine Ornithology* 29: 115–118.

Langrand, O. & Sinclair, J.C. 1994. Additions and supplements to the Madagascar avifauna. *Ostrich* 65: 302–310.

Lansley, P.S., Watson, J. & Farnes, R.F. 2016. Brown Shrike *Lanius cristatus* on Christmas Island, Indian Ocean. *Australian Field Ornithology* 20: 139–141.

Las-Casas, F.M. & de Azevedo-Júnior, S.M. 2010. Ocorrencia de *Knipolegus nigerrimus* (Vieillot, 1818) (Aves, Tyrannidae) no Distrito do Para, Santa Cruz do Capibaribe, Pernambuco, Brasil. *Ornithologia* 3: 18–20.

Lavretsky, P., McCracken, K.G. & Peters, J.L. 2014. Phylogenetics of a recent radiation in the mallards and allies (Aves: *Anas*): inferences from a genomic transect and the multispecies coalescent. *Molecular Phylogenetics and Evolution* 70: 402–411.

Ławicki, Ł., 2012. Azure Tits and hybrids Azure x European Blue Tit in Europe. *Dutch Birding* 34: 219–231.

Ławicki, L. 2013 Japanese Waxwing vagrancy to Europe. *Birding World* 26: 43–44.

Ławicki, Ł., Khil, L. & de Vries, P.P. 2012. Expansion of Pygmy Cormorant in central and western Europe and increase of breeding population in southern Europe. *Dutch Birding* 34: 27–3.

Ławicki, Ł. & van den Berg, A.B. 2015a. WP reports. *Dutch Birding* 37: 340–353.

Ławicki, Ł. & van den Berg, A.B. 2015b. WP reports. *Dutch Birding* 37: 403–420.

Ławicki, Ł. & van den Berg, A.B. 2016a. WP reports. *Dutch Birding* 38: 102–116.

Ławicki, Ł. & van den Berg, A.B. 2016b. WP reports. *Dutch Birding* 38: 183–193.

Ławicki, L. & van den Berg, A.B. 2016c. WP reports. *Dutch Birding* 38: 240-254.

Ławicki, Ł. & van den Berg, A.B. 2016d. WP reports. *Dutch Birding* 38: 322–336.

Ławicki, L. & van den Berg, A.B. 2017a. WP reports. *Dutch Birding* 39: 43–62.

Ławicki, L. & van den Berg, A.B. 2017b. WP reports. *Dutch Birding* 39: 118–132.

Ławicki, L. & van den Berg, A.B. 2017c. WP reports. *Dutch Birding* 39: 200–214.

Ławicki, L. & van den Berg, A.B. 2017d. WP reports. *Dutch Birding* 39: 257–273.

Ławicki, L. & van den Berg, A.B. 2017e. WP reports. *Dutch Birding* 39: 393–415.

Ławicki, L. & van den Berg, A.B. 2018a. WP reports. *Dutch Birding* 40: 46–59.

Ławicki, L. & van den Berg, A.B. 2018b. WP reports. *Dutch Birding* 40: 178–198.

Ławicki, L. & van den Berg, A.B. 2018c. WP reports. *Dutch Birding* 40: 407–423.

Ławicki, L. & van den Berg, A.B. 2019a. WP reports. *Dutch Birding* 41: 51–62.

Ławicki, L. & van den Berg, A.B. 2019b. WP reports. *Dutch Birding* 41: 186–203.

Ławicki, Ł. & Perlman, Y., 2017. Black-winged Kite in the WP: increase in breeding population, vagrancy and range. *Dutch Birding* 39: 1–12.

Leathers, D.J. & Palecki, M.A. 1992. The Pacific/North American teleconnection pattern and United States climate. Part II: temporal characteristics and index specification. *Journal of Climate* 5: 707–716.

Le Corre, M., Salamolard, M. & Portier, M.C. 2003. Transoceanic dispersion of the Red-tailed Tropicbird in the Indian Ocean. *Emu* 103: 183–184.

Le Floch, A., Ropars, G., Lucas, J., Wright, S., Davenport, T., Corfield, M. & Harrisson, M., 2013. The sixteenth century Alderney crystal: a calcite as an efficient reference optical compass? *Proceedings of the Royal Society of London. Series A: Mathematical, Physical and Engineering Sciences* 469: 20120651.

Le, M.H., Nguyen, Q.H., Low, G.W.J., Sadanandan, K.R., Ng, W.Q. & Rheindt, F.R. 2017. The first record of Wedge-tailed Shearwater *Ardenna pacifica* for Vietnam. *BirdingASIA* 27: 122–123.

Leblond, G. 2007. Observation d'une Cicogne blanche *Ciconia ciconia* en Martinique (Petites Antilles). *Alauda* 75: 244–245.

Lee, D.S. 2000. Photograph of Black-bellied and White-faced Whistling-Ducks from North Carolina, with comments on other extralimital waterfowl. *Chat* 64: 93–99.

Lee, D.S. & Walsh-McGee, M. 1998. White-tailed Tropicbird (*Phaethon lepturus*), version 2.0. In *The Birds of North America* (Poole, A. F. & Gill, F. B. Eds). Cornell Lab of Ornithology, Ithaca, NY, USA. https://doi.org/10.2173/bna.353

Lees, A.C. 2016. Evidence for longitudinal migration by a "sedentary" Brazilian flycatcher, the Ash-throated Casiornis. *Journal of Field Ornithology* 87: 251–259.

Lees, A.C. & Ball, A. 2011. Shades of grey: 'eastern' Skylarks and extralimital subspecies identification. *British Birds* 104: 660–666.

Lees, A.C. & Gilroy, J.J. 2004. Pectoral Sandpipers in Europe. *British Birds* 97: 638–646.

Lees, A.C. & Gilroy J.J. 2009. Vagrancy Mechanisms in Passerines and Near-Passerines. In *Rare Birds, Where and When: An analysis of status and distribution in Britain and Ireland*, Slack, R. ed. Volume 1: Sandgrouse to New World Orioles. Rare Bird Books, York.

Lees, A.C. & Gilroy, J.J. 2014. Vagrancy fails to predict colonization of oceanic islands. *Global Ecology and Biogeography* 23: 405–413.

Lees, A.C. & Mahood, S.P. 2011. Jacobin Cuckoo in Finland in September 1976: a plea for reassessment. *Dutch Birding* 33: 325–328.

Lees, A.C. & Martin, R.W. 2015. Exposing hidden endemism in a Neotropical forest raptor using citizen science. *Ibis* 157: 103–114.

Lees, A.C. & Peres, C.A. 2009. Gap-crossing movements predict species occupancy in Amazonian forest fragments. *Oikos* 118: 280–290.

Lees, A.C. & VanderWerf, E.A. 2011. First record of Blyth's Pipit *Anthus godlewskii* for Micronesia. *Bulletin of the British Ornithologists' Club* 131: 212–217.

Lees, A.C., Zimmer, K.J., Marantz, C.A., Whittaker, A., Davis, B.J. & Whitney, B.M. 2013a. Alta Floresta revisited: an updated review of the avifauna of the most intensively surveyed locality in south-central Amazonia. *Bulletin of the British Ornithologists' Club* 133: 165–252.

Lees, A.C., Moura, N.G., Andretti, C.B., Davis, B.J., Lopes, E. V., Henriques, L.M.P., Aleixo, A., Barlow, J., Ferreira, J. & Gardner, T.A. 2013b. One hundred and thirty-five years of avifaunal surveys around Santarém, central Brazilian Amazon. *Revista Brasileira de Ornitologia* 21: 16–57.

Lees, A.C., Thompson, I. & Moura, N.G. 2014. Salgado Paraense: an inventory of a forgotten coastal Amazonian avifauna. *Boletim do Museu Paraense Emílio Goeldi. Ciências Naturais* 9: 135–168.

Lees, A.C., Rosenberg, K.V., Ruiz-Gutierrez, V., Marsden, S., Schulenberg, T.S. & Rodewald, A.D. 2020. A roadmap to identifying and filling shortfalls in Neotropical ornithology. *The Auk* 137: p.ukaa048.

Lefranc, N. & Worfolk, T. 1997. *Shrikes: a Guide to the Shrikes of the World.* Pica Press, Robertsbridge, UK.

Lefroy, J.H. 1981. *Memorials of the Discovery and Early Settlement of the Bermudas or Somers Islands 1515–1685. 5. Compiled from the Colonial Records and Other Original Sources.* Volume 1 with revised chronology and erratum 1511–1652. Second reprinting. Bermuda Historical Society, Hamilton, Bermuda.

Legal, E. 2019. Primeiro registro de *Thamnophilus doliatus* (Aves, Passeriformes, *Thamnophilidae*) para o estado de Santa Catarina, sul do Brasil. *Acta Biológica Catarinense* 6: 19–25.

Legge, W.V. 1905. List of birds observed at the Great Lake in the month of March. *Emu* 4: 103–109.

LeGrand Jr, H.E., Camburn, K.E., Cooper, S., Davis, R.J., Dean, E.V., Forsythe, W.K. & Tyndall, R.L. 2005. 2004 Annual Report of the North Carolina Bird Records Committee. *The Chat* 69: 29–34.

Lehikoinen, P. & Forsman, D. 2020. Chestnut-winged Cuckoo at Ayn Hamran, Oman in December 2019. *Dutch Birding* 42: 341–344.

Lehman, P. 1989. The Changing Seasons Autumn 1988. *American Birds* 43: 50–54.

Lehman, P.E. 2018. River Warbler (*Locustella fluviatilis*) at Gambell, Alaska: First record for North America. *Western Birds* 49: 136–141.

Leitão, A.H. 2011. First sighting of a Brown Noddy *Anous stolidus* for Portugal. *Anuário Ornitológico* 8: 141–142.

Le Nevé, A. & Manzione, M. 2011. First record of the Lesser Sand Plover (*Charadrius mongolus*) in Argentina: a new species for the country and for South America. *Hornero* 26: 177–180.

Lequette, B., Berteaux, D. & Judas, J. 1995. Presence and First Breeding Attempts of Southern Gannets *Morus capensis* and *M. senator* at Saint Paul Island, Southern Indian Ocean. *Emu* 95: 134–137.

Lerner, H.R., Meyer, M., James, H.F., Hofreiter, M. & Fleischer, R.C. 2011. Multilocus resolution of phylogeny and timescale in the extant adaptive radiation of Hawaiian honeycreepers. *Current Biology* 21: 1838–1844.

Lethaby, N. 1998. Identification of Chestnut-flanked and Japanese White-eyes. *Dutch Birding* 20: 57–61.

Lévêque, R., Bowman, R.I. & Billeb, S.L. 1966. Migrants in the Galápagos area. *The Condor* 68: 81–101.

Lewington, I., Alström, P. & Colston, P. 1991. *A Field Guide to the Rare Birds of Britain and Europe*. HarperCollins, London.

Liechti F. & Schaller E. 1999. The use of low-level jets by migrating birds. *Naturwissenschaften* 86: 549–551.

Lima, P.C., Grantsau, R., Lima, R.C.F.R. & dos Santos, S.S. 2002. Notas sobre os registros brasileiros de *Calonectris edwardsii* (Oustalet, 1883) e *Pelagodroma marina hypoleuca* (Mouquin-Tandon, 1841) e primeiro registro de *Phalacrocorax bransfieldensis* (Murphy, 1936) para o Brasil. *Ararajuba* 10: 263–265.

Lima, B. & Kamada, B. 2009. Registros de corvo-bicolor *Corvus albus* (Passeriformes: Corvidae) em territorio brasileiro. *Atualidades Ornitológicas* 150: 10–11.

Lin, D.L., Tsai, Y.L., Ting, T.C., Chan, F.T., Lin, K.H. & Tsai, C.F. 2018. The first wildlife rescue case of a Snowy Sheathbill (*Chionis albus*) from Antarctica. *Journal of the National Taiwan Museum* 71: 19–26.

Lindsell, J. & Fisher, D. 2009. East African Rarities Committee report and change of remit. *Scopus* 29: 23–26.

Lindström, Å. & Alerstam, T. 1986. The adaptive significance of reoriented migration of chaffinches *Fringilla coelebs* and bramblings *F. montifringilla* during autumn in southern Sweden. *Behavioral Ecology and Sociobiology* 19: 417–424.

Lindström, Å. 1989. Finch flock size and risk of hawk predation at a migratory stopover site. *The Auk* 106: 225–232.

Linsley, M.D., Jones, M.J. & Marsden, S.J. 2011. A review of the Sumba avifauna. *Kukila* 10: 60–90.

Liu, Y., Wei, Q., Dong, L. & Lei, J.-Y. 2013. On an update of recent new bird records in China. *Chinese Journal of Zoology* 48: 750–758.

Lobo, P., Kehoe, C.V., Thomason, P.C., Stocker, M.C., Watt, J.E., Rai, B. & Mekhola, R. 2018. Chestnut-flanked White-eye *Zosterops erythropleurus* from Mishmi Hills: An addition to the avifauna of South Asia. *Indian Birds* 14: 82.

Lobo-y-Henriques, J.C.Y., Bates, J. & Willard, D. 2011. First record for the Black-and-white Tanager *Conothraupis speculigera* in Colombia. *Conservación Colombiana* 17: 45–51.

Lockwood, M.W. Texas Bird Records Committee Report for 2007. *Bulletin of the Texas Ornithological Society* 41: 37–72.

Long, A.M., Finn, D.S., Grzybowski, J.A., Morrison, M.L. & Mathewson, H.A. 2019. First documented observation of the federally endangered golden-cheeked warbler (*Setophaga chrysoparia*) in Oklahoma. *Bulletin of the Oklahoma Ornithological Society* 47: 9–16.

Longo, C.A. 2015. WA1874812, *Heliornis fulica* (Boddaert, 1783)]. http://www.wikiaves.com/1874812

Lopes, E.V. 2017. WA2707218, *Oxyruncus cristatus* Swainson, 1821 http://www.wikiaves.com/2707218

Lopes, A. & Kasambe, R. 2016. Recovery of a Red-footed Booby *Sula sula* from the Maharashtra coast, India. *Indian Birds* 12: 86.

Lovas-Kiss, Á., Vincze, O., Löki, V., Pallér-Kapusi, F., Halasi-Kovács, B., Kovács, G., Green, A.J. & Lukács, B.A. 2020. Experimental evidence of dispersal of invasive cyprinid eggs inside migratory waterfowl. *Proceedings of the National Academy of Sciences* 117: 15397–15399.

Loveridge, A. 1969. A sheathbill, *Chionis alba* (Gmelin) on St. Helena. *Bulletin of the British Ornithologists' Club* 89: 48–49.

Lovett, R. & Brina, G. 2001. First documented record of Long-tailed (Mountain) Wagtail *Motacilla clara* in Botswana. *Babbler* 39: 49.

Lowther, P.E. & Nocedal, J. 2012. Olive Warbler (*Peucedramus taeniatus*), version 2.0. In *The Birds of North America* (Poole, A. F., Ed.). Cornell Lab of Ornithology, Ithaca, NY, USA. https://doi.org/10.2173/bna.310

Luca, A. C. 2018. WA2993488, *Lessonia rufa* (Gmelin, 1789)]. http://www.wikiaves.com/2993488

Lucero, F. 2013. Nuevas aves, primeras evidencias y localidades para las provincias de San Juan y Catamarca, Argentina. *Ecoregistros* 3: 14.

Luna, J.C., Ellery, T., Knudsen, K. & McMullan, M. 2011. First confirmed records of Yellow-bellied Sapsucker *Sphyrapicus varius* from Colombia and South America. *Conservación Colombiana* 15: 29–30.

Luz, G.M. 2019. WA3562220, *Euphonia cyanocephala* (Vieillot, 1818)]. http://www.wikiaves.com/3562220

Ma Ming. 1999. Greater Flamingo *Phoenicopterus ruber* and Rufous-tailed Scrub Robin *Cercotrichas galactotes*: two new species for China. *Forktail* 15: 105–106.

Maftei, M. 2013. First record of Franklin's Gull *Leucocephalus pipixcan* from Antarctica. *Marine Ornithology* 41: 49–150.

Malbrant, R. & Maclatchy, A. 1958. A propos de l'occurrence de deux oiseaux d' Afrique australe au Gabon: le manchot du Cap, *Spheniscus demersus* Linné et la grue couronnée, *Balearica regulorum* Bennett. *L'Oiseau et la revue française d'ornithologie* 28: 84–86.

Maclean, G. L. & Kirwan, G. M. 2020. Egyptian Plover (*Pluvianus aegyptius*), version 1.0. In *Birds of the World* (J. del Hoyo, A. Elliott, J. Sargatal, D. A. Christie, and E. de Juana, Editors). Cornell Lab of Ornithology, Ithaca, NY, USA. https://doi.org/10.2173/bow.egyplo1.01

Maftei, M. 2014. The Ross's Gull (*Rhodostethia rosea*) in North America. Masters thesis, Memorial University of Newfoundland. https://research.library.mun.ca/12087/

Mahood, S. P. 2016. Chestnut-cheeked Starling *Agropsar philippensis*: first unequivocal record for Cambodia. *BirdingASIA* 25: 118–119.

Mallalieu, M. & Kaestner, P. 2015. Red-backed Shrikes *Lanius collurio*: first record(s) for Afghanistan. *Sandgrouse* 37: 169–170.

Malling Olsen, K. & Larsson, H. 2003. *Gulls of Europe, Asia and North America*. Christopher Helm, London.

Manekara, S. 2016. Blue-throated Bee-eater *Merops viridis* in Kerala: An escapee, or a wild vagrant? *Indian Birds* 12: 76–78.

Manzoor, M., Shah, I. & Mian, A. 2017. Black-legged Kittiwake (*Rissa tridactyla*) in Karakoram: Record on Vagrancy. *Journal of Bioresource Management* 3: 21–23.

Marantz, C.A. & Remsen, J.V. Jr. 1991. Seasonal distribution of the Slaty Elaenia, a little-known austral migrant of South America (distribución estacional de *Elaenia strepera*, migratorio austral suramericano). *Journal of Field Ornithology* 62: 162–172.

Marchant, S. & Higgins, P.J. 1990. *Handbook of Australian, New Zealand and Antarctic Birds*. Volume 1: Ratites to Ducks. Oxford University Press, Melbourne.

Marcondes, R.S., Del-Rio, G., Rego, M. A. & Silveira, L.F. 2014. Geographic and seasonal distribution of a little-known Brazilian endemic rail (*Aramides mangle*) inferred from occurrence records and ecological niche modeling. *The Wilson Journal of Ornithology* 126: 663–672.

Marin, M., Díaz, F., González, R., Garrido, M. & Beck, J. 2017. First South American record of the Northern Fulmar *Fulmarus glacialis*. *Marine Ornithology* 45: 121–122.

Marinero, N.V., Cortez, R.O., Alfredo Sanabria, E. & Quiroga, L.B. 2012. Ampliación de la distribución de *Sephanoides sephaniodes* (Trochilidae: Aves) en Argentina. *Revista Peruana de Biología* 19: 217–218.

Marshall, H., Collar, N.J., Lees, A.C., Moss, A., Yuda, P. & Marsden, S.J. 2020. Spatio-temporal dynamics of consumer demand driving the Asian Songbird Crisis. *Biological Conservation* 241: 108237

Martin, J.P. 1997. The first Southern Lapwing *Vanellus chilensis* in Mexico. *Cotinga* 8: 52–54.

Martínez, O., Cotayo, L., Kirkconnell, A. & Wiley, J.W. 2016. First record of Lapland Longspur *Calcarius lapponicus* in the Caribbean. *Bulletin of the British Ornithologists' Club* 136: 295–299.

Martuscelli, P., Silva, S.R., Olmos, F. 1997. A large prion *Pachyptila* wreck in south-east Brazil. *Cotinga* 8: 55–57.

Martínez-Vilalta, A., Motis, A. & Kirwan, G.M. 2020. Chinese Pond-Heron (*Ardeola bacchus*), version 1.0. In *Birds of the World* (J. del Hoyo, A. Elliott, J. Sargatal, D. A. Christie, and E. de Juana, Editors). Cornell Lab of Ornithology, Ithaca, NY, USA. https://doi.org/10.2173/bow.chpher1.01

Martin, P.R. & Di Labio, B.M. 2011. Yellow-nosed Albatross. *Ontario Birds* 28: 58–79.

Marzolf, N., Smith, C. & Golladay, S. 2019. Limpkin (*Aramus guarauna*) establishment following recent increase in nonnative prey availability in Lake Seminole, Georgia. *Wilson Journal of Ornithology* 131: 179–184.

Mason, V. 1997. Sooty Oystercatcher a new species for Indonesia. *Kukila* 9: 179–180.

Mason, V. 2011. A revised checklist for the birds of Bali, with notes on recent additions to the avifauna. *Kukila* 15: 1–30.

Mason, P. & Allsop, J. 2009. *The Golden Oriole*. T & AD Poyser, London.

Mather, J.R. 2010. Pacific Diver in Yorkshire: new to Britain and the Western Palearctic. *British Birds* 103: 539–545.

Mathys, B.A. 2011. First record of Aplomado Falcon (*Falco femoralis*) for the West Indies. *The Wilson Journal of Ornithology* 123: 179–180.

Matías, E. & Eisermann, K. 2018. First record of Red-throated Pipit *Anthus cervinus* in Central America. *Bulletin of the British Ornithologists' Club* 138: 383–385.

Matias, R., Catry, P., Pearman, M. & Morrison, M. 2009. Vagrancy of Northern Rockhopper Penguins *Eudyptes moseleyi* to the Falkland Islands. *Marine Ornithology* 37: 287–289.

Matus, R. & Jaramillo, A. 2008. Range extensions and vagrant bird species in the XII Region of Magallanes, Chile. *Cotinga* 30: 34–40.

Maulick, R. & Adhurya, S. 2017. A report of the Dollarbird *Eurystomus orientalis* from southern West Bengal. *Indian Birds* 13: 84.

Maumary, L. & Knaus, P. 2000. Marbled Murrelet in Switzerland: a Pacific Ocean auk new to the Western Palearctic. *British Birds* 93: 190–199.

Maumary, L. & Baudraz, M. 2000. Seltene Vogelarten und ungewöhnliche Vogelbeobachtungen in der Schweiz im Jahre 1999. *Der Ornithologische Beobachter* 97: 307–333.

Mayr, G. 2004. Old World fossil record of modern-type hummingbirds. *Science* 304: 861–864.

McAllan, I.A., Curtis, B.R., Hutton, I. & Cooper, R.M. 2004. The Birds of the Lord Howe Island Group: A Review of Records. *Australian Field Ornithology* 21(Suppl): 1–82.

McAllan, I.A. & Lindsay, K.J. 2016. Stranger in the valley: Expansion in the range of the Spiny-cheeked Honeyeater in central eastern New South Wales. *Australian Field Ornithology* 33: 125.

McAlpine, D.F., Orchard, S.A., Sendall, K.A. & Palm, R. 2004. Status of marine turtles in British Columbia waters: a reassessment. *The Canadian Field-Naturalist* 118: 72–76.

McCaskie, G. 2007. A Ross's Gull reaches southern California. *Western Birds* 38: 137–140.

McCloskey, S. E., Uher-Koch, B. D., Schmutz, J. A. & Fondell, T.F. 2018. International migration patterns of Red-throated Loons (*Gavia stellata*) from four breeding populations in Alaska. *PLOS One* 13: e0189954.

McCrary, J.K. & Young, D.P. Jr. 2008. New and noteworthy observations of raptors in southward migration in Nicaragua. *Ornitologia Neotropical* 19: 573–580.

McGowan, P.J.K., Kirwan, G.M., de Juana, E. & Boesman, P. 2020. Common Quail (*Coturnix coturnix*), version 1.0. In *Birds of the World* (J. del Hoyo, A. Elliott, J. Sargatal, D. A. Christie, and E. de Juana, Editors). Cornell Lab of Ornithology, Ithaca, NY, USA. https://doi.org/10.2173/bow.comqua1.01

McKinlay, G. 2015. First record of Black-faced Spoonbill *Platalea minor* for Palau: an identification challenge. *BirdingASIA* 23: 135–136.

McKinnon, E.A., Artuso, C., Artuso, C. & Love, O.P. 2017. The mystery of the missing warbler. *Ecology* 98: 1970–1972.

McLaren, I.A., Lees, A.C., Field, C. & Collins, K.J. 2006. Origins and characteristics of Nearctic landbirds in Britain and Ireland in autumn: a statistical analysis. *Ibis* 148: 707–726.

McLean, E.B., White, A.M. & Matson, T.O. 1995. Brief Note: Smooth-billed Ani (*Crotophaga ani* L.), a new species of bird for Ohio. *The Ohio Journal of Science* 95: 335–336.

McMaster, D., Rayner, T.S. & McMaster, C.A. 2015. Additional records of Christmas Frigatebird *Fregata andrewsi* in the Northern Territory, Australia. *Australian Field Ornithology* 32: 113–117.

McNair, D. 2000. The status of magnificent frigatebirds in the interior of Florida. *North American Birds* 54: 11–15.

McNair, D.B. & Gore, J.A. 1998. Assessment of occurrences of flamingos in northwest Florida, including a recent record of the Greater Flamingo (*Phoenicopterus ruber*). *Florida Field Naturalist* 26: 40–43.

McNair, D.B. & Post, W. 2001. Review of the occurrence of vagrant Cave Swallows in the United States and Canada. *Journal of Field Ornithology* 72: 485–503.

McNeil, R. & Cadieux, F. 1972. Numerical formulae to estimate flight range of some North American shorebirds from fresh weight and wing length. *Bird-Banding* 43: 107–113.

McNeill, R. & Lambaihang, J. 2013. First record of Eurasian Hoopoe *Upupa epops* for Wallacea. *Kukila* 17: 39–40.

Meadows, B.S. 2003. Additional distributional records from the central Hejaz, western Arabia – an addendum to Baldwin & Meadows (1988). *Bulletin of the British Ornithologists' Club* 123: 154–177.

Mearns, B. & Mearns, R. 1998. *The Bird Collectors*. Academic Press, San Diego and London.

Melling, T. 2009. Should Red-necked Nightjar be on the British List? *British Birds* 102: 110–115.

Melling, T., McGowan, R.Y. & Lewington, I. 2008. The Dorset Yellow Bittern. *British Birds* 101: 137–141.

Melling, T., Dudley, S. & Doherty, P. 2008b. The Eagle Owl in Britain. *British Birds* 101: 478–490.

Mellone, U., López-López, P., Limiñana, R. & Urios, V. 2013. Summer pre-breeding movements of Eleonora's Falcon *Falco eleonorae* revealed by satellite telemetry: implications for conservation. *Bird Conservation International* 23: 487–494.

Menkhorst, P., Rogers, D., Clarke, R., Davies, J., Marsack, P. & Franklin, K. 2017. *The Australian Bird Guide*. CSIRO Publishing, Clayton South, Australia.

Menq, W. 2014. WA1700484, *Chauna torquata* (Oken, 1816). http://www.wikiaves.com/1700484

Menzie, S. 2014. Some notes on ageing and sexing Watercocks *Gallicrex cineria* with specific reference to a vagrant bird in Oman. *Sandgrouse* 36: 176–180.

Merkel, F.W. & Wiltschko, W. 1965. Magnetismus und richtungsfinden zugunruhiger rotkehlchen (*Erithacus rubecula*). *Vogelwarte* 23: 71–77.

Messemaker, R. 2004. First record of Sociable Lapwing *Vanellus gregarius* for Cameroon and western Africa. *Bulletin of the African Bird Club* 11: 34–35.

Messemaker, R. 2009. Eyebrowed Thrush at Merzouga, Morocco, in December 2008. *Dutch Birding* 31: 29–31.

Meyburg, B.U., Meyburg, C. & Pretorius, R. 2017. Year-round satellite tracking of Amur Falcon (*Falco amurensis*) reveals the longest migration of any raptor species across the open sea. In From avian tracking to population processes, British Ornithologists' Union Annual Conference University of Warwick.

Miguel, M.S. & McGrath, T. 2004. First record of the Manx Shearwater for Mexico. *Western Birds* 35: 211–214.

Mikkola, H. 1983. *Owls of Europe*. T & AD Poyser, Calton, UK.

Miller, M.J., Bermingham, E. & Ricklefs, R.E. 2007. Historical biogeography of the new world solitaires (*Myadestes* spp). *The Auk* 124: 868–885.

Milne, P. & McAdams, D.G. on behalf of the Irish Rare Birds Committee 2008. Irish rare bird report 2007. *Irish Birds* 8: 395–416.

Mills, M.S.L. 2015. First record of Pacific Golden Plover *Pluvialis fulva* for Angola. *Bulletin of the African Bird Club* 22: 223–224.

Miranda, L.R. 2012. WA744492, *Thamnophilus doliatus* (Linnaeus, 1764 http://www.wikiaves.com/744492

Miyajima, H., Yamashiro, M. & Tanaka, K. 2012. A record of European Golden Plover *Pluvialis apricaria* at Kin-cho, Kunigami-gun, Okinawa Prefecture; the first record for Japan. *Japanese Journal of Ornithology* 61: 310–313.

Miskelly, C.M., Bester, A.J. & Bell,M. 2006. Additions to the Chatham Islands' bird list, with further records of vagrant and colonising bird species. *Notornis* 53: 215–230.

Miskelly, C.M. 2020. Rose-crowned fruit-dove. In Miskelly, C.M. (ed.) *New Zealand Birds Online*. www.nzbirdsonline.org.nz

Miskelly, C.M., Dowding, J.E., Elliott, G.P., Hitchmough, R.A., Powlesland, R.G., Robertson, H.A., Sagar, P.M., Scofield, R.P. & Taylor, G.A. 2008. Conservation status of New Zealand birds, 2008. *Notornis 55*: 117–135.

Miskelly, C.M., Scofield, R.P., Sagar, P.M., Tennyson, A.J., Bell, B.D. & Bell, E.A. 2011. Vagrant and extra-limital bird records accepted by the OSNZ Records Appraisal Committee 2008-2010. *Notornis* 58: 64–70.

Miskelly, C.M., Crossland, A.C., Sagar, P.M., Saville, I., Tennyson, A.J. & Bell, E.A. 2013. Vagrant and extra-limital bird records accepted by the OSNZ Records Appraisal Committee 2011-2012. *Notornis* 60: 296–306.

Miskelly, C.M., Crossland, A.C., Sagar, P.M., Saville, I., Tennyson, A.J. & Bell, E.A. 2015. Vagrant and extra-limital bird records accepted by the Birds New Zealand Records Appraisal Committee 2013-2014. *Notornis* 62: 85–95.

Miskelly, C.M., Crossland, A.C., Sagar, P.M., Saville, I., Tennyson, A.J. & Bell, E.A. 2017. Vagrant and extra-limital bird records accepted by the Birds New Zealand Records Appraisal Committee 2015–2016. *Notornis* 64: 57–67.

Miskelly, C.M., Crossland, A.C., Saville, I., Southey, I., Tennyson, A.J. & Bell, E.A. 2019. Vagrant and extra-limital bird records accepted by the Birds New Zealand Records Appraisal Committee 2017–2018. *Notornis* 66: 150–163.

Mitchell, D. 2012. Woodchat Shrike on Santa Maria – a new species for the Azores archipelago. *Anuário Ornitológico* 9: 88–89.

Mitra, S.S. & Bochnik, M. 2001. Vagrant Hummingbirds in New York State. *Kingbird* 52: 106–113.

Mlíkovský, J. 2009. Waterbirds of Lake Baikal, eastern Siberia, Russia. *Forktail* 25: 13–70.

Mlodinow, S.G. 1997. The Long-billed Murrelet (*Brachyramphus perdix*) in North America. *Birding* 29: 461–475.

Mlodinow, S.G. 2001. Possible anywhere: Sharp-tailed Sandpiper. *Birding* 33: 330–341.

Mlodinow S.G. 2004. Manx Shearwaters in the North Pacific Ocean. *Birding* 36: 608–615.

Mlodinow, S.G. 2006. Five new species of birds for Aruba, with notes on other significant sightings. *Journal of Caribbean Ornithology* 19: 31–35.

Mlodinow, S.G. & Radamaker, K. 2007. First record of Yellow-browed Warbler (*Phylloscopus inornatus*) for Mexico. *North American Birds* 61: 358–362.

Mlodinow, S.G. & Karlson, K.T. 1999. Anis in the United States and Canada. *North American Birds* 53: 237–45.

Mobakken, C. 2005. Lesser Kestrel at sea off Svalbard. *Dutch Birding* 27: 206.

Mohsanin, S., Dymond, N., Khan, T. & Pierce, A.J. 2014. First records of Siberian Blue Robin *Luscinia cyane* for Bangladesh. *BirdingASIA* 22: 114–115.

Monson, G. & Phillips, A.R. 1981. *Annotated Checklist of the Birds of Arizona*. 2nd ed. University of Arizona Press, Tucson.

Montalti, D., Orgeira, J.L. & Di Martino, S. 1999. New records of vagrant birds in the South Atlantic and in the Antarctic. *Polish Polar Research* 20: 347–354.

Monticelli, D. & Aalto, J. 2011. White-tailed Tropicbird on Flores and Corvo, Azores. *Dutch Birding* 34:100–104.

Monticelli, D., Alfrey, P., Jones, J., Joynt, G., Didner, E., Dufourny, H., Legrand, V. & Crochet, P. A. 2018. Lincoln's Sparrows, Prairie Warbler, Yellow-throated Warbler and Eastern Wood Pewees on Corvo, Azores, in 2010-16. *Dutch Birding* 40: 285–296.

Montgomerie, R. & Holder, K. 2008. Rock Ptarmigan (*Lagopus muta*), version 2.0. In *The Birds of North America* (Poole, A. F., ed.). Cornell Lab of Ornithology, Ithaca, NY, USA. https://doi.org/10.2173/bna.51

Montevecchi, W.A. & Stenhouse, I.J. 2002. Dovekie (*Alle alle*), version 2.0. In *The Birds of North America* (Poole, A. F. & Gill, F. B., eds.). Cornell Lab of Ornithology, Ithaca, NY, USA. https://doi.org/10.2173/bna.701

Moore, R.P., Robinson, W.D., Lovette, I.J. & Robinson, T.R. 2008. Experimental evidence for extreme dispersal limitation in tropical forest birds. *Ecology Letters* 11: 960–968.

Morales Sanchez, J. 1988. Confimacion de la presencia de *Spheniscus humboldti* Meyen (Aves: Sphenisddae) para Columbia. *Trianea* 1: 141–143.

Morris, A.K. 1979. The inland occurrence of tropicbirds in New South Wales during March 1978. *Australian Birds* 13: 51–54.

Morrison, R.I.G. 1984. Migration system of some New World shorebirds. Pages 125–202 in *Behavior of marine animals.* Volume 6. Shorebird: migration and foraging behavior. (Burger, J. & Olla, B. L., Eds.) Plenum Press, New York.

Morrison, K.W. & Sagar, P.M. 2014. First record of interbreeding between a Snares crested (*Eudyptes robustus*) and erect-crested penguin (*E. sclateri*). *Notornis* 61: 109–112.

Moss, S. 1995. *Birds and Weather: A Birdwatchers' Guide.* Hamlyn, London.

Moss, S. 1998. Predictions of the effects of global climate change on Britain's birds. *British Birds* 91: 307–325.

Mueller, T., O'Hara, R.B., Converse, S.J., Urbanek, R.P. & Fagan, W.F. 2013. Social learning of migratory performance. *Science* 34: 1999–1002.

Muheim, R., Moore, F.R. & Phillips, J.B. 2006. Calibration of magnetic and celestial compass cues in migratory birds-a review of cue-conflict experiments. *Journal of Experimental Biology* 209: 2–17.

Muheim, R., Boström, J., Åkesson, S. & Liedvogel, M. 2014. Sensory mechanisms of animal orientation and navigation. Pp. 179–194 in *Animal Movement Across Scales,* L. A. Hansson & Åkesson S. (eds). Oxford University Press, Oxford.

Murphy, R.C. 1936. *Oceanic Birds of South America.* Volume 1. Macmillan Co., American Museum of Natural History, New York.

Murphy, R.C. 1938. The Wandering Albatross in the Bay of Panama. *The Condor* 40: 126.

Murphy, T.G. 2008. Lack of assortative mating for tail, body size, or condition in the elaborate monomorphic Turquoise-browed Motmot (*Eumomota superciliosa*). *The Auk* 125: 11–19.

Naing, T.Z., Zöckler, C., Tun, K.Z., Win, L., Kyaw, M., Lin, N. & Lwin, N., Aung, N.N., Tizard R., Moses, S., Aye, S.N., Zaw, T. & Aung, T.D.W. 2020. New and interesting avifaunal records for Myanmar, 2005– 2019. *BirdingASIA* 33: 118–127.

Naka, L.N., Cohn-Haft, M., Mallet-Rodrigues, F., Santos, M.D. & Torres, M.D.F. 2006. The avifauna of the Brazilian state of Roraima: bird distribution and biogeography in the Rio Branco basin. *Revista Brasileira de Ornitologia* 14: 197–238.

Nakamura, M., Yata, S. & Ikenaga, H. 2013. A record of Eurasian Golden Plover *Pluvialis apricaria* from Kahokugata reclaimed land, Ishikawa Prefecture, the first record for mainland Honshu, Japan. *Japanese Journal of Ornithology* 62: 189–191.

Nandgaonkar, P. S. 2013. Woodchat Shrike *Lanius senator* from Alibaug, Maharashtra: A first record for India. *Indian Birds* 8: 164.

Naniwadekar, R., Viswanathan, A., Kumar, R. & Dalvi, S. 2013. First record of Tristram's Bunting *Emberiza tristrami* from India. *Indian Birds* 8: 134–135.

Narayanan, S.P., Sajith, K.M., Pillai, A.P., Narendran, M.M. & Sreekumar, B. 2008. Records of European Roller *Coracias garrulus* from southern Peninsular India, including the first sighting from Kerala. *Indian Birds* 4: 2–5.

Neate-Clegg, M.H., Román, J.R., Demir, B. & Şekercioğlu, Ç.H. 2019. Endangered Basra Reed-warbler (*Acrocephalus griseldis*) recorded for the first time in Turkey (Aves: Acrocephalidae). *Turkish Journal of Zoology* 43: 250–253.

Nelder, J.A. 1962. A statistical examination of the 'Hastings Rarities'. *British Birds* 55: 283–298.

Nelson, K.N. & Pyle, P. 2013. Distribution and movement patterns of individual Crested Caracaras in California. *Western Birds* 44: 45–55.

Németh, Á. & Vadász C.S. 2008. The first record of the Zitting Cisticola *Cisticola juncidis* (Rafinesque 1810) in Hungary. *Opuscula Zoologica Budapest* 37: 89–90.

Newell D. 2008. Recent records of southern skuas in Britain. *British Birds* 101: 439–441.

Newton, I. 2002. Population limitation in Holarctic owls. Pp. 3-29 in *Ecology and conservation of owls,* Newton, I., Kavanagh, R., Olsen, J. & Taylor, I. Eds.

Newton, I. 2007. Weather-related mass-mortality events in migrants. *Ibis* 149: 453–467.

Newton, I. 2008. *The migration ecology of birds*. Academic Press, London.

Newton, I. & Rothery, P. 2001. Estimation and limitation of numbers of floaters in a Eurasian Sparrowhawk population. *Ibis* 143: 442–449.

Nightingale, B. & Allsopp. K. 1994. Invasion of Red-footed Falcons in spring 1992. *British Birds* 87: 223–231.

Nilsson, C. & Sjöberg, S. 2016. Causes and characteristics of reverse bird migration: an analysis based on radar, radio tracking and ringing at Falsterbo, Sweden. *Journal of Avian Biology* 47: 354–362.

Nilsson, C., Klaassen, R.H. & Alerstam, T. 2013. Differences in speed and duration of bird migration between spring and autumn. *The American Naturalist* 181: 837–845.

Nisbet, I.C.T. 1959. Wader migration in North America and its relation to transatlantic crossings. *British Birds* 52: 205–215.

Nisbet, I.C.T. 1962. South-eastern rarities at Fair Isle. *British Birds* 55: 74–86.

Norbu, N., Wikelski, M.C., Wilcove, D.S., Partecke, J., Tenzin, U. & Tempa, T. 2013. Partial altitudinal migration of a Himalayan forest pheasant. *PLOS One* 8: p.e60979.

Norton, R.L., White, A., Dobson, A. & Massiah, E. 2010. West Indies and Bermuda [Fall 2009 regional report]. *North American Birds* 64: 169–172.

Norton, T., Atkinson, P., Hewson, C. & Garcia-del-Rey, E. 2018. Geolocator study reveals that Canarian Plain Swifts *Apus unicolor* winter in equatorial West Africa. *African Bird Club & Sociedad Ornitologica Canaria*. 15 pp.

Novaes, F.C. 1959. *Quiscalus lugubris* in Brazil. *The Auk* 76: 242–242.

Novaes, F.C. 1978. Sobre algumas aves pouco conhecidas da Amazonia brasileira II. *Boletim do Museu Paraense Emílio Goeldi, série Zoologia* 90: 1–15.

Nunes, G.T., Hoffmann, L.S., Macena, B.C.L., Bencke, G.A. & Bugoni, L. 2015. A Black Kite *Milvus migrans* on the Saint Peter and Saint Paul archipelago, Brazil. *Revista Brasileira de Ornitologia* 23: 31–35.

Nye, E.R., Dickman, C.R. & Kingsford, R.T. 2007. A wild goose chase–temporal and spatial variation in the distribution of the Magpie Goose (*Anseranas semipalmata*) in Australia. *Emu* 107: 28–37.

Oatman, G.F. 2015. Sighting of Hooded Crane *Grus monacha* in Bhutan – December 1989. *BirdingASIA* 24: 10.

Obando-Calderón, G., Camacho-Varela, P., Chaves-Campos, J., Garrigues, R., Montoya, M., Ramírez-Alán, O. & Zook, J. 2014. Lista Oficial de las Aves de Costa Rica Actualización 2014. *Zeledonia* 18: 33–50.

O'Brien, M., Patteson, J.B., Armistead, G.L. & Pearce, G.B. 1999. Swinhoe's Storm-Petrel. *North American Birds* 53: 6–10.

O'Connell, B.M. 2019. Life History of Elegant/Sandwich tern hybrid. *Biodiversity Observations* 10: 10–13.

O'Donnell, E. 2013. The Ruby-crowned Kinglet in County Cork. *Birding World* 26: 425–426.

Ogilvie, M. & the Rare Breeding Birds Panel. 1994. Rare breeding birds in the United Kingdom in 1994. *British Birds* 89: 387–417.

Oliver, W.R.B. 1955. *New Zealand Birds*. 2nd edition. A.H. & A.W. Reed, Wellington, New Zealand.

Ollé, À., Trabalon, F. & Bertran, M. 2015. A review of occurrences of the Pallid Harrier *Circus macrourus* in the Western Mediterranean: a new migrant and wintering species. *Revista Catalana d'Ornitologia* 31: 7–14.

Olmos, F., Eugênio, M., Saviolli, J., Muscat, E., Costa, A., Rotenberg, E. & Chagas, C. 2014. Birds of the Alcatrazes archipelago and surrounding waters, São Paulo, southeastern Brazil. *Check List* 10: 729–738.

Olsen, K., Nielsen, H. H. & Amstrup, O, 2008. Spotless Starling *Sturnus unicolor* in Denmark. *Dansk Ornitologisk Forenings Tidsskrift* 102: 298–300.

Olsen, A. K., Christiansen, S.S., Göller, O.Z., Hansen, M.B., Kauppinen, S. & Sø, R. 2018. Sjældne fugle i Danmark og Grønland i 2016. *Fugleåret* 12: 138–164.

Olsen, A.K., Christiansen, S.S., Göller, O.Z., Hansen, M. B., Kauppinen, S. & Sø, R. 2019. Sjældne fugle i Danmark og Grønland i 2018. *Fugleåret* 13: 140–167.

Olson, S.L. 1971. Two vagrants to Ascension Island. *Bulletin of the British Ornithologists' Club* 91: 90–92.

Olson, S.L. 1975. Paleornithology of St. Helena Island, South Atlantic Ocean. *Smithsonian Contributions to Paleobiology* 23: 1–49.

Olson, S.L. & Hearty, P.J. 2003. Probable extirpation of a breeding colony of Short-tailed Albatross (*Phoebastria albatrus*) on Bermuda by Pleistocene sea-level rise. *Proceedings of the National Academy of Sciences* 100: 12825–12829.

Olson, S.L. 2012. A new species of small owl of the genus *Aegolius* (Aves: Strigidae) from Quaternary deposits on Bermuda. *Proceedings of the Biological Society of Washington* 125: 97–105.

Olson, S.L. 2017. Species rank for the critically endangered Atlantic Lesser Frigatebird (*Fregata trinitatis*). *The Wilson Journal of Ornithology* 129: 661–675.

Olsson, P. 2015. The first Ruddy-breasted Crake *Porzana fusca* for Oman and the Middle East. *Sandgrouse* 37: 57–59.

Onley, D.J. & Schweigman, P. 2004. First record of Franklin's gull (*Larus pipixcan*) in New Zealand. *Notornis* 51: 49–50.

Onton, K., Hassell, C.J., Hulzebosch, M. & Keates, A. 2012. Eurasian Hoopoe (*Upupa epops*) at Roebuck Plains Roadhouse, Western Australia, the first record for Australian Territory. *Amytornis* 4: 5–8.

Oosthuizen, W.C., Dyer, B.M. & de Bruyn, P.J.N. 2009. Vagrant birds ashore at the Prince Edward Islands, southern Indian Ocean, from 1987 to 2009. *African Journal of Marine Science* 31: 445–450.

Ortiz, D. & Capllonch, P. 2008. Dos fruteros migrantes de Argentina. *Ornitologia Neotropical* 19: 473–479.

Oschadleus H.D. 2011. Chestnut Weaver movements in southern Africa. *Afring News* 40: 1–3.

Otgonbayar, B., Buyandelger, S. & Purevsuren, T. 2017. First record of Collared Pratincole *Glareola pratincola* for Mongolia. *BirdingASIA* 27: 119–120.

Otobed, D., Olsen, A.R., Eberdong, M., Ketebengang, H., Etpison, M.T., Pratt, H.D., McKinlay, G.H., Wiles, G.J., VanderWerf, E.A., O'Brien, M. & Leidich, R. 2018. First report of the Palau Bird Records Committee. *Western Birds* 49: 192–205.

Ottema, O. 2004. First sight record of Alpine Swift *Tachymarptis melba* for South America, in French Guiana. *Cotinga* 21: 70–71.

Ottens, G. 2007. Lost and found: Laughing Gull "Atze" in Europe. *Dutch Birding* 29: 288–291.

Ottenburghs, J., van Hooft, P., van Wieren, S.E., Ydenberg, R.C. & Prins, H.H. 2016. Hybridization in geese: a review. *Frontiers in Zoology* 13: 20.

Ouni, R. 2007. Observation d'un Tantale Ibis *Mycteria ibis* et d'un Héron Garde-bœufs Oriental *Bubulcus ibis coromandus* au Cap Bon (Tunisie). *Alauda* 75: 181.

Outlaw, D.C., Voelker, G., Mila, B. & Girman, D.J. 2003. Evolution of long-distance migration in and historical biogeography of *Catharus* thrushes: a molecular phylogenetic approach. *The Auk* 120: 299–310.

Owen, R.P. 1977. A checklist of the birds of Micronesia. *Micronesia* 13: 65–81.

Pacheco, C., Pereira, J., Marques, A., Cardador, D. & Encarnação, P. 2011. Primeiro registo de felosa-boreal *Phylloscopus borealis* em Portugal. *Anuário Ornitológico* 8: 143–147.

Päckert, M., Hering, J., Fuchs, E., Barthel, P. & Heim, W. 2014. Genetic barcoding confirms first breeding record of the Yellow Bittern, *Ixobrychus sinensis* (Aves: Pelecaniformes, Ardeidae) in the Western Palearctic. *Vertebrate Zoology* 64: 251–260.

Palliser, T. & Carter, M. 2013. Rare birds: the 2012 BARC report. *Australian Birdlife* 2: 50–53.

Paradis, E., Baillie, S.R., Sutherland, W.J. & Gregory, R.D. 1998. Patterns of natal and breeding dispersal in birds. *Journal of Animal Ecology* 67: 518–536.

Parasharya, B.M., Rank, D.N., Harper, D.M., Crosa, G., Zaccara, S., Patel, N. & Joshi, C.G. 2015. Long-distance dispersal capability of Lesser Flamingo *Phoeniconaias minor* between India and Africa: genetic inferences for future conservation plans. *Ostrich* 86: 221–229.

Paradis, E., Baillie, S.R., Sutherland, W.J. & Gregory, R.D. 1998. Patterns of natal and breeding dispersal in birds. *Journal of Animal Ecology* 67: 518–536.

Park, J.-Y. & Oh, M.R. 2018. [First record of Grey-headed Canary Flycatcher (*Culicicapa ceylonensis*) in Korea]. *Korean Journal of Ornithology* 25: 62–64.

Parkes, K.C., Kibbe, D.P. & Roth, E.L. 1978. First records of the Spotted Rail (*Pardirallus maculatus*) for the United States, Chile, Bolivia and western Mexico. *American Birds* 32: 295–299.

Parkin, D.T. & Shaw, K.D. 1994. Asian Brown Flycatcher, Mugimaki Flycatcher and Pallas's Rosefinch. Three recent decisions of the British Ornithologists' Union Records Committee. *British Birds* 87: 247–252.

Patrial, L.W., Pessoa, A.S.P. & Pereira, G.A. 2011. Primeiro registro do pelicano-peruano *Pelecanus thagus* no Brasil e registro documentado do pelicano-pardo *P. occidentalis* na costa leste brasileira. *Revista Brasileira de Ornitologia* 19: 539–540.

Patten, M.A.& Marantz, C.A. 1996. Implications of vagrant southeastern vireos and warblers in California. *The Auk* 113: 911–923.

Patten, M.A. & Minnich, R.A. 1997. Procellariiformes occurrence at the Salton Sea and Sonoran Desert. *The Southwestern Naturalist* 42: 302–311.

Patten, M.A., McCaskie, G. & Unitt, P. 2003. *Birds of the Salton Sea: Status, Biogeography, and Ecology*. University of California Press, Berkeley, California.

Patteson, J.B., Patten, M.A. & Brinkley, E.S. 1999. The Black-browed Albatross in North America. First photographically documented record. *North American Birds* 53: 228–231.

Patteson, J.B., Hass, T. & Brinkley, E.S. 2013. Zino's Petrel (*Pterodroma madeira*) off North Carolina: first for North America. *North American Birds* 67: 28–30.

Pavey, C.R., Nano, C.E., Cole, J.R., McDonald, P.J., Nunn, P., Silcocks, A. & Clarke, R.H. 2014. The breeding and foraging ecology and abundance of the Princess Parrot (*Polytelis alexandrae*) during a population irruption. *Emu* 114: 106–115.

Payne, R.B. 1967. *Vidua obtusa* in the Transvaal, South Africa. *Bulletin of the British Ornithologists' Club* 87: 93–95.

Pearman, M & Areta, J. I. 2020. *Birds of Argentina and the South-west Atlantic*. Helm, London.

Pecl, G.T., Araújo, M.B., Bell, J.D., Blanchard, J., Bonebrake, T.C., Chen, I.C., Clark, T.D., Colwell, R.K., Danielsen, F., Evengård, B. & Falconi, L. 2017. Biodiversity redistribution under climate change: Impacts on ecosystems and human well-being. *Science* 355: eaai9214.

Pedler, R.D. & Lynch, C.E. 2016. An unprecedented irruption and breeding of Flock Bronzewings *Phaps histrionica* in central South Australia. *Australian Field Ornithology 33*: 1–13.

Pender, J. 2010. First records of Red-necked Grebe *Podiceps grisegena* from Bangladesh. *BirdingASIA* 14: 86–87.

Pennington, M., Osborn, K., Harvey, P., Riddington, R., Okill, D., Ellis, P. & Heubeck, M. 2004. *The Birds of Shetland*. Christopher Helm, London.

Pennington, M.G., Riddington, R. & Miles. W.T.S. 2012. The Lapland Bunting influx in Britain & Ireland in 2010/11. *British Birds* 105: 654–673.

Pennycuick, C.J. 1971. Soaring flight of the White-backed Vulture *Gyps africanus*. *Journal of Experimental Biology* 55: 13–38.

Perdeck, A.C. 1958. Two types of orientation in migrating starlings, *Sturnus vulgaris* L., and chaffinches, *Fringilla coelebs* L., as revealed by displacement experiments. *Ardea*, 46: 1–37.

Pereira, G.A. & Brito, M.T. 2005. Diversidade de aves silvestres brasileiras comercializadas nas feiras livres da Região Metropolitana do Recife, Pernambuco. *Atualidades Ornitológicas* 126: 7.

Pereira, G.A., Dantas, S.D.M., Silveira, L.F., Roda, S.A., Albano, C., Sonntag, F.A., Leal, S., Periquito, M.C., Malacco, G.B. & Lees, A.C. 2014. Status of the globally threatened forest birds of northeast Brazil. *Papéis Avulsos de Zoologia* 54: 177–194.

Perlut, N.G., Klak, T.C. & Rakhimberdiev, E. 2017. Geolocator data reveal the migration route and wintering location of a Caribbean Martin (*Progne dominicensis*). *The Wilson Journal of Ornithology* 129: 605–610.

Pernetta, J.C. & Watling, D. 1978. The introduced and native terrestrial vertebrates of Fiji. *Pacific Science* 32: 223–244.

Perry, P. 2006. First record of Blue Quail *Coturnix adansonii* for Swaziland. *Bulletin of the African Bird Club* 13: 215–216.

Peterjohn, B.G. 1989. *The Birds of Ohio*. Indiana University Press, Bloomington, USA.

Petersen, E.D.S., Rossi, L.C. & Petry, M.V. 2015. Records of vagrant bird species in Antarctica: new observations. *Marine Biodiversity Records* 8: e61.

Peucker, A.J., Dann, P. & Burridge, C.P. 2009. Range-wide phylogeography of the little penguin (*Eudyptula minor*): evidence of long-distance dispersal. *The Auk* 126: 397–408.

Pfeifer, R., Stadler, J. & Brandl, R. 2007. Birds from the Far East in Central Europe: a test of the reverse migration hypothesis. *Journal of Ornithology* 148: 379–385.

Pfützke, S. & Halley, A. 1995. An Abyssinian Roller in Egypt. *Birding World* 8: 419.

Phillips, J. & Phillips, V. 2005. First record of Bimaculated Lark *Melanocorypha bimaculata* for Seychelles. *Bulletin of the African Bird Club* 12: 40–41.

Pierce, A.J., Hansasuta, C. & Sutasha, K. 2015. The first record of Common Swift *Apus apus* for Thailand and South-East Asia. *BirdingAsia* 24: 139–140.

Pierce, R.J. & Kirwan, G.M. 2019. Pied Avocet (*Recurvirostra avosetta*), version 1.0. In *Birds of the World* (J. del Hoyo, A. Elliott, J. Sargatal, D. A. Christie, and E. de Juana, Editors). Cornell Lab of Ornithology, Ithaca, NY, USA. https://doi.org/10.2173/bow.pieavo1.01

Piers, H. 1923. Accidental Occurrence in Nova Scotia of the Rock Ptarmigan. *Transactions of the Nova Scotian Institute of Science* 16: 1–8.

Piñones C., C. & Bravo N.V. 2017. El Carau (*Aramus guarauna*), una nueva especie para Chile: relato de una experiencia pedagógica. *La Chiricoca* 22: 19–27.

Piot, B. & Lecoq, M. 2018 First record of Magnificent Frigatebird *Fregata magnificens* for Senegal. *Bulletin of the African Bird Club* 25: 216–217.

Pitches, A. 2011. News & Comment: First record of Dwarf Bittern in Malta. *British Birds* 104: 469.

Pitocchelli, J., Jones, J., Jones, D. & Bouchie, J. 2020. Connecticut Warbler (*Oporornis agilis*), version 1.0. In *Birds of the World* (Poole, A. F., Ed.). Cornell Lab of Ornithology, Ithaca, NY, USA. https://doi.org/10.2173/bow.conwar.01

Pomeroy, D., Walsh, F., Flint, P., Hellicar, M. & Shaw, P., 2016. A sustained decline in Cyprus Warbler *Sylvia melanothorax* numbers in western Cyprus, coinciding with the colonisation of its breeding range by the Sardinian Warbler *S. melanocephala*. *Bird Conservation International* 26: 436–450.

Port, D. & Fisch, F. 2013. Primeiro registro de pomba-de-bando, *Zenaida auriculata* (Des Murs, 1847), (Columbiformes: Columbidae) na Ilha da Trindade, Brasil. *Ornithologia* 7: 42–44.

Porter, D.M., 1976. Geography and dispersal of Galapagos Islands vascular plants. *Nature* 264: 745–746.

Porter, R. & Aspinall, S. 2010. *Birds of the Middle East*. 2nd edition. Christopher Helm, London.

Posada, A.K., Kirkconnell, A. & Kirwan, G.M. 2018. First record of White-faced Ibis *Plegadis chihi* in the West Indies. *Bulletin of the British Ornithologists' Club* 138: 272–274.

Post, P.W. 2007. Observations of prion (*Pachyptila*) wrecks on the west coast of South America. *Notornis* 54: 220–225.

Powlesland, R.G. & Pickard, C.R. 1992. Seabirds found dead on New Zealand beaches in 1988, and a review of *Puffinus* species recoveries, 1943 to 1988. *Notornis* 39: 27–46.

Pranty, B., Cooper, L., Menk, G. & Powell, P. 1996. Winter report: December 1995–February 1996. *Florida Field Naturalist* 24: 83–92.

Pranty, B. 2012. Population growth, spread, and persistence of Purple Swamphens (*Porphyrio porphyrio*) in Florida. *Florida Field Naturalist* 40: 1–12.

Pranty, B., Dunn, J.L., Heinl, S., Kratter, A.W., Lehman, P., Lockwood, M.W., Mactavish, B. & Zimmer, K.J. 2007. Annual report of the ABA Checklist Committee: 2006. *Birding* 39: 24–31.

Pranty, B., Barry, J., Dunn, J.L., Garrett, K.L., Gibson, D.D., Lockwood, M.W., Pittaway, R. & Sibley, D.A. 2014. 25th Report of the ABA Checklist Committee 2013–1014. *Birding* 46: 26–36.

Pranty, B., Barry, J., Gustafson, M., Johnson, T., Garrett, K. L., Lang, A., Lockwood, M. W., Pittaway, R., Pyle, P. & Sibley, D. 2016. 27th Report of the ABA Checklist Committee. *Birding* 48: 30–36.

Prater, S.H. 1926. On the occurrence of Pallas' Sandgrouse *Syrrhaptes paradoxurus* within Indian limits. *Journal of the Bombay Natural History Society* 31: 522.

Prater, T. 2012. Important Bird Areas: St Helena. *British Birds* 105: 638–653.

Pratt, H.D., Bruner, P.L. & Berrett, D.G. 1977. Ornithological observations on Yap, Western Caroline Islands. *Micronesica* 13: 49–56.

Pratt, H.D., Retter, M.L., Chapman, D., Ord, W.M. & Pisano, P. 2009. An Abbott's booby *Papasula abbotti* on Rota, Mariana Islands: first historical record for the Pacific Ocean. *Bulletin of the British Ornithologists' Club* 129: 87–91.

Pratt, T.K. & Beehler, B. M. 2015. *Birds of New Guinea*. Second edition. Princeton University Press, Princeton, New Jersey.

Praus, L. 2016. First winter record of Little Bunting (*Emberiza pusilla*) in Morocco. *Go-South Bulletin* 13: 210–212.

Praveen, J., Karuthedathu, D., Prince, M., Palot, M.J. & Dalvi, S. 2013. Identification of South Polar Skuas *Catharacta maccormicki* in the Arabian Sea and Indian Ocean. *BirdingASIA* 19: 83–88.

Praveen, J., Jayapal, R. & Pittie, A. 2013. Notes on Indian rarities 1: Seabirds. *Indian Birds* 8: 113–125.

Priddel, D., Carlile, N. & Wheeler, R. 2008. Population size, breeding success and provenance of a mainland colony of Little Penguins (*Eudyptula minor*). *Emu* 108: 35–41.

Prince, P.A. & Croxall, J.P. 1983. Birds of South Georgia: new records and re-evaluations of status. *Bulletin British Antarctic Survey* 59: 15–27.

Prommer, M., Bagyura, J., Chavko, J. & Uhrin, M. 2012. Migratory movements of Central and Eastern European Saker Falcons (*Falco cherrug*) from juvenile dispersal to adulthood. *Aquila* 119: 111–134.

Pugnali, G. & Pearman, M. 2001. Confirmación de la presencia del Colibrí Rubí (*Chrysolampis mosquitus*) en Argentina. *Hornero* 16: 93–95.

Pyle, P. 1993. A Markham's Storm-Petrel in the northeastern Pacific. *Western Birds* 24: 108–110.

Pyle, P. & Howell, S.N.G. 1993. An Arctic Warbler in Baja California, Mexico. *Western Birds* 24: 53–56.

Pyle, P., Welch, A.J. & Fleischer, R.C. 2011. A new species of shearwater (*Puffinus*) recorded from Midway Atoll, northwestern Hawaiian Islands. *The Condor* 113: 518–527.

Pyle, P., David, R., Eilerts, B.D., Amerson, A.B., Borker, A. & Mckown, M. 2014. Second record of Bryan's Shearwater *Puffinus bryani* from Midway Atoll, with notes on habitat selection, vocalizations and at-sea distribution. *Marine Ornithology* 42: 5–8.

Pyle, P., Keiffer, R.J., Dunn, J.L. & Moores, N. 2015. The Mendocino Shrike: Red-backed Shrike (*Lanius collurio*) x Turkestan Shrike (*L. phoenicuroides*) hybrid. *North American Birds* 69: 4–35.

Pyle, R.L. & Pyle, P. 2017. The Birds of the Hawaiian Islands: Occurrence, History, Distribution, and Status. B.P. Bishop Museum, Honolulu, HI, U.S.A. Version 2 (1 January 2017) http://hbs.bishopmuseum.org/birds/rlp-monograph/

Pyle, P., Gustafson, M., Johnson, T., Kratter, A.W., Lang, A., Lockwood, M.W., Pittaway, R. & Sibley, D. 2017. 28th report of the ABA Checklist Committee 2017. *Birding* 49: 28–35.

Pyle, P., Gustafson, M., Johnson, T., Kratter, A.W., Lang, A., Nelson, K., Lockwood, M.W. & Sibley, D. 2018. 29th report of the ABA Checklist Committee-2018. *Birding* 50: 30–40.

Pyle, P., Gustafson, M., Johnson, T., Kratter, A.W., Lang, A., Nelson, K., Lockwood, M.W. & Sibley, D. 2019. 30th report of the ABA Checklist Committee-2019. *Birding* 51: 37–42.

Pyle, P. 2020. ABA Checklist Committee Mid-Year 2020 Report. https://www.aba.org/aba-checklist-committee-mid-year-2020-report/

Quantrill, R. 2017. Red-billed Queleas at sea off Senegal. *Bulletin of the African Bird Club* 24: 216.

Rabau, T. & Löwe, D. 2018. Primer registro de Aruco *Anhima cornuta* (Linnaeus 1766) para Uruguay. *Achará* 5: 1–8.

Rabeil, T. & Wacher, T. 2011. First record of Greater Kestrel *Falco rupicoloides* for Niger and western Africa. *Bulletin of the African Bird Club* 18: 221–222.

Rabøl, J. & Noer, H. 1973. Spring migration in the skylark (*Alauda arvensis*) in Denmark: influence of environmental factors on the flocksize and correlation between flocksize and migratory direction. *Vogelwarte* 27 50–65.

Radamaker, K.A. & Powel, D.J. 2010. A Little Bunting reaches Baja California Sur. *Western Birds* 41: 55–58.

Ragyov, D.N., Popova-Wightman, L.G., Popov, K.S., Dalakchieva, S.Y., Nikolov B.P. & Nikolov, I.P. 2003. The first Yellow-billed Stork *Mycteria ibis* in Bulgaria. *Sandgrouse* 25: 63–64.

Rahimi, M.M., Mousavi, S.B., Nabiyan, S.M., Joolaee, L. & Qashqaei, A. 2019. Additional records of White-crowned Wheatear *Oenanthe leucopyga* in Iran after more than a century. *Sandgrouse* 41: 211–213.

Raines, R.J. & Bell, A.A. 1967. Penduline Tit in Yorkshire: a species new to Britain and Ireland. *British Birds* 60: 517–520.

Rajeevan, P.C. & Thomas, J. 2011. Buff-breasted Sandpiper *Tryngites subruficollis* from northern Kerala: a third record for India. *Indian Birds* 7: 143–144.

Ralph, C.J., Hunt, G.L., Raphael, M.G. & Piatt, J.F. 1995. Ecology and conservation of the Marbled Murrelet in North America: an overview. Pp. 3–22 in Ralph, C. J., Hunt, G.L., Raphael, M.G. & Piatt, J.F. eds. *Ecology and Conservation of the Marbled Murrelet*. Albany.

Ralph, C.J. & Wolfe, J. 2018. Factors affecting the distribution and abundance of autumn vagrant warblers in northwestern California and southern Oregon. *PeerJ* 6: e5881.

Ramos, R., Carlile, N. & Madeiros, J. 2017. It is the time for oceanic seabirds: Tracking year-round distribution of gadfly petrels across the Atlantic Ocean. *Diversity and Distributions* 23: 794–805.

Ramzy, A. 2016. Rare visit to Taiwan by Siberian Crane is a bird-watcher's dream. *New York Times*, June 2.

Rasmussen, P.C. & Anderton, J.C. 2005. *Birds of South Asia: the Ripley Guide*. Lynx Edicions, Barcelona.

Rasmussen, J.L., Sealy, S.G. & Cannings, R.J. 2008. Northern Saw-whet Owl (*Aegolius acadicus*), version 2.0. In *The Birds of North America* (Poole, A. F. Ed.). Cornell Lab of Ornithology, Ithaca, NY, USA. https://doi.org/10.2173/bna.42

Rastogi, A., Kumari, Y., Rani, S. & Kumar, V. 2011. Phase inversion of neural activity in the olfactory and visual systems of a night-migratory bird during migration. *European Journal of Neuroscience* 34: 99–109.

Rattenborg, N.C., Voirin, B. & Cruz, S.M. 2016. Evidence that birds sleep in mid-flight. *Nature Communications* 7: 1–9.

Reading, R., Harris, S. & Bräunlich, A. 2011. Rufous-bellied Woodpecker *Dendrocopos hyperythrus*: first records for Mongolia. *BirdingASIA* 15: 104–105.

Rebelato, M., Fontana, C. S., Repenning, M. & Hartmann, P.A. 2011. First documented record of Sharp-billed Canastero *Asthenes pyrrholeuca* in Brazil. *Bulletin of the British Ornithologists' Club* 131: 134–136.

Rebstock, G.A., Agüero, M.L. & Boersma, P.D. 2010. Repeated observations of a Cape Gannet *Morus capensis* on the coast of Patagonia, Argentina. *Ostrich* 81: 167–169.

Redinov, K.A., Panchenko, P.S., Formanyuk, O.A. & Petrovich, Z.O. 2014. The Great Skua (*Stercorarius skua*) in Ukraine. *Berkut* 23: 19–23.

Redman, N., Stevenson, T. & Fanshawe, J. 2009. *Birds of the Horn of Africa – Ethiopia, Eritrea, Dijbouti, Somalia and Socotra*. Christopher Helm, London.

Reeber, S. 2002. Nidification de la Spatule d'Afrique *Platalea alba* au lac de Grand-Lieu (Loire-Atlantique). *Ornithos* 9: 42–43.

Reid, K. 1998. Franklin's Gull *Larus pipixcan* at South Georgia. *Bulletin of the British Ornithologists' Club* 118: 55–56.

Remsen Jr, J.V. & Parker III, T.A. 1983. Contribution of river-created habitats to bird species richness in Amazonia. *Biotropica* 15: 223–231.

Renaudier, A. & Comité d'Homologation de Guyane. 2010. Rare birds in French Guiana in 2005–07. *Cotinga* 32: 75–83.

Renfrow, F. 2003. Notes on vagrancy in Brown-headed Nuthatch, with attention to recent range expansion and long-term habitat changes. *North American Birds* 57: 423–428.

Reynolds, M., Nash, S. & Courtot, K. 2015. Peregrine Falcon predation of endangered Laysan Teal and Laysan Finches on remote Hawaiian Atolls. Hawai`i Cooperative Studies Unit Technical Report HCSU-065. University of Hawai'i at Hilo, Hawaii.

Ribas, C.C., Aleixo, A., Nogueira, A.C., Miyaki, C.Y. & Cracraft, J. 2011. A palaeobiogeographic model for biotic diversification within Amazonia over the past three million years. *Proceedings of the Royal Society of London. Series B: Biological Sciences* 279: 681–689.

Richardson, C. & Porter, R. 2020. *Birds of Cyprus.* Helm, London.

Riddiford, N. 1983. Sandhill Crane: new to Britain. *British Birds* 76: 105–109.

Rieta Reig, A. 1969. Sobre un ejemplar de *Aechmophorus major* conservado en Valencia. *Ardeola* 13: 233–235.

Rijmenams, G. & Ebels, E.B. 2011. Eleonora's Valk in Oostvaardersplassen in September 2011. *Dutch Birding* 33: 377–381.

Ritz, T., Wiltschko, R., Hore, P.J., Rodgers, C.T., Stapput, K., Thalau, P., Timmel, C.R. & Wiltschko, W. 2009. Magnetic compass of birds is based on a molecule with optimal directional sensitivity. *Biophysical Journal* 96: 3451–3457.

Rivalan, P., Barbraud, C., Inchausti, P. & Weimerskirch, H. 2010. Combined impacts of longline fisheries and climate on the persistence of the Amsterdam Albatross *Diomedia amsterdamensis. Ibis* 152: 6–18.

Roadhouse, A. 2016. *The Birds of Spurn.* Spurn Bird Observatory.

Robb, R.R., Arendt, D., Larsen, K. & Sherrell, P. 2009. First North American record of Crowned Slaty Flycatcher *Griseotyrannus aurantioatrocristatus*, at Cerro Azul, Panama. *Cotinga* 31: 50–52.

Robbins Jr, S.D. 1991. *Wisconsin Birdlife: Population and Distribution, Past and Present.* University of Wisconsin Press, Madison, USA.

Robertson, I.S. 1994. The first sighting of Black-tailed Gull (*Larus crassirostris*) for Thailand. *Natural History Bulletin of the Siam Society* 42: 294.

Robertson, C.J.R. & Stephenson, B.M. 2005. Cape gannet (*Sula capensis*) breeding at Cape Kidnappers, New Zealand. *Notornis* 52: 238–242.

Robillard, A., Therrien, J. F., Gauthier, G., Clark, K.M. & Bêty, J. 2016. Pulsed resources at tundra breeding sites affect winter irruptions at temperate latitudes of a top predator, the snowy owl. *Oecologia* 181: 423–433.

Robinson, P. 2003. *The Birds of the Isles of Scilly.* Christopher Helm, London.

Robson, C. 2002. *A Field Guide to the Birds of Thailand.* New Holland, London.

Robson, C. 2005. *Field Guide to the Birds of South-East Asia.* New Holland, London.

Robson, C., Roddis, S., Sykes, M. & Sykes, B. 2015. From the field. *BirdingASIA* 23: 138–144.

Robson, C., Roddis, S. & Loseby, T. 2016. From the field. *BirdingASIA* 25: 122–128.

Roca, R.L. 1994. Oilbirds of Venezuela: ecology and conservation. *Publications – Nuttall Ornithological Club* 24: 1–83.

Rodríguez, Y., Navarro, N. & Fernández Ordóñez, J.C. 2017. First record of Eurasian Blackcap (*Sylvia atricapilla*) for Cuba and the West Indies. Pp. 31–32 in Navarro, N. & Reyes, E. 2017. *Annotated Checklist of the Birds of Cuba.* Number 1. 2017 edition. Ediciones Nuevos Mundos, Saint Augustine, USA.

Rodríguez, B. 2018. Daurian Shrike on Tenerife, Canary Islands, in November 2017. *Dutch Birding* 40: 109–111.

Rodríguez, Y., Garrido, O.H., Wiley, J.W. & Kirkconnell, A. 2005. The Common Kingfisher (*Alcedo atthis*): an exceptional first record for the West Indies and the Western Hemisphere. *Ornitologia Neotropical* 16: 141–142.

Rodríguez, G. & Elorriaga, J. 2016. Identification of Rüppell's Vulture and White-backed Vulture and vagrancy in the WP. *Dutch Birding* 38: 349–375.

Rodríguez, G.A., Matheus, J. & Raffali, G. 2017. Primer registro de *Vireo griseus* para Venezuela y registros adicionales de dos migratorios accidentales en la Reserva Biológica de Montecano, estado Falcón. *Revista Venezolana de Ornitología* 7: 38–41.

Rogers, C. 2008. Bird report, 2004. *South Australian Ornithologist* 35: 86–95.

Rogers, M.J.1996. Report on rare birds in Great Britain in 1995. *British Birds* 89: 481–531.

Rogers, M.J. 1988. Report on rare birds in Great Britain in 1987. *British Birds* 81: 535–596.

Rogers, M.J. 2000. Report on rare birds in Great Britain in 1999. *British Birds* 93: 513–567.

Rogers, M.J. 2002. Report on rare birds in Great Britain in 2001. *British Birds* 95: 476–528

Rollinson, D.P., Cardwell, P., De Blocq, A. & Nicolau, J. 2017. Out-of-range sighting of a South Georgian Diving Petrel *Pelecanoides georgicus* in the southeast Atlantic Ocean. *Marine Ornithology* 45: 21–22.

Romano, H., Correia-Fagundes, C., Zino, F.J. & Biscoito, M. 2010. Birds of the archipelagos of madeira and the Selvagens. II-New records and checklist update (1995–2010). *Boletim do Museu Municipal do Funchal* 60: 5–44.

Romay, C.D. & Roselaar, C.K. 2013. American Hawk-Owl caught off Las Palmas, Gran Canaria, Canary Islands, in October 1924. *Dutch Birding* 35: 1–6.

Rose, M.D. & Polis, G.A. 2000. On the insularity of islands. *Ecography* 23: 693–701.

Rosenberg, G.H., Lehman, P.E., Lang, A.J., Stoll, V. & Stoll, R. 2018. Thick-billed Warbler (*Iduna aedon*) at Gambell, Alaska: First record for North America. *Western Birds* 49: 226–230.

Rosenberg, G., Pluym, D.V. & Halsey, L. 2019. Arizona Bird Committee Report, 2015–2017 Records. *Western Birds* 50: 150–175.

Rosenberg, K.V., Dokter, A.M., Blancher, P.J., Sauer, J.R., Smith, A.C., Smith, P.A., Stanton, J.C., Panjabi, A., Helft, L., Parr, M. & Marra, P.P. 2019. Decline of the North American avifauna. *Science* 366: 120–124.

Rottenborn, S.C., McCaskie, G., Daniels, B.E. & Garrett, J. 2016. The 39th annual report of the California Bird Records Committee: 2013 records. *Western Birds* 47: 2–26.

Rottenborn, S.C., Terrill, S.B. & Harvey, H.T. 2018. The 42nd annual report of the California Bird Records Committee: 2016 records. *Western Birds* 49: 238–257.

Round, P.D., Hansson, B., Pearson, D.J., Kennerley, P.R. & Bensch, S. 2007. Lost and found: the enigmatic large-billed reed warbler *Acrocephalus orinus* rediscovered after 139 years. *Journal of Avian Biology* 38: 133–138.

Round, P.D. & Fisher, T.H. 2009. Records of Black-browed Reed Warbler *Acrocephalus bistrigiceps* from Luzon, Philippines. *Forktail* 25: 159–160.

Round, P.D. & Kunsorn, A. 2009. Fire-tailed Sunbirds *Aethopyga ignicauda* in north-west Thailand and notes on their identification. *BirdingASIA* 11: 51–52.

Round, P.D., Ul Haque, E., Dymond, N., Pierce, A.J. & Thompson, P.M. 2014. Ringing and ornithological exploration in north-east Bangladesh wetlands. *Forktail* 30: 109–121.

Roux, J.P. & Martinez, J. 1987. Rare, vagrant and introduced birds at Amsterdam and Saint Paul Islands, southern Indian Ocean. *Marine Ornithology* 14: 3–19.

Rowlands, B.W., Trueman, T., Olson, S.L., McCulloch, M.N. & Brooke, R. K. 1998. *The Birds of St. Helena. An Annotated Check-list.* BOU Check-list 16. British Ornithologists' Union, Tring, UK.

Rowlands, A., Kidner, P. & Condon, P. 2010. Yellow-nosed Albatross: new to Britain. *British Birds* 103: 376–384.

Roy, M. S., Torres-Mura, J.C. & Hertel, F. 1998. Evolution and history of hummingbirds (Aves: Trochilidae) from the Juan Fernandez islands, Chile. *Ibis* 140: 265–273.

Roy, K.J. & Pawlicki, J.M. 2014. Elegant Tern (*Thalasseus elegans*): new to Ontario. *Ontario Birds* 32: 114–136.

Rozemeijer, P.J. 2011. First record of Sungrebe *Heliornis fulica* on Bonaire, Netherlands Antilles. *Cotinga* 33: 123–124.

Rufray, V., Claessens, O. & le Comité d'Homologation de Guyane 2019. *Rapport du Comité d'Homologation de Guyane.* http://www.chnfrance.org/upload_content/Rapport%20CHG%202012-2013-2014.pdf

Ruschi, P.A. & Simon, J.E. 2012. Hummingbirds of Santa Teresa, State of Espírito Santo, Southeastern Brazil. *Boletim do Museu de Biologia Mello Leitão* 29: 31–52.

Rusk, C.L., Walters, E.I. & Koenig, W.D. 2013. Cooperative breeding and long-distance dispersal: A test using vagrant records. *PLOS One* 8: e58624.

Rustamov, E.A., Menliev, S. & Agryzkov, E. 2016. Common Koel *Eudynamys scolopaceus* new for Turkmenistan. *Sandgrouse* 3: 197–198.

Ryall, C., 2016. Further records and updates of ranges expansion of House Crow *Corvus splendens. Bulletin of the British Ornithologists' Club* 136: 39–45.

Ryan, P.G. 1989. Common Nighthawk *Chordeiles minor* and new records of seabirds from Tristan da Cunha and Gough Islands. *Bulletin of the British Ornithologists' Club* 109: 147–149.

Ryan, P.G., Avery, G., Rose, B., Ross, G.B., Sinclair, J.C. & Vernon, C.J. 1989. The Southern Ocean seabird irruption to South African waters during winter 1984. *Marine Ornithology* 17: 41-55.

Ryan, P. 2020. White-winged Apalis (*Apalis chariessa*), version 1.0. In *Birds of the World* (J. del Hoyo, A. Elliott, J. Sargatal, D. A. Christie, and E. de Juana, Editors). Cornell Lab of Ornithology, Ithaca, NY, USA. https://doi.org/10.2173/bow.whwapa1.01

Safford, R.J. & Basque, R. 2007. Records of migrants and amendments to the status of exotics on Mauritius in 1989–93. *Bulletin of the African Bird Club* 14: 26–35.

Safford, R.J. & Hawkins, A.F.A. eds. 2013. *The Birds of Africa.* Volume 8. The Malagasy region. Christopher Helm, London.

Sagar, P.M., Tennyson, A.J. & Scofield, R.P. 2001. Birds of the Snares Islands, New Zealand. *Notornis* 48: 1–40.

Sagar, P.M. 2013. Darter. In Miskelly, C.M. (ed.) New Zealand Birds Online. www.nzbirdsonline.org.nz

Sagot-Martin, F., Lima, R.D., Pacheco, J.F., Irusta, J.B., Pichorim, M. & Hassett, D.M., 2020. An updated checklist of the birds of Rio Grande do Norte, Brazil, with comments on new, rare, and unconfirmed species. *Bulletin of the British Ornithologists' Club* 140: 218–298.

Sagvik, J. 2009. A record of Rose-coloured Starling *Sturnus roseus* in southern Africa. *Bulletin of the African Bird Club* 16: 89–90.

Saikia, P.J. & Gaswami, V.P. 2017. Pied Crow *Corvus albus* at Jodhpur, India: Where did it come from? *Indian Birds* 13: 147–149.

Salewski, V., Bargain, B., Diop, I. & Flade, M. 2009. Quest for a phantom–the search for the winter quarters of the Aquatic Warbler *Acrocephalus paludicola. Bulletin of the African Bird Club* 16: 61–66.

Salewski, V., Flade, M., Lisovski, S., Poluda, A., Iliukha, O., Kiljan, G., Malashevich, U. & Hahn, S. 2019. Identifying migration routes and non-breeding staging sites of adult males of the globally threatened Aquatic Warbler *Acrocephalus paludicola. Bird Conservation International* 29: 503–514.

Salim, M.A., Yassir, W.S., Abed, S.A, Porter, R, Jabbar, M.T., Al-Obeidi, L.A., Hadi, H. & Harbi, Z.S. 2020. Observations of White-faced Whistling Duck *Dendrocygna viduata* in Iraq. *Sandgrouse* 42: 115–117.

Sanders, J. 2010. Glaucous-winged Gull in Gloucestershire: new to Britain. *British Birds* 103: 53–59.

Sandoval, L., Sánchez, C., Biamonte, E., Zook, J. R., Sánchez, J. E., Martínez, D., Loth, D. & O'Donahoe, J. 2010. Recent records of new and rare bird species in Costa Rica. *Bulletin of the British Ornithologists' Club* 130: 237–245.

Santoro, S., Champagnon, J., Kharitonov, S., Zwarts, L., Oschadleus, D., Mañez, M., Samraoui, B., Nedjah, R., Volponi, S. & Cano-Alonso, L. 2019. Long-distance dispersal of the Afro-Eurasian Glossy Ibis from ring recoveries. *Stork, Ibis and Spoonbill (SIS) Conservation,* IUCN Species Survival Commission Stork, Ibis and Spoonbill Specialist Group.

Santos, C.R., Petry, M. V. & Agne, C. E. 2017. Primeiro registro de *Muscisaxicola capistratus* (Passeriformes: Tyrannidae) no Brasil. *Atualidades Ornitológicas* 197: 24.

Santos, M.M., Montalti, D, Juares, M., Coria, N.R. & Archuby, D. 2007. First record of the austral thrush (*Turdus falcklandii*) from the South Shetland Islands, Antarctica. *Notornis* 54: 231–232.

Sawan, M., 2020. First record of Lichtenstein's Sandgrouse *Pterocles lichtensteinii* in Lebanon, 2020. *Ornis Hungarica* 28: 212–214.

Schärer, J. & Cavaillès, S. 2019. Egyptian Nightjar at Chorokhi delta, Georgia, in August 2017 and status in Europe. *Dutch Birding 41*: 36–40.

Schmaljohann, H. & Thoma, M. 2005. First record of American Golden Plover *Pluvialis dominica* for Mauritania, and its status in western Africa. *Bulletin of the African Bird Club* 12: 158–161.

Schmelzer, I. & Phillips, F. 2004. First record of a Barred Owl, *Strix varia*, in Labrador. *The Canadian Field-Naturalist* 118: 273–276.

Schmitt, F. & Pariset, P. 2005. Third record of Rufous-chested Plover *Charadrius modestus* in Peru. *Cotinga* 24: 111.

Schodde, R. & van Tets, G.F. 1981. First record of the Brown Hawk-Owl *Ninox scutulata* from Australasia. *Emu* 81: 171.

Schollaert, V. 2006. First record of Rose-coloured Starling *Sturnus roseus* for Ethiopia and sub-Saharan Africa. *Bulletin of the African Bird Club* 13: 75.

Schroeder, M. A. & Braun, C.E. 1993. Partial migration in a population of Greater Prairie-Chickens in Northeastern Colorado. *The Auk* 110: 21–28.

Schulenberg, T.S., Stotz, D.F., Lane, D.F., O'Neill, J.P. & Parker III, T.A. 2010. *Birds of Peru: revised and updated edition*. Princeton University Press, Princeton, New Jersey.

Schwertner, T.W., Mathewson, H.A., Roberson, J.A. & Waggerman, G.L. 2002. White-winged Dove (*Zenaida asiatica*), version 2.0. In *The Birds of North America* (Poole, A. F. & Gill, F. B., Eds). Cornell Lab of Ornithology, Ithaca, NY, USA.

Schwertner, C.A., Fenalti, P.R. & Fenalti, O.A. 2011. Um novo passeriforme para o Brasil: *Muscisaxicola maclovianus* (Passeriformes: Tyrannidae). *Revista Brasileira de Ornitologia* 19: 453–454.

Schulz-Neto, A. 2004. Aves insulares do arquipélago de Fernando de Noronha, Pp. 147–168. In *Aves marinhas e insulares brasileiras: bioecologia e conservação*, Branco, J. O., ed.). Editora da UNIVALI, Itajaí.

Schunck, F. 2018. WA3101002, *Podiceps occipitalis* Garnot, 1826. http://www.wikiaves.com/3101002

Schweigman, P., Cash, W.F. & Thompson, M.P. 2014. Seasonal movements and survival of royal spoonbill (*Platalea regia*) breeding in New Zealand. *Notornis* 61: 177–187.

Scofield, R.P. & Wiltshire, A. 2004. Snow petrel (*Pagodroma nivea*) records from Macquarie Island. *Notornis* 51: 168–169.

Scott, D.A. 2008. Rare birds in Iran in the late 1960s and 1970s. *Podoces* 3: 1–30.

Scordino, J. J. & Akmajian, A. M. 2012. No it's Not a Murre, it's a Penguin. *Northwestern Naturalist* 93: 232–235.

Scuderi, A. & Corso, A. 2011. Crested Honey Buzzard in Europe. *Birding World* 24: 252–256.

Sealy, S.G. 1974. Breeding phenology and clutch size in the Marbled Murrelet. *The Auk* 91: 10–23.

Sealy, S.G. & Carter, H.R. 2012. Rare inter-ocean vagrancy in Crested Auklet and Parakeet Auklet. *Waterbirds* 35: 64–74.

Senfeld, T., Shannon, T.J., van Grouw, H., Paijmans, D.M., Tavares, E.S., Baker, A.J., Lees, A.C. & Collinson, J.M. 2019. Taxonomic status of the extinct Canary Islands Oystercatcher *Haematopus meadewaldoi*. *Ibis* 162: 1068–1074.

Senzaki, M. 2014. A spring record of White-rumped Sandpiper *Calidris fuscicollis* from Hokkaido. *Japanese Journal of Ornithology* 63: 55–58.

Serpa, G.A., Malacco, G.B., Aleixo, A., Darski-Silva, B. & Madeira, S. 2014. Range extension of the known distribution of the Black-backed Grosbeak, *Pheucticus aureoventris* (Passeriformes: Cardinalidae) in Brazil, with the first records for the states of Rondônia, Amazonas and Goiás. *Revista Brasileira de Ornitologia* 22: 38–41.

Servat, G. & Pearson, D.L. 1991. Natural history and records for seven poorly known bird species from Amazonian Peru. *Bulletin of the British Ornithologists' Club* 111: 92–95.

Seutin, G. & Chartier, B. 1989. The Rock Wren, *Salpinctes obsoletus*, breeding at Churchill, Manitoba. *Canadian Field-Naturalist* 103: 416–417.

Shagir, K.J. & Iqbal, M. 2015. White-eyed Buzzard *Butastur teesa*, a new species for Greater Sundas and Wallacea. *BirdingASIA* 23: 124–125.

Shamoun-Baranes, J. & van Gasteren, H. 2011. Atmospheric conditions facilitate mass migration events across the North Sea. *Animal Behaviour* 81: 691–704.

Shapiro, B., Sibthorpe, D., Rambaut, A., Austin, J., Wragg, G.M., Bininda-Emonds, O.R., Lee, P.L. & Cooper, A. 2002. Flight of the dodo. *Science* 295: 1683–1683.

Sharma, M. & Sangha, H. S. 2012. Purple-backed Starling *Sturnus sturninus* in the Andaman Islands, India, and its status in the Indian subcontinent. *Indian Birds* 64: 172.

Sharrock, J.T.R. & Davies C. 2000. The European Bird Report Non-passerines, including near-passerines. *British Birds* 93: 114–128.

Sharrock, J.T.R. & Davies, C. 2002. The European Bird Report. *British Birds* 95: 174–188.

Shaughnessy, P.D., Kemper, C.M. & Ling, J.K. 2012. Records of vagrant phocid seals (family Phocidae) in South Australia. *Australian Mammalogy* 34: 55–169.

Shema, S. & Njoroge, P. 2018. Desert Wheatear *Oenanthe deserti* at Lake Turkana – the fourth record of the species for Kenya. *Scopus* 38: 18.

Sherony, D.F. 1999. The Fall Migration of Jaegers on Lake Ontario. *Journal of Field Ornithology* 70: 33–41.

Shirihai, H. 1996. *The Birds of Israel*. Academic Press, London

Shirihai, H. 1999. Fifty species new to Israel, 1979–1998: their discovery and documentation, with tips on identification. *Sandgrouse* 21: 45–105.

Shirihai, H. 2007. *A Complete Guide to Antarctic Wildlife. The Birds and Mammals of the Antarctic Continent and the Southern Ocean*. 2nd edition. A.&C. Black, London.

Shirihai, H., Khoury, F., Al-Jbour, S. & Yosef, R. 2000. The first Pink-backed Pelican in Jordan. *Sandgrouse* 22: 127–130.

Shoch, D.T. & Howell, S.N.G. 2013. Occurrence and identification of vagrant 'orange-billed terns' in eastern North America. *North American Birds* 67: 188–209.

Short, L. & Horne, J.F.M. 2001. *Toucans, Barbets, and Honeyguides*. Oxford University Press, Oxford.

Sick, H. 1997. *Ornitologia Brasileira*. Nova Fronteira, Rio de Janeiro.

Sikora, A. & Ławicki, Ł.2019. What brought about the influx of Siberian Accentors to Europe in autumn 2016? A contribution to the debate. *British Birds* 112: 760–765.

Silva, M.C., Matias, R., Ferreira, V., Catry, P. & Granadeiro, J. P. 2016. Searching for a breeding population of Swinhoe's Storm-petrel at Selvagem Grande, NE Atlantic, with a molecular characterization of occurring birds and relationships within the Hydrobatinae. *Journal of Ornithology* 157: 117–123.

Silva, M.A. 2018. WA3216358, *Florisuga mellivora* (Linnaeus, 1758)]. http://www.wikiaves.com/3216358

Silva, M.P., Coria, N.R., Favero, M. & Casaux, R.J. 1995. New records of cattle egret *Bubulcus Ibis*, black necked swan *Cygnus melancorhyphus* and white-rumped sandpiper *Calidris fuscicollis* from the South Shetland Islands, Antarctica. *Marine Ornithology* 23: 65–66.

Silva, M.R. 2019. WA3338245, *Cinclodes fuscus* (Vieillot, 1818)]. http://www.wikiaves.com/3338245

Silva e Silva, R., Olmos, F. & Lima, P.C. 2002. *Catharacta chilensis* (Bonaparte, 1857) no Brasil. *Ararajuba* 10: 275–277.

Silva e Silva, R. & Olmos, F. 2006. Noteworthy bird records from Fernando de Noronha, northeastern Brazil. *Ararajuba* 14: 470–473.

Silva e Silva, R.S. & Carlos, C.J. 2019. A Great Frigatebird *Fregata minor* at Fernando de Noronha archipelago, equatorial Atlantic Ocean. *Bulletin of the British Ornithologists' Club* 139: 333–337.

Simmons, R.E., Mills, M.S.L. & Dean, W.R.J. 2009. Oystercatcher *Haematopus* records from Angola. *Bulletin of the African Bird Club* 16: 211–212.

Simon, J.E., Lima, S.R., Novaes, T.D. & Alves, A. 2005. Primeiro registro de *Podicephorus major* (Boddaert, 1783) (Aves: Podicipedidae) para o estado do Espírito Santo, Brasil. *Boletim do Museu de Biologia Mello Leitão* 18: 59-63.

Simons, A.M. 2004. Many wrongs: the advantage of group navigation. *Trends in Ecology & Evolution* 19: 453–455.

Simpson, R. & Simpson, E. 2011a Registro documentado da Batuíra-de-peito-tijolo *Charadrius modestus* (Charadriiformes: Charadriidae) em Paraty, Rio de Janeiro. *Atualidades Ornitologicas* 162: 18–19.

Simpson, R. & Simpson, E. 2011b. Record of Austral Negrito, *Lessonia rufa* (Passeriformes: Tyrannidae) for the municipality of Ubatuba, north coast of São Paulo. *Atualidades Ornitológicas* 163: 10–11.

Sinclair, I., Hockey, P., Tarboton, W. & Ryan, P. 2011. *Birds of Southern Africa*. 4th edition. Struik Nature, Cape Town, South Africa.

Sinclair, P. H., Nixon, W. A., Eckert, C. D.& Hughes, N. L., eds. 2003. *Birds of the Yukon Territory*. Univ. of Br. Columbia Press, Vancouver.

Singh, D. K. & Panda, A. C. 2014. Sighting of Yellow-bellied Warbler *Abroscopus superciliaris* in Odisha, India. *Indian Birds* 9: 167.

Skerrett, A. 2003. Three new species for Seychelles: Sociable Lapwing *Vanellus gregarius*, Spotted Redshank *Tringa erythropus* and Chiffchaff *Phylloscopus collybita*. *Bulletin of the African Bird Club* 10: 47–49.

Skerrett, A. 2008. The proliferation of records of Amur Falcon *Falco amurensis* in Seychelles since 1995. *Gabar* 19: 23–26.

Skerrett, A. 2017. First record of Greater Painted-snipe *Rostratula benghalensis* for Seychelles. *Bulletin of the African Bird Club* 24: 92–93.

Skerrett, A., Betts, M., Bullock, I., Fisher, D., Gerlach, R., Lucking, R., Phillips, J. & Scott, B. 2006. Third report of the Seychelles Bird Records Committee. *Bulletin of the African Bird Club* 13: 65–72.

Skerrett, A., Betts, M., Bowler, J., Bullock, I., Fisher, D., Lucking, R. & Phillips, J. 2011. Fourth report of the Seychelles Bird Records Committee. *Bulletin of the African Bird Club* 18: 182–192.

Skerrett, A., Betts, M., Bowler, J., Bullock, I., Fisher, D., Lucking, R. & Phillips, J. 2017. Fifth report of the Seychelles Bird Records Committee. *Bulletin of the African Bird Club* 24: 63–75.

Slack, R. 2009. *Rare Birds, where and when: An Analysis of Status & Distribution in Britain and Ireland*. Sandgrouse to New World Orioles. Rare Birds Books, York.

Slaterus, R. 2013 Glanstroepiaal over Kamperhoek. *Dutch Birding* 35: 214–218.

Smiles, M. 2014. Pheasant-tailed Jacana *Hydrophasianus chirurgus* in the United Arab Emirates: the first and second national records. *Sandgrouse* 36: 237–239.

Smiles, M. 2016. Large-billed Leaf Warbler *Phylloscopus magnirostris* in the United Arab Emirates – first record for Arabia. *Sandgrouse* 38: 201–204.

Smith, J. P. 2004. The first Franklin's Gull *Larus pipixcan* in Israel and the Middle East. *Sandgrouse* 26: 65–67.

Smith, P. & Easley, K. 2019. Status and distribution of Paraguayan 'Black Tyrants' (Tyrannidae: *Knipolegus*) with a remarkable new country record. *Avocetta* 43: 149–158.

Smith R.P., Rand, P.W., Lacombe, E.H., Morris, S.R., Holmes, D.W. & Caporale, D.A. 1996. Role of bird migration in the long-distance dispersal of *Ixodes dammini*, the vector of Lyme disease. *Journal of Infectious Diseases* 174: 221–224.

Snow, D.W. 1985. Affinities and recent history of the avifauna of Trinidad and Tobago. *Ornithological Monographs* 36: 238–246.

Snow, D.W. & Perrins, C.M. eds. 1998. *The Birds of the Western Palearctic*. Concise Edition. Oxford University Press, Oxford & New York.

Snow, D. 2020. Sapayoa (*Sapayoa aenigma*), version 1.0. In *Birds of the World* (J. del Hoyo, A. Elliott, J. Sargatal, D. A. Christie, and E. de Juana, Editors). Cornell Lab of Ornithology, Ithaca, NY, USA. https://doi.org/10.2173/bow.sapayo1.01

Snyder, N.F. & Russell, K. 2002. Carolina Parakeet (*Conuropsis carolinensis*), version 2.0. In *The Birds of North America* (Poole, A. F. & Gill, F. B., Eds). Cornell Lab of Ornithology, Ithaca, NY, USA. https://doi.org/10.2173/bna.667

Soldaat, E., Leopold, M.F., Meesters, E.H. & Robertson, C.J. 2009. Albatross mandible at archeological site in Amsterdam, The Netherlands, and WP records of *Diomedea* albatrosses. *Dutch Birding* 31: 1–16.

Somenzari, M., Amaral, P.P.D., Cueto, V.R., Guaraldo, A.D.C., Jahn, A.E., Lima, D.M., Lima, P.C., Lugarini, C., Machado, C.G., Martinez, J. & Nascimento, J.L.X.D. 2018. An overview of migratory birds in Brazil. *Papéis Avulsos de Zoologia* 58: e20185803.

Sorokin A.G. 2017. The record of the rufous hummingbird *Selasphorus rufus* in Ratmanov Island in the Bering Strait (In Russian). *Russian Journal of Ornithology* 26: 3534–3535.

Spar, R. & Bruderer, B. 1996. Soaring flight of Steppe Eagles *Aquila nipalensis* in southern Israel: flight behaviour under various wind and thermal conditions. *Journal of Avian Biology* 27: 289–301.

Spasov, S. 2006. The first record of White-throated Kingfisher *Halcyon smyrnensis* for Bulgaria. *Sandgrouse* 28: 76–77.

Spear, L.B., Lewis, M.J., Myres, M.T. & Pyle, R.L. 1988. The recent occurrence of Garganey in North America and the Hawaiian Islands. *American Birds* 42: 385–392.

Spearpoint, J.A. 1981. A Long-tailed Skua *Stercorarius longicaudus* inland in the Kalahari Gemsbok National Park, South Africa. *Cormorant* 9: 45–46.

Spenneman, D.H.R. 2004. The occurrence of owls in the Marshall Islands. *Notornis* 51: 147–151.

Spierenburg, P. 2005. *Birds in Bhutan: Status and Distribution*. Oriental Bird Club, Bedford, UK.

Sridharan, B., Viswanathan, A., Chimalakonda, D., Singal, R., Subramanya, S. & Prince, M., 2016. New records of Swinhoe's Minivet *Pericrocotus cantonensis*, and Ashy Minivet *Pericrocotus divaricatus* in Bengaluru city, Karnataka, India. *Indian Birds* 12: 70–72.

Srinivasan, U., Dalvi, S. & Yobin, K. 2009. First records of 'white-headed' Black Bulbul *Hypsipetes leucocephalus* from India. *Indian Birds* 5: 28–30.

Stahl, J.C., Weimerskirch, H. & Ridoux, V. 1984. Observations récentes d'oiseaux marins et terrestres visiteurs dans les îles Crozet, sud-ouest de l'Océan Indien. *Gerfaut* 74: 39–46.

Steadman, D.W. 1995. Prehistoric extinctions of Pacific island birds: biodiversity meets zooarchaeology. *Science* 267: 1123–1131.

Steffe, B. 2010. African Openbill at Luxor, Egypt, in May 2009. *Dutch Birding* 32: 254–256.

Steffen, W., Crutzen, P. J. & McNeill, J.R. 2007. The Anthropocene: are humans now overwhelming the great forces of nature. *AMBIO: A Journal of the Human Environment* 36: 614-621.

Steijn, L. & Vries, K.D. 2009. Black-throated Loon in Assam, India, in January 2008. *Dutch Birding* 31: 301–302.

Steiof, K., De Silva, C., Jayarathna, J., Madlow, W., Pohl, M., Puschel, W. & Zerning, M. 2017. Whinchat *Saxicola rubetra* in Sri Lanka in February 2015: First record for the island and the Indian Subcontinent. *Indian Birds* 13: 108–111.

Stevenson, H.M. & Anderson, B.H. 1994. *The Birdlife of Florida*. University Press of Florida, Gainesville, USA.

Stevenson, T. & Fanshawe, J. 2002. *Field Guide to the Birds of East Africa*. T & AD Poyser, London.

Stinson, D.W., Wiles, G.J. & Reichel, J.D. 1997. Migrant land birds and water birds in the Mariana Islands. *Pacific Science* 51: 314–327.

Stoddart, A. 2018. Siberian Accentors in Europe in autumn 2016 and the first British records. *British Birds* 111: 69–83.

Storr, G.M. 1985. *Birds of the Gascoyne Region, Western Australia*. Western Australian Museum, Perth.

Stouffer, P.C. & Bierregaard Jr, R.O. 1993. Spatial and temporal abundance patterns of Ruddy Quail-Doves (*Geotrygon montana*) near Manaus, Brazil. *The Condor* 95: 896–903.

Strewe, R., Villa-De León, C., Navarro, C., Alzate, J. & Utría, G. 2015. Primer registro documentado de la Torcaza aliblanca (*Zenaida asiatica*) en América del Sur. *Ornitologia Colombiana* 15: e75–e78.

Strong, C., Zuckerberg, B., Betancourt, J.L. & Koenig, W.D. 2015. Climatic dipoles drive two principal modes of North American boreal bird irruption. *Proceedings of the National Academy of Sciences* 112: E2795–E2802.

Sueur, F. & Siblet, J.-P. 2010. Observations ornithologiques en Égypte (juillet-août 2008 et mars 2009). *Alauda* 78: 69–73.

Sullivan, B.L., Aycrigg, J.L., Barry, J.H., Bonney, R.E., Bruns, N., Cooper, C.B., Damoulas, T., Dhondt, A.A., Dietterich, T., Farnsworth, A. & Fink, D., 2014. The eBird enterprise: an integrated approach to development and application of citizen science. *Biological Conservation* 169: 31–40.

Sutherland, W.J. 1998. Evidence for flexibility and constraint in migration systems. *Journal of Avian Biology* 29: 441–446.

Szabo, I., Walters, K., Rourke, J. & Irwin, D.E. 2017. First record of House Swift (*Apus nipalensis*) in the Americas. *The Wilson Journal of Ornithology* 129: 411–416.

Tacha, T.C., Nesbitt, S.A. & Vohs, P.A. 1992. Sandhill Crane (*Grus canadensis*), *The Birds of North America Online* (Poole, A. Ed.). Ithaca: Cornell Lab of Ornithology; Retrieved from the Birds of North America Online: http://bna.birds.cornell.edu.libraryproxy.amnh.org:9000/bna/species/031

Tamarozzi, B. 2013. WA1594045, *Rhopospina fruticeti* (Kittlitz, 1833)]. http://www.wikiaves.com/1594045

Tamini, L.L. & Chavez, L.N. 2014. First record of Buller's Albatross (*Thalassarche bulleri*) from a fishing vessel in the south-western Atlantic Ocean off Southern Patagonia (Argentina). *Polar Biology* 37: 1209–1212.

Tavares, D., Boldrini, M., Moura, J., Amorim, C. & Siciliano, S. 2012. Aves, Stercorariidae, Chilean Skua *Stercorarius chilensis* Bonaparte, 1857: First documented record for the state of Espírito Santo, southeastern Brazil. *Check List* 8: 560–562.

Taylor, B. 1998. *Rails: A Guide to Rails, Crakes, Gallinules and Coots of the World*. Pica Press, Robertsbridge, UK.

Taylor, B. 2020. Fairy Flycatcher (*Stenostira scita*), version 1.0. In *Birds of the World* (J. del Hoyo, A. Elliott, J. Sargatal, D. A. Christie, and E. de Juana, Editors). Cornell Lab of Ornithology, Ithaca, NY, USA. https://doi.org/10.2173/bow.faifly1.01

Taylor, B. & Kirwan, G.M. 2019. Buff-spotted Flufftail (*Sarothrura elegans*), version 1.0. In *Birds of the World* (J. del Hoyo, A. Elliott, J. Sargatal, D. A. Christie, and E. de Juana, Editors). Cornell Lab of Ornithology, Ithaca, NY, USA. https://doi.org/10.2173/bow.busflu1.01

Tebb, G., Veron, P.K. & Craig, M. 2003. Laughing Gull *Larus atricilla* in Malaysia: the first record for Asia. *Forktail* 19: 131–132.

Teixeira, D.M., Oren, D. & Best, R. C. 1986. Notes on Brazilian seabirds 2. *Bulletin of the British Ornithologists' Club* 106: 74–77.

Teixeira, R.S.C., Otoch, R. & Raposo, M.A. 2016. First record of Northern Gannet *Morus bassanus* in the Southern Hemisphere. *Bulletin of the British Ornithologists' Club* 136: 151–152.

Telino-Jr, W.R., Neves, R.M.L., Farias, G.B., Brito, M.T., Pacheco, G. & Siqueira, S.A.O. 2006. Ocorrência e aspectos comportamentais da pomba-antártica, *Chionis alba* (Charadriiformes: Chionididae), em Pernambuco, Brasil. *Tangara* 1: 26–29.

Tello, N.S., Stojan-Dolar, M. & Heymann, E.W. 2008. A sight and video record of the oilbird, *Steatornis caripensis*, in Peruvian lowland Amazonia. *Journal of Ornithology* 149: 267–269.

Thévenot, M., Vernon, R. & Bergier, P. 2003. *The Birds of Morocco. An Annotated Check-list. BOU Check-list 20.* British Ornithologists' Union, Tring, UK.

Thiollay, J.-M. 1977. Importance des populations de rapaces migrateurs au Méditerranée occidentale. *Alauda* 45: 115–121.

Thoma, M. & Täschler, A., 2013. Vagrant Bimaculated Larks in Europe and the first record for Switzerland. *British Birds* 106: 101–108.

Thomas, B.T. & Kirwan, G.M. 2019. Oilbird (*Steatornis caripensis*). In del Hoyo, J., Elliott, A., Sargatal, J., Christie, D.A. & de Juana, E., eds. *Handbook of the Birds of the World Alive.* Lynx Edicions, Barcelona.

Thompson, A.L. 1923. The migrations of some British ducks: results of the marking method. *British Birds* 16: 262–276.

Thompson, P.M. & Johnson, D.L. 2003. Further notable bird records from Bangladesh. *Forktail* 19: 85–102.

Thompson, P.M., Chowdhury, S.U., Haque, E.U., Khan, M.M.H. & Halder, R. 2014. Notable bird records from Bangladesh from July 2002 to July 2013. *Forktail* 30: 50–65.

Thorne, R. & Thorne, S. 2014. 'Asian Red-rumped Swallow' in Orkney and Highland: new to Britain. *British Birds.* 107: 405–412.

Thorup, K. 1998. Vagrancy of Yellow-browed Warbler *Phylloscopus inornatus* and Pallas's Warbler *Ph. proregulus* in north-west Europe: Misorientation on great circles? *Ringing & Migration* 19: 7–12.

Thorup, K., 2004. Reverse migration as a cause of vagrancy. *Bird Study* 51: 228–238.

Thorup, K. & Rabøl, J. 2007. Compensatory behaviour after displacement in migratory birds. *Behavioral Ecology and Sociobiology* 61: 825–841.

Thorup, K., Ortvad, T.E., Holland, R.A., Rabøl, J., Kristensen, M.W. & Wikelski, M. 2012. Orientation of vagrant birds on the Faroe Islands in the Atlantic Ocean. *Journal of Ornithology* 153: 1261–1265.

Tigreros A., A.-F., Dávila, S.G. & Londoño, G. 2019. First record of the Clay-colored Sparrow (Passerellidae: *Spizella pallida*) for South America. *Ornitología Neotropical* 30: 85–87.

Tiwari, J.K. 2008. Grey Hypocolius *Hypocolius ampelinus* in Kachchh, Gujarat, India. *Indian Birds* 4: 12–13.

Tobalske, B.W., Hedrick, T.L., Dial, K.P. & Biewener, A.A. 2003. Comparative power curves in bird flight. *Nature* 421: 363–366.

Tobgay, T. 2017. First record of Oriental Pratincole *Glareola maldivarum* for Bhutan. *BirdingASIA* 27: 120–121.

Tobias, J.A. & Seddon, N. 2007. Ornithological notes from southern Bolivia. *Bulletin of the British Ornithologists' Club* 127: 293–300.

Tomkovich, P.S. 1982. Osobennosti osenneiy migratsii ostrokhvostykh pesochnikov [Peculiarities of the autumn migration of the Sharp-tailed Sandpiper]. *Byulleten Moskovskogo Obschestva ispytatelei prirodny otdel biologicheskii* 87: 56–61.

Tonkin, S. & Gonzalez, J.M. 2019. Ringing recovery of Yellow-browed Warbler in Andalucía confirms overwintering in consecutive winters. *British Birds* 112: 686–687.

Toochin, R. 1997. A Xantus's Hummingbird in British Columbia: A first Canadian record. *Birders Journal* 6: 293–297.

Torés, N., Johow, P. & Johow, F. 2013. Nuevo registro para la avifauna de Chile: el Cuclillo de pico amarillo. *La Chiricoca* 16: 16–19.

Torres-Mura, J.C., Lemus, M.L. & Garín, C. 2003. Registro del Turpial Norteño (*Icterus galbula*, Aves: Icteridae) en Calama, II Región, Chile. *Noticiario Mensual del Museo Nacional de Historia Natural, Chile* 351: 9–12.

Torres-Vivanco, A., Vital-García, C. & Moreno-Contreras, I. 2015. A new record of Lewis's Woodpecker, *Melanerpes lewis* (Gray, 1849) (Aves, Picidae) from Chihuahua, with comments on its status in Mexico. *Check List* 11: 1792.

Tosoni, A.C. & Piot, B. 2014. *Sturnus unicolor* en Suisse. *Nos Oiseaux* 61: 13–16.

Towers, J.R., Atkins, R., Howard, K., Sykes, J. & Dunstan, G. 2015. The first confirmed record of a Blue-footed Booby in Canada. *British Columbia Birds* 25: 13–16.

Turner, D.A. 1997. Family Musophagidae (Turacos). In del Hoyo, J., Elliott, A. & Sargatal, J. (Eds) *Handbook of the Birds of the World,* Volume 4. Lynx Edicion, Barcelona.

Tyler, S.J., Randall, R.D. & Brewster, C.A. 2008. New bird records for Botswana and additional information on some rarities. *Bulletin of the African Bird Club* 15: 36–52.

Underhill, L.G. 2015. Four records of Citrine Wagtail *Motacilla citreola* in South Africa: the bigger picture. *Biodiversity Observations* 6: 65–67.

Unitt, P. 1984. The birds of San Diego County. *San Diego Society of Natural History Memoirs* 13: 1–276.

Unitt, P. 2004. San Diego County Bird Atlas. *Proceedings of the San Diego Society of Natural History* 39. San Diego Natural History Museum, San Diego.

Ussher, R.J. & Warren, R. 1900. *The birds of Ireland: an account of the distribution, migrations and habits of birds as observed in Ireland, with all additions to the Irish list.* Gurney and Jackson.

Urban, E.K., Fry, C.H. & Keith, S. 1986. *The Birds of Africa.* Volume 2. Academic Press, London.

Valera, F., Rey, P., Sanchez-Lafuente, A.M. & Muñoz-Cobo, J. 1990. The situation of penduline tit (*Remiz pendulinus*) in southern Europe: A new stage of its expansion. *Journal für Ornithologie* 131: 413–420.

van Balen, B., Trainor, C. & Noske, R. 2013. Around the Archipelago. *Kukila* 17: 41–72.

van Bemmelen, R.S. & Wielstra, B. 2008. Vagrancy of Brünnich's Guillemot *Uria lomvia* in Europe. *Seabird* 21: 16–31.

van Bemmelen, R.S., van Spanje. T & Camphuysen, C.J. 2017. Olive-backed Pipits and Yellow-browed Warblers at Cap Blanc peninsula, Mauritania, in November. *Dutch Birding* 39: 103–105.

van Bemmelen, R.S., Lehmhus, J. & Mlodinow, S.G., 2018. Hybrid Northern Shoveler x Blue-winged Teal on Schiermonnikoog, Netherlands, in May 2014, and identification and WP occurrence. *Dutch Birding* 40: 71–81.

van Buren, A. N. & Boersma, P.D. 2007. Humboldt Penguins (*Spheniscus humboldti*) in the northern hemisphere. *The Wilson Journal of Ornithology* 119: 284–288.

van Els, P. & Brady, M.L. 2014. Specimen of Indian Pitta *Pitta brachyura* from the Islamic Republic of Iran. *Bulletin of the British Ornithologists' Club* 134: 160–162.

van den Berg, A.B. & Bosman, C.A.W. 1999. *Rare Birds of the Netherlands.* Pica Press, Robertsbridge, UK.

van den Berg, A.B. 2014. WP reports. *Dutch Birding* 36: 402–420.

van den Berg, A.B. 2015. WP reports. *Dutch Birding* 37: 42–54.

van den Berg, A.B. & Haas, M. 2010. WP Reports. *Dutch Birding* 32: 266–276.

van den Berg, A.B. & Haas, M. 2012. WP reports. *Dutch Birding* 34: 178–194.

van den Berg, A.B. & Haas, M. 2013a. WP reports. *Dutch Birding* 35: 129–138.

van den Berg, A.B. & Haas, M. 2013b. WP reports *Dutch Birding* 35: 194–208.

van den Berg, A.B. & Ławicki, Ł. 2015. WP reports. *Dutch Birding* 37: 261–275.

van der Ploeg, J. & Minter, T. 2004. Cinereous Vulture *Aegypius monachus*: first record for the Philippines. *Forktail* 20: 109–1

van der Veen, S.R. 2011. Bruce's Green Pigeon at Luxor, Egypt, in January 2011. *Dutch Birding* 33: 121–122.

Van Doren, B.M. & Horton, K.G. 2018. A continental system for forecasting bird migration. *Science* 361: 1115–1118.

Van Doren, B.M., Horton, K.G., Stepanian, P.M., Mizrahi, D.S. & Farnsworth, A. 2016. Wind drift explains the reoriented morning flights of songbirds. *Behavioral Ecology* 27: 1122–1131.

VanderWerf, E., O'Brien, M., Leidich, R., Basilius, U.i & Yalap, Y. 2018. First Report of the Palau Bird Records Committee. *Western Birds.* 49: 192–205.

Vaughan, R. 1988. Birds of the Thule District, northwest Greenland. *Arctic* 41: 53–58.

Veit, R.R. 1997. Long-distance dispersal and population growth of the Yellow-headed Blackbird *Xanthocephalus xanthocephalus. Ardea* 85: 135–143.

Veit, R.R. 2000. Vagrants as the expanding fringe of a growing population. *The Auk* 117: 242–246.

Veit, R.R. 2004. The provenance of vagrants and their evaluation by rarities committees. *Bird Observer* 32: 377–38.

Veit, R.R., Zawadzki, L.C. & Manne, L.L. 2016. Vagrancy and colonization of St. Thomas and St. John, US Virgin Islands, by Adelaide's Warblers (*Setophaga adelaidae*). *Journal of Caribbean Ornithology* 29: 47–50.

Venturini, A.C., Paz, R.P. & Jacomelli-Jr, J.A. 2007. Registro do corta-ramos-de-rabo-branco *Phytotoma rutila* para o sudeste do Brasil: Linhares, Espírito Santo. *Atualidades Ornitológicas* 136: 1–3.

Verhelst, J., Salaman, P., Donegan, T., Ellery, T. & Pacheco, J. 2018. Revision of the status of bird species occurring or reported in Colombia 2018. *Conservacion Colombiana* 25: 1–47.

Veronese, C.L. 2013. WA1104832, *Mimus triurus* (Vieillot, 1818). http://www.wikiaves.com/1104832

Villagómez, S., Gámez, E. & Molina, D. 2017. Primer registro del colimbo mayor (*Gavia immer* Brünnich, 1764) en Oaxaca, México. *Huitzil* 18: 180–184.

Villalobos, J. & Sandoval, L. 2012. Ten new bird species for Isla del Coco, Costa Rica. *Check List* 8: 568–571.

Villers, A., Millon, A., Jiguet, F., Lett, J.M., Attié, C., Morales, M. B. & Bretagnolle, V. 2010. Migration of wild and captive-bred Little Bustards *Tetrax tetrax*: releasing birds from Spain threatens attempts to conserve declining French populations. *Ibis* 152: 254–261.

Vinicombe, K. 1994. The Welsh Monk Vulture. *British Birds* 87: 613–622.

Vinicombe, K. 2007a. Vagrancy and melting ice. *Birdwatch* 177: 42.

Vinicombe, K.E. 2007b. The status of Red-headed Bunting in Britain. *British Birds* 100: 540–551.

Visser, M.E., Perdeck, A.C., van Balen, J.H. & Both, C. 2009. Climate change leads to decreasing bird migration distances. *Global Change Biology* 15: 1859–1865.

Vizcarra, J.K. & Vicetti, R. 2013. Primer registro documentado de la Parina Chica (*Phoenicoparrus jamesi*) en los Humedales de Ite, costa sur del Perú. *Boletín Informativo Unión de Ornitólogos del Perú* 8: 21–26.

Vooren, C.M. & Fernandes, A.C. 1989. *Guia de albatrozes e petréis do sul do Brasil.* Porto Alegre, Sagra.

Vooren, C.M. 2004. The first two records of *Sula capensis* in Brazil. *Ararajuba* 12: 76–77.

Voous, K.H. 1960. *Atlas of European Birds.* Nelson, London.

Voous, K.H., 1985. Additions to the avifauna of Aruba, Curaçao, and Bonaire, south Caribbean. *Ornithological Monographs* 36: 247–254.

Votier, S.C., Harrop, A.H. J. & Denny, M. 2003. A review of the status and identification of American Wigeon in Britain & Ireland. *British Birds* 96: 2–22.

Votier, S.C., Kennedy, M. & Bearhop, S. 2007. Supplementary DNA evidence fails to confirm presence of Brown Skuas *Stercorarius antarctica* in Europe: a retraction of Votier *et al.* (2004). *Ibis* 149: 619–621.

Waheed, A. 2016. Egyptian record of Wahlberg's Eagle *Hieraaetus wahlbergi* is the first in the Western Palearctic and Middle East. *Sandgrouse* 38: 96–98.

Wainwright, C.E., Stepanian, P.M. & Horton, K.G. 2016. The role of the US Great Plains low-level jet in nocturnal migrant behavior. *International Journal of Biometeorology* 60: 1531–1542.

Walbridge, G., Small, B. & McGowan, R.Y. 2003. Ascension Frigatebird on Tiree – new to the Western Palearctic. *British Birds* 96: 58–73.

Walker, D. 2004. Audouin's Gull: new to Britain. *British Birds* 97: 534–541.

Wallace, A.R. 1852. On the monkeys of the Amazon. *Proceedings of the Zoological Society of London* 1852: 107–110.

Wallace, I. 2004. *Beguiled by Birds.* Christopher Helm, London.

Wallraff, H.G. 1972. Homing of pigeons after extirpation of their cochleae and lagenae. *Nature New Biology* 236: 223–224.

Wang, K.-F., Tain, N.-C. & Zhang, G.-P. 2011. Black-throated Accentor (*Prunella atrogularis*) was found in Qinling Mountain. *Chinese Journal of Zoology* 46: 147–149.

Wang, Y., Xue, W. & Wang, H. 2019. Save China's yellow-breasted bunting. *Science* 365: 651–651.

Wanless, R.M., Aguirre-Muñoz, A., Angel, A., Jacobsen, J.K., Keitt, B.S. & McCann, J. 2009. Birds of Clarion Island, Revillagigedo archipelago, Mexico. *The Wilson Journal of Ornithology* 121: 745–751.

Warburton, T. 2009. The Philippine Owl Conservation Programme: why is it needed? *Ardea*, 97: 429–438.

Warning, N. 2016. Rock wren transport in railroad boxcars. *The Southwestern Naturalist* 61: 203–209.

Waschkies, I., Schmidt, C. & Djerf, J. 2005. First record of Pin-tailed Sandgrouse *Pterocles alchata* for Egypt since 1917? *Sandgrouse* 27: 163–165.

Wassink, A., Ahmed, R., Bussutil, S. & Salemgareev, A. 2011. Oriental Plover, Franklin's Gull, Syrian Woodpecker and Masked Shrike new to Kazakhstan. *Dutch Birding* 33: 239–244.

Wassink, A. 2013. Birds of Kazakhstan: new and interesting data, part 4. *Dutch Birding* 35: 30–34.

Watanabe, Y. & Kagoshima, K. 2016. A record of Asian Paradise Flycatcher *Terpsiphone paradisi* on Yonaguni-jima, Okinawa Prefecture, Japan. *Japanese Journal of Ornithology* 65: 43–45.

Weenink, R., van Duivendijk, N. & Ebels, E. B. 2011. Spaanse Keizerarend over Loozerheide in mei 2007. *Dutch Birding* 33: 94–102.

Weidensaul, S. & Lockerman, S. 2013. Bahama Woodstar (*Calliphlox evelynae*) in Lancaster County. *Pennsylvania Birds* 27: 78–79.

Weimerskirch, H., Chastel, O., Barbraud, C. & Tostain, O. 2003. Frigatebirds ride high on thermals. *Nature* 421: 333–334.

Weimerskirch, H., Tarroux, A., Chastel, O., Delord, K., Cherel, Y. & Descamps, S. 2015. Population-specific wintering distributions of adult south polar skuas over three oceans. *Marine Ecology Progress Series* 538: 229–237.

Weintraub, J.D. & San Miguel, M. 1999. First record of the Ivory Gull (*Pagophila eburnea*) in California. *Western Birds* 30: 39–43.

Welch, G. & Welch, H. 1988. The autumn migration of raptors and other soaring birds across the Bab-el-Mandeb straits. *Sandgrouse* 10: 26–50.

Welch, G. & Welch, H. 2017. First record of Oriental Honey Buzzard *Pernis ptilorhynchus* for Djibouti and Africa, in 1987. *Bulletin of the African Bird Club* 24: 96–97.

White, R. 2011. First record of Lesser Kestrel *Falco naumanni* for the Comoros. *Bulletin of the African Bird Club* 18: 79–80.

White, N.E., Phillips, M.J., Gilbert, M.T.P., Alfaro-Núñez, A., Willerslev, E., Mawson, P.R., Spencer, P.B. & Bunce, M. 2011. The evolutionary history of cockatoos (Aves: Psittaciformes: Cacatuidae). *Molecular Phylogenetics and Evolution* 3: 615–622.

White, G. & Hayes, F.E. 2002. Second report of the Trinidad and Tobago Rare Bird Committee. *Living World, Journal of the Trinidad and Tobago Field Naturalists' Club* 2002: 51–56.

White, S. & Kehoe, C. 2016. Report on scarce migrant birds in Britain in 2013. Part 1: non-passerines. *British Birds* 109: 21–45.

White, A., Cummins, R.H. & Boardman, M.R. 2014. A White Tern (*Gygis alba*) in the Bahamas. *North American Birds* 67: 384–385.

Whitfield, S.M., Frezza, P., Ridgley, F. N., Mauro, A., Patterson, J. M., Pernas, A. & Lorenz, J. J. 2018. Status and trends of American Flamingos (*Phoenicopterus ruber*) in Florida, USA. *The Condor* 120: 291–304.

Whittaker, A., Antoine-Feill, S.A.H. & Scheiele, Z.R. 2004. First confirmed record of Oilbird *Steatornis caripensis* for Brazil. *Bulletin of the British Ornithologists' Club* 124: 106–108.

Whittaker, A., da Silva, J.P.F., Lucio, B. & Kirwan, G.M. 2019. Old World vagrants on Fernando de Noronha, including two additions to the Brazilian avifauna, and predictions for potential future Palearctic vagrants. *Bulletin of the British Ornithologists' Club* 139: 189–204.

Whitecross, M.A., Retief, E.F. & Smit-Robinson, H.A. 2019. Dispersal dynamics of juvenile Secretarybirds *Sagittarius serpentarius* in southern Africa. *Ostrich* 90: 97–110.

Whitney, B.M. 2003. Family Conopophagidae (Gnateaters). Pp. 732–748 in del Hoyo, J., Elliot, A. & Christie, D.A. (Eds.). *Handbook of the birds of the world,* Volume 8. Lynx Edicions, Barcelona.

Whittington, B. 1992. Red-tailed Tropicbird – A First for Canada. *Birders Journal* 1: 309.

Whittington M. 2011. Chestnut Weaver *Ploceus rubiginosus*, a brand new visitor to South Africa. *Ornithological Observations* 2: 32–33.

Wiedenfeld, D.A. 2006. Aves, the Galapagos Islands, Ecuador. *Check List* 2: 1–27.

Wilkinson, C.P., Esmonde-White, D.A., Underhill, L.G. & Whittington, P.A. 1999. African Penguins *Spheniscus demersus* along the Kwazulu-Natal coast, 1981–1997. *Marine Ornithology* 27: 111–113.

William, B. 1974. Two new southern migrants for Brazil. *The Auk* 91: 820–821.

Williams, R. 2016. Neotropical notebook. *Neotropical Birding* 18: 45–52.

Williams, R.S.R., Jurado Zevallos, M.A., Fernandez Gamarra, E. & Flores Balarezo, L. (2011) First record of Buff-necked Ibis *Theristicus caudatus* for Peru. *Cotinga* 33: 92–93.

Williams, S.M., Weber, S.B. & Oppel, S. 2017. Satellite telemetry reveals the first record of the Ascension Frigatebird (*Fregata aquila*) for the Americas. *Wilson Journal of Ornithology* 129: 600–604.

Williams, S.O. & Daw, M.J. 2016. A Rufous-necked Wood-Rail (*Aramides axillaris*) in New Mexico. *North American Birds* 69: 190–197.

Williams, T.C., Williams, J.M., Ireland, L.C. & Teal, J.M. 1977. Autumnal bird migration over the western North Atlantic Ocean. *American Birds* 31: 251–267.

Williams, T.J. 1996. Double-crested Cormorant in Cleveland: new to the Western Palearctic. *British Birds* 89: 163–169.

Willemoes, M., Strandberg, R., Klaassen, R.H., Tøttrup, A.P., Vardanis, Y., Howey, P.W., Thorup, K., Wikelski, M. & Alerstam, T. 2014. Narrow-front loop migration in a population of the common cuckoo *Cuculus canorus*, as revealed by satellite telemetry. *PLOS One* 9: e83515.

Wiltschko, R., Walker, M. & Wiltschko, W. 2000. Sun-compass orientation in homing pigeons: compensation for different rates of change in azimuth? *Journal of Experimental Biology* 203: 889–894.

Wiltschko, W. & Wiltschko, R., 1972. Magnetic compass of European robins. *Science* 176: 62–64.

Wiltschko, W., Daum, P., Fergenbauer-Kimmel, A. & Wiltschko, R. 1987. The development of the star compass in garden warblers, *Sylvia borin*. *Ethology* 74: 285–92.

Wiltschko, W., Weindler, P. & Wiltschko, R. 1998. Interaction of magnetic and celestial cues in the migratory orientation of passerines. *Journal of Avian Biology* 29: 606–617.

Wiltschko, R. & Wiltschko, W. 2015. Avian navigation: a combination of innate and learned mechanisms. *Advances in the Study of Behavior* 47: 229–310.

Wingate, D.B. 1973. *A Checklist and Guide to the Birds of Bermuda*. Island Press, Hamilton, Bermuda.

Wingate, D.B. 1983. A record of the Siberian flycatcher (*Muscicapa sibirica*) from Bermuda: An extreme extralimital vagrant. *The Auk* 100: 212– 213.

Wingate, D.B. & Watson, G.E. 1974. First North Atlantic record of the White Tern. *The Auk* 91: 614–617.

Winkler, D.W., Gandoy, F.A., Areta, J.I., Iliff, M.J., Rakhimberdiev, E., Kardynal, K.J. and Hobson, K.A., 2017. Long-distance range expansion and rapid adjustment of migration in a newly established population of Barn Swallows breeding in Argentina. *Current Biology* 27: 1080–1084.

Winkler, H. & Christie, D.A. 2020. African Grey Woodpecker (*Chloropicus goertae*), version 1.0. In *Birds of the World* (J. del Hoyo, A. Elliott, J. Sargatal, D. A. Christie, and E. de Juana, Editors). Cornell Lab of Ornithology, Ithaca, NY, USA. https://doi.org/10.2173/bow.grywoo1.01

Winklhofer, M., Dylda, E., Thalau, P., Wiltschko, W. & Wiltschko, R. 2013. Avian magnetic compass can be tuned to anomalously low magnetic intensities. *Proceedings of the Royal Society of London. Series B: Biological Sciences* 280: 20130850.

Witt, C., Baumann, M., Bautista, E. & Beckman, E. 2015. Long-distance dispersal of a sedentary Andean flycatcher species with a small geographic range, *Ochthoeca piurae* (Aves: Tyrannidae). *Check List* 11: 1795.

Witt, H.H. 1977. Zur Biologie der Korallenmöwe *Larus audouinii*, Brut usnd Ernährung. *Journal of Ornithology* 118: 134–155.

Woehler, E.J. & Gilbert, C.A. 1990. Hybrid rockhopper-macaroni penguins, interbreeding and mixed species pairs at Heard and Marion Islands. *Emu* 90: 198–201.

Wolfson, D.W., Fieberg, J.R. & Andersen, D.E. 2020. Juvenile Sandhill Cranes exhibit wider ranging and more exploratory movements than adults during the breeding season. *Ibis* 162: 556–562.

Wood, R. & Wood, H. 2012. Black Skimmer – a first confirmed record for South Africa. *Ornithological Observations* 3: 213–217.

Woodall, P.F. 2020. Common Kingfisher (*Alcedo atthis*), version 1.0. In *Birds of the World* (J. del Hoyo, A. Elliott, J. Sargatal, D. A. Christie, and E. de Juana, Editors). Cornell Lab of Ornithology, Ithaca, NY, USA. https://doi.org/10.2173/bow.comkin1.01

Woods, R.W. 2017. *The Birds of the Falkland Islands: an annotated checklist.* BOC Checklist Series: 25. British Ornithologists' Club, Tring, UK.

Woodward, P.W. 1972. The natural history of Kure Atoll, northwestern Hawaiian Islands. *Atoll Research Bulletin.*

Woolfenden, G.E. & van Deventer, M. 2006. First record of the Varied Bunting from Florida. *Florida Field Naturalist* 34: 1–3.

Wu, J., Wilcove, D.S., Robinson, S.K. & Powell, G. 2015. White-rumped Sandpiper *Calidris fuscicollis* in Sichuan, China. *BirdingASIA* 23: 93.

Yamamoto Y., Kaneda, K. & Kikuchi, K. 2015. The first record of a Blue-tailed Bee-eater *Merops philippinus* in Japan. *Strix* 31: 147–151.

Yang, D-S., Rottenborn, S.C., Terrill, S., Searcy, A.J. & Villablanca, F.X. 2016 First California records of the Little Stint and Nazca Booby confirmed through molecular analysis. *Western Birds* 47: 58–66.

Yates, B. 2010. Least Tern in East Sussex: new to Britain and the Western Palearctic. *British Birds* 103: 339–347.

Yerger, J.C. & Mohlmann, J.D. 2008. First North American record of Brown Hawk Owl (*Ninox scutulata*) on Saint Paul Island, Alaska. *North American Birds* 62: 4–8.

Young, B.E., Easley, K., Garrigues, R., Mactavish, B., Murgatroyd, P. & Zook, J. R. 2010. Swallow-tailed Gull *Creagrus furcatus* in Costa Rica. *Cotinga* 32: 24–26.

Zannetos, S.P., Zevgolis, Y. & Akriotis, T. 2018. First record of Crested (or Crested-type) Honey Buzzard *Pernis ptilorhynchus* for Greece. *Bulletin of the British Ornithologists' Club* 138: 386–388.

Zavalaga, C.B. & Paredes, R. 2009. Records of Magellanic penguins *Spheniscus magellanicus* in Peru. *Marine Ornithology* 37: 281–282.

Zelaya, P.V., Salvador, S. A., Giraudo, H. & Klavins, J. 2013. Primeros registros de tersina (*Tersina viridis*) para la ecoregión chaqueña de Argentina. *Nuestras Aves* 58: 52–54.

Zenzal, Jr, T.J. & Moore, F.R. 2016. Stopover biology of Ruby-throated Hummingbirds (*Archilochus colubris*) during autumn migration. *The Auk* 133: 237–250.

Zimmer, K. & Isler, M.L. 2003. Family Thamnophilidae (typical antbirds). In del Hoyo J., Elliott A. & Christie D. (Eds.). *Handbook of the Birds of the World*, Volume 8. Lynx Editions, Barcelona.

Zimmerman, C.E., Hillgruber, N., Burril, S.E., St. Peters, M.A. & Wetzel, J.D. 2005. Offshore marine observation of Willow Ptarmigan, including water landings, Kuskokwim Bay, Alaska. *The Wilson Bulletin* 117: 12–14.

Zonfrillo, B. 1986. Diet of Bulwer's Petrel *Bulweria bulwerii* in the Madeiran Archipelago. *Ibis* 128: 570–572.

GENERAL INDEX

SPECIES INDEX

▲ Cape May Warbler *Setophaga tigrina*, Southeast Farallon Island, San Francisco, California, United States, 29 September 2008 (*Matt Brady*).